Arsenic Exposure
and Health Effects

Arsenic Exposure and Health Effects

Proceedings of the Third International Conference on Arsenic
Exposure and Health Effects, July 12-15, 1998, San Diego, California

Editors:

Willard R. Chappell
University of Colorado at Denver
Denver, CO, USA

Charles O. Abernathy
U.S. Environmental Protection Agency
Washington, DC, USA

and

Rebecca L. Calderon
U.S. Environmental Protection Agency
Research Triangle Park, NC, USA

1999

ELSEVIER
Amsterdam-Lausanne-New York-Oxford-Shannon-Singapore-Tokyo

ELSEVIER SCIENCE Ltd
The Boulevard, Langford Lane
Kidlington, Oxford OX5 1GB, UK

First edition 1999

Library of Congress Cataloging in Publication Data
A catalog record from the Library of Congress has been applied for.

British Library Cataloguing in Publication Data

Arsenic exposure and health effects : proceedings of the
   Third International Conference on Arsenic Exposure and
   Health Effects, July 12-15, 1998, San Diego, California L.
   Calderon
   1.Arsenic - Toxicology - Congresses 2.Arsenic - Health
   aspects - Congresses
   I.Chappell, Willard R. II.Abernathy, Charles O.
   III.Calderon, R. L. (Rebecca L.)
   571.9'54715

   ISBN 008043648X

ISBN: 0 08 043648 X

⊛ The paper used in this publication meets the requirements of ANSI/NISO Z39.48-1992 (Permanence of Paper).
Printed in The Netherlands.

KURT J. IRGOLIC
1938–1999

In July 1999, Professor Kurt Irgolic died in an alpine hiking accident. Professor Irgolic was a well-liked and admired member and colleague of the arsenic research community. He was a key member of the Society for Environmental Geochemistry and Health Arsenic Task Force and contributed substantially to all three of the International Arsenic Exposure and Health Effect Conferences. These Proceedings are dedicated to the memory of our cherished colleague and friend.

Professor Irgolic received his Ph.D. in Inorganic and Analytical Chemistry from Karl-Franzens-University Graz, Austria in 1964. He started his academic career at Texas A&M University with Professor R.A. Zingaro. He was at Texas A&M from 1964 to 1989 and served as Chairman of the Inorganic Chemistry Division from 1986 to 1989. Dr. Irgolic returned to Karl-Franzens-University Graz in 1989 as Professor and Head of the Institute from Analytical Chemistry. Using the experience gained during his 25 years in the US, Kurt built the Office for Foreign Relations at Graz University to an internationally recognized group. In addition to his teaching and research activity, he served in several high administrative posts, culminating in the position of Dean of the Faculty of Natural Sciences. He was very proud of the long and distinguished history of his alma mater. He remained at Karl-Franzens until his untimely death in 1999.

Kurt's research interests were broad. He published 261 articles and 16 books. It would be hard to determine what he felt were his most important scientific contributions because he never distinguished among them. For him they were all important. He was equally interested in analyzing archeological materials from South America and the chemistry of arsenic in various media. He was widely regarded as one of the world's pre-eminent experts on arsenic analyses. His intellectual legacy and integrity will be carried on by the many students that he mentored during his academic career. The high standard of Kurt's character serves as a worthy guidepost for all of us to aspire to in our personal and professional lives.

While his scientific accomplishments brought him international recognition as one of the best chemists in the world and his teaching abilities trained some of the world's foremost chemists, his humanity endeared him to those of us who were fortunate to have known him. He brightened many tedious meetings with his humor and common sense. He was always cheerful and energetic. It was an honor to have known him and a great pleasure to have called Kurt a friend.

To his wife Gerlinde and to his daughter Brigit, we share in your pride of Kurt as a chemist, teacher and family man. He and his contributions will not be forgotten.

Preface

The Society of Environmental Geochemistry and Health (SEGH) Third International Conference on Arsenic Exposure and Health Effects was held July 12–15, 1998 in San Diego, California. The conference was sponsored by both public and private groups. In addition to SEGH and the University of Colorado at Denver, other sponsors included the US Environmental Protection Agency (US EPA), the Agency for Toxic Substances and Disease Registry (ATSDR), the Atlantic Richfield Company (ARCO), the Electric Power Research Institute (EPRI), the International Council on Metals in the Environment (ICME), Elf Atochem, and the Environmental Arsenic Council (EARC). The funding was about equally spread between the public and private sectors.

The attendees included scientists from academia, industry, local government, state government, the US federal government and several government scientists from outside the US. The disciplines represented and the content of the presentations included geochemistry, chemistry, molecular biology, biochemistry, epidemiology and medicine. Several outstanding papers and posters, which were presented at the conference, generated lively discussion and debate not only about scientific issues but also about policy and regulatory issues. The interactions between the attendees both during and outside the sessions resulted in many new contacts between scientists.

Panel discussions were held after each session to encourage discussion and debate. The final session was a panel discussion on the shape of the dose–response curve. This panel, needless to say, generated considerable debate and controversy. One of the most notable comments during the discussion was made by a physician from Bangladesh where there is probably the biggest arsenic public health problem in the world involving perhaps 80 million people at risk from elevated arsenic in their drinking water. Having observed the heated debate over how to extrapolate the dose–response relation to doses corresponding to less than 10 micrograms per liter or less, this scientist noted the irony when in Bangladesh there are perhaps tens of millions of people whose drinking water contains more than ten times that amount. While developed countries are considering spending perhaps billions of dollars per year to reach concentrations of 10 micrograms per liter or less, countries like Bangladesh, India and China are trying to deal with much more severe, epidemic scale, arsenic problems with millions of dollars or less.

The seeds of the conference(s) were sown at a meeting of the SEGH Executive Board in December, 1991. They agreed that it should form an Arsenic Task Force similar to the SEGH Lead in Soil Task Force that had been formed in the 1980s. It was clear that there was a growing controversy regarding the proposed changes in the US EPA Maximum Contaminant Level (MCL) for arsenic in drinking water. This is the enforceable standard for drinking water. In addition to impacting on water utilities, the development of the standard would also have the possibility for significant economic impacts on the cleanup of superfund sites and on the electric power industry (because of arsenic in fly ash).

The Task Force was formed in 1992 and chaired by Willard Chappell. An international conference seemed to be an excellent way to begin to compile the data and to determine what needed to be done in this area. The First SEGH International Conference on Arsenic Exposure and Health Effects was held in New Orleans in 1993. This conference was successful in attracting the top arsenic researchers in the world. It was followed by the

Workshop on Epidemiology and Physiologically-Based Pharmacokinetics that was held in Annapolis, MD in 1994. Perhaps the most significant outcome of the Workshop was the realization that there are many arsenic "hot spots" in the world. Although now widely recognized by the scientific community, significant public health problems existed in countries such as India, Thailand and China. The Second SEGH International Conference on Arsenic Exposure and Health Effects was held in 1995 in San Diego with one of its main purposes being to highlight the global aspects of the problem and most of the impact countries were represented. At that time, the biggest problem was in West Bengal, India where an estimated 30 million people are at risk from arsenic exposure in the ground water.

The Third SEGH International Conference on Arsenic Exposure and Health Effects (1998) was also held in San Diego and continued the theme of global impact of arsenic. In addition, two new countries with significant arsenic problems, Inner Mongolia and Bangladesh, were represented. The attendees were to learn that the Bangladesh problem could be larger than the one in West Bengal with a possible 80 million people (two-thirds of the population) at risk. This situation caught the attention of the media later in 1998 with the publication of a front-page article in the Nov. 16, 1998 *New York Times*. The article was syndicated and published in newspapers around the world.

Over 200 people, including the speakers and poster presenters, attended the conference. Of these, approximately one-third (64) were non-US citizens; the largest groups were from Asia (24) and Europe (22). There were 47 platform and 100 poster presentations. Twenty-two of the platform speakers were from outside the USA and represented 17 countries.

In addition to the continued focus on global impact, the conference also featured a session on mechanisms of cancer carcinogenesis. Several scientists presented their work on this important issue which is central to considerations of such questions as the shape of the dose–response curve at low doses. This latter issue was featured in the final session of the conference. Another session that was new and of great interest was on the treatment of victims of chronic arsenic poisoning.

We believe this was the most dynamic conference to date and this resulting monograph represents the state-of-the-art in arsenic research on a world-wide basis. We believe that it will contribute to the solution of the many problems existing throughout the world and are very grateful to the authors for their diligence and fine work and to the sponsors for the support that made it possible.

We are also deeply appreciative of the fine efforts of Rosemary Wormington of the Environmental Sciences Program of the University of Colorado at Denver who put in long hours as conference coordinator. She kept the entire Conference "going" and, more than anyone else, is responsible for the success of the conference.

Co-Editors:
Willard R. Chappell, *University of Colorado at Denver*
Charles O. Abernathy, *U.S. Environmental Protection Agency*
Rebecca L. Calderon, *U.S. Environmental Protection Agency*

Contents

List of Contributors

Abernathy, Charles
Office of Drinking Water
WH 550 D RM. 1037 East Tower
U.S. EPA, 401 M St., S.W.
Washington, DC 20460
USA

Aikawa, H.
Tokai University,
Tokai, Japan

Ajjimangkul, Sirinporn
Center of Primary Health Care
Southern Region, Ampur Muang
Nakorn Sri Thammarat Province 80000
Thailand

Andersen, Melvin E.
KS Crump Group/ICF Kaiser
P.O. Box 14348
Research Triangle Park, NC 27709
USA

Anderson, Henry
Wisconsin Bureau of Public Health
P.O. Box 2659
Madison, WI 53701
USA

Angle, C.R.
University of Nebraska Medica Center
Omaha, NE
USA

Aposhian, Mary M.
Dept. of Molecular and Cellular Biology
Life Sciences South Bldg.
University of Arizona
P.O. Box 210106
Tucson, AZ 85721-0106
USA

Aposhian, H. Vasken
Dept. of Molecular and Cellular Biology
Life Sciences South Bldg.
University of Arizona
P.O. Box 210106
Tucson, AZ 85721-0106
USA

Arif, Ashraf Islam
Dhaka Community Hospital
1089 Malibagh Chowdhury para
Dhaka
Bangladesh

Arnold, Lora L.
University of Nebraska Medical Center
Pathology/Microbiology Dept.
Omaha, NE 68198-3135
USA

Ayala-Fierro, Felix
Department of Pharmacology and Toxicology
College of Pharmacy, COP 228
University of Arizona, P.O. Box 21-0207
Tucson, AZ 85721
USA

Barber, David S.
Virginia-Maryland Reg. College of Vet. Medicine
Blacksburg, VA 24061
USA

Bencko, Vladimír
Institute of Hygiene & Epidemiology
Charles University of Prague
CZ 128 00 Praha 2, Studnickova 7
Czech Republic

Biggs, Mary Lou
School of Public Health
University of California, 140 Warren Hall
Berkeley, CA 94720-7360
USA

Biswas, Bhajan K.
School of Environmental Studies
Jadavpur University
Calcutta 700032
India

Bogdon, Gregory M.
Dept. of Molecular and Cellular Biology
University of Arizona
P.O. Box 210106
Tucson, AZ 85721-0106
USA

Brown, Kenneth
Kenneth G. Brown, Inc.
P.O. Box 16608
Chapel Hill, NC 27516-6608
USA

Browning, Steven R.
Dept.Preventive Medicine and Environ. Health
University of Kentucky
1141 Red Mile Rd., Ste. 201
Lexington, KY 40502
USA

Buchet, Jean
  Industrial Toxicology and Occupational Med. Unit
  Catholic University of Louvain
  Clos Chapelle-aux-Champs 30 - BTE 30.54
  B-1200 Brussels
  Belgium
Cano, Martin
  University of Nebraska Medical Center
  Pathology/Microbiology Dept.
  600 South 42nd St.
  Omaha, NE 68198-3135
  USA
Carter, Dean E.
  Department of Pharmacology and Toxicology
  University of Arizona
  College of Pharmacy
  Tucson, AZ 85721
  USA
Casarez, Elizabeth A.
  Department of Pharmacology and Toxicology
  University of Arizona
  P.O. Box 21-0207
  Tucson, AZ 85721
  USA
Cebrian, Mariano
  CINVESTAV-IPN
  AV. Instituto Politecnico Nacional
  2508. San Pedro Zacatenco, D.F.C.P. 07300
  Mexico
Centeno, J.A.
  Armed Forces Institute of Pathology
  Washington DC
  USA
Chakraborti, Dipankar
  Director, School of Environmental Studies
  Jadavpur University
  Calcutta 700032
  India
Chakraborty, D.
  University of Nebraska Medica Center
  Omaha, NE
  USA
Chao Yan, Chong
  Nelson Institute of Environmental Medicine
  New York University School of Medicine
  Tuxedo, NY 1098
  USA
Chatterjee, A.
  B.C. Roy Institute of Basic Medical Sciences
  University College of Medicine
  Calcutta
  India
Chen, Chien-Jen
  Graduate Institute of Epidemiology
  National Taiwan University
  1 Jen-Ai Road, Section 1
  Taipei 10018
  Taiwan
Chen, Tian-xin
  First Department of Pathology
  Osaka City University Medical School
  1-4-54 Asahi-machi
  Abeno-ku
  Osaka 545-8585
  Japan
Chiou, Hung-Yi
  National Taiwan University
  Taipei
  Taiwan
Chiswell, Barry
  Chemistry Department
  The University of Queensland
  Brisbane 4007
  Australia
Choprapawon, Chanpen
  Office of the Permanent Scretary, moph
  Thivanondh Road
  Ampur Muang
  Nonthaburi, 11000
  Thailand
Chouchane, Salem
  Nelson Institute of Environmental Medicine
  New York University School of Medicine
  Tuxedo, NY 10987
  USA
Chowdhury, Uttam K.
  School of Environmental Studies
  Jadavpur University
  Calcutta 700032
  India
Clewell, Harvey J.
  KS Crump Group
  ICF Kaiser International
  602 E. Georgia Ave.
  Ruston, LA 71270
  USA
Clifford, Dennis A.
  Department of Civil & Environmental Engineering
  University of Houston
  Houston, TX 77204-4791
  USA
Cohen, Samuel M.
  University of Nebraska Medical Center
  Pathology/Microbiology Dept.
  Omaha, NE 68198-3135
  USA
Cragin, D.W.
  Elf Atochem North America Inc.
  2000 Market Street
  Philadelphia, PA 19103-3222
  USA
Crecelius, E.A.
  Battelle Marine Sciences,
  1529 West Sequim Bay Road
  Sequim, WA 98382
  USA

Crump, Kenny S.
    KS Crump Group, ICF Kaiser International
    602 E. Georgia Ave.
    Ruston, LA 71270
    USA
Cullen, William R.
    Department of Chemistry
    University of British Columbia, 2036 Main Mall
    Vancouver
    Canada V6T 1Z1
Dai, G.J.
    School of Public Health
    China Medical University
    Shenyang 110001
    China
Dasgupta, J.
    Inst. Post Graduate Medical Education and Res.
    Calcutta
    India
De, B.K.
    Inst. Post Graduate Medical Education and Res.
    Calcutta
    India
Dekerkhove, Diane
    Department of Nuclear Physics
    Oxford University, Oxford
    UK
DeSesso, John M.
    Mitretek Systems, Inc., 7525 Colshire Drive
    McLean, VA 22102
    USA
Dhar, Ratan K.
    School of Environmental Studies
    Jadavpur University
    Calcutta 700032
    India
Donohue, Joyce M.
    Office of Science and Technology
    U.S. EPA, 401 M St., S.W.
    Washington 20460-0001
    USA
Dutta, S.
    B.C. Roy Institute of Basic Medical Sciences
    University College of Medicine,
    Calcutta
    India
Eickhoff, J.
    Environ, 4350 North Fairfax Drive
    Arlington, VA 22203
    USA
Eldan, Michal
    Luxembourg Industries (PAMOL) Ltd.
    27 Hamered St., P.O. Box 13
    Tel-Aviv 61000
    Israel
Fabianova, Eleonora
    Specialized State Institute of Public Health
    USA

Farago, Margaret
    Royal School of Mines, Imperial College
    Prince Consort Road
    London SW7 2BP
    UK
Farr, Craig F.
    Elf Atochem North America, Inc.
    2000 Market Street
    Philadelphia, PA 19103
    USA
Ferreccio, Catterina
    Pontificia Universidad Catolica de Chile/Gredis,
    Fleming 9840-Casa 4
    Las Condes, Santiago
    Chile
Focazio, Michael J.
    U.S. Geological Survey
    333 W. Nye Lane
    Carson City, NV 89706
    USA
Friedle, Matthias
    Institute of Sanitary Engineering, Water Quality
    and Solid Waste Management
    University of Stuttgart, Bandtäle 1
    70569 Stuttgart
    Germany
Froines, John
    Center for Occupational & Environmental Health,
    School of Public Health
    University of California, Los Angeles
    10833 LeConte Ave.
    Los Angeles, CA 90024-1772
    USA
Fukushima, Shoji
    First Department of Pathology
    Osaka City University Medical School
    1-4-54 Asahi-machi, Abeno-ku
    Osaka 545-8585
    Japan
Gailer, Jürgen
    Dept. of Molecular and Cellular Biology
    University of Arizona
    P.O. Box 210106
    Tucson, AZ 85721-0106
    USA
García-Vargas, Gonzalo G.
    CINVESTAV-IPN
    AV. Instituto Politecnico Nacional
    2508. San Pedro Zacatenco
    D.F.C.P. 07300
    Mexico
Garnett, Corrine M.
    National Research Centre for Environ. Toxicology
    The University of Queensland
    39 Kessels Rd.
    Coopers Plains
    Brisbane 4108
    Australia

Germolec, Dori
  Environmental Immunology Laboratory
  National Institute of Environmental Health
  Sciences, P.O. Box 12233
  Research Triangle Park, NC 27709
  USA
Ghosh, N.
  Inst. of Post Graduate Medical Education and Res.
  Calcutta
  India
Ghoshal, U.C.
  Inst. of Post Graduate Medical Education and Res.
  Calcutta
  India
Ghurye, Ganesh
  Department of Civil & Environmental Engineering
  University of Houston
  Houston, TX 77204-4791
  USA
Goessler, Walter
  Institut for Analytische Chemie
  Karl-Franzens-Universitat Graz
  Universitatsplatz 1, A-8010 Graz
  Austria
Gómez-Muñoz, Aristides
  CINVESTAV-IPN
  AV. Instituto Politecnico Nacional
  2508. San Pedro Zacatenco
  D.F.C.P. 07300
  Mexico
Götzl, Miloslav
  Department of Oncology
  District Hospital of Bohnice
  Slovak Republic
Grime, Geoffrey
  Department of Nuclear Physics
  Oxford University
  Oxford
  UK
Grissom, R.E.
  Division of Health Assessment and Consultation
  1600 Clifton Road N.E., ATSDR
  Atlanta, GA 30333
  USA
Haque, Reina
  School of Public Health
  University of California, 140 Warren Hall
  Berkeley, CA 94720-7360
  USA
Haufroid, Vincent
  Industrial Toxicology and Occupational Medicine
  Unit, Catholic University of Louvain
  Clos Chapelle-aux-Champs 30 - BTE 30.54
  B-1200 Brussels
  Belgium
Healy, Sheila M.
  Dept. of Molecular and Cellular Biology
  Life Sciences South Bldg.

University of Arizona
  P.O. Box 210106
  Tucson, AZ 85721-0106
  USA
Helsel, Dennis R.
  U.S. Geological Survey
  333 W. Nye Lane
  Carson City, NV 89706
  USA
Hernández, Maria C.
  Facultad de Medicina
  Universidad Autonoma de Coahuila
  AP 70228
  Mexico
Hertz-Picciotto, Irva
  University of North Carolina
  Department of Epidemiology
  Chapel Hill, NC 27599
  USA
Hicks, Jeffrey B.
  Geomatrix Consultants
  USA
Hindmarsh, J.T.
  Division of Biochemistry
  Ottawa General Hopital
  Ottawa, Ontario
  Canada K1H 8L6
Hoet, Perine
  Industrial Toxicol. & Occupational Medicine Unit
  Catholic University of Louvain
  Clos Chapelle-aux-Champs 30, BTE 30.54
  B-1200 Brussels
  Belgium
Holson, Joseph F.
  WIL Research Laboratories, Inc.
  1407 George Road
  Ashland, OH 44805
  USA
Hopenhayn-Rich, Claudia
  Dept. Preventive Medicine and Environ. Health
  University of Kentucky
  1141 Red Mile Rd., Ste. 201
  Lexington, KY 40502
  USA
Hsu, Lin-I
  National Taiwan University
  Taipei
  Taiwan
Hsueh, Yu-Mei
  National Taiwan University
  Taipei
  Taiwan
Hu, Yu
  Nelson Institute of Environmental Medicine
  New York University School of Medicine
  57 Old Forge Road
  Tuxedo, NY 10987
  USA

Irgolic, Kurt J.
    Institut for Analytische Chemie
    Karl-Franzens-Universitat Graz
    A-8010 Graz, Universitatsplatz 1
    Austria
Kabir, Saiful
    Dhaka Community Hospital
    Dhaka
    Bangladesh
Kavanagh, Peter,
    Centre for Environmental. Technology
    Royal School of Mines, Imperial College
    of Science, Technology and Medicine
    Prince Consort Road
    London SW7 2BP
    UK
Knobeloch, Lynda
    Wisconsin Bureau of Public Health
    P.O. Box 2659
    Madison, WI 53701
    USA
Kosnett, Michael J.
    University of Colorado Health Sciences Center
    c/o 1630 Welton St., Suite 300
    Denver, CO 80202
    USA
Kuehnelt, Doris
    Institut for Analytische Chemie
    Karl-Franzens-Universitat Graz
    A-8010 Graz, Universitatsplatz 1
    Austria
Kuo, Tsung-Li
    Department of Forensic Medicine
    College of Medicine
    National Taiwan University
    Taipei 10018
    Taiwan
Lai, Vivian W.-M.
    Department of Chemistry
    University of British Columbia
    2036 Main Mall
    Vancouver BC
    Canada V6T 1Z1
Le, X. Chris
    Department of Public Health Science,
    Faculty of Medicine
    13-103 CSB, University of Alberta
    Edmonton, AB
    Canada T6G 2G3
Lee, Chyi Chia R.
    First Department of Pathology
    Osaka City University Medical School
    Osaka 545-8585
    Japan
Leininger, Joel
    Nat. Institute of Environmental Health Sciences
    Research Triangle Park, NC
    USA

Lewis, Denise Riedel
    Epidemiology and Biomarkers Branch/HSD
    MD-58A, NHEERL/US EPA
    Research Triangle Park, NC 27711
    USA
Li, F.J.
    School of Public Health, China Medical University
    Shenyang 110001
    China
Lison, Dominique
    Industrial Toxicology and Occupational Med. Unit
    Catholic University of Louvain
    Clos Chapelle-aux-Champs 30 - BTE 30.54
    B-1200 Brussels
    Belgium
Luster, Michael I.
    Center for Disease Control and Prevention
    National Inst. for Occupational Safety and Health
    1095 Willowdale Road
    Morgantown, WV 26505-2888
    USA
Ma, Heng Z.
    Institute of Endemic Disease for Prevention and
    Treatment in Inner Mongolia
    Huhhot, Inner Mongolia
    China 010020
Ma, Mingsheng
    Department of Public Health Science
    13-103 CSB, University of Alberta
    Edmonton, AB
    Canada T6G 2G3
Mandal, Badal K.
    School of Environmental Studies
    Jadavpur University
    Calcutta 700032
    India
Maria Del Razo, Luz
    Section of Environmental Toxicology
    Department of Pharmacology and Toxicology
    CINVESTAV-IPN, PO. Box 14-740
    Mexico City
    Mexico
Mazumder, D.N. Guha
    Department of Gastroenterology
    Inst. Post Graduate Medical Education & Research
    244 Acharya Jagadish Chandra Bose Rd
    Calcutta, 700020
    India
Meacher, O.M.
    Department of Community and Environmental
    Medicine, University of California-Irvine,
    Irvine, CA 92696-1825
    USA
Menzel, O.B.
    Dept. of Community and Environmental Medicine
    University of California-Irvine,
    Irvine, CA 92696-1825
    USA

Moore, Lee
School of Public Health
University of California
Berkeley, CA 94720-7360
USA

Moore, Michael R.
Nat. Research Centre for Environmental Toxicology
The University. of Queensland
39 Kessels Rd., Coopers Plains
P.O. Box 594, Archerfield 4108
Brisbane 4108
Australia

Morales, Knashawn H.
Department of Biostatistics
Harvard School of Public Health
655 Huntington Avenue
Boston, MA 02115
USA

Mumford, Judy L.
NHEERL, U.S. Environmental Protection Agency
Research Triangle Park, NC 27711
USA

Na, Yifei
First Department of Pathology
Osaka City University Medical School
1-4-54 Asahi-machi, Abeno-ku
Osaka 545-8585
Japan

Ng, Jack C.
Nat. Research Centre for Environmental Toxicology
The University. of Queensland
39 Kessels Rd.
Coopers Plains
Queensland 4108
Australia

Okoji, Russel S.
Center for Occupational & Environmental Health
University of California, Los Angeles
10833 LeConte Ave.
Los Angeles, CA 90024-1772
USA

Peralta, Cecilia
Dept Preventive Med. and Environmental Health
University of Kentucky
1141 Red Mile Rd., Ste. 201
Lexington, KY 40502
USA

Peraza, Marjorie A
Department of Pharmacology and Toxicology
College of Pharmacy, COP 228
University of Arizona, P.O. Box 21-0207
Tucson, AZ 85721
USA

Powell, LaTanya A.
Dept. of Molecular and Cellular Biology
University of Arizona, P.O. Box 210106
Tucson, AZ 85721-0106
USA

Powell, Jonathan
Department of Gastroenterology
St. Thomas' Hospital
London
UK

Pradipasen, M.
Faculty of Public Health
Mahidol University
Bangkok 10400
Thailand

Qi, Lixia
Nat. Research Centre for Environmental Toxicology
The University. of Queensland
39 Kessels Rd., Coopers Plains
P.O. Box 594, Archerfield 4108
Brisbane
Australia

Quamruzzaman, Quazi
Dhaka Community Hospital
1089 Malibagh Chowdhury para
Dhaka
Bangladesh

Radabaugh, Timothy R.
Dept. of Molecular and Cellular Biology
Life Sciences South Bldg.
University of Arizona
P.O. Box 210106
Tucson, AZ 85721-0106
USA

Rahman, Mahmuder Rahman
Dhaka Community Hospital
1089 Malibagh Chowdhury para
Dhaka
Bangladesh

Rames, Jiri
Institute of Hygiene & Epidemiology
Charles University of Prague
CZ 128 00 Praha 2
Studničkova 7
Czech Republic

Rossman, Toby G.
Nelson Institute of Environmental Medicine
NYU Medical Center
57 Old Forge Road
Tuxedo, NY 10987
USA

Rott, Ulrich
Institute of Sanitary Engineering, Water Quality
and Solid Waste Management
University of Stuttgart
Bandtäle 1
70569 Stuttgart
Germany

Roy Chowdhury, Tarit
School of Environmental Studies
Jadavpur University
Calcutta 700032
India

Roy, Shibtosh
  Dhaka Community Hospital
  1089 Malibagh Chowdhury para
  Dhaka
  Bangladesh
Roy, Sibtosh
  School of Environmental Studies
  Jadavpur University
  Calcutta 700032
  India
Roy, B.K.
  Inst. of Post Graduate Medical Education and Res.
  Calcutta
  India
Ryan, Louise
  Department of Biostatistics
  Harvard School of Public Health and Dana Farber
  Cancer Institute, 44 Binney Street
  Boston, MA 02115
  USA
Saha, J.
  Inst. of Post Graduate Medical Education and Res.
  Calcutta
  India
Samanta, Gautam
  School of Environmental Studies
  Jadavpur University
  Calcutta 700032
  India
Sancha, Ana Maria
  Department of Civil Engineering,
  University of Chile, Blanco Encalada 2120
  Santiago
  Chile
Santra, A.
  Inst. of Post Graduate Medical Education and Res.
  Calcutta
  India
Schoof, Rosalind A.
  Exponent, Inc.
  15375 SE 30th Place,
  Bellevue, WA 98007
  USA
Scialli, Anthony R.
  Georgetown University Medical Center
  3800 Reservoir Road NW
  Washington, DC 20007
  USA
Seawright, Alan A.
  Nat. Research Centre for Environ. Toxicology
  The University. of Queensland
  39 Kessels Rd., Coopers Plains
  P.O. Box 594, Archerfield 4108
  Brisbane
  Australia
Shipp, Annette M.
  KS Crump Group
  ICF Kaiser International, 602 E. Georgia Ave.

Ruston, LA 71270
  USA
Siripitayakunkit, Unchalee
  Division of Epidemiology, Ministry of Public Health
  Thiwanon Rd., Muang District
  Nonthaburi 11000
  Thailand
Smith, Allan
  School of Public Health
  University of California
  140 Warren Hall
  Berkeley, CA 94720-7360
  USA
Snow, Elizabeth T.
  Nelson Institute of Environmental Medicine,
  New York University School of Medicine,
  57 Old Forge Road,
  Tuxedo, NY 10987
  USA
St. John, Margaret K.
  University of Nebraska Medical Center
  Pathology/Microbiology Dept., 600 South 42nd St.
  Omaha, NE 68198-3135
  USA
Steinmaus, Craig
  School of Public Health
  University of California, 140 Warren Hall
  Berkeley, CA 94720-7360
  USA
Styblo, Miroslav
  Department of Pediatrics,
  University of North Carolina CB# 8180
  Chapel Hill, NC 27599
  USA
Sun, Gui Fan
  School of Public Health
  China Medical University
  Shenyang 110001
  China
Sun, Tian Z.
  Institute of Endemic Disease for Prevention and
  Treatment in Inner Mongolia
  Huhhot, Inner Mongolia
  China 010020
Susten, A.S.
  Division of Health Assessment and Consultation
  1600 Clifton Road N.E., ATSDR
  Atlanta, GA 30333
  USA
Thomas, David J.
  Pharmacokinetics Branch
  Experimental Toxicology Division
  NHEERL, U.S. Environmental Protection Agency
  Research Triangle Park, NC 27711
  USA
Thornton, Iain
  Imperial College of Science,
  Technology and Medicine

Royal School of Mines
Prince Consort Road
London, SW7 2BP
UK

Tripp, Anthony R. Tripp
Department of Civil & Environmental Engineering
University of Houston
Houston, TX 77204-4791
USA

Tseng, Chin-Hsiao
National Taiwan University
Taipei
Taiwan

Vahter, Marie
Institute of Environmental Medicine
Karolinska Institutet, Box 210
S-171 77Stockholm
Sweden

Vega, Libia
Environmental Immunology Laboratory
Nat. Institute of Environmental Health Sciences
P.O. Box 12233
Research Triangle Park, NC 27709
USA

Visudhiphan, P.
Ramathibodi Hospital
Mahidol University
Bangkok 10400
Thailand

Vorapongsathorn, T.
Faculty of Public Health, Mahidol University
Bangkok 10400
Thailand

Wang Changqing
Department of Chemistry
University of British Columbia
Vancouver
Canada V6T 1Z1

Wanibuchi, Hideki
First Department of Pathology
Osaka City University Medical School
1-4-54 Asahi-machi
Abeno-ku
Osaka 545-8585
Japan

Warzecha, Charles
Wisconsin Bureau of Public Health
P.O. Box 2659
Madison, WI 53701
USA

Watkins, Sharon A.
U.S. Geological Survey
Carson City, NV 89706
USA

Welch, Alan H.
U.S. Geological Survey
333 W. Nye Lane

Carson City, NV 89706
USA

Wildfang, Eric K.
Dept. of Molecular and Cellular Biology
University of Arizona
Tucson, AZ 85721-0106
USA

Winski, Shannon L.
University of Colorado Health Science Center
School of Pharmacy
Denver, CO 80262
USA

Wu, Ke G.
Institute of Endemic Disease for Prevention and
Treatment in Inner Mongolia
Huhhot, Inner Mongolia
China 010020

Wu, Meei-Maan
Institute of Biomedical Sciences
Academia Sinica
Taipei 11529
Taiwan

Xia, Ya J.
Institute of Endemic Disease for Prevention and
Treatment in Inner Mongolia
Huhhot, Inner Mongolia
China 010020

Yager, Janice W.
Electric Power Research Institute
3412 Hillview Avenue
Palo Alto, CA 93404
USA

Yamamoto, Shinji
First Department of Pathology
Osaka City University Medical School
1-4-54 Asahi-machi
Abeno-ku
Osaka 545-8585
Japan

Yamauchi, H.,
St. Marianna Medical University
Japan

Yoshida, T.,
Tokai University
Japan

Yost, L.J.
Exponent, Inc. (formerly PTI Environmental
Services)
15375 SE 30th Place
Bellevue, WA 98007
USA

Zakharyan, Robert
Dept. of Molecular and Cellular Biology
Life Sciences South Bldg.
University of Arizona
Tucson, AZ 85721-0106
USA

Arsenic Exposure and Health Effects
W.R. Chappell, C.O. Abernathy and R.L. Calderon (Editors)

Arsenic in the Global Environment: Looking Towards the Millennium

Iain Thornton

ABSTRACT

Health problems associated with exposure to arsenic continue to command world attention. This paper focuses on a number of recent investigations and research developments, and attempts to provide a bridge with those ideas covered in the 2nd International Conference on Arsenic Exposure and Health Effects in 1995. The locations at which natural or anthropogenic sources of arsenic are considered to be of concern continue to grow and include those associated with the burning of arsenic-rich coal, and with mining and smelting activities. Issues to be considered in health-based risk assessment for arsenic include the essentiality of the element, speciation and bioavailability, and the nature of the dose–response curve. Recent developments in the regulatory process are briefly discussed. The full extent of arsenic-related health problems has still to be fully identified and quantified.

Keywords: arsenic geochemistry, location-specific studies, essentiality, exposure assessment, speciation, bioavailability, environmental regulation, remediation

INTRODUCTION

Worldwide concern with arsenic and its influence on human health has increased markedly since the Second International Conference held in San Diego in 1995. The debate has been stimulated by numerous publications in the scientific press, including the review article of the Arsenic Task Force of the Society for Environmental Geochemistry and Health which focused on future research needs to provide better scientific underpinning of the risk assessment process (Chappell et al., 1997). This paper aims to build on a series of ideas that are introduced both in that publication and in the closing contribution to the 1995 Conference (North, Gibb and Abernathy, 1997).

Three main themes are covered (i) *places/locations* in which either natural or anthropogenic sources of arsenic are considered to be sufficiently large and/or extensive to be of concern; (ii) *scientific issues* that influence arsenic mobility, pathways, exposure and effects and (iii) *actions*, including the need for remediation and development of remedial strategies, necessary to protect potentially exposed populations.

LOCATIONS

A listing and brief descriptions of location-specific studies were presented to the 1995 Conference (Thornton and Farago, 1997). This earlier paper covered the geochemistry of arsenic and included data on concentrations reported in rocks, soils and plant materials in both uncontaminated and contaminated environments.

There have also been several reports covering diffuse emissions of arsenic into the atmosphere on global, regional and national scales, classified into various categories of both natural and anthropogenic sources. For example, Nriagu (1989) listed data for worldwide natural emissions of arsenic into the atmosphere in the 1980s totalling 1.1 to 23.5 tonnes per year, derived mostly from volcanoes, wind-born soil particles, sea spray and biogenic processes. However, the uncertainty on this 'estimated emission' is large and reflects the limited data on which it is made. Corresponding values for anthropogenic emissions were mostly attributed to pyrometallurgical non-ferrous metal mining and production, iron and steel manufacturing and coal combustion (Nriagu and Pacyna, 1988). It has since been proposed that coal combustion alone accounts for 20 percent of the atmospheric emission, and also that arsenic from coal ash may be leached into soils and waters (Alloway and Ayres, 1993). Pacyna (1996, 1998) later provided more detailed estimates of atmospheric emissions of arsenic in Europe, peaking in the period 1960–65 at around 11,500 t per year and falling to 4570 t per year by 1985. At this later date, over half of the emission was from USSR (Europe) of which some 75 per cent was attributed to non-ferrous metal manufacturing. Forward projections by Pacyna (1996), estimate emissions for Europe to fall to ca. 1900 t As in the year 2000 and ca. 1600 t in 2010, with Russia accounting for around 30% of the total.

These emission data are of limited application, but can be used to give broad indications of the countries in which arsenic deposition may be expected, and, on a more local scale, those industrial locations around which arsenic deposition may be expected to occur and where land contamination may present a hazard to health.

In developing further the list of places/locations where arsenic is of concern, this may be based on (i) those countries or districts in which human health problems have already been related to exposure and (ii) those where elevated concentrations of arsenic in the environment have caused scientists and medical clinicians to look for possible health effects.

Warner North (1997) suggested we made an arbitrary division of these two groups into:
(a) high dose regions where people ingested from one to several milligrams of arsenic per day and where various skin and internal cancers and other health problems have been found;
(b) medium or intermediate dose regions, where intake ranged from 100 microgrammes to 1 milligram per day, where health outcomes were less certain or not as yet proven and in which epidemiological studies are essential; and

(c) low dose regions with an intake of below or around 100 micrograms per day, corresponding to the level of the present US drinking water standard.

The list of high and medium dose regions continues to grow and I would foresee that it will continue to grow further into the new millenium. Pressures to meet the requirements of the rapidly growing world population for both drinking water and food will put pressure on both ground and surface water resources, sometimes interacting with geological beds of uncertain chemical stability. Strata enriched in arsenic include some black shales, pyritic rich rocks, volcanic sediments and geothermal deposits. Geological strata have been shown to contaminate drinking water in many countries, ranging from Minnesota in the USA, Kutahya Province in Turkey, to Inner Mongolia, China as noted in posters shown at the 1998 Conference in San Diego. Recent discoveries of arsenic-rich deposits that may impinge on water and/or soil quality have been made in the Caribbean with areas of arsenic-rich soil derived from transported volcanic materials in Jamaica, and rivers with geothermal inputs in the Island of Dominique. As yet there is no direct information on possible human exposure routes. Similar geologic materials are to be expected in other islands within the Lesser Antilles, including the geologically unstable island of Monserrat.

Emissions from the burning of arsenic-rich coal have continued to attract attention with epidemiological research linked to environmental monitoring. A new EU funded research project co-ordinated by Imperial College with partners in Germany, Austria, the Czech Republic and Slovakia comprises a case-control epidemiological study of skin cancer around coal-fired power plant in the county of Prievidza, Slovakia, where, for many years, arsenic-rich coal was burnt and where emissions have led to contamination of nearby urban areas and to elevated concentrations in hair, urine and blood in local children (Bencko and Symon, 1977).

Attention has previously been drawn to the presence of arsenic in mining and smelting residues, with examples from the United Kingdom, Greece, Thailand and Ghana (Thornton and Farago, 1997). The association between arsenic and gold mineralisation has been long recognised and arsenic continues to be used as a pathfinder element in geochemical exploration for gold (Nichol et al., 1994). Elevated concentrations of arsenic in mining wastes and in land, surface and groundwaters contaminated from gold and copper mining and processing are now widely recognised and by their frequency imply a global significance. For example, in North America (a) toxic inorganic arsenic species found in elevated concentrations in freshwater biota have been attributed to pollution from two gold mines near Yellowknife, North West Territories (Koch et al., 1998), and (b) arsenic associated with goethite and jarosite (sulphide minerals) in tailings from mine wastes and processed gold ores in the Mother Lode district of California has been leached due to seasonal flooding into the lake waters of the Don Pedro Reservoir, where mineralogical studies are now focusing on solubility and potential bioavailability (Savage et al., 1998). Recent studies in the Iberian Copper Belt in South-east Portugal have focused on the San Domingos mine, where copper ores and pyrite were worked in the late 19th century. Local soils containing 1 per cent or more of arsenic were found in a small urban community, where mine wastes had been used in the foundations of houses and as a material covering for roads (Fehily, 1998). A heavily contaminated river draining this mining area was characterised by the large amounts of precipitated iron (ochre) in river sediments which had acted as a sink for arsenic dispersed from the mine tailings (Thornton, 1998). This river drains into a large reservoir used for drinking water and irrigation purposes; dilution with uncontaminated water would seem to have removed potential risk of elevated human exposure. It is however recognised that environmental monitoring has been limited in this and in similar mineralised areas of Southern Europe.

Arsenic enrichment in soils (up to 430 μg/g) together with Cu and Pb has been recorded at the site of the Avoca mine in Ireland. However, little was released by partial extraction procedures in the laboratory and it was concluded that the arsenic was bound within the lattice of sulphide minerals (Gallagher and O'Connor, 1997).

Agricultural land. Reference has been made in discussions by the Arsenic Task Force to the large areas of land in the United States and elsewhere contaminated with arsenic used (a) as a growth promoter in pigs, where phenyl arsenic acid and its derivatives are found in the litter, (b) as an active ingredient in copper–chrome-arsenate wood preservative (c) for many years as a pesticide to treat fruit and other crops and (d) as a defoliant for cotton. It is difficult to quantify the extent and degree of such contamination, though many thousands of hectares of land are thought to be affected.

ISSUES

Many of the issues in the arsenic debate are generic and of fundamental importance when considering the process of human health based risk assessment for arsenic.

Essentiality of Arsenic

The evidence would seem to be compelling that arsenic has a physiological role affecting methionine metabolism in animals. Essentiality of arsenic has been shown in studies with rats, hamsters, chickens, goats and minipigs (Anke et al., 1976; Uthus, 1994). It has also been suggested that arsenic may impose a risk to hemodialysis patients (Mayer et al., 1993), and that this establishes a role for its essentiality to the human (Uthus, 1994). However, it is likely that under normal environmental conditions, natural ambient concentrations of arsenic will be sufficient to meet the nutritional needs of man.

However, in considering a deficiency–sufficiency–toxicity scenario, if we assume a linear dose–response curve for arsenic in man, it is feasible that a risk assessment process, including the incorporation of safety factors, will result in the calculation of a "safe" level of arsenic intake below the requirement for healthy physiological function. Similar scenarios have been portrayed for the essential micronutrients copper and zinc. The different status of arsenic is however that such a calculation would forecast "safe" levels of arsenic intake/exposure below those to which the population are already exposed.

Exposure Assessment

As with other potentially toxic elements, it is important to stress the importance of determining all sources of exposure and not only the prime route (i.e. water in West Benghal, soil and dust ingestion in South West England). Total exposure will be the sum of exposures from the diet, drinking water, direct ingestion of soil and dust, inhalation and percutaneous absorption. Measurements of arsenic in materials comprising the prime source may well underestimate total exposure and in doing so, confuse the interpretation of the dose–response curve in the risk assessment paradigm.

Speciation and Bioavailability

To a geochemist and environmental chemist, this is perhaps the most important missing contribution to present-day risk assessment. In my presentation to the 1995 Conference (Thornton and Farago, 1997), I stressed the urgency to develop and accept methods for the identification and quantification of the mineral and chemical forms of arsenic in rocks, soils, sediments and atmospheric particles that control its solubility and pathways into waters, the food chain and man.

The solubilities of arsenic species play a major part in controlling its pathways in the soil–plant; soil–plant–animal (human); soil (dust)–human; rock–soil–water–human pathways. Over 200 mineral species of arsenic have been identified. The geochemistry of arsenic has been reviewed by Thornton and Farago (1997) and its behaviour in soil by O'Neill (1990).

Although it is realised that both mineral and chemical forms of arsenic in environmental media can change over time, the rates of change or ageing are not understood. Such changes can be important, particularly when the ageing process may lead to the formation of in-soluble arsenic species, such as ferric arsenate, which has a low bioavailability—to both plants and animals/man. The formation of this insoluble compound/mineral by the chemical weathering of other arsenic species in the surface environment may be regarded as a process of natural remediation (Thornton, 1996). Its presence may also explain the low uptake and the apparent lack of adverse health effects in people living on very heavily contaminated soils and mine/smelter wastes, such as those in the historical arsenic mining areas of South-west England (Farago et al., 1997). Examination of soils and mine wastes by a sequential extraction procedure (Woolson et al., 1973) showed water-extractable As in agricultural soils to be as little as 0.05–0.03 percent, and in mine wastes to range from 0.02–1.2 percent. Ninety-three percent of the total arsenic in mine spoil and nearby soils was present in the Fe—organic and residual fractions, accounting for the low bioavailability (Kavanagh et al., 1997). At the Anaconda smelting site in Montana, low bioaccessibility and bioavailability of arsenic has been attributed to the presence of sparingly soluble As-bearing phases in soils and house dusts, including mainly metal-arsenic oxides and phosphates (Davis et al., 1996). Further, a proportion of the arsenic minerals was shown to be encapsulated by carbonate and silicate rinds.

Dose Response

The debate on the shape of the dose–response curve for arsenic in relation to cancer prevalence and whether this is linear or non-linear was addressed by North et al. (1997) and is detailed in later chapters. It is possible that this may vary on a location-specific basis. Further, one should not ignore the possibility of differences between populations in relation to their genetic susceptibility or tolerance to arsenic exposure and to the possibility of adaptation or the development of tolerance over a period of time.

ACTIONS

The Regulatory Process

The current position in the US regarding the *drinking water* directive has been, and continues to be, one of the main driving forces in the arsenic debate, the focus of the SEGH Task Force and the underpinning of the International Conferences on Arsenic Exposure and Health Effects. The question still remains in the US—should the drinking water standard for arsenic be reduced from 50 µg/l and if so, to what level? On a precautionary basis, the WHO have already introduced a provisional guideline value of 10 µg/l (WHO, 1993).

Guidelines/regulations for soil continue to cause problems. In the UK, guidelines pub-lished for the redevelopment of contaminated land are at present referred to as "tentative threshold trigger concentrations". Values for arsenic are 10 µg/g for urban development and gardens and 40 µg/g for parks and amenity areas (ICRCL, 1987). Below these values there is no cause for concern, above them—local government etc. must use professional judgement as to whether remediation may be necessary. There is at present no agreed action value above which remediation is required and at least one local authority has imposed its own action value of 140 µg/g As.

These threshold trigger values are unrealistic and would, for instance, question further redevelopment over some 700 km² of the counties of Devon and Cornwall in the south-west of England where extensive arsenic contamination from historical mining and smelting of arsenopyrite has been recorded (Abrahams and Thornton, 1987).

Somewhat different approaches have been taken in Australia where a site-specific risk assessment has been applied to contaminated land by the National Environmental Health

Forum and in the Netherlands where, until recently, the principle of multi-functionality has driven the regulatory process.

Now in the UK, the CLEA (Contaminated Land Exposure Model) has been developed, and proposed action values requiring remediation are aimed at "fitness for purpose" (Ferguson and Denner, 1993). Using this approach, preliminary soil action guideline values have been proposed and cited in a National (Australian) Environmental Health Forum Monograph (Taylor and Langley, 1996), as below:

	As (μg/g)
Residential with garden	175
Residential without garden	300
Allotment	250
Parks, open space	300
Commercial/industrial	1000

Remediation

Technologies for partial removal of arsenic in drinking water supplies are discussed elsewhere in this and the previous volumes and include coagulation, with ferric chloride and alum, anion exchange with chloride-form strong base resins, the use of novel iron-impregnated absorbent, and the *in-situ* treatment of groundwater.

Treatment of contaminated soil and mine waste presents greater difficulty and remediation usually requires removal and disposal. However, *in-situ* methods have been tested on waste materials in Southwest England using a range of industrial minerals; the most suitable candidates for amendments were found to be natural zeolites and diatomaceous earth (kieselguhr) (Atkinson et al., 1990).

THE FUTURE

Contamination of the environment with arsenic from both natural and anthropogenic sources is widespread, occurs in many parts of the world and may be regarded as a global issue. I would predict that we will continue to find many more situations where contamination of surface and sub-surface waters and/or soils will result in the intermediate exposure scenario. New occurrences will be found particularly in Central and Eastern Europe and the developing world.

The response to these by national government and international bodies will depend on the results of a number of current site-specific studies to determine exposure and factors influencing exposure, and the more generic issue of the dose–response curve to predict risk. It is emphasised that bioaccessibility and exposure will be influenced by the nature of the contamination, including the chemical and mineral forms of arsenic and their solubility.

In conclusion, the arsenic issue is a global one and will be with us, probably, at an increasing level, into the millennium. The full significance of related health outcomes has still to be fully identified and quantified. Whether adverse health effects will be limited to those situations where exposure is high, or whether long term low-level exposure can give rise to ill health, has still to be decided. The need for remedial action will depend on the outcome of such studies.

REFERENCES

Abrahams, P.W. and Thornton, I. 1987. Distribution and extent of land contaminated by arsenic and associated metals in mining regions in Southwest England. *Trans. Inst. Mining Metall. (Section B: Appl. Earth Sci.),* **96**, 131–138.

Alloway, B.J. and Ayres, D.C. 1993. *Chemical Principles of Environmental Pollution.* Blackie Academic and Professional, London, 291 pp.

Atkinson, K., Edwards, R.P., Mitchell, P.B., and Waller, C. 1990. Roles of industrial minerals in reducing the impact of metalliferous mine waste in Cornwall. *Trans. Inst. Mining Metall.* (Sect. A: Min. Ind.), **99**, A158–172.

Bencko, V. and Symon, K. 1977. Health aspects of burning coal with a high arsenic content. 1. Arsenic in hair, urine and blood in children residing in a polluted area. *Env. Res.*, **13**, 378–385.

Chappell, W.R., Beck, B.D., Brown, K.G., Chaney, R., Cothern, C.R., Irgolic, K.J., North, D.W., Thornton, I., and Tsongas, T.A. 1997. Inorganic arsenic: a need and an opportunity to improve risk assessment. *Environ. Health Perspect.*, **105**, 1060–1067.

Davis, A., Ruby, M.V., Bloom, M., Schoof, R., Freeman, G., and Bergstrom, P.D. 1996. Mineralogic constraints on the bioavailability of arsenic in smelter-impacted soils. *Environ. Sci. Technol.*, **30**, 392–399.

Farago, M.E., Thornton, I., Kavanagh, P., Elliott, P., and Leonardi, G.S. 1997. Health aspects of human exposure to high arsenic concentrations in soil in south-west England. In: *Arsenic Exposure and Health Effects* (eds. C.O. Abernathy, R.L. Calderon and W.R. Chappell). Chapman and Hall, London, pp. 210–226.

Fehily, L.J. 1998. Assessment of Contamination from Sao Domingos Copper Mine, South East Portugal. Unpublished MSc Thesis. Imperial College of Science, Technology and Medicine.

Ferguson, C.C. and Denner, J. 1993. Soil guideline values in the UK: new risk-based approach. In: *Contaminated Soil '93* (eds. F. Arendt, G. J. Annokkee, R. Bosman and W. J. van den Brink). Kluwer Academic Publishers, Dordrecht, pp. 365–372.

Gallagher, V. and O'Connor, P. 1997. Characterisation of the Avoca Mine Site: Geology, Mining Features, History and Soil Contamination Study. Geological Survey of Ireland Technical Report MS97/1, April 1997.

ICRCL, 1987. Guidance on the Assessment and Redevelopment of Contaminated Land. ICRCL 59/83, Second Edition. Department of the Environment, London.

Kavanagh, P.J., Farago, M. E., Thornton, I., and Braman, R.S. 1997. Bioavailability of arsenic in soil and mine wastes of the Tamar valley, SW England. *Chem. Speciation Bioavail.*, **9**, 77–81.

Koch, I., Wang, L., Cullen, W.R., Ollson, C.A. and Reimer, K.J. 1998. Arsenic speciation in Yellowknife biota: impact on the terrestial environment. In: *Third International Conference on Arsenic Exposure and Health Effects*. Book of Posters, University of Colorado at Denver, 1998.

Nichol, I., Hale, M. and Fletcher, W.K. 1994. Drainage geochemistry in gold exploration. In: *Drainage Geochemistry* (eds. M. Hale and J.A. Plant), Elsevier Science, Amsterdam, pp. 499–559.

North, D.W., Gibb, H.J. and Abernathy, C.O. 1997. Arsenic: past present and future considerations. In: *Arsenic Exposure and Health Effects* (eds. C.O. Abernathy, R.L. Calderon and W.R. Chappell). Chapman and Hall, London, pp. 406–423.

Nriagu, J.O. and Pacyna, J.M. 1988. Quantitative assessment of worldwide contamination of air, water and soils by trace elements. *Nature*, **333**, 134–139.

Nriagu, J.O. 1989. *Nature*, **338**, 47–49.

Pacyna, J.M. 1996. Atmospheric Emissions of Heavy Metals for Europe. A Final Report for the International Institute of Applied Systems Analysis, Hagan Monday, January 1996.

Pacyna, J.M. 1998. Source inventories for atmospheric trace metals. In: *Atmospheric Particles* (eds. R.M. Harrison and R. Van Grieken). John Wiley and Sons, pp. 385–423.

Savage, K.S., Bird, D.K., O'Day, P. and Waychunas, G.A. 1998. Atomic environment of arsenic in mine waste rock, tailings and their weathering products at San Pedro Reservoir, Tuolumne County, California. In: *Third International Conference on Arsenic Exposure and Health Effects*. Book of Posters, University of Colorado at Denver, 1998.

Taylor, R. and Langley, A. 1996. *Exposure Scenarios and Exposure Settings*. National Environmental Health Forum Monograph: Soil Series No. 2.

Thornton, C.F. 1998. Drainage Reconnaissance Survey of the San Domingos Mine Area, Portugal. Unpublished MSc Thesis. Imperial College of Science, Technology and Medicine.

Thornton, I. 1996. Sources and pathways of arsenic in the geochemical environment: health implications. In: *Environmental Geochemistry and Health*. Geological Society Special Publication, No. 113, (eds. J. D. Appleton, R. Fuge and G. J. McCall), The Geological Society of London, pp. 153–161.

Thornton, I. and Farago, M. E. 1997. The geochemistry of arsenic. In: *Arsenic Exposure and Health Effects* (eds. C.O. Abernathy, R.L., Calderon and W.R. Chappell). Chapman and Hall, London, pp. 1–16.

Woolson, E.A., Axley, J.H., and Kerney, P.C. 1973. The chemistry in phytotoxicity of arsenic in soils, II. Effects of time and phosphorus. *Proc. Soil Sci. Soc. Am.*, **37**, 254.

World Health Organisation (WHO) 1993. Recommended Guidelines for Drinking Water.

Arsenic Exposure and Health Effects
W.R. Chappell, C.O. Abernathy and R.L. Calderon (Editors)
1999 Elsevier Science B.V.

Arsenic in Ground Water Supplies of the United States

Alan H. Welch, Dennis R. Helsel, Michael J. Focazio, Sharon A. Watkins

ABSTRACT

High arsenic concentrations in ground water have been documented in many areas of the United States. Within the last decade, parts of Maine, Michigan, Minnesota, South Dakota, Oklahoma, and Wisconsin have been found to have widespread arsenic concentrations exceeding 10 µg/L. These high concentrations most commonly result from: (1) upflow of geothermal water, (2) dissolution of, or desorption from, iron-oxide, and (3) dissolution of sulfide minerals. Because the MCL for arsenic is currently being evaluated, estimating the exceedance frequency for different arsenic concentrations in regulated water supplies is particularly timely. Estimates of the frequency of exceedance, which are based on analyses of about 17,000 ground water samples, suggest that about 40% of both large and small regulated water supplies have arsenic concentration greater than 1 µg/L. The frequency of exceedance decreases for greater arsenic concentrations—about five percent of systems are estimated to have arsenic concentration greater than 20 µg/L. Comparison of these estimates with previously published work, based on 275 samples collected from regulated water supplies, shows very good agreement for the United States as a whole, although the two approaches yield somewhat different results for some parts of the nation.

Keywords: arsenic, water supply, ground water

INTRODUCTION

Arsenic can impact human health through the ingestion of ground water used for water supply. An understanding of arsenic concentrations in ground water can: (1) assist water managers and users in overcoming adverse health effects through avoidance or treatment, (2) assist epidemiologists interested in evaluating the intake of arsenic from drinking water, which can contribute much of the human exposure to inorganic arsenic (Borum and Abernathy, 1994), and (3) provide a basis for evaluating the costs of adopting a particular value for a drinking-water standard (or MCL —Maximum Contaminant Level). Estimating the frequency of exceedance for arsenic in regulated ground water supplies is the focus of this manuscript. The estimates were made for values in the range being considered for a revised MCL. These estimates are timely because the U.S. Environmental Protection Agency (EPA) must issue a proposed and a final regulation for arsenic in drinking water by 2000 and 2001, respectively.

ARSENIC IN GROUND WATER

Within the conterminous United States, widespread high arsenic concentrations in ground water most commonly result from: (1) upflow of geothermal water, (2) dissolution of, or desorption from, iron-oxide, (3) dissolution of sulfide minerals, and (4) evaporative concentration. Figure 1 and Table 1 indicate areas where these processes appear to be important. Concentrations of naturally occurring arsenic in ground water vary regionally due to a combination of climate and geology. At a broad regional scale, arsenic concentrations exceeding 10 µg/L appear to be more frequently observed in the western U.S. than in the east (Welch et al., in press). Investigations of ground water in Maine, Michigan, Minnesota, South

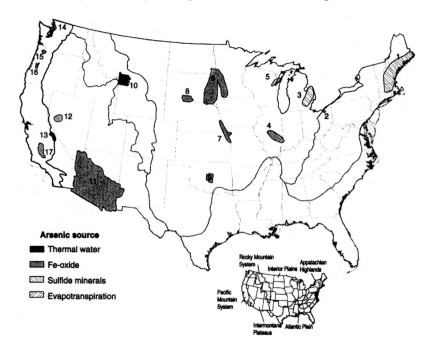

Fig. 1. Areas where high arsenic concentration have been documented in ground-water systems. Physiographic provinces modified from Fenneman (1931).

TABLE 1

Areas with high arsenic concentrations derived from natural sources

Source of arsenic [1]	Hydrologic units and number of area shown on Figure 1	References
Sulfide minerals and Fe-oxide	Bedrock (1)[2]	Zuena and Keane, 1985; Boudette et al., 1985; Marvinney et al., 1994; Ayotte et al., 1998; Peters et al., 1998
Fe-oxide (D)	Paleozoic sandstone (2)	Matisoff et al., 1982
Sulfide minerals	Glacial deposits, sandstone[3] and shale (3)	Westjohn et al., 1998; Kolker et al., 1998
Fe-oxide (D)	Glacio-fluvial deposits[4] (4)	Voelker, 1986; Holm and Curtiss, 1988; Panno et al., 1994; Holm, 1995; Korte, 1995
Sulfide minerals	Ordovician carbonate and clastic rocks (5)	Simo et al., 1996
Fe-oxide (D, P)	Glacial deposits and shale (6)	Roberts et al., 1985; Kanivetsky, in press
Fe-oxide (D)	Alluvium (7)	Ziegler et al., 1993; Korte, 1991
Fe-oxide (P)	Volcanic ash (8)	Carter et al., 1998
Fe-oxide (P)	Sandstone and mudstone (9)	Schlottmann and Breit, 1992; Norvell, 1995
Black shale lithic fragments	Glacio-fluvial deposits	Yarling, 1992
Geothermal water	Volcanic rocks (10)	Stauffer and Thompson, 1984; Ball et al., 1998
Fe-oxide (P)	Basin fill sediments, including volcanic, alluvial, and lacustrine deposits (11)	Owen-Joyce and Bell, 1983; Owen-Joyce, 1984; Robertson, 1989
Fe-oxide (D, P) and evaporative concentration	Basin fill sediments, including alluvial and lacustrine deposits (12)	Welch and Lico, 1998
Geothermal water	Volcanic rocks (13)	Mariner and Willey, 1976; Eccles, 1976; Wilkie and Hering, 1998
Fe-oxide[5] (P)	Alluvium (14)	Goldstein, 1988; Ficklin et al. 1989; Davies et al., 1991
Fe-oxide (D)	Basin-fill deposits (15)	Hinkle, 1997
Fe-oxide (P)	Felsic-volcanic tuff (16)	Goldblatt et al., 1963; Nadakavukaren et al., 1984
Fe-oxide and evaporative concentration	Basin-fill sediments , including alluvial and lacustrine deposits (17)	Fujii and Swain, 1995; Swartz, 1995; Swartz et al., 1996

[1] Known or inferred. For areas with Fe-oxide as a source of arsenic, dissolution of the oxide and desorption are important processes that can release arsenic to ground water. The letters 'D' and 'P' in parentheses refer to the processes of dissolution and pH-influenced desorption of arsenic, respectively.
[2] Arsenic concentrations in ground water are generally higher in bedrock aquifers compared with overlying glacial aquifers.
[3] The sandstone contains arsenic-rich pyrite, which may be a source of the arsenic in the overlying glacial aquifer. Pyrite has not been identified in the glacial deposits.
[4] May include a contribution of arsenic from underlying coal-bearing units. Arsenic-rich ground water may extend into the upper Kankakee River basin within Indiana, as suggested by high arsenic in surface water, sediment and biota (Fitzpatrick et al., 1998; Schmidt and Blanchard, 1997).
[5] Arsenopyrite has been mentioned as a possible source of arsenic. However, high pH (the median pH of 11 samples with arsenic >50 µg/L is 8.25) and generally low sulfate concentrations (<25 mg/L; Ficklin et al., 1989) imply that sulfide mineral oxidation is limited, suggesting that the arsenic may be from Fe-oxide that was formed from the oxidation of arsenopyrite. The ground water with the highest arsenic concentration (15,000 µg/L) also had the highest pH (9.23).

Dakota, Oklahoma, and Wisconsin within the last decade suggest that arsenic concentrations exceeding 10 µg/L are more widespread and common than previously recognized (Table 1).

Arsenic release from iron oxide appears to be the most common cause of widespread arsenic concentrations exceeding 10 µg/L in ground water. This can occur in response to different geochemical conditions, including release of arsenic to ground water through reaction of iron oxide with organic carbon. Iron oxide also can release arsenic to alkaline ground water, such as that found in some felsic volcanic rocks, including areas 8, 14, and 16 (Figure 1) in the western U.S. Geothermal water and high evaporation rates also are associated with arsenic concentrations greater than 10 µg/L in ground water, particularly in the west. Geothermal systems occur throughout much of the western United States,

TABLE 2

Comparison of arsenic concentrations in public and non-public water supplies, by physiographic province. A total of 18,468 sites were used.

Physiographic Province	Number of sites	Median arsenic in µg/L	Significance
	Public supply	Public Supply	$(p > |z|)$
	Non-public supply	Non-public Supply	
Appalachian Highlands	376	≤1	0.6552
	2212	≤1	
Atlantic Plain	646	≤1	0.0067
	2047	≤1	
Interior Plains, including the Interior	342	≤1	0.3289
Highlands, and Laurentian Upland	3947	≤1	
Intermontane Plateaus	458	3	0.1389
	4640	3	
Pacific Mountain System	303	2	0.7159
	2401	2	
Rocky Mountain System	74	≤1	0.6444
	1022	≤1	

although most affect relatively small regions. A notable exception is the Yellowstone geothermal system, which causes arsenic concentrations as high as 360 µg/L in the Madison River at the park boundary, and as high as 19 µg/L arsenic in the Missouri River at a point 470 km downstream (Nimick, 1994; Nimick et al., 1998). These studies clearly demonstrate the existence of ground water with arsenic concentrations that exceed both the current and possible new MCL for arsenic.

ESTIMATING ARSENIC IN REGULATED WATER SUPPLIES

The arsenic content in ground water used to supply water for regulated systems has been estimated from arsenic data from across the United States. The basic approach consisted of combining arsenic data for ground water with the locations of water supplies that use ground water (Figure 2). The arsenic data were retrieved from the U.S. Geological Survey's National Water Information System (NWIS). These data, along with other water-quality measurements and ancillary data are collectively referred to as the USGS Arsenic Database.

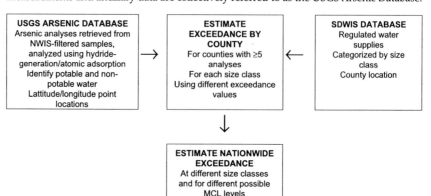

Fig. 2. Approach used to estimate arsenic exceedance in regulated supplies using ground water.

Within this database, geothermal water (water with a temperature >50°C) and slightly saline water (dissolved solids >3,000 mg/L or specific conductance >4,000 µS/cm) are considered to be unlikely to be a source for a regulated water supply. Accordingly, analyses of this ground water were excluded from the evaluation.

The USGS integrated the arsenic occurrence data with information on public suppliers and population served. This was accomplished by relating arsenic concentrations on a county level from the USGS data base with public water supplier data from US EPA's Safe Drinking Water Information System (SDWIS) database. Data were retrieved from SDWIS for all (surface water, ground water, and purchased water) community public suppliers and their sources during late summer of 1997. Exceedance frequencies were estimated for public water supplies for two system sizes, based on population served. The size classification corresponds to that used by Frey and Edwards (1997) to allow direct comparison with their estimates. The size classes are large systems—those that serve populations greater than 10,000—and small systems, which serve 1,000 to 10,000 people.

COMPARISON OF PUBLIC AND NON-PUBLIC GROUND WATER SAMPLES

Ancillary data within the USGS Arsenic Database includes information on the use of the water. These data allow a comparison of arsenic concentrations in water that is used for public supply with water that is not used for public supply. The classification as public supply was based on primary use of water listed in NWIS as bottling, commercial, medicinal, public supply, or institutional; all other primary use categories were considered non-public supply. It is worth emphasizing that the use of the term 'public water supply' is not synonymous with 'regulated water supply', where the latter refers to systems that are subject to the Safe Drinking Water Act. The term 'public water supply' is an informal term used here and refers to the water uses listed above.

The arsenic analyses for public and non-public supply sources were compared using non-parametric tests employing the procedure NPAR1WAY in the SAS/STAT® software system (SAS Institute Inc., 1990, p. 1195–1210.). The statistical comparisons indicated that arsenic concentrations in non-public supply samples tended to be higher than concentrations in public supply samples. However, the magnitude of the differences is not very large—the results are so significant due to the large amount of data used. The medians of both the non-public and public supply samples are ≤1 µg/L. The test results are summarized below.

Test and test statistic	Probability of more extreme test statistic
Wilcoxon two-sample test, normal approximation, z	0.0001
Kruskal-Wallis, chi-square approximation, χ^2	0.0001

When the data are split into Physiograhic Provinces, only the Atlantic Plain had a significant difference in arsenic concentration between public supplies and non-public ground water sources (Table 2). Yet, the medians for both groups are ≤1 µg/L.

These data suggest that the magnitude of difference in arsenic concentrations between public supply and other data represented in the USGS Arsenic Database is very small. Although the USGS data were not collected with the specific intent of developing national occurrence estimates in drinking water, these data appear to be appropriate for character-izing arsenic concentrations in ground water used for public supply.

ESTIMATES OF THE FREQUENCY OF EXCEEDANCE

A common way to summarize national occurrence data for use in the regulatory process is to present the data in terms of the frequency with which systems exceed specified concen-trations of a contaminant. The systems are categorized by size classes and the concentrations are often selected to correspond with ranges of potential drinking water standards that are being assessed as part of the regulatory process. Estimates of arsenic exceedance in regulated water supplies are based on 17,496 arsenic analyses for 595 counties that have five or more sites represented in the USGS Arsenic Database (Figure 3). About 47% of large systems are located in these counties. About 32% of all small systems are located in those same 595 counties.

Estimates based on the USGS Arsenic Database suggest that an MCL of 20 µg/L would increase the frequency of exceedance over that estimated for the current MCL of 50 µg/L, although less than five percent of both the small and large systems would be affected at either level. Decreasing the MCL below 10 would result in a substantial increase in the frequency of exceedance, with about forty percent of the systems exceeding the standard if an MCL of 1 µg/L were adopted. Generally, the frequency of exceedance for the large and small systems at the various levels shown on Figure 4 are similar.

Estimates of exceedance based on the USGS Arsenic Database are broadly similar to previous estimates (Figure 4) using the NAOS (National Arsenic Occurrence Survey) by Frey and Edwards (1997). (Comparisons were not possible at concentrations of 20 and 50 µg/L because they were not published for the NAOS data.) This broad agreement is somewhat surprising considering the differences in the two databases. NAOS includes 275 filtered samples of ground water used by regulated water supplies, whereas over 17,000 analyses were used in the approach described above. The exceedance estimates based on the NAOS database also were adjusted for treatment, which resulted in lower arsenic values. This adjustment is one factor that could account for the generally lower frequency of exceedance compared with the estimates based on the USGS Database. The adjustment was not large, however, ranging from zero to ten percent (Frey and Edwards, 1997).

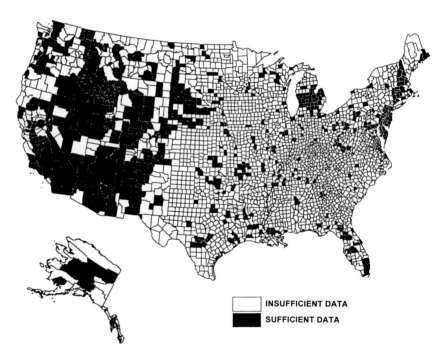

☐ INSUFFICIENT DATA
■ SUFFICIENT DATA

Fig. 3. Counties with five or more sites with ground-water samples analyzed for arsenic.

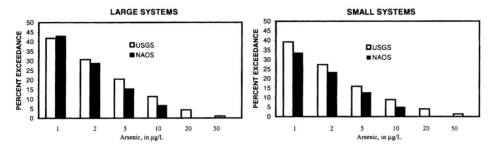

Fig. 4. Exceedance frequency of arsenic concentrations in small and large regulated water supply systems.

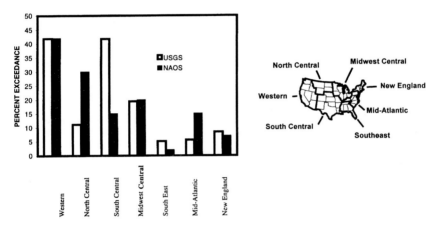

Fig. 5. Exceedance frequencies of arsenic at 5 μg/L for seven regions. Regions and exceedance frequency data for NAOS are from Frey and Edwards, 1997. The western region includes Alaska and Hawaii, although the USGS Arsenic database has less than 5 samples for the state.

The agreement between the approach discussed above with that of Frey and Edwards (1997) ranges from very good to poor at 5 μg/L for regions of the United States (Figure 5). The exceedance estimates for the Western, Midwest, and New England regions show very good agreement. Although the estimates for the South East region differ by more than a factor of two, both suggest that this region has the lowest frequency of exceedance at the 5 μg/L level. The poor agreement for the North Central, South Central, and Mid-Atlantic regions may, in part, be due to combination of sparse data and the approach described above. For instance, only a few counties are represented in the USGS estimates for several of the states in the Mid-Atlantic region (Figure 3). One weakness of the current approach is that counties are relatively small in the central and eastern United States compared to the west. An undesirable artifact of the approach is that the data density for the east must be much greater to include a county in the frequency estimates. An alternative approach is to include data within a search radius greater than the size of most counties in the central and eastern United States.

DISCUSSION

Estimates of the frequency of exceedance based on the USGS Arsenic Database suggest that about 40% of both large and small water supplies have arsenic concentration greater than 1 µg/L. The frequency of exceedance decreases for greater arsenic concentrations—about five percent of systems are estimated to have arsenic concentration greater than 20 µg/L. Comparison of these with the estimates based on the NAOS Database shows very good agreement for the United States as a whole, although the two approaches yield somewhat different results for some parts of the nation. This agreement suggests that the USGS Arsenic Database may be used to estimate the frequency of exceedance for the large number of systems that supply small populations.

REFERENCES

Ayotte, J. D., M. G. Nielson, and G. R. Robinson. 1998. Relation of arsenic concentrations in ground water to bedrock lithology in eastern New England. 1998 Geol. Soc. of Am. Annual meeting abstracts with programs. pp. A-58.

Ball, J. W., D.K. Nordstrom, E.A. Jenne, and D.V. Vivit. 1998. Chemical Analyses of Hot Springs, Pools, Geysers, and Surface Waters from Yellowstone National Park, Wyoming, and Vicinity, 1974–1975.U.S. Geol. Surv. Open-File Rep. 98-182. p. 45.

Borum, D.R. and C.O. Abernathy. 1994. Human oral exposure to inorganic arsenic. In: W.R. Chappell, C.O. Abernathy, and C.R. Cothern (eds.), Arsenic exposure and health. *Sci. Tech. Letters, Northwood*, **16**, pp. 31–30.

Boudette, E. L., F. C. Canney, J. E. Cotton, R. I. Davis, W.H. Ficklin, and J.M. Motooka. 1985. High levels of arsenic in the groundwater of southeastern New Hampshire. A Geochemical Reconnaissance. U.S. Geol. Surv. Open-File Rep. pp. 25.

Carter, J.M., S.K. Sando, T.S. Hayes, and R.H. Hammond. 1998. Source Occurrence, and Extent of Arsenic Contamination in the Grass Mountain Area of the Rosebud Indian Reservation, South Dakota. U.S. Geol. Surv. Wat. Res. Inv. Rep. 97-4286. p. 90.

Davies, J., R. Davis, D. Frank, F. Frost, D. Garland, S. Milham, R.S. Pierson, R.S. Raasina, S. Safioles, and L. Woodruff. 1991. Seasonal study of arsenic in ground water: Snohomish County, Washington. Snohomish Health District and Washington State Dept. of Health unpublished report. p. 18.

Eccles, L.A. 1976. Sources of arsenic in streams tributary to Lake Crowley California. U. S. Geol. Surv. Wat. Res. Inv. Rep. 76-36. pp. 76–86.

Fenneman, N.M. 1931. *Physiography of Western United States*. McGraw-Hill, New York. p. 534.

Ficklin, W.H., D.G. Frank, P.K. Briggs, and R.E. Tucker. 1989. Analytical results for water, soil, and rocks collected near the vicinity of Granite Falls, Washington as part of an arsenic-in-groundwater study. U. S. Geol. Surv. Open-File Rep. 89-148. p. 9.

Fitzpatrick, F., T.L. Arnold, and J.A. Colman. 1998. Surface-Water-Quality Assessment of the Upper Illinois River Basin in Illinois, Indiana, and Wisconsin-Spatial distribution of geochemicals in the fine fraction of streambed sediment, 1987. U.S. Geol. Surv. Wat. Res. Inv. Rep. 98-4109. p. 89.

Frey, M.M. and M.A. Edwards. 1997. Surveying arsenic occurrence. *J. Am. Water Works Assoc.* **89**, 105–117.

Fujii R. and W.C. Swain. 1995. Areal distribution of trace elements, salinity, and major ions in shallow ground water, Tulare basin, Southern San Joaquin Valley, California. U.S. Geol. Surv. Wat. Res. Inv. Rep. 95-4048. p. 67.

Goldblatt, E.L., S.A. Van Denburgh, and R.A. Marsland. 1963. The unusual and widespread occurrence of arsenic in well waters of Lane County, Oregon. Lane County Health Dept. Rept. p. 24.

Goldstein, L. 1988. A review of arsenic in ground water with an emphasis on Washington state. Unpublished M.S. thesis, Evergreen State College, Olympia, Washington. p. 102.

Hinkle, S.R. 1997. Quality of shallow ground water in alluvial aquifers of the Willamette Basin, Oregon, 1993-95. U.S. Geol. Surv. Wat. Res. Inv. Rep. 97-4082-B. p. 48.

Holm, T.R. 1995. Ground-water quality in the Mahomet Aquifer, McLean, Logan, and Tazewell Counties. Illinois State Water Survey Contract Report 579. p. 42.

Holm, T.R. and C.D. Curtiss. 1988. Arsenic contamination in east-central Illinois ground waters. Illinois Dept. of Energy and Natural Res., Energy and Environmental Affairs Division. p. 63.

Kanivetsky, R. in press. Arsenic in ground water of Minnesota: Hydrogeochemical modeling and characterization. Minnesota Geol. Surv. Rep.

Kolker, A., W.F. Cannon, D.B. Westjohn, and L.G. Woodruff. 1998. Arsenic-rich pyrite in the Mississippian Marshall Sandstone: Source of anomalous arsenic in southeastern Michigan ground water. 1998 Geol. Soc. of Am. Annual meeting abstracts with programs. pp. A-59.

Korte, N. E. 1991. Naturally Occurring Arsenic in Groundwaters of the Midwestern United States. Oak Ridge National Laboratory Environmental Tech. Section Informal Rep. p. 20.

Korte, N.E. 1995. Naturally Occurring Arsenic in the Groundwater at Chanute Air Force Base, Rantoul, Illinois. Environmental Sciences Division Publication no. 3501. p. 54.

Mariner, R.H. and L.M. Willey. 1976. Geochemistry of Thermal Waters in Long Valley, Mono County, California. *J. Geophys. Res.*, **81** (5), 792–800.

Marvinney, R.G., M.C. Loiselle, J.T. Hopeck, D. Braley, and J.A. Krueger. 1994. *Arsenic in Maine Groundwater: An Example From Buxton, Maine*. 1994 Focus Conference on Eastern Regional Ground Water Issues. pp. 701–714.

Matisoff, G., C.J. Khourey, J.F. Hall, A.W. Varnes, and W.H. Strain. 1982. The nature and source of arsenic in northeastern Ohio Ground Water. *Ground Water*, **20** (4), 446–456.

Nadakavukaren, J.J., R.L. Ingermann, and G. Jeddeloh. 1984. Seasonal Variation of Arsenic Concentration in Well Water in Lane County, Oregon. *Bull. Environ. Contam. Toxicol.*, **33** (3), 264–269.

Nimick, D.A. 1994. Arsenic transport in surface and ground water in the Madison and upper Missouri River Valleys, Montana. EOS. pp. 247.

Nimick, D.A., J.N. Moore, C.E. Dalby, and M.W. Savka. 1998. The fate of arsenic in the Madison and Missouri Rivers, Montana and Wyoming. *Water Resour. Res.*, **34** (11), 3051–3067.

Norvell, J.L.S. 1995. Distribution of, sources of, and processes mobilizing arsenic, chromium, selenium, and uranium in the central Oklahoma aquifer. Unpublished M.S. thesis, Colorado School of Mines, Golden, Colorado. p. 169.

Owen-Joyce, S.J. and C.K. Bell. 1983. Appraisal of water resources in the Upper Verde River area, Yavapai and Coconino Counties, Arizona. Arizona Dept. of Water Res. Bull. 2. p. 219.

Owen-Joyce, S.J. 1984. Hydrology of a Stream-Aquifer System in the Camp Verde Area, Yavapai County, Arizona. Arizona Dept. of Water Res. Bull. 3. p. 60.

Panno, S. V., K.C. Hackley, K. Cartwright, and C.L. Liu. 1994. Hydrochemistry of the Mahomet Bedrock Valley Aquifer, East-Central Illinois: Indicators of Recharge and Ground-Water Flow. *Ground Water*, **32** (4), 591–604.

Peters, S.C., J.D. Blum, B. Klaue, and M.R. Karagas. 1998. Arsenic Occurrrence in New Hampshire Ground Water. 1998 Geol. Soc. of Am. Annual meeting abstracts with programs. pp. A-58.

Roberts, K., B. Stearns, and R. L. Francis. 1985. Investigation of arsenic in southeastern North Dakota ground water, A Superfund Remedial Inv. Rep. North Dakota State Dept. of Health. p. 225

Robertson, F.N. 1989. Arsenic in ground-water under oxidizing conditions, south-west United States. *Environ. Geochem. Health*, **11** (3/4), 171–186.

SAS Institute Inc., 1990, SAS/STAT® User's Guide, Version 6, Fourth Edition, Volume 2: Cary, North Carolina, SAS Institute Inc., p. 1686.

Schlottmann, J.L. and G.N. Breit. 1992. Mobilization of As and U in the central Oklahoma aquifer. In: Y.K. Kharaka and A.S. Maest (eds.), *Water–Rock Interaction*. Balkema, Rotterdam. pp. 835–838.

Schmidt, A.R. and S.F. Blanchard. 1997. Surface-water-quality assessment of the upper Illinois River Basin in Illinois, Indiana, and Wisconsin. Results of Investigations through April 1992. U.S. Geol. Surv. Wat. Res. Inv. Rep. 96-4223. p. 63.

Simo, J.A., P.G. Freiberg, and K.S. Freiburg. 1996. Geologic constraints on arsenic in groundwater with applications to groundwater modeling: Groundwater Research Rept. WRC GRR 96-01, University of Wisconsin. p. 60.

Stauffer, R.E. and J.M. Thompson. 1984. Arsenic and antimony in geothermal waters of Yellowstone National Park, Wyoming, USA. *Geochim. Cosmochim. Acta*, 48 (11), 2547–2561.

Swartz, R.J. 1995. A study of the occurrence of arsenic on the Kern Fan element of the Kern Water Bank, southern San Joaquin Valley, California. Unpublished M.S. thesis, California State University, Bakersfield, California, p. 138.

Swartz, R.J., Thyne, G.D., and Gillespie, J.M. 1996. Dissolved arsenic in the Kern Fan San Joaquin Valley, California: Naturally occurring or anthropogenic. *Environ. Geosci.*, 3 (3), 143–153.

Voelker, D.C. 1986. Observation-well network in Illinois, 1984. U.S. Geol. Surv. Open-File Rep. 86-416. p. 108.

Welch, A.H. and Lico M.S. 1998. Factors controlling As and U in shallow ground water, southern Carson Desert, Nevada. *Appl. Geochem.*, 13 (4) 521–539.

Welch, A.H., D.B. Westjohn, D.R. Helsel, and R.B. Wanty. in press. Arsenic in ground water of the United States: Occurrence and geochemistry. Accepted for publication in *Ground Water*.

Westjohn, D.B., A. Kolker, W.F. Cannon, and D.F. Sibley. 1998. Arsenic in ground water in the "Thumb Area" of Michigan. The Mississippian Marshall Sandstone Revisited, Michigan: Its Geology and Geologic Resources, 5th symposium. pp. 24–5.

Wilkie, J.A., and J.G. Hering. 1998. Rapid oxidation of geothermal arsenic(III) in streamwaters of the eastern Sierra Nevada. *Environ. Sci. Technol.*, **32**, 657–662.

Yarling, M. 1992. Anomalous concentrations of arsenic in the groundwater at Wakarusa, Indiana: A byproduct of chemical weathering of shales. Indiana Dept. of Environmental Management. p. 38.

Ziegler, A.C., W.C. Wallace, D.W. Blevins, and Maley, R.D. 1993. Occurrence of Pesticides, Nitrite Plus Nitrate, Arsenic, and Iron in Water From Two Reaches of the Missouri River alluvium, northwestern Missouri, July 1988 and June–July 1989. U.S. Geol. Surv. Open-File Rep. 93-101. p. 30.

Zuena, A.J. and P.E. Keane. 1985. Arsenic Contamination of Private Potable Wells. *EPA National Conference on Environmental Engineering Proceedings*, Northeastern University Boston, MA. pp. 717–725.

Arsenic Exposure and Health Effects
W.R. Chappell, C.O. Abernathy and R.L. Calderon (Editors)

Airborne Exposure to Arsenic Occurring in Coal Fly Ash

Janice W. Yager, Harvey J. Clewell, III, Jeffrey B. Hicks,
Eleonora Fabianova

ABSTRACT

The relationship between respiratory tract deposition of airborne arsenic and urinary excretion of total arsenic was investigated in five workers during and following a 9-day maintenance activity involving exposure to high concentrations of arsenic in fly ash. Full-shift air samples were collected in the breathing zone of each worker during each day of exposure. The resulting data on arsenic mass versus particle size were analyzed using the U.S. EPA particle deposition model to estimate the deposited dose of arsenic associated with the exposures. The total deposited dose of arsenic over the exposure period was then compared with the total excretion of arsenic in 24 hr urine samples collected during, and for two days following, the exposures to estimate the bioavailability of arsenic in fly ash. The observed particle distributions consisted primarily of large, non-respirable particles, greater than 6 μm in aerodynamic diameter, which would be expected to deposit in the naso-pharynx or tracheo-bronchial region rather than in the deep lung. The average deposition efficiency was on the order of 75%, and the calculated bioavailabilities (fraction of deposited arsenic excreted in urine) for each subject ranged from 1% to 25% , with an average value of 11%. These results support the conclusions of a previous study performed at the same location, which found that the relative urinary excretion of arsenic from fly ash was much lower than had been reported for other exposures to airborne arsenic, such as at copper smelters.

Keywords: arsenic, fly ash, occupational exposure, particulates, respiratory deposition

INTRODUCTION

Arsenic is considered a human carcinogen based principally on results from epidemiological studies that include investigations of occupational exposure as well as ecological studies of exposure to naturally occurring arsenic in drinking water (U.S. EPA, 1984). Occupational epidemiological studies have been conducted principally in copper smelting operations wherein past high airborne exposures have been associated with increased risk of lung cancer (Enterline et al., 1987, 1995; Enterline and Marsh, 1982; IARC, 1980; Jarup and Pershagen, 1991; Jarup et al., 1989; Pinto et al., 1977; Welch et al., 1982). Similarly, most occupational studies to date relating arsenic air exposure to urinary excretion have been conducted in non-ferrous smelters (Hakala and Pyy, 1995; Offergelt et al., 1992; Pinto et al., 1976; Smith et al., 1977; Vahter et al., 1986; Yamauchi et al., 1989). It is of interest to examine other work settings to assess potential impacts of differences in exposure parameters. For example, differences in arsenic compound(s), physical form(s), particle size distributions, and matrix compositions in different work settings may lead to differences in arsenic uptake, distribution, metabolism and excretion. Further, estimation of differences in the relationship between air monitoring and urinary excretion in different settings can be useful in refining and validating physiological-based pharmacokinetic (PBPK) models for arsenic (Wyzga and Yager, 1993).

A previous study was undertaken in this coal-fired power plant in Slovakia during a routine maintenance outage and described, in general, the relationship between airborne occupational exposure to inorganic arsenic in coal fly ash and urinary excretion of arsenic metabolites (Yager et al., 1997). In that study, it was noted, based on a limited number of size-fractionated air samples collected, that approximately 90% of particle mass and arsenic in this work setting were present in particle size fractions ≥ 3.5 μm. Further, the mean concentration of urinary arsenic metabolites excreted was approximately one-third that described for an equivalent arsenic breathing-zone concentration observed in a number of copper smelter studies.

The purpose of this current work was to refine observations of the relationship between airborne arsenic exposure and urinary excretion of speciated metabolites in a small number of workers. Such information is valuable in refining knowledge concerning lung deposition, and, in addition, can lend insight into an approximation of bioavailability in the absence of a comprehensive arsenic PBPK model.

METHODS AND MATERIALS

Power Plant

The coal-fired power plant at which the study was carried out is located in the central Slovak Republic and burns local low-grade brown lignite coal containing a mean concentration of approximately 800 ppm arsenic (maximum 1350 ppm) as the principal fuel (Proceedings of Study on Health Impact of Environmental Pollution, 1995). In comparison with coals used in US coal-fired power plants, Slovak lignite coal contains on average about 30–300 times higher arsenic concentrations. Technological improvements, as well as reduction of the total output from 620 MW to 250 MW, has resulted in a reduction of arsenic emissions to ambient air from 90 tons of arsenic emitted per year in 1980 to 2.7 tons emitted in 1993 (Proceedings of Study on Health Impact of Environmental Pollution, 1995).

Work activities during the maintenance outage studied at this coal-fired power plant were very similar to those observed in coal fired power plants in the United States (Hicks, 1993). During the initial stages of a maintenance outage, work activities are routinely directed toward removal of accumulated fly ash and clinker from inside the boiler structure as well as the electrostatic precipitators by use of vacuum systems and manual methods later followed by manual wet methods. Maintenance and repair inside the boiler is often performed in and around residual fly ash that is not completely removed, and this ash becomes airborne when agitated.

Subjects

Workers were initially interviewed to determine interest in study participation and general health background. Five healthy power plant workers participated in this study with informed consent. Three of the workers were cleaning the boiler on the first days of the task when exposure is highest whereas two of the workers started after the major cleaning had been completed. A questionnaire was administered to all subjects to obtain information on work history, diet (including consumption of seafood in the week preceding the start of the study), smoking, alcohol intake, medical history and other lifestyle factors.

Air Sampling

Time-integrated full shift air samples were collected in the breathing zone of each worker during each day of the work week. For daily personal samples, airborne particle size distribution samples were collected by use of a battery-powered personal air sampling pump (Gilian® Model HFS 113A) connected to a six-stage personal cascade impactor (Marple® Model 296). Cascade impactor filters were desiccated and weighed on the microbalance prior to sampling. Gravimetric and arsenic analyses were carried out on each filter as described below.

Analysis of Air Samples

Gravimetric Method for Total Dust

Filters were weighed on a six-place microbalance (Cahn Instruments) following a 48-hr desiccation period. Filters were weighed within 0.01 mg. Differences between matched-weight filters were calculated and results were expressed as total mass of particulate (mg) per m³ of air. Quality control included daily instrument calibration with standard weights replicate analysis and analysis of field blanks which constituted approximately 10% of samples.

Arsenic Analysis

Filters containing collected particulate were digested according to NIOSH Method 7900 in closed Teflon beakers. Analysis was conducted using graphite furnace–atomic absorption spectrometry (Varian® Spectre Model AA-30 with GTA-96 graphite tube atomizer). Daily analytical quality control procedures consisted of analysis of three separate reagent blanks and instrument calibration at three different known arsenic concentrations. Weekly quality control analyses for arsenic were carried out on the National Bureau of Standards (NBS) Trace Elements in Coal Fly Ash, 1633a, Standard Reference Material (SRM) sample. Results of these analyses ($n = 19$) showed a mean arsenic recovery of 110% (range 91.5–145) of the target concentration in the SRM coal fly ash. Replicate analyses of 10% of all samples were also carried out on a weekly basis.

Daily Work Activity Diary

A daily diary of work activities for each worker was completed by a member of the research team, based on brief employee interviews throughout the day and observations made in the workplace. Information included time of day, work location, work activities, and respirator usage. Standard respirators at this plant consisted of washable fabric dust masks held in place by tie strings.

Respirator Fit Test

The respirator Fit Factor (FF) was determined for each individual worker using a TSI Model 8020 Portacount Plus (TSI, St. Paul, MN). The FF is the measured concentration outside the mask divided by the measured concentration inside the mask. The average Fit Factor value for each worker was determined by utilizing software provided with the Portacount Plus® in which five separate measurements were recorded while the respirator wearer performed pre-set timed exercises (e.g., normal breathing, deep breathing, side to side head motion, up and down head motion, talking, etc.).

Urine Sampling

Twenty-four hour urine samples were collected for up to 9 consecutive days of exposure beginning pre-shift on the first morning of boiler cleaning. Samples were also collected for an additional two days after the last workplace exposure had occurred. Urine was collected in clean containers provided by the laboratory.

Urine Analysis

Creatinine concentration in urine samples was determined using the standard Jaffe colorimetric method. Arsenic species consisting of inorganic arsenic (As_i), monomethylarsonic acid (MMA) and dimethylarsinic acid (DMA) were quantified in urine by the hydride generation method coupled with atomic absorption spectrometry (Crecelius, 1978; Irgolic, 1987). Daily quality control procedures included triplicate blank reagent analyses, daily calibration checks using three different known concentrations of standard solutions of As_i, MMA and DMA, and daily spiked urine samples containing standard addition quantities of known arsenic species. Additionally, results of weekly analysis of the NBS SRM 2670 standard urine sample "Toxic Metals in Freeze Dried Urine" showed a mean recovery of 105.4% (range 80–143) with a CV of 26.5%. Analytic results from this laboratory compared favorably with other laboratories participating in an inter-laboratory comparison study (Crecelius and Yager, 1997).

Arsenic Lung Deposition

To estimate the amount of arsenic deposited, each stage of the impactor was treated as a monodisperse aerosol with a mass median aerodynamic diameter (MMAD) equal to the stated cut size. The U.S. EPA particle deposition model (U.S. EPA, 1994), which is based on the work of Menache et al. (1995), was used to estimate the mass of arsenic deposited in each region of the respiratory tract for the aerosol MMAD representing each impactor stage. The breathing rate was assumed to be 20 L/min, which represents an average occupational ventilation rate for a moderate activity level (Astrand, 1983; Dankovic and Bailer, 1993). Once the fractional deposition of aerosol was determined for each MMAD, this fraction was then multiplied by the mass of arsenic on the impactor stage for that cut-size. The total mass of deposited arsenic was then determined by summing the masses across all the impactor stages for each day's sample.

Bioavailability

The sum of concentrations of arsenic species (As_i and the methylated metabolites) in the pre-shift, first-day urine sample for each individual was assumed to represent the background total urinary arsenic concentration for that subject, and was subtracted from the samples obtained during and following the exposure period. The exposure-related mass of arsenic excreted in urine for each worker was determined by multiplying the total concentration of arsenic species (adjusted for background) times the volume for each urine sample, and then summing the mass for all the individual samples. This value of total excreted As was then divided by the estimated total amount of As deposited in the respiratory tract to obtain an estimate of the systemic bioavailability (percent systemic uptake) of the deposited arsenic.

RESULTS AND DISCUSSION

Particle Deposition

Analysis of the mass of aerosol from the different stages of the cascade impactors revealed a highly polydisperse aerosol. Figure 1 shows the average cumulative distribution of arsenic as a function of particle size for three workers on the two days associated with the highest exposures. Consistent with previous analyses of high-arsenic fly ash (Buchet et al., 1997; Yager et al., 1997), the particle distribution was dominated by relatively large particles: less than 10% of the arsenic was associated with particles less than 3.5 μm in diameter, while nearly 50% was associated with particles greater than 10 μm. As shown in Figure 2, particle size distributions measured by all five individuals' personal samplers were heavily skewed toward the larger particles. As noted earlier, the personal exposure for Subjects 1 through 3 was much higher than for Subjects 4 and 5, who only worked during the later stages of the cleaning process.

Individual mean respirator Fit Factors ranged from 1.7 to 2.3 for four of the subjects, with a slightly higher value of 3.5 for Subject 3, indicating that the masks provided very little protection from the aerosol exposure. In addition, estimates obtained from the workers regarding the percent of time each day that they wore the masks indicated that the masks were used only about half of the time (generally during the highest exposures). Due to the low efficiency of the masks, together with the reports of the workers that they often did not wear the masks, it was decided that personal exposure would not be adjusted to account for the effect of the masks.

Based on the particle size distribution of the aerosol, calculations with the USEPA deposition model indicate that the major site of deposition was the naso-pharynx, with much lesser deposition in the tracheo-bronchial and pulmonary regions for all 5 workers (Figure 3).

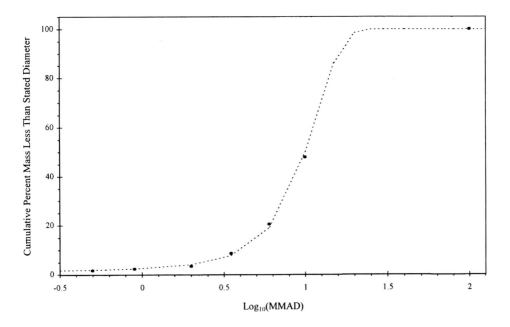

Fig. 1. Cumulative distribution of the percentage of arsenic as a function of particle diameter. Points represent cut-points for impactor stages. Dotted line is best-fit curve assuming normal distribution.

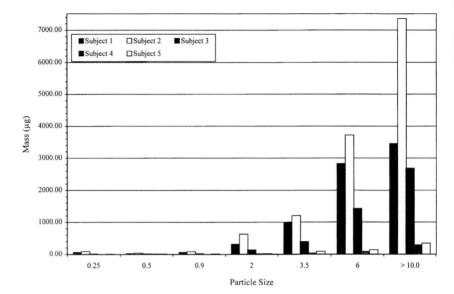

Fig. 2. Mass of arsenic collected in each impactor stage of personal samplers. Numbers on *x*-axis identify cut points for impactor stages.

Aerosol deposited in the naso-pharynx is primarily cleared to the gastro-intestinal (GI) tract or blown out of the nose, and the surface area for direct absorption into the blood is small compared to the other portions of the respiratory tract. Therefore, it is reasonable to expect that systemic exposure to arsenic in these workers may have resulted to a large extent from oral absorption of particles deposited in the nose and cleared to the GI tract.

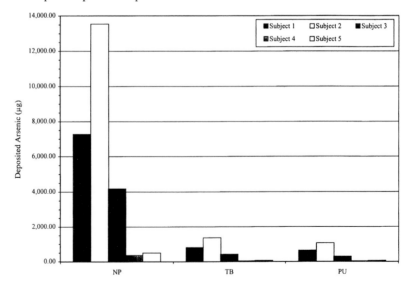

Fig. 3. Mass of arsenic deposited in each region of respiratory tract: NP = naso-pharyngeal, TB = tracheo-bronchial, PU = pulmonary (alveolar).

Urinary Excretion

The daily urinary volume represented by the samples provided by the workers ranged from 0.8–1.5 L/day. The temporal relationship between the exposure to airborne arsenic and urinary excretion of total arsenic was highly variable, as shown for each of the workers in Figure 4a–4e. Nevertheless, the correlation depicted in Figure 5 indicates that, except for Subject 2, total urinary excretion was a roughly linear function of total arsenic deposited in the respiratory tract. Subject 2, who had by far the highest measure of exposure to large particles, is clearly an outlier for this relationship, and was not included in the calculation of the correlation shown in this figure. It can only be speculated that in the case of this individual a much greater proportion of the aerosol was coughed or blown out due to the irritant effect of very large particles and high mass load.

The mean pre-shift, first-day sum of arsenic metabolites in urine was 8 μg/L whereas 2 days following the last exposure, the mean was 18 μg/L. Urinary concentrations during the exposures, on the other hand, ranged as high as 30–200 μg/L. Thus it appears that the fraction of the inhaled arsenic which was available to the systemic circulation was for the most part cleared by two days after the last exposure. This conclusion is also supported by the relatively rapid and complete excretion of arsenic in the urine of workers exposed to airborne arsenic in another study (Mann et al., 1996).

Bioavailability

As described earlier, an estimate of systemic bioavailability (BA), defined as the total urinary excretion above background divided by the total respiratory tract deposition, was calculated for each of the five workers. The results, annotated in Figure 5, ranged from 1% to 25%, with an average value of 11%. The value calculated for Subject 5 (25%) is less certain than those calculated for the other subjects, because no pre-exposure urine sample was obtained from this worker and the average pre-exposure concentration for the other four workers was used as a background estimate. A low bioavailability for the exposures received by these workers is also consistent with the regression performed on four of the Subjects: the slope of the regression, shown in Figure 5, is 0.076 μg arsenic in urine per μg arsenic deposited, or roughly 8% bioavailability.

The relatively low systemic bioavailability of arsenic from fly ash in this study is supported by a previous study (Yager et al., 1997). A regression of urinary arsenic concentration versus airborne arsenic concentration performed previously at the same location as the current study resulted in a slope of 0.1 μg arsenic in urine per μg/m³ arsenic in air, with an intercept (background urine concentration) of 12.2 μg/g creatinine. This slope was an order of magnitude smaller than those obtained in studies of exposures to arsenic-containing dust other than fly ash (Yager et al., 1997), yet it is entirely consistent with the current results. The background urinary arsenic estimate from the previous regression (12.2 μg/g creatinine) is also in good agreement with the average pre-exposure concentration measured in that study (10.8 μg/g creatinine), as well as in the present study (8 μg/L).

The reason for the unusually low arsenic bioavailability that has been estimated in these studies of Slovak fly ash is not known with certainty. However, it seems likely that the apparent difference in bioavailability between copper smelter dust and fly ash noted in the earlier study may result from different particle deposition characteristics. In particular, the particle size distribution associated with exposures at copper smelters may be significantly smaller than in the case of fly ash. "Respirable" particles less than 6 μm in diameter can reach the pulmonary region, where dissolution and systemic absorption occur efficiently. Non-respirable particles larger than 6 μm, on the other hand, deposit almost exclusively in the naso-pharyngeal and tracheo-bronchial regions and are rapidly cleared to the gastro-intestinal tract. Systemic bioavailability of arsenic could therefore be lower for the larger particle distribution, if the oral bioavailability of arsenic in large particles cleared to the

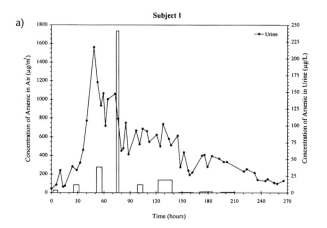

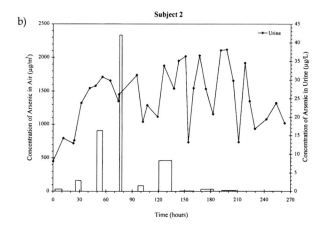

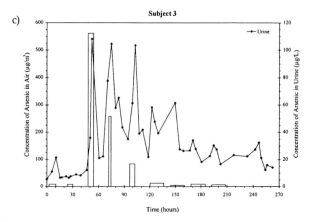

Fig. 4. Time-course of concentration of arsenic in air (rectangles, left axis) and urine (points, right axis) over the course of the study: (a) subject 1, (b) subject 2, (c) subject 3, (d) subject 4, (e) subject 5.

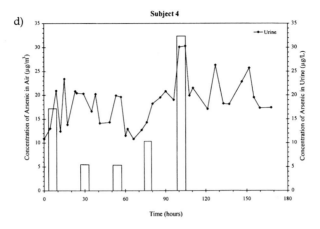

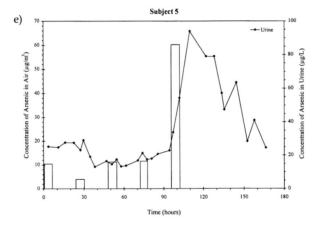

Fig. 4. Continued.

gastrointestinal tract is significantly less than the bioavailability of arsenic from small particles deposited in the deep lung.

The particle size distribution measured in the case of the Slovak fly ash, both in this study and previously (Yager et al., 1997) was dominated by relatively large particles. Less than 10% of the arsenic was associated with particles less than 3.5 μm in diameter, and less than 20% was associated with particles less than 6 μm. In contrast, a measurement of the particle size distribution for dust at a copper smelter (Smith et al., 1977) found that nearly 50% of the particles were less than 5 μm in diameter. The roughly 3-fold greater fraction of respirable particles in the case of the smelter study is consistent with the higher apparent bioavailability of airborne arsenic observed in studies at smelters. Thus differences in particle size distribution may well account for the observed difference in bioavailability between the studies of fly ash exposure and those at smelters.

Another possible reason for a low bioavailability is the effect of protective respirators. Since the particle measurements were obtained from personal breathing zone samplers, any particles trapped by the masks would reduce the actual worker exposure relative to the sampler. However, the respirators in this exposure provided very little protection, with fit

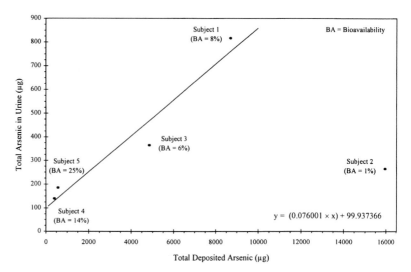

Fig. 5. Relationship of total arsenic excreted in urine over the study period to total arsenic deposited in the respiratory tract during the exposures. Equation represents best fit to data from subjects 1, 3, 4, and 5. Bioavailability (BA) for each subject is shown in parentheses.

factors of around 2, and were often not even used by the workers. At most, the use of the masks could have reduced the personal exposures by 30 to 50%. This effect would probably not serve to explain the difference noted between the bioavailability of copper smelter dust and fly ash. Rudimentary forms of respiratory protection (masks, neckerchiefs, etc.) were undoubtedly used in the exposures at smelters, too.

Apparent bioavailability might also be reduced if the subjects failed to collect all of their urine. Daily urine volumes, which ranged from 0.8 to 1.5 L/day, were all within the normal range, arguing that losses due to uncollected urine were not substantial. Some extent of incomplete collection is, of course, likely, but would also be expected in the case of the smelter studies.

Another of the differences between the studies performed on exposures to fly ash and the studies, primarily at copper smelters, performed previously (Enterline et al., 1987, 1995; Enterline and Marsh, 1982; Hakala and Pyy, 1995; Jarup et al., 1989; Jarup and Pershagen, 1991; Offergelt et al., 1992; Pinto et al., 1976, 1977; Smith et al., 1977; Vahter et al., 1986; Welch et al., 1982; Yamauchi et al., 1989) relate to the chemical composition of the dusts and the duration of the exposures. A comparative study in hamsters found that while the form of arsenic in copper smelter dust was much less soluble than the arsenic in fly ash, the lung clearance and urinary excretion of arsenic following intratracheal instillation of the different dusts were still markedly similar. Thus differences in solubility do not appear to be reflected in the duration of retention in the lung.

With respect to exposure duration, the smelter studies typically involved the measurement of air and urinary concentrations of arsenic in workers who had been exposed chronically. A steady-state relationship between exposure and urinary excretion had undoubtedly been reached in most of the workers. In the case of the fly-ash exposures, there is no indication that they were in a steady-state condition. In fact, the first-morning pre-exposure urinary total arsenic concentrations for the subjects at the beginning of the two-week maintenance exposure were not significantly different from those of community referents (Yager et al., 1997). Urinary excretion of total arsenic in the present workers over the 11-day period of the fly-ash studies might underestimate systemic bioavailability for two reasons: (1) insufficient time for the complete absorption of deposited arsenic, and (2) insufficient time for the complete excretion of absorbed arsenic.

However, in a recent study in hamsters (Buchet et al., 1997), more than 40% of arsenic from a high-arsenic fly ash was excreted in the urine within two days after intratracheal instillation of the fly ash into the lungs. At the same time-point, less than 20% of the instilled arsenic remained in the lungs. The arsenic content and particle size distribution of the high-arsenic fly ash used in that study were very similar to the fly ash inhaled by the present workers. The rapid lung clearance and high urinary excretion of this high-arsenic fly indicates that most of the arsenic from the fly ash deposited in the respiratory tract of the subjects in the current study would be excreted by the end of the sampling period. Consistent with this expectation, the urinary total arsenic concentrations in the workers at the end of the collection period had returned to within a factor of two of the background concentration prior to the exposure.

In summary, the arsenic bioavailability (fraction of deposited arsenic appearing in the urine) observed in this and a previous study (Yager et al., 1997) of workers exposed to Slovak fly ash is about 10%. The bioavailability of arsenic from fly ash in these studies is much lower than observed in other exposures to airborne arsenic, such as for workers at copper smelters (Yager et al., 1997). The lower bioavailability of arsenic from fly ash observed in these studies, as compared to copper smelter studies, is consistent with the observation of a smaller fraction of fly ash particles in the respirable size range (less than 6 microns), and cannot readily be explained by other differences from studies at copper smelters. Additional studies of the particle size distribution and bioavailability of airborne arsenic for different occupational exposures are needed to corroborate this conclusion.

REFERENCES

Astrand, I. 1983. Effect of physical exercise on uptake, distribution and elimination of vapors in man. In: *Modeling of Inhalation Exposure to Vapors: Uptake, Distribution, and Elimination* (ed. V. Fiserova-Bergerova), Vol. II. CRC Press, Boca Raton, FL, pp. 107–130.

Buchet, J.P., Lauwerys, R., Fabries, J.F., and Yager, J.W. 1997. Factors affecting the retention in hamster lung of arsenic present in fly ash and copper smelter dust. In: Arsenic. Exposure and Health Effects (eds. C.O. Abernathy, R.L. Calderon and W.R. Chappell). Chapman & Hall, New York, pp. 272–282.

Crecelius, E.A. 1978. Modification of the arsenic speciation technique using hydride generation. *Anal. Chem.*, 50, 826–827.

Crecelius, E. and Yager, J.W. 1997. Interlaboratory Comparison of Analytical Methods for Arsenic Speciation in Human Urine. *Environ. Health Perspect.*, 105 (6), 650–653.

Dankovic, D.A. and Bailer, A.J. 1993. The impact of exercise and intersubject variability on dose estimates for dichloromethane derived from a physiologically based pharmacokinetic model. *Fundam. Appl. Toxicol.*, 22, 20–25.

Enterline, P.E. and Marsh, G.M. 1982. Cancer among workers exposed to arsenic and other substances in a copper smelter. *Am. J. Epidemiol.*, 116, 895–911.

Enterline, P.E., Henderson, V.L. and Marsh, G.M. 1987. Exposure to arsenic and respiratory cancer. A reanalysis. *Am. J. Epidemiol.*, 125, 929–938.

Enterline, P.E., Day, R. and Marsh, G.M. 1995. Cancers related to exposure to arsenic at a copper smelter. *Occup. Environ. Med.*, 52, 28–32.

Hakala, E. and Pyy, L. 1995. Assessment of exposure to inorganic arsenic by determining the arsenic species excreted in urine. *Toxicol. Letters*, 77, 249–258.

Hicks, J.B. 1993. Toxic constituents of coal fly ash. In: *Managing Hazardous Air Pollutants. State of the Art.* (eds. W. Chow and K.K. Connor). CRC Press, Inc., Boca Raton, FL, pp. 262–275.

IARC. Arsenic and arsenic compounds. Vol. 23. Some metals and metallic compounds. IARC Monographs on the Evaluation of the Carcinogenic Risk of Chemicals to Man. International Agency for Research on Cancer, Lyon, 1980, pp. 39–141.

Irgolic, K.J. 1987. Analytical procedures for determination of organic compounds of metals and metalloids in environmental samples. *Sci. Total Environ.*, 64, 61–73.

Jarup, L., Pershagen, G. and Wall, S. 1989. Cumulative arsenic exposure and lung cancer in smelter workers: a dose–response study. *Am. J. Ind. Med.*, 15, 31–41.

Jarup, L. and Pershagen, G. 1991. Arsenic exposure, smoking, and lung cancer in smelter workers—a case-control study. *Am. J. Epidemiol.*, 134, 545–551.

Mann, S., Droz, P., and Vahter, M. 1996. A physiologically based pharmacokinetic model for arsenic exposure. II. Validation and application in humans. *Toxicol. Appl. Pharmacol.*, **140**, 471–486.

Menache, M.G., Miller, F.J. and Raabe, O.G. 1995. Particle inhalability curves for humans and small laboratory animals. Ann. Occup. Hyg., 19, 317–328.

Offergelt, J.A., Roels, H., Buchet, J.P., Boeckx, M. and Lauwerys, R. 1992. Relation between airborne arsenic trioxide and urinary excretion of inorganic arsenic and its methylated metabolites. *Brit. J. Ind. Med.*, **49**, 387–393.

Pinto, S.S., Enterline, P.E., Henderson, V., and Varner, M.O. 1977. Mortality experience in relation to a measured arsenic trioxide exposure. *Environ. Health Perspect.*, **19**, 127–130.

Pinto, S.S., Varner, M.O., Nelson, K.W., Labbe, A.L. and White, L.D. 1976. Arsenic trioxide absorption and excretion in industry. *J. Occupat. Med.*, **18**, 677–680.

Proceedings of Study on Health Impact of Environmental Pollution. Specialized State Health Institute, Banska Bystrica, Slovakia; State Health Institute, Prievidza, Slovakia; and International Centre of Pesticides Safety, Busto Garolfo, Italy. PHARE Project EC/91/HEA/18. September 29, 1995.

Smith, T.J., Crecelius, E.A. and Reading, J.C. 1977. Airborne arsenic exposure and excretion of methylated arsenic compounds. *Environ. Health Perspect.*, **19**, 89–93.

U.S. EPA. 1984. Final Report. Health Assessment Document for Inorganic Arsenic. EPA-600/8-83-021.

U.S. EPA. 1994. Methods for Derivation of Inhalation Reference Concentrations and Application of Inhalation Dosimetry. EPA/600/8-90/066F. Office of Research and Development, Washington, DC. October.

Vahter, M., Friberg, L., Rahnster, B., Nygren, A. and Nolinder, P. 1986. Airborne arsenic and urinary excretion of metabolites or inorganic arsenic among smelter workers. *Int. Arch. Occup. Environ. Health*, **57**, 79–91.

Welch, K., Higgins, I., Oh, M. and Burchfield, C. 1982. Arsenic exposure, smoking, and respiratory cancer in copper smelter workers. *Arch. Environ. Health*, **37**, 325–335.

Wyzga, R.E. and Yager, J.W. 1993. Carcinogenic risks of arsenic in fly ash. In: *Managing Hazardous Air Pollutants. State of the Art.* (eds. W. Chow and K.K. Connor). CRC Press, Inc. Boca Raton, FL, pp. 255–261.

Yager, J.W., Hicks, J.B. and Fabianova, E. 1997. Airborne arsenic and urinary excretion of arsenic metabolites during boiler cleaning operations in a Slovak coal-fired power plant. *Environ. Health Perspect.*, **105**, 836–842.

Yamauchi, H., Takahashi, K., Mashiko, M. and Yamamura, Y. 1989. Biological monitoring of arsenic exposure of gallium arsenide- and inorganic arsenic-exposed workers by determination of inorganic arsenic and its metabolites in urine and hair. *Am. Ind. Hyg. Assoc. J.*, **50**, 606–612.

Arsenic Exposure and Health Effects
W.R. Chappell, C.O. Abernathy and R.L. Calderon (Editors)
© 1999 Elsevier Science B.V. All rights reserved.

Consistency of Biomarkers of Exposure to Inorganic Arsenic: Review of Recent Data

Jean P. Buchet, Perrine Hoet, Vincent Haufroid, Dominique Lison

ABSTRACT

Health impairments can be observed following undue absorption of inorganic arsenic (As) which mainly occurs through inhalation in occupational settings or ingestion in the general population. Assessment of exposure based on the determination of the external dose often lacks precision due to variations in chemical species involved, in concentrations or due to varying toxicokinetic parameters. Therefore, on the basis of As metabolism data, three biomarkers of internal exposure are generally proposed: the urinary excretion of the element and its concentration in hair and nails (blood levels are generally too low and transient). This paper has as its main purpose the review of recent studies of the relationship between external and internal As doses to assess their consistency among investigators. To avoid difficulties from possibly poor analytical techniques only literature published during the last two decades is considered. Despite some encouraging reports, the value of As levels in hair and nails as indices of internal exposure appears limited and efforts are needed to develop a standardized procedure to solve the problem of external contamination of samples. The agreement on the relationship between airborne As concentration and urinary excretion (sum of inorganic, mono- and dimethylated forms) appears better. The level of As urinary excretion (seafood excluded) as a function of As oral intake through drinking water in steady state conditions has been reported by several authors from different countries. Despite possible ethnic and environmental differences, reported results display a quite satisfactory consistency. Most strikingly, for the same As level in drinking water, a higher urinary excretion of the element is observed in populations with a lower socio-economic status. The physiopathological basis and implications of this differential relationship deserves further study.

Keywords: inorganic arsenic, exposure biomarkers

INTRODUCTION

An adequate assessment of risks for human health resulting from the use of toxic chemicals is tightly bound to a careful characterization of dose/effects and dose/response relationships. Most of the time investigations to collect these data represent costly and time-consuming efforts and data obtained have not always the desired quality or the required precision. There are many possible causes for such a situation. Indeed the selection of effects to take into consideration and of populations in which responses have to be investigated strongly influence the validity of the global risk assessment. Even when the most relevant effect marker for a chemical is known and when an agreement on the tolerable occurrence of this effect is reached, the range of safe doses and thus acceptable exposure levels is still difficult to determine. Doses can be assessed using best estimated contributions of the commonest exposure routes: inhalation, ingestion and dermal uptake ("external dose"). In experimental conditions, the dose absorbed by a subject can be precisely controlled but, generally, imprecision is the rule, due to frequently unavoidable quantitative as well as qualitative variations in the exposure.

To evaluate the dose absorbed by a subject on a more individual basis, biological markers of exposure are useful. They derive from data on the metabolism of the chemical and are often based on the determination of the chemical itself or its metabolite(s) in relevant biological media. Such biomarkers not only take individual biological characteristics and personal habits into consideration but also integrate doses from all possible routes of exposure in the general and in the occupational environment. They allow the assessment of an "internal" dose which is obviously more directly related to health effects than the "external" dose derived from fragmentary or oversimplified information.

This paper aims at reviewing the consistency of current biomarkers of exposure to inorganic As in order to examine how reliably they allow comparisons between studies of dose/effects or dose/response relationships.

METHODOLOGY

Many analytical techniques have been developed to measure As traces in mineral and biological matrices. Their performances have been discussed by Irgolic et al. (1994). As concerns As measurements in biological media and particularly in urine, the most efficient excretion route, analytical specificity is of utmost importance because the presence of arsenicals from marine origin may greatly confound levels relevant for the assessment of exposure to inorganic As. Therefore, only literature data published during the last two decades and involving determinations in urine of chemical species closely related to inorganic As metabolism are considered. In all studies considered, the sum of tri- and pentavalent inorganic species, monomethylarsonic and dimethylarsinic acids was calculated and used in regression models. Furthermore, since toxicokinetics data indicate that As binds to proteins especially in hair and nails and because these media are readily accessible to sampling, studies reporting their As content were also reviewed.

Two main absorption routes of inorganic As were retained in the present review: inhalation of contaminated air occurring most frequently in industrial settings, and ingestion of As-rich drinking water which is prevalent in several countries around the world. Only papers reporting a quantitative relationship between parameters reflecting external and internal As doses were selected in order to assess their agreement.

RESULTS

Relationships between Exposure Parameters in the Case of Airborne Exposure to Inorganic Arsenic

Very low airborne As concentrations are commonly found in the general environment; by contrast, in some industrial settings, significant amounts of inorganic As may be inhaled, a major part of which is bioavailable. The roasting of non ferrous metal ores represents a main source of air contamination by As trioxide but the dispersal of other less soluble components (natural sulfides from ores; calcium arsenate from coal combustion) sometimes makes difficult the assessment of the bioavailability of the inhaled dose. The potential use of As levels in fingernails as a biological indicator of occupational exposure to the element was investigated by Agahian et al.(1990) in gold mine crushing mill operators. These authors developed a digestion procedure allowing the recovery of 70% of nail As after the application of a washing technique to remove 98% of the exogenous element. Although limited in number ($n = 12$) matched air and fingernail As concentrations were said to be well correlated ($r = 0.89$); the equation between air As concentration (y) in $\mu g/m^3$ as function of fingernail As level (x) in $\mu g/g$ was $y = 1.79 * x - 5.9$. This relation is however questionable because it leads to a concentration of 3.3 μg As/g in nails of non exposed subjects, which is one order of magnitude above mean control values commonly cited in the literature.

The As content in hair of workers from a copper smelter and a GaAs crystal production plant was examined by Yamauchi et al.(1989) who found that most of As in hair is in the inorganic form. However these authors were not able to distinguish the contaminant inorganic As from that eliminated from the body and concluded that "As determination in hair appears to be of value only when used for environmental monitoring of As con-tamination". de Peyster and Silvers (1995) also concluded that hair arsenic in the hair of workers from a semiconductor manufacturing facility where safe work practices minimized the air contamination (worst case 8 h TWA of $4 \mu g$ As/m^3) was not useful. It was found that, in that particular environment, non- occupational sources of As (tapwater, tobacco, diet) were the major contributors to hair levels in workers.

In four surveys of workers exposed to As trioxide in copper smelters or sulfuric acid plants the relationship between the urinary excretion of inorganic As and its methylated meta-bolites and As levels at the workplaces was studied. Mean parameter values were reported in two studies and regression equations in the others (Table 1). When the theoretical urinary excretions of As corresponding to an air exposure of 30 $\mu g/m^3$ (as in the study by Lagerkvist) are calculated on the basis of these regression equations, values amounting to 45 and 94 μg As/L are found for the Offergelt et al.(1992) and Vahter et al (1986) studies respectively, indicating some disagreement. According to Hakala and Pyy (1995) the ingestion of dimethylarsinic acid in the diet may confound the relationship between As in air and its urinary excretion. These authors found the best correlation ($r = 0.78$) with the sum of the concentrations of the tri- and pentavalent inorganic forms in urine collected up to 8 h after the exposure. A concentration of 5 $\mu g/L$ inorganic As in urine was found to correspond to a 8 h TWA exposure of 10 μg As/m^3. In a study by Landrigan et al.(1982) among workers in a lead-acid battery plant a close correlation ($n = 47$; $r = 0.84$) was found between urinary As concentration and exposure to arsine. In the range up to 30 $\mu g/m^3$ each μg arsine in air increased the urinary concentration by 2.43 μg As/L above a background level of 12 $\mu g/L$,

TABLE 1

Literature data related to the relationship between arsenic in air and the urinary excretion of the element.

Survey	Subject number	As-air ($\mu g/m^3$)	As-urine ($\mu g/L$)
		Mean values	
Lagerkvist et al. (1994)	31	30	27
Yamauchi et al. (1989)	11	330	240
		Regression equations	
Offergelt et al. (1992)	82	Log As-urine = 1.13 + 0.353 * log As-air	
Vahter et al. (1986)	17	As-urine = 34 + 2 * As-air	

leading to the excretion in urine of 85 μg As/L for an exposure of 30 μg As/m^3. This seems to indicate a more efficient absorption of the gaseous arsenical by the respiratory route by comparison with the results of studies involving As species found in non ferrous metal smelters.

Relationships Between Exposure Parameters in the Case of Chronic Consumption of Arsenic-rich Drinking Water

Undue chronic ingestion of inorganic As is most of the time associated with the consumption of drinking water contaminated by the element naturally present in many minerals and rocks in contact with the aquifer. Observations by Tseng in Taiwan (Tseng et al., 1968) prompted many other studies in several countries around the world where high As levels in drinking water are prevalent. It is obvious that, for a correct estimation of the ingested dose of As, both the element concentration and the volume of water consumed (as beverage as well as water soaking food items during cooking) must be known. Environmental (climate, food types) and physiological (sex, age) factors may induce variations of the ingested water volume which may be estimated to vary by a factor 2 to 3. In addition, variations resulting from differences in As levels in water are significant as concentrations from below 10 μg/L up to several mg/L may be found. As most studies report results related to populations without sex or age selection it seems reasonable to use the As level in drinking water as a parameter reflecting the external dose.

As in hair and nails versus As in water

Mean As levels in hair (136 subjects) associated with exposure via drinking water in California and Nevada were reported by Valentine et al. (1979); on average an increase of 0.26 μg As/g hair was found to correspond with a 100 μg As/L increment in drinking water. Similar investigations were performed recently in other countries and lead to some discrepancy. Indeed for the same increment of 100 μg As/L in drinking water, increases of As content in hair (μg As/g hair) were 0.09 in 115 Taiwanese subjects (Chiou et al., 1997), 0.94 in 293 subjects from Bangladesh (Biswas et al., 1998) and 1.0 in 42 subjects from Finland (Kurttio et al., 1998).

Few studies report the relationship between nail and drinking water As levels. The first significant correlation between these parameters was observed in New Hampshire by Karagas et al. (1996). However, the number of subjects ($n = 21$) and the range of As levels in drinking water (non detectable to 137 μg/L) were limited. Based on the regression analysis performed in subjects who were drinking water with detectable levels of As, a 10-fold increase in well water concentration was reflected by about a 2-fold increase in toenail concentration ranging from 0.073 to 2.25 μg/g. Two more recent observations have been made over a wider range of As level in drinking water (up to several hundreds μg As/L) but, as for observations in hair (see above), the increase in nail concentration corresponding to an increase of 100 μg As/L in drinking water varies by a factor of 10. It amounts to 0.27 μg/g in the Chiou et al. study (1997) and 2.5 μg/g in that of Biswas et al. (1998). Although no correlation study was performed in the survey by Olguin et al. (1983), a mean increase of As in nails of 4 μg/g was found in subjects consuming water with 400 as compared to 5 μg As/L, indicating a 1 μg/g increase of As in nails for a 100 μg As/L increase in drinking water.

Discrepancies between the range of control values for hair and nail As contents are not observed in the studies. Further investigations including analytical procedures for sample treatment and measurement are needed to examine the possible causes of disagreement.

On the basis of the growth rate of hair (10 mm/month) and nails (3 mm/month) a return to the baseline concentration of As is to be expected only when the consumption of As-rich water is discontinued for a long period which may be several years. (Chowdury et al., 1997; Lin et al., 1997).

Relationship between hair and nail As contents

Hair and nails (as well as skin) are efficient binding tissues for As and thus their As content can be expected to be closely related in the same subject. Two sets of data collected in India are presented in Figure 1; one relates to 17 family members drinking water with 276 μg As/L (Chowdury et al., 1997), the other to 7 subjects in a family using water with 320 μg As/L (Das et al., 1995). No relationship between hair and nail As levels can be observed. A high correlation between the same parameters ($r = 0.84$) was however reported by Hewitt et al. (1995) among 35 employees at a pesticide manufacturing plant which had formerly produced arsenical pesticides. A close examination of the scatter plot of data reveals, however, that a severe bias is introduced by a single observation and that other data form a much more randomly distributed cluster of points. Thus, if one accepts the hypothesis of a related affinity of As for hair and nails, it seems that severe analytical problems remain to be solved.

Relationship between As Level in Drinking Water and the Urinary Excretion of the Element

The urinary excretion of As is the parameter most frequently used to assess environmental exposure. In many surveys aiming at characterizing health risks resulting from the consumption of As-rich water, data are available on the basis of groups of subjects. These are presented in Table 2 with references to the original publications. Urinary concentrations are expressed in μg/L, if necessary after an adjustment taking into account a mean daily production of 1.5 L of urine containing 1.5 g creatinine. Observations made at high exposure levels in Belgium (Buchet et al., 1981b) correspond to experimental conditions in volunteers after the ingestion of 5 consecutive daily doses ranging between 0.125 and 1.0 mg As; assuming a water consumption of 2 L/day this would be equivalent to water concentrations ranging between 62.5 and 500 μg As/L. A scatter plot is presented in Figure 2A, in which different symbols are used to identify countries where data were collected. The goodness of fit of a linear regression model applied to the total data set (Figure 2A) is moderate: $r^2 = 0.59$; the slope of the regression line equals 0.59 (95% CI: 0.49–0.81). For As levels in water below 100 μg/L (Figure 2B), the fit is worse($r^2 = 0.1$) and the slope, which is equal to 0.44 (95% CI: 0.03-0.84), becomes not significantly different from 0 when a single extreme observation is not considered. Better fits of the linear model to the data are obtained when they are considered in subsets corresponding to observations made in Alaska, California, Nevada and Belgium (Figure 3A),in Argentina and Chile (Figure 3B), and in India and Mexico (Figure 3C). Percentages of variance explained by the linear model and slopes of the regression lines (with 95% CI) are 97%, 0.52 (0.48–0.56), 99%, 0.94 (0.88–1.00) in groups A and B respectively. The linear regression model does not fit well observations in group C (Figure 3C). If one excepts a single observation in India (that with the highest As in water), r^2 equals 0.72 and the slope is 1.62 (1.07–2.17). Slopes and r^2 for linear regressions between levels of As in urine (y) and in drinking water (x) were recently published by Kurttio et al.(1998) in Finland, Biswas et al. (1998) in India and Chiou et al. (1997) in Taiwan. Observations made in the first two studies are in agreement with the values reported above: slopes/r^2 are 0.8/0.74 and 1.27/0.89 respectively; this is not the case in the Taiwanese study (slope is 0.18; no r^2 reported).

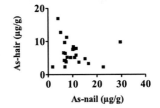

Fig. 1. Relationship between hair and nail As levels
in subjects drinking As-rich water.

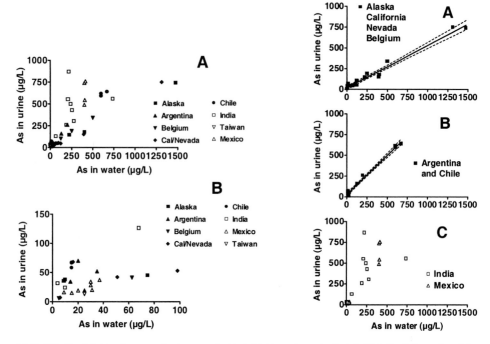

Left: Fig. 2. Scatter plot of As urinary excretion in groups of subjects drinking water contaminated with As (A: all observations; B: low exposure range; see Table 2 for details).

Right: Fig. 3. Relationship between As urinary excretion and As level in drinking water in different populations.

It has been shown (Buchet et al., 1981a) that after a single moderate dose of inorganic arsenic (0.5 mg As), the background urinary excretion of the element is restored after one week and that a few weeks are necessary after repeated moderate doses or a single acute dose such as in a suicide attempt (Mahieu et al., 1981). When As has been ingested chronically for years, longer periods of clean water use appear necessary to restore a background urinary excretion. Intervention studies (Hopenhayn-Rich et al. 1996; Moore et al., 1997) show that the consumption during 2 months of water with 45 μg As/L instead of 600 μg/L reduces the urinary excretion from 636 to 166 μg/L. Even when water with less than 1 μg/L is used, a two month period does not appear sufficient to obtain a steady state equilibrium (Kurttio et al., 1998). As observed in West Bengal (Chowdury et al., 1997) the restoration of a normal urinary excretion of As may require a few years.

DISCUSSION AND CONCLUSIONS

The use of As levels in hair and nails as indices of internal exposure intensity is of potential interest but there is a clear need for a well standardized analytical procedure for these determinations. In addition the rates of As elimination in urine and hair or nails differ so strongly that urinary As is considered representative of the recent (previous 1 or 2 days) exposure while hair and nail levels reflect the earlier (several months before) exposure intensity. It is thus not surprising that neither As in hair nor in nails has ever been found correlated with a urinary As level determined only on a single occasion. To make sound comparisons these biological exposure parameters may thus relate to a steady state situation

TABLE 2

Literature data related to the relationship between arsenic in drinking water and the urinary excretion of the element (mean values)

Countries	Authors	Drinking water As level					
		Low			High		
			μg As/L in			μg As/L in	
		n	water	urine	n	water	urine
Alaska	Harrington et al. (1978)	67	10	38	30	75	45
					49	401	178
	Kreiss et al. (1983)	95	9	35	39	222	143
					13	1475	742
Argentina	Vahter et al. (1995)	5	25	19	5	35	52
		5	14	34	11	200	260
	Lerda et al. (1994)	155	20	70	282	130	160
Belgium	Buchet et al. (1981b, 1996)	135	<5	6	1	62.5	41
		2			1	125	102
					1	250	190
					1	500	341
California Nevada	Valentine et al. (1979)	32	<6	7	28	51	42
					46	98	53
					22	123	47
					35	393	153
	Warner et al. (1994)	18	16	68	18	1312	750
Chile	Hopenhayn- Rich et al. (1996)	98	15	58	122	600	592
	Moore et al. (1997)	55	15	66	70	600	616
	Biggs et al. (1997)	108	15	67	122	670	640
India	Chatterjee et al. (1995)				47	278	303
					58	737	555
					45	241	499
					40	193	258
					58	257	428
					52	210	550
	Dahr et al. (1997)	62	4	31			
	Chakraborti et al. (1998)				8	68	126
	Mandal et al. (1996)	250	<10	24	1166	220	865
Mexico	Gonsebatt et al. (1994)	34	30	34	35	400	740
		30	37	37	33	412	758
	Del Razo et al. (1997)	34	31	20	35	404	544
	Garcia Vargas et al. (1994)	31	20	19	36	400	490
	Wyatt et al.(1998)	10	9	16			
		10	15	15			
		10	30	28			
Taiwan	Lin et al. (1998)	25	25	13			

i.e. after at least one year at the same exposure level. Other difficulties in interpreting data are worth mentioning: (1) hair and nails can both adsorb As from external sources and without appropriate washing their As content may not accurately reflect the internal dose but rather indicate the presence of the element in the environment; (2) a wide range (0.01–8 μg/g) is reported in the scientific literature for the As concentration in hair and nails of non exposed subjects; in our experience, these levels are below 1 $\mu g/g$ and such values were recently proposed as reference values (Hewitt et al., 1995).

In the work environment, the urinary excretion of As is most often correlated with the air concentration; however the ratio between both parameters differs in studies reported in the literature. Although the urine sampling time after exposure and analytical techniques for As measurements in air and in urine are comparable, generally, no information is available on the airborne As species. These may, however, exhibit highly different bioavailabilities according to the chemical composition, particle size and solubility in water. In addition, pre-exposure levels of As in urine often differ sufficiently to interfere with the increment caused by exposure, particularly at levels below 50 μg As/m³. Therefore comparisons between various settings should be made with caution and the measurements are mainly useful to monitor exposure under reproducible conditions, e.g. in a follow-up programme of industrial hygiene.

The bioavailability of As in drinking water is probably very similar in different countries; however different urinary excretion rates are observed in populations living in different geographical areas but with similarly high exposure through drinking water. These geographical differences may of course be ascribed to some extent to differences in water volume consumed daily (precise data are generally not available). One may also wonder to what extent the differences are related to the socio-economic (or nutritional status) of the populations. Indeed Biswas et al. (1998) observed in many As-affected villages in West Bengal that poor people are more affected than villagers consuming similarly contaminated water but in a better condition of health.

It seems worth investigating whether a possible protein deficiency induced by a poor food supply can lead to a depletion of binding sites for As and its metabolites and thus facilitate their urinary elimination.

REFERENCES

Agahian, B., Lee, J.S., Nelson, J.H. and Johns, R.E. 1990. Arsenic levels in fingernails as a biological indicator of exposure to arsenic. *Am. Ind. Hyg. Assoc. J.*, **51**, 646–651.

Biggs, M.L., Kalman, D.A., Moore, L.E., Hopenhayn-Rich, C., Smith, M.T. and Smith, A.H. 1997. Relationship of urinary arsenic to intake estimates and a biomarker of effect, bladder cell micronuclei. *Mutat. Res.* **386**, 185–195.

Biswas, B.K., Dhar, R.K., Samanta, G., Mandal, B.K., Chakraborti, D., Faruk, I., Islam, K.S., Chowdhury, M.M., Islam, A. and Roy, S. 1998. Detailed study report of Samta, one of the arsenic-affected villages of Jessore district, Bangladesh. *Current Science*, **74**, 133–145.

Buchet, J.P., Lauwerys, R. and Roels, H. 1981a. Comparison of the urinary excretion of arsenic metabolites after a single oral dose of sodium arsenite, monomethylarsonate, or dimethylarsinate in man. *Int. Arch. Occup. Environ. Health*, **48**, 71–79.

Buchet, J.P., Lauwerys, R. and Roels, H. 1981b. Urinary excretion of inorganic arsenic and its metabolites after repeated ingestion of sodium metaarsenite by volunteers. *Int. Arch. Occup. Environ. Health*, **48**, 111–118.

Buchet, J.P., Staessen, J.,Roels, H., Lauwerys, R. and Fagard, R. 1996. Geographical and temporal differences in the urinary excretion of inorganic arsenic: a Belgian population study. *Occup. Environ. Med.*, **53**, 320–327.

Chakraborti, D., Samanta, G., Mandal, B.K., Chowdhury, T.R., Chanda, C.R., Biswas, B.K., Dhar, R.K., Basu, G.K. and Saha, K.C. 1998. Calcutta's industrial pollution: groundwater arsenic contamination in a residential area and sufferings of people due to industrial effluent discharge—an eight-year study report. *Current Science*, **74** (4), 346–355.

Chatterjee, A., Das, D., Mandal, B.K., Chowdhury, T.R., Samanta, G. and Chakraborti, D. 1995. Arsenic in ground water in six districts of West Bengal, India: the biggest arsenic calamity in the world. Part I. Arsenic species in drinking water and urine of the affected people. *Analyst*, **120**, 643–650.

Chiou, H.-Y., Hsueh, Y.-M., Hsieh, L.-L., Hsu, L.-I., Hsu, Y.-H., Hsieh, F.-I., Wei, M.-L., Chen, H.-C., Yang, H.-T., Leu, L.-C., Chu, T.-H., Chuan, C.-W., Yang, M.-H. and Chen, C.-J.1997. Arsenic methylation capacity, body retention, and null genotypes of glutathione S-transferase M1 and T1 among current arsenic-exposed residents in Taiwan. *Mutat. Res.*, **386**, 197–207.

Chowdury, T.R., Mandal, B.K., Samanta, G., Basu, G.K., Chowdury, P.P., Chanda, C.R., Karan, N.K., Lodh, D., Dahr, R.K., Das, D., Saha, K.C. and Chakraborti, D. 1997. Arsenic in groundwater in six districts of West Bengal, India: the biggest arsenic calamity in the world: the status report up to August, 1995. In *Arsenic Exposure and Health Effects* (eds. C.O. Abernathy, R.L. Calderon, and W.R. Chappell), Chapman and Hall, pp. 93–111.

Das, D., Chatterjee, A., Mandal, B.K., Samanta, G., Chakraborti, D. and Chanda, B. 1995. Arsenic in groundwater in six districts of West Bengal, India: the biggest arsenic calamity in the world. Part 2. Arsenic concentration in drinking water, hair, nails, skin-scale and liver tissue (biopsy) of the affected people. *Analyst*, **120**, 917–924.

Del Razo, L.M., Garcia-Vargas, G.G., Vargas, H., Albores, A., Gonsebatt, M.E., Montero. R., Ostrosky-Wegman, P., Kelsh, M. and Cebrián, M.E. 1997. Altered profile of urinary arsenic metabolites in adults with chronic arsenicism. A pilot study. *Arch. Toxicol.* **71**, 211–217.

De Peyster, A. and Silvers, J.A. 1995. Arsenic levels in hair of workers in a semiconductor fabrication facility. *Am. Ind. Hyg. Assoc. J.*, **56**, 377–383.

Dhar, R.K., Biswas, B.K., Samanta, G., Mandal, B.K., Chakraborti, D., Roy, S., Jafar, A., Islam, A., Ara, G., Kabir, S., Khan, A.W., Akther Ahmed, S. and Hadi, S.A. 1997. Groundwater arsenic calamity in Bangladesh. *Current Science*, **73**, 47–59.

Garcia-Vargas, G.G., Del Razo, L.M., Cebrián, M.E., Albores, A., Ostrosky-Wegman, P., Montero, R., Gonsebatt, M.E., Lim, C.K. and De Matteis, F. 1994. Altered urinary prophyrin excretion in a human population chronically exposed to arsenic in Mexico. *Hum. Exper. Toxicol.*, **13**, 839–847.

Gonsebatt, M.E., Vega, L., Montero, R., García-Vargas, G., Del Razo, L.M., Albores, A., Cebrián, M.E. and Ostrosky-Wegman, P. 1997. Lymphocyte replicating ability in individuals exposed to arsenic via drinking water. *Mutat. Res.*, **313**, 293–299.

Gonsebatt, M.E., Vega, L., Salazar, A.M., Montero, R., Guzmán, P., Blas, J., Del Razo, L.M., García-Vargas, G., Albores, A., Cebrián, M.E., Kelsh, M. and Ostrosky-Wegman, P. 1997. Cytogenetic effects in human exposure to arsenic. *Mutat. Res.*, **386**, 219–228.

Hakala, E. and Pyy, L. 1995. Assessment of exposure to inorganic arsenic by determining the arsenic species excreted in urine. *Toxicol. Lett.*, **77**, 249–258.

Harrington, J.M., Middaugh, J.P., Morse, D.L. and Housworth, J. 1978. A survey of a population exposed to high concentrations of arsenic in well water in Fairbanks, Alaska. *Am. J. Epid.*, **108** (5), 377–385.

Hewitt, D.J., Millner, G.C., Nye, A.C. and Simmons, H.F. 1995. Investigation of arsenic exposure from soil at a superfund site. *Env. Res.*, **68**, 73–81.

Hopenhayn-Rich, C., Biggs, M.L., Smith, A.H., Kalman, D.A. and Moore, L.E. 1996. Methylation study of a population environmentally exposed to arsenic in drinking water. *Environ. Health Perspect.*, **104** (6), 620–628.

Irgolic, K.J. 1994. Determination of total arsenic and arsenic compounds in drinking water, in *Arsenic Exposure and Health* (eds. W.R. Chappell, C.O. Abernathy, and C.R. Cothern), Science and Technology Letters, Northwood, pp. 51–60.

Hopenhayn-Rich, C., Biggs, M.L., Kalman, D.A., Moore, L.E. and Smith, A.H. 1996. Arsenic methylation patterns before and after changing from high to lower concentrations of As in drinking water. *Environ. Health Perspect.*, **104** (11), 1202–1206.

Karagas, M.R., Morris, J.S., Weiss, J.E., Spate, V., Baskett, C. and Greenberg, E.R. 1996. Toenail samples as an indicator of drinking water arsenic exposure. *Cancer Epid., Biom. Prev.*, **5**, 849–852.

Kreiss, K., Zack, M.M., Feldman, R.G., Niles, C.A., Chirico-Post, J., Sax, D.S., Landrigan, P.J., Boyd, M.H. and Cox, D.H. 1983. Neurologic evaluation of a population exposed to arsenic in Alaskan well water. *Arch. Environ Health*, **38** (2), 116–121.

Kurttio, P., Komulainen, H., Hakala, E., Kahelin, H. and Pekkanen, J. 1998. Urinary excretion of arsenic species after exposure to arsenic present in drinking water. *Arch. Environ. Contam. Toxicol.*, **34**, 297–305.

Lagerkvist, B.J. and Zetterlund, B. 1994. Assessment of exposure to arsenic among smelter workers: a five-year follow-up. *Am. J. Ind. Med.*, **25**, 477–488.

Landrigan, P.J., Costello, R.J. and Stringer, W.J. 1982. Occupational exposure to arsine. *Scand. J. Work Environ. Health*, **8**, 169–177.

Lerda, D. 1994. Sister-chromatid exchange (Sce) among individuals chronically exposed to arsenic in drinking water. *Mutat. Res.*, **312** (2), 111–120.

Lin, T.-H. and Huang, Y.-L. 1998. Arsenic species in drinking water, hair, fingernails, and urine of patients with blackfoot disease. *J. Toxicol. Environ. Health*, **53**, 85–93.

Mandal, B.K., Chowdhury, T.R., Samanta, G., Basu, G.K., Chowdhury, P.P., Chanda, C.R., Lodh, D., Karan, N.K., Dhar, R.K., Tamili, D.K., Das, D., Saha, K.C. and Chakraborti, D. 1996. Arsenic in groundwater in seven districts of West Bengal, India—The biggest arsenic calamity in the world. *Curr. Sci.*, **70** (11), 976–986.

Mahieu, P., Buchet, J.P., Roels, H. and Lauwerys, R. 1981. The metabolism of arsenic in humans acutely intoxicated by As_2O_3. Its significance for the duration of the BAL therapy. *Clin. Toxicol.*, **18**, 1067–1075.

Moore, L.E., Smith, A.H., Hopenhayn-Rich, C., Biggs, M.L., Kalman, D.A. and Smith, M.T. 1997. Micronuclei in exfoliated bladder cells among individuals chronically exposed to arsenic in drinking water. *Cancer Epid. Biom. Prev.*, **6**, 31–36.

Moore, L.E., Smith, A.H., Hopenhayn-Rich, C., Biggs, M.L., Kalman, D.A. and Smith, M.T. 1997. Decrease in bladder cell micronucleus prevalence after intervention to lower the concentration of arsenic in drinking water. *Cancer Epid. Biom. Prev.*, **6**, 1051–1056.

Offergelt, J.A., Roels, H., Buchet, J.P., Boeckx, M. and Lauwerys, R. 1992. Relation between airborne arsenic trioxide and urinary excretion of inorganic arsenic and its methylated metabolites *Br. J. Ind. Med.*, **49**, 387–393.

Olguin, A., Jauge, P., Cebrian, M. and Albores, A. 1983. Arsenic levels in blood, urine, hair and nails from a chronically exposed human population. *Proc. West. Pharmacol. Soc.*, **26**, 175–177.

Tseng, W.P., Chu, H.M., How, S.W., Fong, J.M., Lin, C.S. and Yeh, S. 1968. Prevalence of skin cancer in an endemic area of chronic arsenicism in Taiwan. *J. Natl. Cancer Inst.*, **40**, 453–463.

Vahter, M., Friberg, L., Rahnster, B., Nygren, A. and Nolinder, P. 1986. Airborne arsenic and urinary excretion of metabolites of inorganic arsenic among smelter workers. *Int. Arch. Occup. Environ. Health*, **57**, 79–91.

Vahter, M., Concha, G., Nermell, B., Nilsson, R., Dulout, F. and Natarajan, A.T. 1995. A unique metabolism of inorganic arsenic in native Andean women. *Eur. J. Pharmacol.*, **293**, 455–462.

Valentine, J.L., Kang, H.K. and Spivey, G. 1979. Arsenic levels in human blood, urine and hair in response to exposure via drinking water. *Environ. Res.*, **20**, 24–32.

Warner, M.L., Moore, L.E., Smith, M.T., Kalman, D.A., Fanning, E. and Smith, A.H. 1994. Increased micronuclei in exfoliated bladder cells of individuals who chronically ingest arsenic-contaminated water in Nevada. *Cancer Epid. Biom. Prev.*, **3**, 583–590.

Wyatt, C.J., Lopez Quiroga, V., Olivas Acosta, R.T. and Mendez, R.O. 1998. Excretion of arsenic in urine of children, 7–11 years, exposed to elevated levels of As in the city water supply in Hermosillo, Sonora, Mexico. *Environ. Res.* **78**, 19–24.

Yamauchi, H., Takahashi, K., Mashiko, M. and Yamamura, Y. 1989. Biological monitoring of arsenic exposure of gallium arsenide- and inorganic arsenic-exposed workers by determination of inorganic arsenic and its metabolites in urine and hair. *Am. Ind. Hyg. Assoc. J.*, **50**, 606–612.

Arsenic Exposure and Health Effects
W.R. Chappell, C.O. Abernathy and R.L. Calderon (Editors)
© 1999 Elsevier Science B.V. All rights reserved.

Hair Arsenic as an Index of Toxicity

J. Thomas Hindmarsh, Diane Dekerkhove, Geoffrey Grime,
Jonathan Powell

ABSTRACT

Hair arsenic levels are a useful indicator of chronic arsenic poisoning in forensic cases, provided that external contamination of the hair by arsenic can be excluded. Unfortunately its lack of clinical precision limits its usefulness in assessing severity of poisoning in subjects exposed to arsenic in their drinking water. External contamination of the hair from washing in this water may also be a confounding factor. Our preliminary experiments have shown that the locations in the hair of external arsenic contamination and of arsenic derived from ingestion are identical.

Keywords: arsenic, hair, toxicity, electromyography, urine, blood, neuropathy

INTRODUCTION

There are few biochemical indicators of arsenic toxicity. It is rapidly cleared from the blood, little remaining 10 hours after an oral dose (Mealey et al., 1959), thus blood levels are of little value in assessing human arsenic intoxication, except perhaps for identifying arsenic as a cause of acute poisoning in patients whose renal function is seriously impaired. Because of the rapid and efficient excretion of arsenic in the urine, analysis of this is a reliable index of recent (1 to 3 days previous) exposure but little remains 96 hours after exposure has ceased (Buchet et al., 1981) except after heavy and prolonged arsenic ingestion when the urine may take longer to become normal (Dewar and Lenihan, 1956; Valentine et al., 1979). Thus, urine is best for identifying ongoing exposure and cannot be reliably used for the retrospective diagnosis of chronic arsenic poisoning.

Nails and hair have similar affinities for arsenic but data on nails is somewhat limited because hair is more convenient to work with. Nails can take up and concentrate arsenic *in vitro* (Smith and Hendry, 1934) and therefore like hair, are subject to external arsenic contamination. Subsequent discussion will be confined to hair. Hair arsenic can be employed to identify chronic arsenic poisoning provided that external contamination can be excluded. However, the relationship between hair arsenic levels and degree of toxicity is only very approximate.

NORMAL REFERENCE INTERVAL

Recent studies agree that hair arsenic levels in persons not subjected to significant arsenic exposure are less than 1 μg/g (Hindmarsh et al., 1977). However, levels can be higher in people living in polluted environments. Liebscher and Smith (1968) analysed 1250 hair samples from persons living in Glasgow, Scotland—a large industrial city—and quoted a range of 0.02 to 8.17 μg/g. Their data were log-normally distributed with the tail to the right and applying limits of plus and minus two Standard Derivations of the mean to the log-normal distribution gave a range of approximately 0.06 to 3 μg/g. Sky-Peck obtained similar results when he analysed 987 hair samples from residents of Chicago in the 1980s (Sky-Peck, 1990). Also Bencko and Symon (1977) have reported levels of up to 10 μg/g in children living near an electric power plant burning soft coal. Goldsmith et al. (1972) have reported that hair arsenic becomes elevated above 0.4 μg/g when drinking water arsenic exceeds 50 μg/L and Valentine et al. (1979) support this finding.

SAMPLE COLLECTION AND PREPARATION

At least one gram of hair should be collected, from several sites around the nape of the neck (to minimize external contamination) cutting close to the scalp so that the whole hair shaft is obtained. The sample should be washed in a non-ionic detergent, distilled deionized water, then ethanol, and then chopped into small pieces and thoroughly mixed and dried before analysis. Some analysts prefer pubic hair to minimize external contamination but dust in factories can be very pervasive and there can be no certainty that this site is not contaminated. There is no need to pull the hairs out to include the roots as the first centimeter of hair is rapidly impregnated after acute arsenic ingestion. It is important when using hair arsenic as an index of toxicity that the mean arsenic content of a large sample collected as described above be used. Much confusion has arisen from attempts to interpret arsenic concentrations from individual hairs or parts of an individual hair as indices of toxicity. Inter- and intra-hair variations are enormous (many fold), thus making interpretation of results from single hairs of limited value (Cornelis, 1973). Also, hair samples from different subjects adsorb arsenic with different degrees of avidity (Ferguson et al., 1983) adding to the difficulties in interpretation.

DEPOSITION ON HAIR

How arsenic in the blood stream reaches the hair is still a matter of conjecture. Young and Rice (1944) found arsenic in the hair of guinea pigs injected with sodium arsenite well beyond where it could reach by growth alone. They concluded it must have reached this point on the hair from sweating or sebaceous secretions, followed by adsorption onto the hair surface. Pearson and Pounds (1971), however, have questioned whether arsenic can be adsorbed onto hair from sweat. Similar results to those of Young and Rice were obtained by Lima (1966) who identified the location of hair arsenic in humans 10 days after starting the daily ingestion of Fowler's solution. He detected high hair arsenic levels four centimeters from the scalp even though hair growth was probably only 3.5 mm since the drug ingestion. He considered whether washing could move arsenic along the hair by capillary action, but showed that although there was some spreading by washing, most of an applied bolus of arsenic remained at the point of deposition on the hair even after washing with various solutions. His findings imply that ingested arsenic is secreted in the sweat and/or sebaceous secretion and becomes deposited on the surface of the hair where it remains firmly attached for the life of the hair.

SOURCES, LOCATION AND NATURE OF HAIR ARSENIC

Arsenic can be deposited in hair from the bloodstream after enteral or parenteral administration; also, it can be deposited on the hair by external contamination from arsenic containing dusts such as may occur in smelter workers. Arsenic from these two sources is avidly bound to the hair and cannot be readily removed by washing. Nor can arsenic from these two sources be differentiated by any known technique. Their location in a cross-section of hair is identical. We have shown by electron probe microscopy (Hindmarsh and McCurdy, 1986) that externally contaminated hair from workers in a gold smelter, not surprisingly demonstrated a greater concentration of arsenic on the outer surface of the hair. In contrast, Smith (1976), using the same technique, also on hairs from workers in a gold smelter, showed arsenic distribution throughout the cross-section of several hairs although one sample showed a greater concentration on the outer surface. Cookson and Pilling (1975), using proton induced x-ray emission (PIXE), showed peaks of arsenic also on the surface of transected hairs in a patient ingesting an arsenic containing medication.

Our recent experiments demonstrate a mixed picture: Figure 1 (a–e) demonstrates the location of arsenic in cross sections of hairs obtained from five gold smelter workers whose hair was contaminated by exposure to arsenic dust in their work environment and it can be seen that, whereas most of the arsenic is found on the outer surface in samples (a) and (b), in samples (c), (d) and (e) most of the arsenic is in the core of the hair. In contrast, Figure 2 (a, b) are samples from a single patient who had been ingesting arsenic (Fowler's solution) for one year and it is obvious that most of the arsenic is located in the core of the hair. Thus it can be seen that the location of arsenic in a cross section of hair cannot distinguish whether the arsenic was ingested or derived from external contamination.

There is always concern that some of the arsenic in the hair of gold-smelter workers might be derived from arsenic ingestion and inhalation rather than external contamination, but we can rest assured that most of the arsenic on the hair depicted in Figure 1 must be from external contamination as most of the levels would be in the lethal range were they due largely to ingestion or inhalation. The samples shown in Figures 1 and 2 have been in storage for 25 years and it is possible that changes in arsenic location may have occurred. Therefore, further experiments on fresh samples are ongoing. The use of hair lacquer has been claimed to prevent external arsenic contamination of hair but this remains to be proved (Barrowcliff, 1971).

Few studies have been made of the chemical nature of hair arsenic but Yamauchi et al. (1989) have reported it to be mostly inorganic with a trace of dimethylarsinic acid. Fish

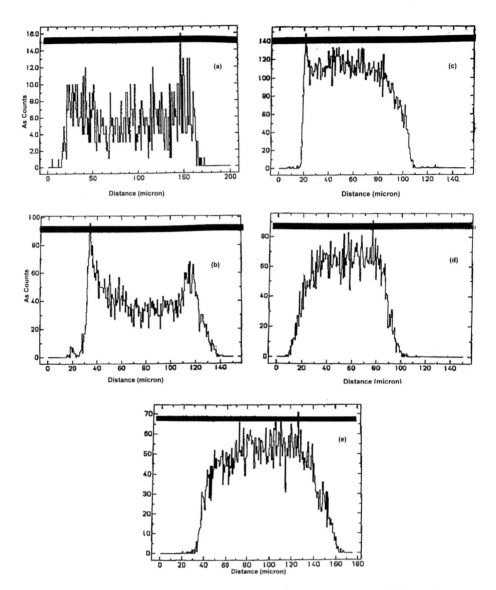

Fig. 1. Location of arsenic in transections of hair from workers in a gold smelter. Arsenic levels: (a) 29, (b) 161, (c) 287, (d) 386, (e) 452 μg/g.

arsenic is not deposited in the hair: arsenocholine and arsenobetaine are rapidly excreted in the urine and are not deposited in the hair (Vahter et al., 1983; Marafante et al 1984).

Hair can concentrate arsenic from a solution in which it is suspended. Smith and Hendry (1935) showed that human hair placed in a solution of sodium arsenite for 10 days (10 μg/ml) increased its arsenic content from 0 to 50 μg/g. 60% of this arsenic remained after 15 days soaking in distilled water (water changed every day). Rapid rinsing with distilled water or dilute sodium hydroxide did not affect the arsenic content significantly. Similar results were

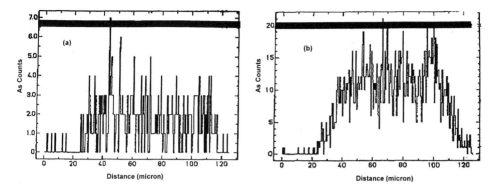

Fig. 2. Location of arsenic in transection of hair from a patient with chronic arsenic poisoning. Arsenic levels (a) 10 and (b) 75 μg/g.

obtained by Young and Rice (1944) who compared hair from guinea pigs injected with sodium arsenite daily for 9 and 13 days, with hair soaked in sodium arsenite (0.2 mg/L) for 17 days. In the latter experiment, hair arsenic content increased from 0.34 to 290 μg/g. Soaking of the hair from the latter experiment in water for 24 hours reduced the arsenic content by 7%. Further soaking in water for 12 days reduced the arsenic content to 100 μg/g. Only a further 8% could be extracted with acid, alkali, ethanol, or ether. Similar results were obtained with the hair samples from the arsenic-injected animals. They concluded that washing techniques, even though they eventually removed most of the arsenic, could not help distinguish between "external" and "internal" arsenic. Atalla et al. (1965) concluded the same after observing the effects of various washing techniques on hair from a rabbit injected with sodium arsenite, human hair soaked in sodium arsenite solution, and arsenic-contaminated hair from workers in an arsenic plant. Hair arsenic levels of 12 μg/g have been reported in corpses buried in arsenic-containing soil (Barrowcliff, 1971).

Numerous authors (Smith, 1961; Erikson, 1966; Shapiro, 1967; Pearson and Pounds, 1971; Barrowcliff, 1971) have shown that the dates of arsenic dosing can be determined by the location of the arsenic peaks on the hair (human hair grows at the rate of approximately 0.35 mm per day or 13 mm a month). However, although this seems to be reliable in some instances, it can be misleading. The uptake of arsenic along the length of a hair shaft can be inconsistent. Thus, peaks and valleys may represent variable uptake as well as intermittent dosing. Also, in acute poisoning, excessive sweating may produce bizarre patterns of arsenic deposition. Lander et al. (1965), studying the location of arsenic in hair in cases of severe acute poisoning, found that four of twelve cases had higher levels in the distal rather than the proximal sections even though three were sampled within 8 hours of arsenic ingestion. They acknowledged that contamination from vomitus might explain their findings, although, in some cases, they also found high levels in the distal toe nails, a less likely site for contamination from vomitus. The spreading of arsenic by the intense sweating associated with acute arsenic poisoning is a more likely explanation.

Some authors have claimed that arsenic in hair derived from ingestion can be distinguished from arsenic derived from external contamination by the fact that the latter would be evenly distributed along the length of the hair shaft whereas the former would demonstrate peaks and valleys corresponding to arsenic doses. However, this is not the case: Maes and Pate (1977) soaked hair in ^{76}As labelled arsenite and demonstrated zones of high and low activity along the length of the hair shaft showing that external contamination can be unevenly distributed.

TABLE 1

Approximate interpretation of hair arsenic levels (μg/g). Mean of $\geq$1 g of hair from several sites on head.

Normal	< 1 (<3 in polluted areas)
Chronic poisoning	$\geq$10 (can be lower)
Compatible with death from chronic arsenic poisoning	$\geq$45 (can be lower, also higher levels have been reported in subjects who survived).
External contamination:	up to 8000

An important question is whether washing hair in arsenic-contaminated water can produce significant hair contamination; that is, in persons drinking arsenic-contaminated water, how much of the hair arsenic is derived from ingestion and how much from washing? Ferguson et al. (1983) have shown that arsenic uptake by hair from arsenic-containing solutions is relatively slow, implying that washing in well water contaminated with arsenic in the amounts that have been described would not significantly elevate hair arsenic levels. However, Harrington et al. (1978) have shown, in a study from Alaska of subjects using arsenic-contaminated wells that 31 persons drinking bottled water but washing in contaminated well water ($\geq$100 μg/L) had a mean hair arsenic of 5.74 μg/g and those who had stopped drinking the contaminated water 3 months previously but who were still washing in it (22 subjects) had a mean hair arsenic level of 2.29 μg/g, compared to a mean hair level of 0.46 μg/g in 84 subjects drinking and washing in water whose arsenic content was < 100 μg/L. Forty-four persons washing in and still drinking contaminated water ($\geq$100 μg/L) had a mean hair arsenic level of 3.29 μg/g. Also, Inamasu et al. (1978) measured hair arsenic levels in 7 persons who used arsenic contaminated water (3.53 mg/L) for washing but not for drinking. Hair levels were elevated in 5 of the 7 subjects, the highest being 5.2 μg/g. These findings imply that washing in contaminated water can produce significant hair contamination.

ACUTE POISONING

In acute poisoning, hair levels are often very elevated 24 hours after arsenic ingestion (Lander et al., 1965). The one case of acute arsenic poisoning that I have seen had a mean hair level of 100 μg/g 25 hours after the ingestion of a cup of copper acetoarsenite (Paris Green). She died 72 hours after ingestion.

INTERPRETATION OF HAIR ARSENIC LEVELS

An approximate interpretation of hair arsenic levels in chronic poisoning is given in Table 1. It must be emphasized, however, that external arsenic contamination must be completely excluded, if hair levels are to be used in assessment of poisoning. The relationships between hair arsenic levels and the extent of poisoning as described in Table 1 are only very approximate and can be very misleading. Thus, the diagnosis of chronic arsenic poisoning must be made clinically by demonstrating the typical dermatologic lesions and neuropathy, reserving the hair arsenic level for confirmation of diagnosis and then only after external contamination has been excluded.

HOW DOES HAIR ARSENIC COMPARE WITH OTHER OBJECTIVE FEATURES OF ARSENIC TOXICITY?

The clinical features of chronic arsenic poisoning (skin lesions, clinical neuropathy) are somewhat crude and imprecise indicators of the severity of chronic arsenic poisoning. Unfortunately, the search for more objective evidence of the severity of chronic arsenic poisoning (such as nerve conduction studies) has been disappointing. In 1977 we (Hind-marsh et al., 1977) studied 32 people in Nova Scotia who had been using arsenic-contaminated wells for many years. Twelve controls whose wells were not contaminated but who were from the same community were also studied. A neurologist who did not know which patients were in the exposed or control groups studied motor and sensory functions of the median, peroneal and sural nerves. We found that hair arsenic was a more reliable predictor of presumably arsenic related electromyographic abnormalities than was well arsenic content; of the nineteen persons with hair arsenic levels of more than 1 $\mu g/g$, ten had electrophysiological abnormalities whereas all 25 of those whose hair arsenic levels were 1 $\mu g/g$ or less were electrophysiologically normal even though 13 of this latter group were using arsenic-contaminated wells. Thus, in our limited experience, these two objective determinants of arsenic toxicity agree quite well and if electromyography were a "gold standard", we could conclude that hair arsenic is an acceptable index of toxicity. We also found an approximate correlation between well water arsenic and hair arsenic levels indicating that hair arsenic increases as exposure increases. However, others have been unable to confirm our findings. In a much larger study from Alaska (Kreiss et al., 1983) of 147 persons exposed to arsenic from domestic wells whose arsenic concentrations ranged up to 4.781 $\mu g/g$, the investigators were unable to demonstrate a dose–response relationship between arsenic ingestion and symptoms or physical findings of peripheral neuropathy including electromyography. Also, Southwick et al. (1981) studied 145 people in Utah whose drinking water arsenic content averaged 0.18 $\mu g/g$ in one community and 0.21 $\mu g/g$ in another community. A matched control group of 105 people whose drinking water arsenic averaged 0.02 $\mu g/g$ was also studied. They found a clear relationship between the amount of arsenic consumed and the amount of arsenic in hair and urine samples. However, electro-myographic abnormalities (slowing of sural nerve conduction rates) were similar in the exposed and control populations. Kreiss et al. (1983) have commented that velocity slowing is a late finding in axonal neuropathies such as those caused by arsenic and whereas amplitude decreases would be an earlier finding, the lack of reproducibility of this parameter makes it difficult to interpret, thus rendering nerve conduction studies of limited value.

Feldman et al. (1979) found electromyography more useful in their study of 70 arsenic-exposed workers and 41 controls in a copper smelter. Electromyographic abnormalities correlated well with the arsenic contents of urine, hair, and nails. However, hair arsenic content in their high exposure group was very high (average 182.6 $\mu g/g$) indicating significant external contamination of the hair by arsenic dust, as this level is well into the lethal range had it been due to inhalation and ingestion alone. Nevertheless, it gave an approximate index of the extent of exposure because those workers whose hair was heavily contaminated were also, no doubt, inhaling similarly large amounts of arsenic dust. In contrast, Dafoe (1978) compared electromyographic assessment of 403 persons (including 26 gold smelters) from Yellowknife, Canada, a city of intense gold mining and smelting, with a control population of 120 persons from Hay River, Canada (no gold mining). Mean hair arsenic level for the gold workers was 27 $\mu g/g$. In his study there was no clear relationship between electromyographic abnormalities and hair arsenic levels although a positive association was demonstrated between electromyographic abnormalities and urine arsenic levels.

It seems, then, that nerve conduction studies have failed to prove themselves useful in the evaluation of chronic arsenic poisoning. However, recent refinements in technique may improve their usefulness and we await reports of their evaluation in the large Asian studies of arsenic poisoning due to well water contamination.

In conclusion, hair arsenic is a useful test for confirming chronic arsenic poisoning in forensic cases, provided external contamination can be excluded; it is, at least, a "smoking gun". However, its lack of clinical precision limits its usefulness in the evaluation of subjects exposed to arsenic from their water supply. Also, the unknown contribution to hair arsenic made from washing in arsenic-contaminated water further compounds the problem in such cases. In our Nova Scotia study (Hindmarsh et al., 1977) two of our worst cases of poisoning, both with obvious skin lesions, clinical neuropathy and grossly abnormal nerve conduction studies, although living in the same house and using the same contaminated well for many years, had hair arsenic levels of 47 and 4.2 μg/g respectively, thus demonstrating the poor precision of hair arsenic in assessing the severity of poisoning. Unfortunately, clinical findings lack precision too. Also, nerve conduction studies have produced variable results and we must search for other ways accurately to assess severity of poisoning.

ACKNOWLEDGEMENT

Our thanks are due to Drs Marie Vahter and Michael Kosnett for advice and to Ms Francine Huot for typing the manuscript.

REFERENCES

Atalla, L.T., Silva, C.M., Lima, F.W. Activation analysis of arsenic in human hair. *Aun. Acad. Bras. Cienc.*, **37**, 433–441.

Barrowcliff, D. 1971. The Stoneleigh Abbey poisoning case. *Medicolegal J. (Cambridge)*, **39**, 79–90.

Bencko, V., Symon, K. 1977. Health aspects of burning coal with a high arsenic content. *Environ. Res.*, **13**, 378–85.

Buchet, J.P., Lauwerys, R., Roels, H. 1981. Comparison of the urinary excretion of arsenic metabolites after a single oral dose of sodium arsenite, monomethylarsonate, or dimethylarsinate in man. *Int. Arch. Occup. Environ. Health.*, **48**, 71–79.

Cookson, J.A., Pilling, F.D. 1975. Trace element distribution across the diameter of human hair. *Phys. Med. Biol.*, **20**, 1015–1020.

Cornelis, R. `1973. Neutron activation analysis of hair, failure of a mission. *J. Radioanal. Chem.*, **15**, 305–316.

Dafoe, G.H. 1978. Canadian Public Health Association final report, electromyography: Yellowknife and Hay River, Northwest Territories.

Dewar, W.A., Lenihan, J.M.A. 1956. A case of chronic arsenical poisoning: examination of tissue samples by activation analysis. *Scot. Med. J.*, **1**, 236–238.

Erikson, N.E. 1966. Arsenic in Hair. Proc. 1st Int. Conf. Forens. Act. Anal., San Diego, pp. 279–286.

Feldman, R.G., Niles, C.A., Kelly-Hayes, M., Sax, D.S., Dixon, W.J., Thompson, D.J., Landau, E. 1979. Peripheral neuropathy in arsenic smelter workers. *Neurology*, **29**, 939–944.

Fergusson, J.E., Holzbecher, J., Ryan, D.E. 1983. The sorption of copper, manganese, zinc and arsenic onto human hair, and their desorption. *Sci. Total Environ.*, **26**, 121–135.

Goldsmith, J.R., Deane, M., Thom, J., Gentry, G. 1972. Evaluation of health implication of elevated arsenic in well water. *Water Res.*, **6**, 1133–1136.

Harrington, J.M., Middaugh, J.P., Morse, D.L., Housworth, J. 1978. A survey of a population exposed to high concentrations of arsenic in well water in Fairbanks, Alaska. *Am. J. Epidemiol.*, **108**, 377–384.

Hindmarsh, J.T., McCurdy, R.F. 1986. Clinical and environmental aspects of arsenic toxicity. *Crit. Rev. Clin. Lab. Sci.*, **23**, 315–347.

Hindmarsh, J.T., McLetchie, O.R., Heffernan, L.P.M., Hayne, O.A., Ellenberger, H.A., McCurdy, R.F., Thiebaux, H.J. 1977. Electromyographic abnormalities in chronic environmental arsenicalism. *J. Anal. Toxicol.*, **1**, 270–276.

Inamasu, T., Ishinishi, N., Kodama, Y. 1978. Environmental health survey of water pollution by arsenic contained hot water from geothermal power station. *Jpn. J. Hygiene*, **33**, 528–531.

Kreiss, K., Zack, M.M., Feldman, R.G., Niles, C.A., Chirico-Post, J., Sax, D.S., Landrigan, P.J., Boyd, M.H., Cox, D.H. 1983. Neurological evaluation of a population exposed to arsenic in Alaskan well water. *Arch. Environ. Health*, **38**, 116–121.

Lander, H., Hodge, P.R., Crisp, C.S. 1965. Arsenic in hair and nails, its significance in acute arsenic poisoning. *J. Forens. Med.*, **12**, 52–67.

Lima, F.W. 1966. Exogenous contamination of hair by capillary action of arsenic solutions. Proc. Ist Int. Conf. Forens. Act. Anal., San Diego, 261–278.

Maes, D., Pate, D.B. 1977. The absorption of arsenic into single human head hairs. *J. Forens. Sci.*, **22**, 89–94.

Marafante, E., Vahter, M., Dencker, L. 1984. Metabolism of arsenocholine in mice, rats and rabbits. *Sci. Total Environ.*, **34**, 223–240.

Mealey, J., Brownell, G.L., Sweet, W.H. 1959. Radioarsenic in plasma, urine, normal tissues and intracranial neoplasms. *Arch. Neurol. Psychiatry*, **81**, 310–320.

Pearson, E.F., Pounds, C.A. 1971. A care involving the administration of known amounts of arsenic and its analysis in hair. *J. Forens. Sci. Soc.*, **11**, 229–234.

Shapiro, H.A. 1967. Arsenic content of human hair and nails, its interpretation. *J. Forens. Med.*, **14**, 65–71.

Sky-Peck, H.H. 1990. Distribution of trace elements in human hair. *Clin. Physiol. Biochem.*, **8**, 70–80.

Smith, H. 1961. Estimation of arsenic in biological tissue by activation analysis. *J. Forens. Med.*, **8**, 165–171.

Smith, R.A. 1976. A method to distinguish between arsenic in and on human hair. *Environ. Res.*, **12**, 171–173.

Smith, S., Hendry, E.B. 1934. Arsenic in its relation to the keratin tissue. *BMJ*, **2**, 675–677.

Southwick, J., Western, A.E., Beck, M.M., Whitley, J., Isaacs, R. 1981. Community Health associated with arsenic in drinking water in Millard County, Utah. U.S. EPA Rep 600/1-18-064, Utah Dept. Health, Salt Lake City, UT.

Vahter, M., Marafante, E., Dencker, L. 1983. Metabolism of arsenobetaine in mice, rats and rabbits. *Sci. Total Environ.*, **30**, 197–211.

Valentine, J.L., Kang, H.K., Spivey, G. 1979. Arsenic levels in human blood, urine, and hair in response to exposure via drinking water. *Environ. Res.*, **20**, 24–32.

Yamauchi, H., Takahashi, K., Mashiko, M., Yamamura, Y. 1989. Biological monitoring of arsenic exposure of Gallium Arsenide and inorganic arsenic-exposed workers by determination of inorganic arsenic and its metabolites in urine and hair. *Am. Ind. Hyg. Assoc. J.*, **50**, 606–612.

Young, E.G., Rice, F.A.H. 1944. On the occurrence of arsenic in human hair and its medicolegal significance. *J. Lab. Clin. Med.*, **29**, 439–446.

Arsenic Exposure and Health Effects
W.R. Chappell, C.O. Abernathy and R.L. Calderon (Editors)
© 1999 Elsevier Science B.V. All rights reserved.   51

Estimating Total Arsenic Exposure in the United States

R.E. Grissom, C.O. Abernathy, A.S. Susten, J.M. Donohue

ABSTRACT

Arsenic is a component of the earth's crust and is ubiquitous in the environment. Consequently, humans are exposed to arsenic from a variety of sources such as food, water, air and soil. For the majority of people in the U.S., the major exposure route is through consumption of food. Ingestion of soil, dust, or water contaminated with arsenic also occurs in areas with naturally elevated arsenic levels or near hazardous waste sites. In addition, there are some people whose cultural practices such as the use of herbal medicines may cause additional exposures to arsenic. Dermal absorption of arsenic may occur if activities such as rice farming require daily exposure to water contaminated with arsenic. Unfortunately, many exposure scenarios only consider one type of exposure, e.g., exposure to arsenic in soil. It is critical to obtain a complete exposure estimate, especially at lower doses, to be certain that the exposure data used for the risk assessment does not underestimate the true exposure and present an unrealistic value.

Keywords: arsenic, risk assessment, routes, pathways, bioavailability

The opinions expressed in this manuscript are those of the authors and do not necessarily reflect the opinions or policies of ATSDR or of U.S. EPA.

INTRODUCTION

Arsenic is a naturally occurring element found in a variety of forms in soil, water, air, and food. Thus, all humans are exposed to arsenic on a daily basis. However, arsenic pollution caused by humans has raised concerns about public health (ATSDR, 1997). When industrial facilities are believed to have released arsenic into the local environment, government health officials are often asked "How much arsenic is too much?" Such an inquiry must be answered, and the public must be assured that the government agencies are doing an adequate public health assessment. When formulating a balanced and accurate risk assessment, one cannot depend solely on the measured levels of arsenic in a single pathway and route of exposure. In this chapter, we examine the need to look at potential exposures from environmental and non-environmental sources, the importance of considering bioavailability from all sources, the role of biological monitoring, the use of sentinel species, and dermal penetration in assessing public health threats from exposures to arsenic.

SOURCES OF EXPOSURE TO ARSENIC

Potential environmental sources of arsenic exposure include household dust, soil, drinking water, foods, and airborne particulates. Since arsenic is a natural substance, most of the arsenic in the above sources may be there because arsenic is part of the natural background. Thus, exposures to the general public and employees through occupational activities may occur as a result of breathing the air, drinking water, eating food, or inadvertently ingesting dust and soil. Data are available on the background levels of arsenic in various environmental media. For example, in "unpolluted soil", arsenic levels range from one to 40 milligrams of arsenic per kilogram of soil (mg As/kg) depending on the region of the country (ATSDR, 1997; Thornton and Farago, 1997). Areas with sulfide ore deposits or geothermal activity may contain concentrations as high as 8,000 mg As/kg soil (Thornton and Farago, 1997). Mean levels are usually around 5 mg As/kg soil (ATSDR, 1997). The levels of arsenic detected in indoor dust are highly variable; levels exceeding 2,000 mg As/kg have been reported (Millette et al., 1995; Lepow et al., 1975; Diaz-Barriga et al., 1993). In one study, the geometric mean arsenic level from 477 composite interior dust samples was 73 mg As/kg (Hwang et al., 1997). Levels in uncontaminated groundwater are generally ≤2 micrograms of arsenic per liter of water (μg As/L); surface water levels of arsenic seldom exceed 10 μg As/L; and ambient air levels usually do not exceed 0.10 micrograms of arsenic per cubic meter of air (μg As/m^3) (ATSDR, 1997). Relatively little is known about the levels of arsenic in food. Table 1 presents a summary of the levels of organic and inorganic arsenicals reported in different foods.

Anthropogenic enterprises may have a dramatic effect on natural environmental arsenic levels (ATSDR, 1997). For example, mining and smelting activities, and use of pesticide and fossil fuels have resulted in environmental levels of arsenic that can be far above background. Around hazardous waste sites, arsenic is most frequently reported in soil followed by groundwater and air (see Table 2) (HazDat, 1997).

Although soil, food, water and air are obvious exposure media, others (e.g., arsenicals in medicinals) are often overlooked. Such arsenic compounds as arsphenamine and Fowler's Solution were often used to treat illnesses (Abernathy and Ohanian, 1992; Hunter, 1995; Sasieni et al., 1994). These medicinals are not currently in use, but some arsenicals are still prescribed in medicine. For example, arsenic trioxide is currently administered in China and India to treat certain forms of leukemia (Treleaven et al., 1993; Chen et al., 1996a; Chen et al., 1996b; Shen et al., 1997). The efficacy of compounds containing arsenic in the control of blood counts from patients with chronic myeloid leukemia has been known for more than 100 years.

Herbal medications have been used for centuries by various populations including Chinese and Indians (Subbarayappa, 1997; Bhopal, 1993; Chan et al., 1993; Tay and Seah,

TABLE 1

Arsenic levels reported in food and medicinals

Media	Total As	Inorganic As	Reference
Potatoes	$2.3\,\mu g\,As^a$	10%	EPA (1988)
Vegetables	$2.8\,\mu g\,As^a$	5%	EPA (1988)
Vegetables: treated areas	≤3 mg As/kg	NR^b	Eisler (1988)
Vegetables: untreated areas	≤1 mg As/kg	NR	Eisler (1988)
Wine	≤2–33 μg As/L	NR	ATSDR (1997)
Milk and dairy products	$12\,\mu g\,As^a$	75%	EPA (1988)
Meat: beef and pork	0.15 mg As/kg	75%	EPA (1988)
Fish	0.1 –64 mg As/kg	DL^c–0.12 mg AS/kg	Chew (1996)
Shellfish	0.2–126 mg As/kg	<0.01–0.6 mg/kg	Chew (1996)
Poultry	NR	65%	EPA (1988)
Fruits	$5\,\mu g\,As^a$	10%	EPA (1988)
Cereals and grains	$6.4\,\mu g\,As^a$	65%	EPA (1988)
Rice	ND	35%	EPA (1988)
Medicinals	25–107,000 mg/kg	100%	Tay and Seah (1975)
Fowler's solution	1% potassium arsenite	100%	ATSDR (1997)

[a]Daily dietary intake derived from EPA (1988); [b]NR = not reported; [c]DL = detection limit.

TABLE 2

ATSDR sites involving arsenic by type of contaminated medium (HazDat, 1997)

Type of Medium	Total Sites (n = 2597)	National Priority List Sites (n = 1430)
Air	47 (1.8%)	37 (2.6%)
Groundwater	199 (7.7%)	166 (11.6%)
Soil	628 (24.2%)	518 (36.2%)

1975; Kew et al., 1993) and are becoming increasingly popular in "Western" countries such as Australia, England, and the U.S. (Ernst, 1998; Anna and Stephen, 1997). Herbal medications and teas may be a significant source of exposure (Espinoza et al., 1995; Tay and Seah, 1975; Wong et al., 1998) but are often overlooked by health risk assessors evaluating the health impacts of environmental arsenic sources. Arsenic levels in herbal medicinals may exceed 100,000 mg As (see Table 1) (ATSDR, 1997). With increasing migration to the United States from Asian and other countries, health risk assessors need to be more aware of cultural practices that result in exposure to arsenic.

The impact on risk from arsenic in water and air has been considered in many publications (ATSDR, 1997; EPA, 1988). In this communication, we will focus on some pathways that are not always considered in risk assessments. Although the contribution of these pathways may be minimal when exposure from a single source is high (e.g., arsenic-contaminated drinking water), they may appreciably contribute to the total risk. For example, imported traditional Chinese herbal-ball medicinals have been shown to contain arsenic at levels that range from 0.1 to 36.6 mg As/ball (Espinoza et al., 1995). Ingestion of these herbal preparations alone could pose a health threat.

ESTIMATING EXPOSURE AND PUBLIC HEALTH IMPACT

The first step in evaluating the public health impact of arsenicals is to determine the extent of exposure. In evaluations of total exposure, environmental risk assessors almost always consider drinking water, air, soil, and sometimes food as sources of exposure. Assessing the

impact on public health of arsenic levels in the environment, however, has routinely
involved the assumption that exposure is occurring through a single pathway and route
(e.g., ingestion of arsenic-contaminated soil). Site-specific concentrations of arsenic are
evaluated and decisions about public health impact are often made on the basis of comparing
estimates of exposure in terms of dose (milligrams of arsenic per kilogram body weight per
day [mg As/kg BW/day]) to health guideline values for cancer and non-cancer endpoints.
Although generally considered a very conservative approach given the uncertainty factors
built into the health guidance values, it is obvious that if the assessor does not include
exposure estimates for all sources of arsenic (e.g., indoor dust and medicinals), there may be a
gross underestimation of total exposure.

The mere presence of arsenic in environmental media does not *a priori* indicate that
overexposure to arsenic will occur. There are a number of factors that need to be considered
in an exposure evaluation. One factor often not considered is the bioavailability of the arsenic
in the various environmental media. It is a critical factor that determines how much of the
arsenic that is taken into the body (external dose) reaches critical target organs and cells (i.e.,
the internal dose).

BIOAVAILABILITY

Bioavailability is a primary factor influencing the relationship between the external dose and
the internal dose. It is the percentage of the external dose that reaches the systemic
circulation, i.e., the fraction of the external dose that is absorbed (Paustenbach et al., 1997).
The internal dose is the arsenic available to interact with receptors and cause adverse health
effects. Bioavailability issues are becoming important in evaluating potential adverse health
effects from exposure to arsenic in the U.S. There are numerous factors that affect the
absorption of arsenic from food, soil, or dust including the following: species of arsenic, solubility, soil pH, presence of clay or organic matter (i.e., cation exchange capacity [CEC]), and
soil type (Pascoe et al., 1994; Chaney and Ryan, 1994; Rabinowitz, 1993; Freeman et al., 1992).

Absorption of arsenic from water has been measured in both humans and animals.
Absorption of water-soluble arsenicals (e.g., sodium arsenate and sodium arsenite) in
humans ranged from approximately 50 to 95 percent (ATSDR, 1997). Similar results have
been observed in animals. The uptake of less water soluble forms such as arsenic triselenide
and lead arsenate are usually lower than the water soluble species, but significant amounts
can be absorbed. Absorption of arsenic triselenide in hamsters was reported to be approximately 12 percent and absorption of lead arsenate was approximately 22 percent (ATSDR,
1997).

Absorption of arsenicals from soil has been measured in laboratory animals. Absorption of
sodium arsenate in soil administered in capsules to rabbits ranged from approximately 17 to
50 percent (Freeman et al., 1993; Griffin and Turck, 1991). Absorption of sodium arsenate in
soil or dust administered to monkeys was approximately 15 and 25 percent, respectively
(Freeman et al., 1995). Differences among species may be partially responsible for variation in
bioavailability. Other factors including variation in the concentration or species of arsenic in
soil and variation in the soil type can also bias the results of bioavailability studies. The
biochemical environment of the gastrointestinal (GI) tract, including the acidic environment
of the stomach, the influence of food, and the influence of normal GI tract flora, can influence
absorption of arsenic (Berti and Lipski, 1995). Studies that do not evaluate both hepatic and
fecal excretion, and enterohepatic recirculation cannot adequately address absorption of
arsenic from the GI tract.

In vitro assays have been developed to estimate bioavailability of arsenic in soil and dust at
sites (Ruby et al., 1996). The relevance of events observed in *in vitro* studies to physiological
events that occur *in vivo* has not been adequately evaluated. If the acid environment in the

stomach causes the arsenic to dissociate from its binding sites on the dust or soil, binding to ions may occur in the small intestine and affect arsenic bioavailability (Ruby et al., 1993; Chaney and Mielke, 1886). Site-specific extrapolations based on soil mineralogy and *in vitro* studies, therefore, are incomplete. Validation of a model may be possible for arsenic under the conditions at a specific site; however, this does not mean that the model has been validated for other sites (Oreskes et al., 1994). Different sites are not identical; therefore, information concerning similarities and differences between sites is needed prior to making extrapolations from one site to another.

RELATIVE BIOAVAILABILITY

Most toxicological data have been obtained from studies involving ingestion of water soluble forms of arsenic and are expressed as mg As/kg BW/day (ATSDR, 1997). The effects of exposure to these arsenicals have been documented. The effects of many species of arsenic, particularly arsenicals that are not water-soluble, have not been adequately evaluated.

Relative bioavailability compares the absolute bioavailability of two different chemical species of arsenic. If there is information concerning the bioavailability of an arsenical in an environmental medium such as soil, the relative bioavailability of that arsenical can be estimated using equation one (Canady et al., 1997) where the absolute bioavailability of arsenic in water is 100%.

$$\frac{\text{Absolute As Bioavailability}_{soil}}{\text{Absolute As Bioavailability}_{water}} = \text{Relative Bioavailability}_{soil}$$

Once the relative bioavailability of arsenicals has been determined, the relative toxicity can be estimated; and the likelihood that adverse health effects will occur can be more accurately estimated. Relative bioavailability, therefore, will facilitate toxicological evaluation of arsenic compounds with different bioavailabilities. There are few studies properly assessing absolute bioavailability; consequently, toxicity of many arsenicals will continue to be estimated using the currently available models and literature.

BIOLOGICAL MONITORING

If arsenic exposure is suspected, biologic monitoring can be conducted to verify that exposure is occurring. For arsenic, biological monitoring assesses current exposure status (CDC, 1987; Buchet et al., 1981).

Soil in the communities surrounding Anaconda, Montana, remains contaminated with arsenic, even though the copper smelter located there has been closed since 1980. Because of concern that exposure to arsenic may be occurring, children between the ages of two and six who lived near the smelter were tested for urinary arsenic in March and again in July 1985 (CDC, 1987; Binder et al., 1987). The average level of urinary arsenic for the six children who were less than eight years of age was 76.0 μg As/L. The families were relocated to protect the health of the children. The average urinary arsenic level after the children were relocated was 15.3 μg As/L. The average pre-move level for persons greater than or equal to eight years of age was 17.2 μg As/L; their average post-move level was 14.6 μg As/L. Although five individuals had levels of urinary arsenic greater than 50 μg As/L prior to the move, none had levels greater than 50 μg As/L after relocation (CDC, 1987).

If people are living in a contaminated environment but do not have arsenic in their urine, additional information is needed to evaluate risks. Evaluation of their hair may provide information about recent past exposures, depending on the length of the hair (ATSDR, 1997). Detecting arsenic in hair, provided the assay was properly conducted, can provide information concerning exposures over the past few months. Finger or toe nails can also be evaluated for arsenic. Analyses of hair and nails, however, are problematic (ATSDR, 1997; Wilhelm and Idel, 1996). They are not reliable at low-level exposure and are subject to bias due to surficial contamination.

Biological monitoring may be useful in some cases. For example, arsenic can only be detected in the urine for a few days after an exposure has occurred; therefore, urinalysis can only address recent exposures. The value of using hair and/or nails for long-term monitoring of arsenic exposure is questionable (Wilhelm and Idel, 1996).

Several studies suggest that site activities and education alter behavior resulting in reduced exposure (Kimbrough et al., 1994; Porru et al., 1993; Kimbrough et al., 1995). Interpretation of biological monitoring, therefore, may be complicated. Failure to detect arsenic in urine may be the result of altered behavior rather than an indicator that arsenic is not present in the environment at levels that pose a health threat.

USE OF SENTINEL SPECIES

Evaluation of pets may be useful in assessing human health threats from exposure to arsenic (LeBlanc and Bain, 1997; Grahl-Nielsen, 1996; Coppock et al., 1996). Detecting arsenic in the urine of pets indicates that exposure is currently occurring. Detecting arsenic in the hair of pets indicates that exposure is currently occurring or has occurred within the past days or weeks. The pets' owners are likely to have access to the same source of arsenic as their pets.

A study investigating exposure of dogs to arsenic was conducted in Mesa de Oro, CA (MdO). The mean arsenic level in soil was 374 mg As/kg soil; the maximum level reported was 1,320 mg As/kg soil. The control dogs used in the study were from a local area that was not contaminated. The urine and fecal levels of arsenic in the MdO dogs were higher than the control dogs; however, the differences were not significant (see Table 3). The surface hair arsenic levels for the MdO dogs were significantly higher than the control dogs. The arsenic that was part of the internal structure of the hair of the MdO dogs was greater than in the control dogs; however, the difference was not significant (see Table 3). The fact that the surface levels of arsenic on the hair of the MdO dogs were significantly higher than the hair of the control dogs shows that analysis of hair can be biased (CA DHS, 1997).

EXPOSURE PATHWAYS AND ROUTE APPROACH

Assessing the impact on public health of arsenic levels in the environment has routinely involved the assumption that exposure is occurring through a single pathway, usually soil exposures, and route, usually ingestion. It is believed that an integrated exposure dose approach should be used. The single exposure pathway and route are valid only when one

TABLE 3

Arsenic levels in sentinel species (CA DHS, 1997)

Media	Control Dogs (N = 4)		MdO Dogs (N = 15 or 16)	
	Mean	Range	Mean	Range
Urine: urinary arsenic levels as μg As/gm creatinine	21.75	5–53	55.00	4–252
Hair: Internal (integral component) hair levels of arsenic as mg As/kg hair	0.33	0.2–0.5	0.44	<0.1–1.0
Hair: Internal + external hair (surficial contaminant) levels of arsenic as mg As/kg hair	0.44	0.2–1.1	2.11*	0.1–7
Feces: as mg As/kg feces	0.76	0.25–2.7	0.95	0.25–4.8

*Statistically different from comparison dogs (Kruskal-Wallis $p = 0.0373$).

source is predominant. In other cases, the single exposure estimate may result in an underestimation of exposure and, consequently, an underestimate of the likelihood that adverse health effects will occur. Specific factors such as identification of all sources of arsenic, level of contamination in each source, chemical form (species) of arsenic, dose estimation, plausible exposure scenarios, toxicological assessment, weight of evidence, and bioavailability must be evaluated to insure that public health assessments are protecting people's health.

UNIQUE EXPOSURE CASE

In some circumstances, exposure to arsenic through a route that is usually unimportant should be considered. For example, there were reports of Blackfoot Disease in Taiwan (Tseng et al., 1968; Tseng, 1977), but it was not found in Thailand (Choprapawon and Rodcline, 1997). One possible explanation for this observation is that exposure to arsenic from dermal exposure may have been a factor.

The most common routes of exposure to arsenic are ingestion and inhalation. Dermal uptake of inorganic arsenic is assumed to be minor compared to the other two routes of exposure. Although the skin serves as a barrier that prevents environmental substances from entering the body, it is not a perfect barrier (Scheuplein, 1977; Berti and Lipski, 1995; Wester et al., 1993; Wepierre and Marty, 1979). There have been reports where dermal exposure to arsenicals appeared to be the primary route of exposure. For example, arsenic-induced skin cancers resulting from the use of arsenical pesticides were reported in Beaujolais wine growers and their families (Thiers et al., 1967). An employee in a chemical manufacturing facility reportedly died when he spilled arsenic trichloride on his leg (Delepine, 1923).

Dermal penetration of arsenic does not generally appear to be an important issue. However, in cases where there is extended contact with water that is contaminated with arsenic through activities such as farming, fishing, or salt production, this may not be the case. Rice farmers, in particular, may have substantial dermal hydration of the feet and lower legs while working in the rice fields. Uptake of arsenic through the skin may explain why Blackfoot disease is prevalent in Taiwan (Tseng, 1977; Tseng et al., 1968) but not in other countries such as India, Bangladesh, Mexico, and Chile (EPA, 1988; Guha Mazumder et al., 1988; Tseng, 1977) which also have high arsenic in the groundwater.

Dramatic increases in *in vitro* dermal permeability have been observed following increased dermal hydration of hairless mouse skin (Lambert et al., 1989). In addition, *in vivo* studies using both laboratory animals and humans have shown that increased skin hydration enhances transcutaneous absorption (Berti and Lipski, 1995; Idson, 1978). Under such specialized conditions, it is possible that dermal penetration of arsenic was a factor in the development of blackfoot disease.

CONCLUSIONS

The above discussion has described how the use of a single route and pathway analysis may provide an inadequate estimate of exposure. In those cases, the total exposure dose, from all pathways, should be addressed. It is also important to consider the bioavailability and the relative bioavailability when assessing the likelihood that an adverse effect will occur. In some cases, the use of a sentinel species may provide a useful indicator of exposure potential, especially to children. In all cases, an integrated exposure approach should be used to evaluate potential health threats at sites containing arsenic.

The current health-based guidelines are based on data from exposures to arsenic in drinking water in Taiwan (ATSDR, 1997; EPA, 1988). The question arises as to whether they are appropriate for assessing health threats in the U.S., particularly if the exposures involve soil or food (Brown and Abernathy, 1997).

REFERENCES

Abernathy, C.O., Ohanian, E.V. 1992. Noncarcinogenic effects of inorganic arsenic. *Environ. Geochem. Health*, **14**, 35–41.

Anna, K.D., Stephen, P.M. 1997. Safety issues in herbal medicine: implications for the health professions. *Med. J. Austral.*, **166**, 538–541.

ATSDR. 1997. Agency for Toxic Substances and Disease Registry. Toxicological Profile for Arsenic, U.S. Public Health Service, Agency for Toxic Substances and Disease Registry, Atlanta, Georgia.

Berti, J.J., Lipski, J.J. 1995. Transcutaneous drug delivery: a practical review. *Mayo Clin. Proc.*, **70**, 581–586.

Bhopal, Raj. 1993. Use of Indian ethnic remedies. *Br. Med. J.*, **306**, 1003–1004.

Binder, S., Forney, D., Kaye, W., Paschal, D. 1987. Arsenic exposure in children living near a former copper smelter. *Bull. Environ. Contam. Toxicol.*, **39**, 114–121.

Brown, K.B., Abernathy, C.O. 1997. The Taiwan skin cancer risk analysis of inorganic arsenic ingestion: Effects of water consumption rates and food arsenic levels. In: C.O. Abernathy, R.L. Calderon and W.R. Chappell (eds.), *Arsenic Exposure and Health Effects*, pp. 260–271. Chapman and Hall, London.

Buchet, J.P., Lauwerys, R., Roels, H. 1981. Urinary excretion of arsenic and its metabolites after repeated ingestion dose of sodium metaarsenite by volunteers. *Int. Arch. Occup. Environ. Health*, **48**, 111–118.

CA DHS (California Department of Health Services). 1997. Health Consultation, Mesa de Oro (a/k/a Central Eureka Mine Site) dog biomonitoring, Sutter Creek, Amador County, CA.

Canady, A.C., Hanley, J.E., Susten, A.S., 1997. ATSDR science panel on bioavailability of mercury in soil: lessons learned. *Risk. Anal.*, **17**, 527–532.

CDC (Centers for Disease Control). 1987. Progress in Chronic Disease Prevention Reduction of Children's Arsenic Exposure Following Relocation — Mill Creek, Montana. Morb Mortal Wkly Rep., 36, 505–507.

Chan, T.Y.K., Chan, J.C.N., Tomlinson, B., Critchley, J.A.J.H. 1993. Chinese herbal medicines revisited: a Hong Kong perspective. *Lancet*, **342**, 1532–1534.

Chaney, R.L., Mielke, H.W. 1986. Standards for lead limitations in the United States. *Trace Subst. Environ. Health*, **20**, 355–577.

Chaney, R.L., Ryan, J.A. 1994. Risk-based standards for arsenic, lead, and cadmium in urban soils. DECHEMA, Deutsche Gesellschaft für Chemisches Apparatewesen, Chemische Technik und Biotechnologie e.V., Frankfurt am Main.

Chen, C.J., Chiou, H.Y., Chiang, M.H., Lin, L.J., Tai, T.Y. 1996a. Dose–response relationship between ischemic heart disease mortality and long-term arsenic exposure. *Arter. Thromb. Vas. Biol.*, **16**, 504–510.

Chen, G.Q., Zhu, J., Shi, X.G., Ni, J.H., Zhong, H.J., Si, G.Y., Jin, X.L., Tang, W., Li, X.S., Xong, S.M., Shen, Z.X., Sun, G.L., Ma, J., Zhang, P., Zhang, T.D., Gazin, C., Naoe, T., Chen, S.J., Wang, Z.Y., Chen, Z. 1996b. *In vitro* studies on cellular and molecular mechanisms of arsenic trioxide (As$_2$O$_3$) in the treatment of acute promyelocytic leukemia—As$_2$O$_3$ induces NB4 cell apoptosis with down regulation of BCL-2 expression and modulation of PML-RAR-Alpha/PML proteins. *Blood*, **88**, 1052–1061.

Choprapawon, C., Rodcline, A. 1997. Chronic arsenic poisoning in Ronpibool Nakhon Sri Thammarat, the southern province of Thailand, In: C.O. Abernathy, R.L. Calderon and W.R. Chappell (eds.), *Arsenic, Exposure and Health*. Chapman and Hall, London, pp. 69–77.

Coppock, R.W., Mostrom, M.S., Stair, E.L., Semalulu S.S. 1996. Toxicopathology of oilfield poisoning in cattle: a review. *Vet. Hum. Toxicol.*, **38**, 36–42.

Chew, C. 1996. Toxicity and Exposure concerns related to arsenic in seafood: An arsenic literature review for risk assessments. Prepared for USEPA Region 10 ESAT, Seattle Washington as Technical Instruction Document 10-9601-815 under Work Unit document 4038.

Delepine, S. 1923. Observations upon the effects of exposure to arsenic trichloride upon health. *J. Indust. Hyg.*, **4**, 410–423.

Diaz-Barriga, F., Santos, M.A., De Jesus Mejia, J., Batres, L., Yanez, L., Carrizales, L., Vera, E., Del Razo, L.M., Cebrian, M.E. 1993. Arsenic and cadmium exposure in children living near a smelter complex in San Luis Potosi, Mexico. *Environ. Res.*, **62**, 242–250.

Eisler, R. 1988. Arsenic hazards to fish, wildlife, and invertebrates: a synoptic review. U.S. Fish Wildl. Serv. Biol. Rep., 85(1.12). 92 pp.

EPA (U.S. Environmental Protection Agency) 1988. Special report on ingested inorganic arsenic. Skin cancer; nutritional essentially. Risk Assessment Forum, U.S. Environmental Protection Agency, Washington, D.C. EPA/625/3-87/013 July 1988.

Ernst, E. 1998. Harmless herbs? A review of the recent literature. *Am. J. Med.*, **104**, 170–178.

Espinoza, E.O., Mann, M.J., Bleasdell, B. 1995. Arsenic and mercury in traditional Chinese herbal balls. *N.E. J. Med.*, **333**, 803–804.

Freeman, G.B., Johnson, J.D., Killinger, J.M., Liao, S.C., Feder, P.I., Davis, A.O., Ruby, M.V., Chaney, R.L., Lovre, S.C., Bergstrom, P.D. 1992. Relative bioavailability of lead from mining waste soil in rats. *Fundam. Appl. Toxicol.*, **19**, 388–398.

Freeman, G.B., Johnson, J.D., Killinger, J.M., Liao, S.C., Davis, A.O., Ruby, M.V., Chaney, R.L., Lovre, S.C., Bergstrom, P.D. 1993. Bioavailability of arsenic in soil impacted by smelter activities following oral administration in rabbits. *Fund. Appl. Toxicol.*, **21**, 83–88.

Freeman, G.B., Schoof, R.A., Ruby, M.V., Davis, A.O. 1995. Bioavailability of arsenic in soil and house dust impacted by smelter activities following oral administration in cynomolgus monkeys. *Fund. Appl. Toxicol.*, **28**, 215–222.

Grahl-Nielsen, O. 1996. Distribution of trace elements from industrial discharges in the Hardangerfjord, Norway: A multivariate data analysis of saithe, flounder and blue mussel as sentinel organisms. *Marine Poll. Bull.*, **32**, 564–571.

Griffin, S., Turck, P. 1991. Bioavailability in rabbits of sodium arsenite adsorbed to soils. *The Toxicologist*, **11**, 195.

Guha Mazumder, D.N., Chakraborty, A.K., Ghose, A., Gupta, J.D., Chakraborty, D.P., Dey, S.B., Chattopadhaya, N. 1988. Chronic arsenic toxicity from drinking tubewell water in rural West Bengal. *Bull. WHO*, **64**, 499–506.

HazDat. 1997. Arsenic Agency for Toxic Substances and Disease Registry. Atlanta, GA.

Hunter, W.N. 1995. Rational drug design: a multidisciplinary approach. *Mole. Med. Today*, **1**, 31– 34.

Hwang, Y.H., Bornschein, R.L., Grote, J., Menrath, W., Roda, S. 1997. Urinary arsenic excretion as a biomarker of arsenic exposure in children. *Arch. Environ. Health*, **52**, 139–147.

Idson, B. 1978. Hydration and percutaneous absorption. *Curr. Prob. Dermatol.*, **7**, 132–141

Kew, J., Morris, C., Aihie, A., Fysh, R., Jones, S., Brooks, D. 1993. Arsenic and mercury intoxication due to Indian ethnic remedies. *BMJ*, **306**, 506–507.

Kimbrough, R.D., LeVois, M., Webb, D.R. 1994. Management of children with slightly elevated blood lead levels. *Pediatrics*, **93**, 188–191.

Kimbrough, R., LeVois, M., Webb, D. 1995. Survey of lead exposure around a closed lead smelter. *Pediatrics*, **95**, 550–554.

Lambert, W.J., Higuchi, W.I., Knutson, K., Krill, S.L. 1989. Effects of long-term hydration leading to the development of polar channels in hairless mouse stratum corneum. *J. Pharm. Sci.*, **78**, 925–928.

LeBlanc, G.A., Bain, L.J. 1997. Chronic toxicity of environmental contaminants: sentinels and biomarkers. *Environ. Health Perspect.*, **105**, 65–80.

Lepow, M.L., Bruckman, L., Gillette, M., Markowits, S., Robino, R., Kapish, J. 1975. Investigations into sources of lead in the environment of urban children. *Environ. Res.*, **10**, 415–426.

Millette, J.R., Brown, R.S., Mount, M.D. 1995. Lead arsenate. *Microscope*, **43**, 187–191.

Oreskes, N., Shrader-Frechette, K., Belitz, K. 1994. Verification, validation, and confirmation of numerical models in the earth sciences. *Science*, **263**, 641–646.

Pascoe, G.A., Blanchet, R.J., Linder, G. 1994. Bioavailability of metals and arsenic to small mammals at a mining waste-contaminated wetland. *Arch. Environ. Contam. Toxicol.*, **27**, 44–50.

Paustenbach, D.J., Bruce, G.M., Chrostowaki, P. 1997. Current views on oral bioavailability of inorganic mercury in soil: implications for health risk assessment. *Risk Anal.*, **17**, 533–544.

Porru, S., Donato, F., Apostoli, P., Coniglio, L., Duca, P., Alessio, L. 1993. The utility of health education among lead workers: the experience of one program. *Am. J. Ind. Med.*, **23**, 473–481.

Rabinowitz, M.B. 1993. Modifying soil lead bioavailability by phosphate addition. *Bull Environ. Contam. Toxicol.*, **51**, 438–444.

Ruby, M.V., Davis, A., Link, T.E., Schoof, R., Chaney, R.L., Freeman, G.B., Bergstrom, P. 1993. Development of an *in vitro* test to evaluate *in vivo* bioaccessibility of ingested mine-waste lead. *Environ. Sci. Technol.*, **27**, 2870–2877.

Ruby, M.V., Davis, A., Schoof, R., Eberle, S., Sellstone, C.M. 1996. Estimation of lead and arsenic bioavailability using a physiologically based extraction test. *Environ. Sci. Tech.*, **30**, 422–430.

Sasieni, P., Evans, S., Cuzick, J. 1994. Long term follow-up of patients treated with medicinal arsenic. In: W.R. Chappell, C.O. Abernathy, and C.R. Cothern (eds.), *Arsenic Exposure and Health*, pp. 101–107. Science and Technology Letters.

Scheuplein, R.J. 1977. Permeability of the skin. In: D.H.K. Lee (ed.), *Handbook of Physiology, Sect 9, Reaction to Environmental Agents*. pp 299–322. American Physiology Society, Bethesda, MD.

Shen, Z.X., Chen, G.Q., Ni, J.H., Li, X.S., Xiong, S.M., Qiu, Q.Y., Zhu, J., Tang, W., Sun, G.L., Yang, K.Q., Chen, Y., Zhou, L., Fang, Z.W., Wang, Y.T., Ma, J., Zhang, P., Zhang, T.D., Chen, S.J., Chen, Z., Wang, Z.Y. 1997. Use of arsenic trioxide (As$_2$O$_3$) in the treatment of acute promyelocytic leukemia (APE). 2. Clinical efficacy and pharmacokinetics in relapsed patients. *Blood*, **89**, 3354–3360.

Subbarayappa, B.V. 1997. Siddha medicine: an overview. *Lancet*, **350**, 1841–1844.

Tay, C.H., Seah, C.S. 1975. Arsenic poisoning from anti-asthmatic herbal preparations. *Med. J. Aust.*, **2**, 424–428.

Thiers, H., Colomb, D., Moulin, G., Colin, L. 1967. Arsenical skin cancer in Beaujolais wine growers. *Ann. Dermatol. Syphiligr.*, **94**, 133–158.

Thornton, I., Farago, M. 1997. The geochemistry of arsenic. In: C.O. Abernathy, R.L. Calderon and W.R. Chappell (eds.), *Arsenic Exposure and Health Effects*, pp. 1–16. Chapman and Hall, London.

Treleaven, J., Meller, S., Farmer, P., Birchall, D., Goldman, J., Piller, G. 1993. Arsenic and Ayurveda. *Leuk. Lymphoma*, **10**, P343–345.

Tseng, W.P. 1977. Effects and dose–response relationships of skin cancer and blackfoot disease with arsenic. *Environ. Health Perspect.*, **19**, 110–119.

Tseng, W.P., Chu, H.M., How, S.W., Fong, J.M., Lin, C.S., Yeh, S. 1968. Prevalence of skin cancer in an endemic area of chronic arsenicism in Taiwan. *J. Natl. Cancer Inst.*, **40**, 453–463.

Wepierre, J., Marty, J.P. 1979. Percutaneous absorption of drugs. *Pharmacol. Sci.*, **1**, 23–26.

Wester, R.C., Maibach, H.I., Sedik, L., Melendres, J., Wade, M. 1993. *In Vivo* and *in vitro* percutaneous absorption and skin decontamination of arsenic from water and soil. *Fund. Appl. Toxicol.*, **20**, 336–340.

Wilhelm, M., Idel, H. 1996. Hair analysis in environmental medicine. *Zentralbl. Hyg. Umweltmed.*, **198**, 485–501.

Wong S.S., Tan, K.C., Goh, C.L. 1998. Cutaneous manifestations of chronic arsenicism—Review of seventeen cases. *J. Am. Acad. Derm.*, **38**, 179–185.

Arsenic Exposure and Health Effects
W.R. Chappell, C.O. Abernathy and R.L. Calderon (Editors)

Arsenic Compounds in Terrestrial Biota

Kurt J. Irgolic, Walter Goessler, Doris Kuehnelt

ABSTRACT

During the past two decades trimethylarsine oxide, the tetramethylarsonium cation, arsenobetaine, arsenocholine, arsenic-containing riboses, and arsenic-containing lipids were identified in marine organisms. Terrestrial biota appeared to be restricted to the simpler organic compounds of arsenic such as methylarsonic acid, dimethylarsinic acid, and perhaps trimethylarsine. The development of an analytical method for the identification and quantification of inorganic and organic arsenic compounds with very low detection limits (extraction, evaporation of extract, dissolution of the residue, chromatography, nebulization in a hydraulic high-pressure nebulizer, detection of arsenic in the column effluent with an inductively coupled argon plasma mass spectrometer) proved that almost all of the arsenic compounds thus far found in marine organisms are also present in terrestrial biota. The detection limits of the analytical system are so low, that arsenic compounds can be quantified in human food derived from terrestrial biota. Knowledge of arsenic compounds in food should improve the assessment of risk posed by ingested arsenic.

Keywords: arsenic compounds, terrestrial biota, human food, risk assessment

Sadly, Kurt Irgolic died before this book was published.

INTRODUCTION

Life developed in the primordial sea. The cells had to find ways to harvest radiant energy from the sun, convert carbon dioxide and water to organic compounds, extract the chemical energy stored in organic compounds, and discover catalysts for the catabolic and metabolic reactions, without which organisms would not have been able to evolve. These catalysts are proteins (enzymes), many of which have metal ions as cofactors. Consequently, the biological entities in the early seas had to collect such essential metal ions and at the same time prevent elements that hamper biochemical reactions from penetrating the cell membranes. If such elements, nevertheless, did reach the inside of a cell, they had to be chemically altered to avoid damage.

Arsenic is quite a rare element (in 51st place in the list of the crustal abundance of elements), the primary minerals of which are sulfides (As_4S_4, Realgar; As_2S_3, Orpiment; FeAsS, Arsenopyrite) or arsenides ($FeAs_2$, Löllingite; NiAs, Niccolite). These primary arsenic minerals were stable in the primordial, reducing (oxygen-free) atmosphere of the earth. Consequently, these minerals should not have weathered and the concentrations of arsenic compounds in the oceans should have been very low. Anions derived from the weak arsenous acid (H_3AsO_3) and the quite strong arsenic acid (H_3AsO_4) should not have been present in the ocean. The concentrations of the oxoanions of arsenic (arsenite, $H_2AsO_3^-$; arsenates, $H_2AsO_4^-$, $HAsO_4^{2-}$) probably began to increase with the enrichment of molecular oxygen in the atmosphere as a consequence of the invention of the oxygen-generating photosynthesis. Arsenic-containing sulfides and arsenides became thermodynamically unstable and were subject to weathering to arsenites and arsenates. The arsenites and arsenates were dissolved by water. The marine organisms had to cope and did cope with these arsenic compounds. Although the early organisms learned to live with and to thrive in spite (or because) of the presence of arsenic compounds in their aqueous surrounding several billion years ago, we still are not certain yet, whether arsenic compounds have an essential biochemical function or are at low doses utterly useless and at higher doses only toxic.

Today seawater contains arsenite and arsenate in a thermodynamic disequilibrium (more arsenite than expected) at a concentration of total arsenic of approximately $2\,\mu g/L$. As much as 100 years ago determinations of arsenic in marine organisms suggested that the concentration of arsenic in the organisms is much higher than in the seawater. During the past three decades high concentrations of arsenic were found with much improved analytical methods in many marine animals and plants. Bioaccumulation factors (concentration of As in fresh organisms/concentration of As in seawater) of 1000 are the rule. In exceptional cases this factor assumes values of 200,000 (Francesconi and Edmonds, 1993). Most of the marine organisms, in which total arsenic had been determined, possess concentrations of arsenic of a few mg/kg and several up to several hundred mg/kg on a wet mass basis.

ARSENIC COMPOUNDS IN MARINE ORGANISMS

Marine organisms are for many people a major part of their food and for many more people a welcome enrichment of their diet. With the consumption of seafood the intake of arsenic increases inexorably. For instance, the average Japanese person with a diet rich in marine products consumes up to $273\,\mu g$ As per day, an intake ~5-times higher than for a land-locked person (Yamauchi and Fowler, 1994). The shrimps and lobsters, coveted delicacies, are known to be rich in arsenic. Because arsenic is popularly known as an insidious poison, concerns about the healthfulness of arsenic-rich seafood arose, although few—if any—ill effects incontrovertibly attributable to arsenic in seafood are on record. These concerns fostered experiments with the goal of identifying and quantifying arsenic compounds in marine organisms. In 1977 arsenobetaine was identified as the major arsenic compound in the Western Rock Lobster (Edmonds and Francesconi, 1977; Cannon et al., 1981). Subsequent to this breakthrough in environmental arsenic chemistry, several research groups devoted their efforts to the identification of arsenic compounds in marine organisms. These efforts were greatly assisted by chromatographic techniques for the separation of the arsenic compounds and by arsenic-specific detection systems, that are unresponsive to all substances co-eluting with an arsenic compound. Such arsenic-specific detection systems listed in order of decreasing detection limits are flame atomic absorption spectrometers, inductively-coupled argon plasma atomic emission spectrometers, graphite furnace atomic absorption spectrometers, and inductively-coupled argon plasma mass spectrometers (Irgolic, 1992; 1994). The importance of these research endeavors is manifested by the founding of the Japanese Arsenic Chemists' Society, which organizes an international meeting every other year. All these efforts uncovered an interesting array of organic arsenic compounds, several of which were unexpected. The inorganic arsenic compounds arsenite and arsenate are the precursors of the organic arsenic compounds. Methylarsonic acid, dimethylarsinic acid, trimethylarsine oxide, the tetramethylarsonium cation, arsenocholine, arsenobetaine, eleven 5-ribosyl(dimethyl)arsine oxides with various aglycons bonded to the oxygen atom on the C-1 carbon atom of the ribose, two 5-ribosyl(trialkyl)arsines, and dimethyl[3-(2'-sulfoethylamidcarbonyl)-1-propyl]arsine oxide (Fig. 1) were identified in marine organisms (Francesconi and Edmonds, 1993; 1997).

ARSENIC COMPOUNDS IN TERRESTRIAL SYSTEMS

Such a rich variety of organic arsenic compounds seemed to exist only in marine systems. Terrestrial biota were limited to arsenite, arsenate, methylarsonic acid, dimethylarsinic acid, and trimethylarsine (Cullen and Reimer, 1989; Braman, 1975). The more complex organic arsenic compounds appeared to be unique to marine biota. Although concentrations of total arsenic can be higher in fresh waters (for example, in hydrothermal areas and regions with sulfidic mineralization) than in sea water, in which the arsenic concentration is uniformly approximately 2 μg/L, they are generally much lower than in sea water (U.S. National Academy of Sciences, 1977). In uncontaminated soils the average concentration of total arsenic is a few mg/kg. Most plants growing on these soils do not accumulate arsenic and, therefore, have in their tissues arsenic at concentrations below mg/kg on a wet mass basis. In contaminated areas tissues may reach a few mg/kg (wet mass). Arsenic accumulators, for instance, the mushroom *Laccaria amethystina*, reach concentrations of total arsenic of approximately 100 mg/kg (dry mass) (Byrne and Tusek-Znidaric, 1983; Slekovec and Irgolic, 1996). Because the concentrations of arsenic compounds generally are much lower in terrestrial biota than in marine organisms, methods with very low detection limits must be used to identify and quantify arsenic compounds formed in terrestrial systems. Solid samples must be extracted with suitable solvents to bring the arsenic compounds into solution, aliquots of which can be chromatographed. The preparation of extracts suitable for chromatography invariably requires dilution and produces solutions, in which the arsenic compounds are present at low concentration, frequently at μg/L levels. The development of high-performance liquid chromatographic (HPLC) techniques, especially cation- and anion-exchange chromatographic methods for the separation of inorganic and organic arsenic compounds that—depending on the pH of the mobile phase—may be neutral, anionic, or zwitterionic and the use of an inductively coupled argon plasma mass spectrometer (ICP-MS) as the on-line arsenic-specific detector to monitor arsenic in the column effluents allowed the identification and quantification of arsenic compounds in terrestrial biota. The HPLC-ICP-MS system, in which the concentric nebulizer had been replaced by a hydraulic high pressure nebulizer (HHPN), reached quantification limits of 0.1 ng arsenic (100 μL of reconstituted extract injected) and detection limits of 0.01 ng arsenic. With these analytical methods the complex organic arsenic compounds (in addition to arsenite, arsenate, methyl-arsonic acid, and/or dimethylarsinic acid) given in Figure 1 were found in terrestrial biota (Table 1).

Many of the terrestrial biota, in which these arsenic compounds were identified, were collected in areas contaminated with inorganic arsenic. Such areas were sites in Styria

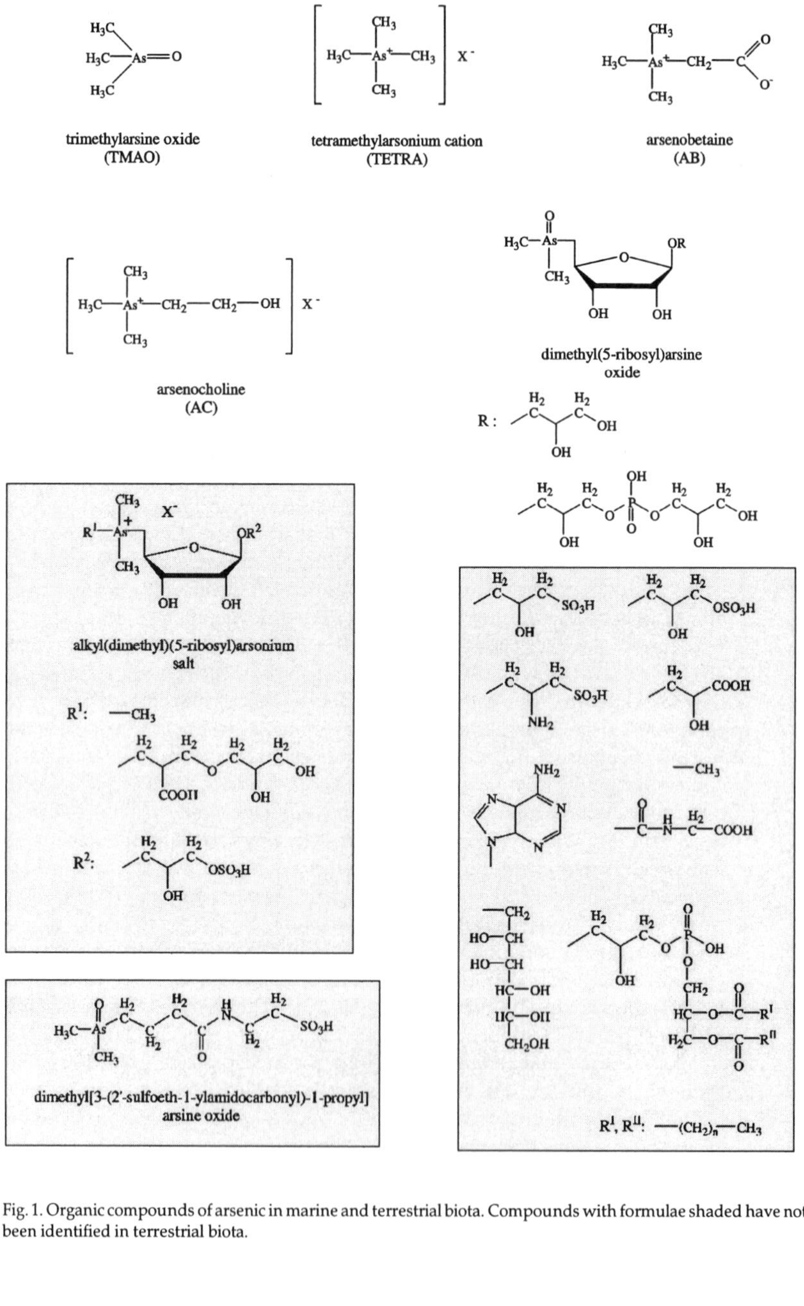

Fig. 1. Organic compounds of arsenic in marine and terrestrial biota. Compounds with formulae shaded have not yet been identified in terrestrial biota.

TABLE 1
Complex organic arsenic compounds found in terrestrial biota

Trimethylarsine Oxide in:		
green plants	*Dactylis glomerata*	Geiszinger (1998)
	Trifolium pratense	Geiszinger (1998)
	Plantago lanceolata	Geiszinger (1998)
	fern	Kuehnelt et al. (to be published)
lichens		Kuehnelt et al. (to be published)
mushrooms	*Laccaria amethystina*	Larsen et al. (1998)
soil		Geiszinger (1998)
ant-hill material		Kuehnelt et al. (1997c)
urine of mice and hamsters exposed to trimethylarsine		Yamauchi et al. (1990)
urine of rats exposed to trimethylarsine		Goessler and Buchet (to be published)
Tetramethylarsonium Ion in:		
green plants	*Dactylis glomerata*	Geiszinger (1998)
	Trifolium pratense	Geiszinger (1998)
	Plantago lanceolata	Geiszinger (1998)
	fern	Kuehnelt et al. (to be published)
lichens		Kuehnelt et al. (to be published)
mushrooms	*Sarcodon imbricatum*	Byrne et al. (1995)
	Amanita sp.	Kuehnelt et al. (1997a,b); Slejkovec et al. (1997)
	Leucocoprinus badhamii	Slejkovec et al. (1997)
	Agaricus campester	Slejkovec et al. (1997)
	Sparassis crispa	Slejkovec et al. (1997)
	Gomphus clavatus	Slejkovec et al. (1997)
	Albatrellus sp.	Slejkovec et al. (1997)
urine of rats exposed to trimethylarsine		Goessler and Buchet (to be published)
Arsenobetaine in:		
green plants	*Dactylis glomerata*	Geiszinger (1998)
	Trifolium pratense	Geiszinger (1998)
	Plantago lanceolata	Geiszinger (1998)
	fern	Kuehnelt et al. (to be published)
lichens		Kuehnelt et al. (to be published)
mushrooms	*Agaricus* sp.	Byrne et al. (1995)
	Sarcodon imbricatum	Byrne et al. (1995)
	Collybia sp.	Kuehnelt et al. (1997a)
	Amanita sp.	Kuehnelt et al. (1997a,b); Slejkovec et al. (1997)
	Calvatia sp.	Slejkovec et al. (1997)
	Lycoperdon sp.	Slejkovec et al. (1997)
	Geastrum sp.	Slejkovec et al. (1997)
	Agaricus sp.	Slejkovec et al. (1997)
	Macrolepiota procera	Slejkovec et al. (1997)
	Leucocoprinus badhamii	Slejkovec et al. (1997)
	Sparassis crispa	Slejkovec et al. (1997)
	Gomphus clavatus	Slejkovec et al. (1997)
	Albatrellus sp.	Slejkovec et al. (1997)
	Ramaria pallida	Slejkovec et al. (1997)
	Tricholoma pardinum	Slejkovec et al. (1997)
	Lyophyllum conglobatum	Slejkovec et al. (1997)
	Volvariella volvacea	Slejkovec et al. (1997)
	Laccaria sp.	Slejkovec et al. (1997); Larsen et al. (1998)
earthworms		Geiszinger et al. (1998); Geiszinger (1998)
ants		Kuehnelt et al. (1997c)
ant-hill material		Kuehnelt et al. (1997c)
soil		Geiszinger (1998)
human urine after exposure to trimethylarsine		Goessler et al. (1997b; 1998)
Arsenocholine in:		
green plants	*Plantago lanceolata*	Geiszinger (1998)
mushrooms	*Amanita muscaria*	Kuehnelt et al. (1997a,b); Slejkovec et al. (1997)
	Leucocoprinus badhamii	Slejkovec et al. (1997)
	Sparassis crispa	Slejkovec et al. (1997)
	Gomphus clavatus	Slejkovec et al. (1997)
	Albatrellus sp.	Slejkovec et al. (1997)
Arsenoriboses in:		
freshwater algae	*Chlorella* sp.	Goessler et al. (1997a; to be published)
earthworms		Geiszinger et al. (1998); Geiszinger (1998)

(Austria) with geologically caused concentrations of total arsenic in soils of approximately 40 mg/kg (Geiszinger et al., 1998) and a former smelting site in the Poella-Valley, Carinthia (Austria), at which arsenopyrite had been converted to arsenic trioxide by roasting (Kuehnelt et al., 1997b). Although this facility ceased operation approximately 100 years ago, the concentration of total arsenic in the humic soils at this site is still ~700 mg/kg (Kuehnelt et al., 1997b). Consequently, the terrestrial biota from these sites possess quite high concentrations of total arsenic, for instance ~22 mg/kg (dry mass) for the mushroom *Amanita muscaria* (Kuehnelt et al., 1997a,b), ~12 mg/kg (dry mass) for *Formica* sp. (ants) (Kuehnelt et al., 1997c), and up to 18 mg/kg (dry mass) for earthworms (Geiszinger et al., 1998).

Such high concentrations on a dry mass basis (the concentration on a fresh or wet mass basis is approximately one hundredth of the concentration on a dry mass basis) are a boon for the analytical chemist, because the diluted extracts prepared from these terrestrial biota still have concentrations of total arsenic much higher than the method detection limits. When the diluted extracts contain several arsenic compounds, the concentration of each arsenic compound in the aliquots prepared for chromatography must be at least ten times higher than the method detection limit to produce quantitatively meaningful results. Conclusions drawn about the (qualitative) presence of an arsenic compound from a chromatographic signal barely above the background are not convincing. Biota from contaminated sites with high concentrations of arsenic avoid such difficulties and allow the presence of complex organic arsenic compounds—formerly thought to exist only in marine organisms—to be established with confidence in terrestrial biota and the concentrations of these arsenic compounds to be determined accurately and precisely.

Although the HPLC-ICP-MS method for the identification and quantification of arsenic compounds in terrestrial biota was successfully applied to arsenic-rich mushrooms, lichens, and earthworms during the initial phase of our efforts to prove that terrestrial organisms are as capable as marine organisms of transforming arsenic compounds, the detection limit of this analytical method at 0.01 ng As (100 μL injected; 0.1 μg As/L) is sufficiently low to be useful with samples containing arsenic at concentrations characteristic of uncontaminated sites.

For an arsenic compound to be detectable in the chromatogram, an amount of an arsenic compound of 0.01 ng arsenic must be contained in the 100 μL of a diluted extract injected onto the column. When one gram of a biological sample is extracted, a particular arsenic compound is completely transferred into the extract, and the residue from the extract is dissolved to 10 mL, then at least 1 ng of arsenic must be present for each arsenic compound to be identifiable. The minimal amount of an arsenic compound in an original sample necessary for identification can be reduced perhaps by two orders of magnitude by increasing the mass of the sample to be extracted, by decreasing the volume of solvent for the dissolution of the residue obtained by evaporating the extract, and by increasing the injection volume. Food destined for human consumption with the exception of coffee extracts, rice, spinach, parsley, and seafood was found to contain total arsenic at concentrations up to 0.1 mg/kg (Nriagu and Azcue, 1990). Even the arsenic compounds present at 0.01 mg/kg should be identifiable provided that the extraction procedure is close to quantitative. The chromatograms (Fig. 2) of a dilute solution of synthetic arsenic compounds (1 ng for each compound, 100 μL injection) and of a reconstituted extract (~3.5 ng As, 100 μL injection) obtained from the edible mushroom *Xerocomus badius* prove that the arsenic compounds in terrestrial biota with concentrations of total arsenic in the range known to exist in food is possible.

The chromatograms obtained, for instance, with extracts from mushrooms contain several signals for arsenic compounds (Kuehnelt et al., 1997b), that could not be identified with the available synthetic arsenic compounds.

Fig. 2. Chromatogram of synthetic arseno-betaine bromide, trimethylarsine oxide, arsenocholine bromide, and tetramethyl-arsonium iodide (1 ng for each compound, 100 μL injected) and of an extract of the edible mushroom *Xerocomus badius* (Supel-cosil LC-SCX column; 20 mM aqueous pyridine solution at pH 2.5; 1.5 mL/min flow rate; injection volume 100 μL; HHPN: heating 140°C, cooling 4°C).

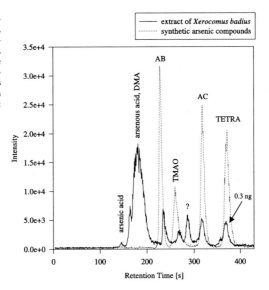

CONCLUSIONS

Improvements in the chromatographic methods for the separation of organic arsenic compounds and in the detection of arsenic in the column effluents allowed many of the complex organic arsenic compounds, that were believed to exist only in marine organisms, to be identified in terrestrial biota. The analytical procedure based on the extraction of the arsenic compounds, chromatographic separation of the arsenic compounds, nebulization of the effluent from the chromatographic column in a hydraulic high pressure nebulizer, and introduction of the aerosol into the argon plasma of an inductively coupled argon plasma mass spectrometer, makes it possible to at least identify an arsenic compound, when this compound contributes 1 μg of arsenic or more to the total arsenic in 1 kg of the biological sample (~1 g fresh or dry mass extracted). Consequently, this method is capable of identifying and quantifying the arsenic compounds (arsenic concentrations at least 0.01 mg/kg, more than 80% extractable) in terrestrial foods. With information about the types of arsenic compounds in terrestrial (and marine) food, the certainly incorrect assumption that all arsenic in terrestrial food is inorganic arsenic becomes obsolete and the assessment of risk posed by arsenic to people can be carried out on the basis of the quite divergent toxicities of inorganic and organic compounds of arsenic (Chappell et al., 1997). Such a change in risk assessment will probably make it unnecessary to lower the maximum contaminant level (MCL) for arsenic in drinking water to less than 10 μg/L and will have the potential of saving not inconsiderable sums of money. The annual cost for the United States of America for compliance with an MCL of 2 μg arsenic/L in drinking water was estimated at 2.1 billion U.S. dollars (Kempic, 1997).

REFERENCES

Braman, R.S. 1975. Arsenic in the environment. In: E.A. Woolson (ed.), *Arsenical Pesticides*, pp. 108–123. ACS Symposium Series 7, Washington D.C.

Byrne, A.R. and Tusek-Znidaric, M. 1983. Arsenic accumulation in the mushroom *Laccaria amethystina*. *Chemosphere*, **12**, 1113–1117.

Byrne, A.R., Slejkovec, Z., Stijve, T., Fay, L., Goessler, W., Gailer, J. and Irgolic, K.J. 1995. Arsenobetaine and other arsenic species in mushrooms. *Appl. Organomet. Chem.*, **9**, 305–313.

Cannon, J.R., Edmonds, J.S., Francesconi, K.A., Raston, C.L., Saunders, J.B., Skelton, B.W. and White, A.H. 1981. Isolation, crystal structure and synthesis of arsenobetaine, a constituent of the western rock lobster, the duskyshark, and some samples of human urine. *Austr. J. Chem.*, **34**, 787–798.

Chappell, W.R., Beck, B.D., Brown, K.G., Chaney, R., Cothern, C.R., Irgolic, K.J., North, D.W., Thornton, I. and Tsongas, T.A. 1997. Inorganic arsenic: a need and an opportunity to improve risk assessment. *Environ. Health Persp.*, **105**, 1060–1067.

Cullen, W.R. and Reimer, K.J. 1989. Arsenic speciation in the environment. *Chem. Rev.*, **89**, 713–764 and references therein.

Edmonds, J.S. and Francesconi, K.A. 1977. Methylated arsenic from marine fauna. *Nature*, **265**, 436.

Francesconi, K.A. and Edmonds, J.S. 1993. Arsenic in the sea. *Oceanogr. Mar. Biol. Annu. Rev.*, **31**, 111–151 and references therein.

Francesconi, K.A. and Edmonds, J.S. 1997. Arsenic and marine organisms. *Adv. Inorg. Chem.*, **44**, 147–189.

Geiszinger, A., Goessler, W., Kuehnelt, D., Francesconi, K.A. and Kosmus, W. 1998. Determination of arsenic compounds in earthworms. *Environ. Sci. Technol.*, **32**, 2238–2243.

Geiszinger, A. 1998. Spurenelemente in Regenwuermern (*Lumbricidae, Oligochaeta*) und ihre oekologische Bedeutung. PhD thesis, Karl-Franzens-University Graz.

Goessler, W., Lintschinger, J., Szakova, J., Mader, P., Kopecky, J., Doucha, J. and Irgolic, K.J. 1997a. *Chlorella* sp. and arsenic compounds: an attempt to prepare an algal reference material for arsenic compounds. *Appl. Organomet. Chem.*, **11**, 57–66.

Goessler, W., Schlagenhaufen, C., Kuehnelt, D., Greschonig, H. and Irgolic, K.J. 1997b. Can humans metabolize arsenic compounds to arsenobetaine? *Appl. Organomet. Chem.*, **11**, 327–335.

Goessler, W., Schlagenhaufen, C., Kuehnelt, D., Greschonig, H. and Irgolic, K.J. 1998. Can humans metabolize arsenic compounds to arsenobetaine? A reply to John Edmonds letter. *Appl. Organomet. Chem.*, **12**, 873–876.

Irgolic, K.J. 1992. Arsenic. In: M. Stoeppler (ed.), *Hazardous Metals in the Environment*, pp. 287–350. Elsevier, Amsterdam.

Irgolic, K.J. 1994. Determination of total arsenic and arsenic compounds in drinking water. In: W.R. Chappell, C.O. Abernathy and C.R. Cothern (eds.), *Arsenic Exposure and Health*, pp. 51–60. Science and Technology Letters, Northwood.

Kempic, J.B. 1997. Arsenic removal technologies: An evaluation of cost and performance. In: C.O. Abernathy, R.L. Calderon and W.R. Chappell (eds.), *Arsenic Exposure and Health Effects*, pp. 393–405. Chapman & Hall, New York.

Kuehnelt, D., Goessler, W. and Irgolic, K.J. 1997a. Arsenic compounds in terrestrial organisms I: *Collybia maculata, Collybia butyracea* and *Amanita muscaria* from arsenic smelter sites in Austria. *Appl. Organomet. Chem.*, **11**, 289–296.

Kuehnelt, D., Goessler, W. and Irgolic, K.J. 1997b. Arsenic compounds in terrestrial organisms II: Arsenocholine in the mushroom *Amanita muscaria*. *Appl. Organomet. Chem.*, **11**, 459–470.

Kuehnelt, D., Goessler, W., Schlagenhaufen, C. and Irgolic, K.J. 1997c. Arsenic compounds in terrestrial organisms III: Arsenic compounds in *Formica* sp. from an old arsenic smelter site. *Appl. Organomet. Chem.*, **11**, 859–867.

Larsen, E.H., Hansen, M. and Goessler, W. 1998. Speciation and health risk considerations of arsenic in the edible mushroom *Laccaria amethystina* collected from contaminated and uncontaminated locations. *Appl. Organomet. Chem.*, **12**, 285–291.

Nriagu, J.O. and Azcue, J.M. 1990. Food contamination with arsenic in the environment. In: J.O. Nriagu and M.S. Simmons (eds.), *Food Contamination from Environmental Sources*, pp. 121–143. John Wiley & Sons, New York.

Slejkovec, Z., Byrne, A.R., Stijve, T., Goessler, W. and Irgolic, K.J. 1997. Arsenic compounds in higher funghi. *Appl. Organomet. Chem.*, **11**, 673–682.

Slekovec, M. and Irgolic, K.J. 1996. Uptake of arsenic by mushrooms from soil. *Chem. Spec. Bioavailab.*, **8**, 67–73.

U.S. National Academy of Sciences. 1977. *Arsenic*, pp. 20–23. National Academy of Sciences, Washington D.C.

Yamauchi, H., Kaise, T., Takahashi, K. and Yamamura, Y. 1990. Toxicity and metabolism of trimethylarsine in mice and hamsters. *Fund. Appl. Toxicol.*, **14**, 399–407.

Yamauchi, H. and Fowler, B.A. 1994. Toxicity and metabolism of inorganic and methylated arsenicals. In: J.O. Nriagu (ed.), *Arsenic in the Environment. Part II*, pp. 35–53. John Wiley & Sons, New York, and references therein.

Arsenic Exposure and Health Effects
W.R. Chappell, C.O. Abernathy and R.L. Calderon (Editors)
© 1999 Elsevier Science B.V. All rights reserved.

Exposure to Arsenosugars from Seafood Ingestion and Speciation of Urinary Arsenic Metabolites

X. Chris Le, Mingsheng Ma, and Vivian W.-M. Lai

ABSTRACT

Ingestion from food and water is the major route of exposure to arsenic by the general population. Most seafoods contain $\mu g/g$ levels of arsenic, arsenobetaine being the major arsenic species in crustaceans and arsenosugars in seaweeds. It is necessary to understand human exposure to these arsenic species from dietary ingestion and their effects on urinary speciation of arsenic, which are the topics of this study. Methods based on high performance liquid chromatography (HPLC) with both inductively coupled plasma mass spectrometry (ICPMS) and hydride generation atomic fluorescence detection (HGAFD) were used for the quantitation of arsenic species in urine and seafood samples. Commercial seaweed products, mussels and shrimp were chosen for ingestion studies because of their high arsenosugar and/or arsenobetaine contents. Detailed speciation of arsenic compounds in urine samples collected before and after the ingestion of shrimp, mussels and seaweed demonstrated that arsenobetaine was rapidly excreted unchanged into the urine and that arsenosugars were completely metabolized to dimethylarsinic acid (DMA) and several other uncharacterized arsenic species, which were also excreted into urine. The ingestion of arsenosugars from food is a confounding factor in assessing inorganic arsenic intake from drinking water. Arsenosugars are abundant in seaweed and marine bivalves. Recent studies also found arsenosugars in some terrestrial organisms, suggesting that they may be present in human diet excluding seafood.

Keywords: arsenosugar, arsenobetaine, food, urine, speciation, metabolism, biomarker

INTRODUCTION

Exposure to arsenic by the general population occurs mainly through arsenic present in drinking water and food (WHO, 1981; NRC, 1999). In some areas of the world, the high natural arsenic content of the drinking water has caused endemic, chronic arsenic poisoning (Chatterjee et al., 1995; Chowdhury et al., 1997; Luo et al., 1997). Epidemiological studies of populations exposed to high levels of arsenic have demonstrated a relationship between elevated arsenic exposure via drinking water and the prevalence of skin and several internal cancers (Tseng, 1977; Chen et al., 1985 and 1992; Bates et al., 1992; Hopenhayn-Rich et al., 1996). However, extrapolation from high level exposure data to predict potential impact at low levels involves many uncertainties. In order to reliably assess possible health impacts of exposure to low levels of arsenic from drinking water ingestion, it is important to understand the relative contributions of arsenic from food sources (Chappell et al., 1997).

Inorganic arsenite and arsenate are the major arsenic species present in drinking water. Minor amounts of monomethylarsonic acid (MMA), dimethylarsinic acid (DMA), and the methylated arsenic(III) species can also be present in natural waters (Braman and Foreback, 1973; Andreae, 1986; Hasegawa, 1997). Most seafoods contain $\mu g/g$ levels of arsenic, arsenobetaine being the major arsenic species in crustaceans and arsenosugars in seaweeds (Edmonds and Francesconi, 1981 and 1997; Cullen and Reimer, 1989). Both arsenobetaine and arsenosugars are present in bivalves (Shibata and Morita, 1992; Le et al., 1994b; Larsen, 1995). Minor amounts of arsenocholine, trimethylarsine oxide (Me$_3$AsO), tetramethyl-arsonium ion (Me$_4$As$^+$), DMA, MMA, arsenate, and unidentified arsenicals have also been found in marine animals (Shiomi et al., 1987; Cullen and Dodd, 1989; Larsen et al., 1993; Edmonds et al., 1997; Goessler et al., 1998). Very little is known about arsenic species in food other than seafood.

Urinary excretion is the major pathway for the elimination of arsenic compounds from the body (Buchet and Lauwerys, 1994; Vahter, 1994; Le et al., 1994a; Stoeppler and Vahter, 1994). Arsenic levels in urine reflect recent exposure. Therefore, a convenient approach to assessing arsenic exposure is through a quantitative measurement of arsenic species in urine samples.

Exposure to inorganic arsenic has commonly been assessed on the basis of the concentration of inorganic arsenite and arsenate as well as their metabolites, DMA and MMA. The ingestion of seafood can also cause a considerable increase in urinary arsenic concentration. To obtain a reliable assessment of arsenic exposure from drinking water, one must ensure that the ingestion of arsenic from non-drinking water sources can be differentiated and its contribution to urinary excretion of these arsenic species can be accounted for. The primary objective of this work was to study how the ingestion of arsenic from the diet affects the urinary arsenic speciation.

METHODOLOGY

Standards, Reagents, and Samples

An atomic absorption arsenic standard solution containing 1000.0 mg As/L as arsenite in 2% KOH (Sigma, St. Louis, MO) was used as the primary arsenic standard. Sodium arsenate, As(O)OH(ONa)$_2$·7H$_2$O) and sodium cacodylate, (CH$_3$)$_2$As(O)ONa were obtained from Sigma, and monomethylarsinate, CH$_3$As(O)OHONa, was obtained from Chem Service (West Chester, PA). Stock solutions (1000 mg As/L) of these arsenicals were prepared by dissolving appropriate amounts of the corresponding arsenic compounds in 0.01M hydro-choloric acid, and standard solutions were prepared by serial dilution with deionized water.

Mussels and seaweed products, nori and kelp powder, were purchased from a local supermarket in Edmonton, Canada. Standard Reference Material (SRM 1566a) oyster tissue was obtained from the National Institute of Standards and Technology (NIST, Gaithersburg,

MD). To extract arsenic species from the samples, a homogenized subsample (2–5 g dry weight for seaweed products and oyster tissue, or 20 g wet weight for mussel meat) was weighed into a test tube, to which was added 20 ml of a methanol/water mixture (1:1, V/V). The sample in the tube was sonicated for 20 min. After centrifugation, the extract was removed and placed in a 150-ml beaker. The extraction process with the aid of sonication was repeated a further four times. The extracts were combined in the beaker, slowly evaporated to dryness, and the residue dissolved in 10 ml of deionized water. After filtration through a 0.45 μm nylon membrane, the sample was subjected to HPLC analysis of arsenic species.

Urine samples were obtained from healthy subjects who volunteered to participate in the seafood ingestion studies. The experimental details and possible health effects concerning the ingestion of seaweed products and mussels associated with the experiments were discussed with the volunteers. All procedures followed were in accordance with the ethical guidelines of the Research Ethics Board, Faculty of Medicine, University of Alberta. The volunteers were asked not to eat any seafood for at least 72 h prior to commencing the mussel/seaweed ingestion experiment. Each volunteer collected two urine samples during the 12 h period prior to the consumption of mussels/seaweed. These samples were used to determine the background level of arsenic species in the urine resulting from a regular diet that excluded any seafood. The volunteers then consumed approximately 250 g (wet weight) of cooked mussels or 9.5 g (dry weight) of seaweed nori in one meal. The time of this meal was referred to as time zero. All urine was completely collected in separate 500-ml polyethylene containers for the subsequent three days. The volunteers did not eat any other seafood during the experiment period. The urine samples were stored at 4°C and were analyzed for arsenic speciation within one week.

Speciation of Arsenic

Two methods were used for the speciation of arsenic compounds. In the first approach, arsenic species were separated on a reversed phase C18 column (ODS-3, 150 mm × 4.6 mm, 3-μm particle size. Phenomenex, Torrance, CA) and detected by hydride generation atomic fluorescence (Le and Ma, 1998). A solution (pH 5.8) containing 5 mM tetrabutylammonium hydroxide, 4 mM malonic acid, and 5% methanol, was used as the HPLC mobile phase, at a flow rate of 1.5 ml/min. Effluent from the HPLC column was mixed at two T-joints, with continuous flows of hydrochloric acid (2 M) and sodium borohydride (1.3%). Arsines generated were separated from liquid waste and carried by a continuous flow of argon to an atomic fluorescence detector (Excalibur 10.003, P.S. Analytical, Kent, UK) for quantitation.

The second method was based on HPLC separation with inductively coupled plasma mass spectrometry (ICPMS) detection (Le et al., 1994b). Separation of arsenic species was carried out on a reversed phase C18 column (Inertsil ODS-2, 250 mm × 4.6 mm, 5-μm particle size. GL Sciences, Tokyo, Japan). An aqueous solution (pH 6.8) containing 10 mM tetra-ethylammonium hydroxide, 4.5 mM malonic acid, and 0.1% methanol was used as the mobile phase, at a flow rate of 0.8 ml/min. The HPLC effluent was directly introduced to an inductively coupled plasma mass spectrometer (PlasmaQuad 2 Turbo Plus. VG Elemental, Fisons Instrument). Signal intensity (counts per second) at m/z 75 was monitored for the quantitation of arsenic. Signals at m/z 77 were also monitored and used to correct for interference from $ArCl^+$.

RESULTS AND DISCUSSION

Major Arsenic Species in Seafood

Arsenic occurs in seawater mainly as inorganic arsenate, at levels of 1–3 μg/L (Andreae, 1978 and 1979; Cullen and Reimer, 1989). Arsenite, MMA, and DMA are present at lower levels, usually under 10% of the total arsenic concentration (Braman and Foreback, 1973; Andreae,

1986). In marine animals the levels of arsenic are much higher than the background concentrations in the surrounding water. Typical arsenic concentrations found in marine animals, range from 0.3 $\mu g/g$ in salmon to a high of 340 $\mu g/g$ in the midgut gland of the carnivorous gastropod *Charonia sauliae* (Cullen and Reimer, 1989; Francesconi and Edmonds, 1993). Arsenobetaine was first isolated and unequivocally characterized in 1977 (Edmonds et al.) from the tail muscle of the western rock lobster, *Panulirus cygnus*. Since then, this compound has been found to be present in almost all marine animals, usually as the major arsenic species (Cullen and Reimer, 1989; Francesconi and Edmonds, 1993, 1997).

The concentration of arsenic in marine algae (seaweed) is also considerably above that in their surrounding water. For example, an arsenic concentration of 95–109 $\mu g/g$ was found in a marine alga, *Laminaria digitata*. A variation of the total arsenic content between 0.4 and 32 $\mu g/g$ was reported for three main classes of macroalgae. Although a high concentration of arsenate (up to 50% of the total arsenic) is present in brown algae, *Sargassum muticum* (Whyte and Englar, 1983), *Laminaria digitata* (Andreae, 1986), and *Hizikia fusiforme* (Edmonds et al., 1987), generally only a small portion (usually under 10%) of the total arsenic present in the marine algae is in inorganic form. The rest is present as organoarsenicals, both water soluble and lipid soluble (Francesconi and Edmonds, 1993).

The major arsenic compounds present in marine algae are arsenosugar derivatives (Edmonds and Francesconi, 1981 and 1997; Shibata and Morita, 1988; Shibata et al., 1990; Le et al., 1994b; Pergantis et al., 1997). Arsenosugars are a group of arsenic-containing ribo-furanosides. Examples of typical arsenosugars are shown below:

$$R - CH_2 - O - O - CH_2\text{-}CH\text{-}CH_2\text{-}Y$$
$$\underset{X}{|}$$
OH   OH

	R	X	Y
X	$(CH_3)_2As(O)-$	$-OH$	$-OH$
XI	$(CH_3)_2As(O)-$	$-OH$	$-OPO_3HCH_2CH(OH)CH_2OH$
XII	$(CH_3)_2As(O)-$	$-OH$	$-SO_3H$
XIII	$(CH_3)_2As(O)-$	$-OH$	$-OSO_3H$
XIV	$(CH_3)_2As(O)-$	$-NH_2$	$-SO_3H$
XV	$(CH_3)_3As^+-$	$-OH$	$-OSO_3H$

Edmonds and Francesconi (1981 and 1983) first identified the water-soluble arsenosugar derivatives X, XI, and XII from the brown kelp, *Ecklonia radiata*, with XII being the major arsenic component (79%). Subsequently, this arsenosugar has been found to be a major arsenic compound in all sources of brown algae. Fifteen arsenosugars have been identified to date, X–XIII being the principal species (Shibata et al., 1992; Le et al., 1994b; Francesconi and Edmonds, 1997).

Figure 1 shows a chromatogram from the speciation analysis of arsenic compounds in a commercial seaweed food product, kelp powder. The principal arsenic compounds appear to be arsenosugars X, XI, XII, and XIII (Lai et al., 1997). Other commonly consumed seaweed products, nori and yakinori, contain arsenosugars X and XI as the major arsenic species, with concentrations as high as 20 $\mu g/g$ (dry weight) (Le et al., 1994a, 1996).

Arsenosugars are not only present in edible seaweed, they have also been found in several other food items, such as oysters, mussels, clams, and some terrestrial organisms. Figure 2 shows a chromatogram from the speciation analysis of arsenic compounds in an oyster tissue standard reference material. It demonstrates the presence of arsenosugars X and XI in

Fig. 1. A chromatogram from the HPLC/ICPMS analysis of an extract from kelp powder. HPLC separation was performed on an Inertsil ODS-2 column (4.6 × 250 mm), with a mobile phase (pH 6.8) containing 10 mM tetraethylammonium hydroxide, 4.5 mM malonic acid and 0.1% methanol, and a flow rate of 0.8 ml/min. X, XI, XII and XIII are four arsenosugars.

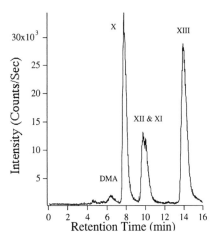

Fig. 2. A chromatogram from the HPLC/ICPMS analysis of an extract from the standard reference material oyster tissue (SRM 1566a). The same conditions as shown in Figure 1 were used. AsB, X and XI stand for arsenobetaine and two arseno-sugars, respectively.

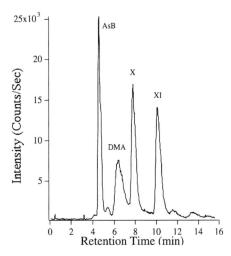

addition to arsenobetaine. Similar co-existence of arsenobetaine and arsenosugars have been observed in mussels and clams (Le et al., 1994b; Shibata and Morita, 1992; Larsen, 1995). Originally arsenosugars X and XIII were isolated from the kidney of the giant clam *Tridacna maxima*, and their presence was attributed to their formation by symbiotic algae which were also in the giant clam. However, bivalves such as *M. lusoria*, which have been found to contain arsenosugar XI, do not contain symbiotic algae, and it has been suggested (Shibata et al., 1992) that the bivalve's food source, phytoplankton, may be the origin of the arsenicals.

Many of the arsenic compounds initially thought of as marine origin have recently been found in the terrestrial environment. For example, arsenobetaine has been found in mushrooms (Byrne et al., 1995; Kuehnelt et al., 1997a), freshwater fish (Shiomi et al., 1995), and ants (Kuehnelt et al., 1997c) arsenosugars in terrestrial alga (Lai et al., 1997), terrestrial plants (Koch et al., 1998), freshwater shellfish (Koch et al., 1998) and earthworms (Geiszinger et al., 1998), and arsenocholine in mushrooms (Kuehnelt et al., 1997b). These new findings suggest that these organoarsenicals may be widely present in human diet. However, little research has been carried out on arsenic speciation in human diet other than seafood.

Urinary Arsenic Speciation

A common approach for the assessment of recent exposure to arsenic is through the measurement of arsenic species in urine because most arsenic compounds are rapidly excreted into urine. For the assessment of exposure to inorganic arsenicals, such as those from drinking water ingestion, determination of urinary inorganic arsenite and arsenate and their metabolites, MMA and DMA, has become a common practice. Exposure to excess amounts of arsenobetaine, which can result from the ingestion of crustacean seafood, does not contribute to these four arsenic species in urine. This is demonstrated in Figure 3, showing a typical pattern of the urinary excretion of arsenobetaine. Following the ingestion of 415 μg arsenobetaine from shrimp, over 66% was excreted within 36 h after the ingestion. Arsenobetaine did not undergo any metabolic change and it did not affect the concentration of inorganic arsenic, MMA and DMA, consistent with previous findings (Crecelius, 1977; Yamauchi et al., 1992; Le et al., 1993).

Unfortunately, the fact that arsenobetaine is rapidly excreted into urine without metabolic change has led to an incorrect perception that all organoarsenicals of seafood origin are excreted unchanged. In fact, arsenosugars undergo metabolism in the body, a behavior distinct from that of arsenobetaine. Figure 4 shows chromatograms obtained from HPLC/ICPMS analyses of a seaweed product, nori, and urine samples from a volunteer before and after the ingestion of nori (9.5 g). An arsenosugar (X) is the major arsenic species in the nori sample (Fig. 4c) and its content is approximately 21 μg/g (dry weight). Other forms of arsenic in the nori are approximately 0.7 μg/g (dry weight). Urine samples collected 13 h before (Fig. 4a) and 13.5 h after (Fig. 4b) the volunteer ate 9.5 g of nori show that the original arsenosugar is not present in the urine samples. Instead, five new arsenic compounds are present in the urine sample collected 13.5 h after the ingestion of nori (Fig. 4b). These compounds do not correspond to the arsenosugar present in nori on the basis of their retention times (Fig. 4c).

Similarly, the ingestion of mussels, which contain arsenobetaine and two arsenosugars (X and XI) as the major arsenic species (Fig. 5a), results in the excretion of arsenobetaine and the metabolites of arsenosugars. No trace of the original arsenosugars is detected in any of the urine samples (Fig. 5 b–e). The metabolism of the arsenosugars results in several, yet uncharacterized, arsenic-containing metabolites. The retention times of these metabolites are different from those of more than a dozen known arsenic species available to us. We have not been able to characterize these metabolites because of their trace levels in the urine samples (<30 ng/ml). Further development of analytical techniques, including preconcentration techniques combined with HPLC, electrospray ionization mass spectrometry, and nuclear magnetic resonance spectroscopy, will be useful for the characterization of the new metabolites.

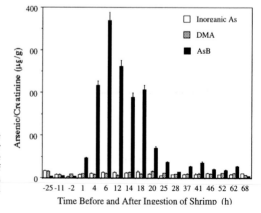

Fig. 3. Speciation of arsenicals in urine samples collected from a male volunteer (34 years old) with respect to the consumption of shrimp. Concentration of MMA was below the limit of quantitation and was not included in the figure (Le and Ma, 1997).

Fig. 4. Chromatograms of a nori extract (c) and of urine samples from a male volunteer (62 years old) 13 h before (a) and 13.5 h after (b) the ingestion of 9.5 g nori. An Inertsil ODS-2 column (4.6 × 250 mm) was used for the separation and ICPMS was used for detection. The major arsenic species in the nori is an arsenosugar (X). Peaks marked * indicate five new arsenic metabolites (Ma and Le, 1998).

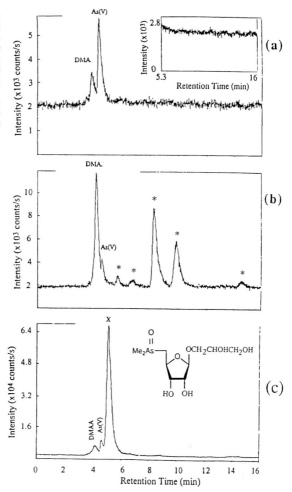

In addition to the presence of the new metabolites, there are substantial increases of DMA in urine samples collected after the ingestion of arsenosugars. Speciation of inorganic arsenic, MMA, and DMA in urine samples from a female volunteer is shown in Figure 6. The concentration of DMA is increased from a normal background level of 10–15 ng/ml to as high as 90 ng/ml after the consumption of 250 g of mussels. Speciation analyses of urine samples from an additional thirteen volunteers (nine males and four females) also showed substantial increases of DMA concentration following the consumption of either seaweed products or mussels.

CONCLUDING REMARKS

The most commonly used biomarkers of exposure (or internal dose) to inorganic arsenic have been based on the measurement of urinary arsenite, arsenate, MMA and DMA. These biomarkers are valid only when other ingested forms of arsenic are not metabolized, such as in the case of arsenobetaine. Ingestion of arsenobetaine, which is the principal arsenic species

Fig. 5. Arsenic speciation of a mussel extract (a) and urine samples from a male volunteer (32 years old) collected 0.5 h before (b), 2 h (c), 17 h (d) and 42.5 h (e) after the ingestion of 250 g mussels. Peaks marked * indicate four unidentified new arsenic species from the metabolism of two arsenosugars, X and XI (Le and Ma, 1998).

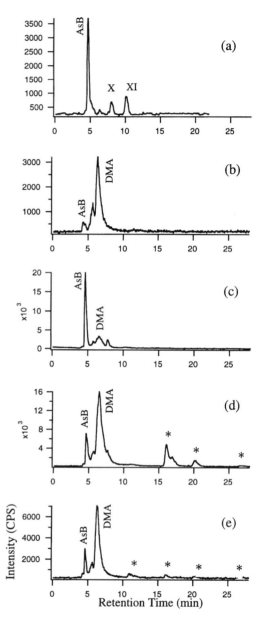

in crustacean seafood, results in urinary excretion of this species unmodified. The presence of arsenobetaine does not affect the determination of inorganic arsenic, MMA and DMA when appropriate speciation techniques are used. Therefore, ingestion of crustacean seafood does not affect the validity of the traditional biomarker of exposure to inorganic arsenic. However, in the case of the ingestion of arsenosugar-containing food, arsenosugars are metabolized. The substantial increase of DMA in urine samples as a result of arsenosugar metabolism is a major confounding factor in assessing inorganic arsenic intake using the

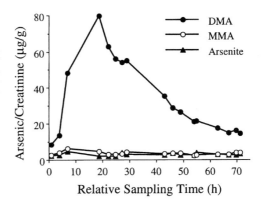

Fig. 6. Urinary arsenic/creatinine concentrations of a female volunteer (31 years old) before and after the ingestion of 9.5 g nori. HPLC separation with hydride generation atomic fluorescence detection was used for arsenic speciation. HPLC separation was carried out on an ODS-3 column (150 × 4.6 mm), with a mobile phase (pH 5.8) containing 5 mM tetrabutylammonium hydroxide, 4 mM malonic acid, and 5% methanol, at a flow rate of 1.5 ml/min.

traditional biomarkers which are strictly based on measurements of urinary arsenite, arsenate, MMA, and DMA.

Arsenosugars are present in seaweeds, oysters, mussels and clams (Francesconi and Edmonds, 1997; Le et al., 1994b), which are common for human consumption. Recent studies also identified arsenosugars in commercial food products of terrestrial algae (Lai et al., 1997), freshwater shellfish (Koch et al., 1998) and earthworms (Geiszinger et al., 1998), confirming that arsenosugars are not limited to the marine environment and may be present in other food items for common human consumption. The ingestion of arsenosugar-containing food in the human diet will invalidate the use of the conventional biomarkers of exposure to inorganic arsenic. One potential approach to dealing with this problem could be the use of the unique arsenosugar metabolites as an indicator of arsenosugar ingestion, which assists the interpretation of exposure data. Thus one should adopt appropriate caution when attempting to establish any correlation between the elevated urinary arsenic levels and the exposure to inorganic arsenic. Possible contribution to urinary arsenic species from dietary sources should be considered.

ACKNOWLEDGMENTS

This work was supported by the American Water Works Association Research Foundation and the Natural Sciences and Engineering Research Council of Canada. The authors thank Drs. W.R. Cullen at the University of British Columbia, K.J. Reimer at Royal Military College, and J. Feldmann at the University of Aberdeen for their contributions.

REFERENCES

Andreae, M.O. 1978. Distribution and speciation of arsenic in natural waters and some marine algae. *Deep-Sea Res.*, **25**, 391–402.

Andreae, M.O. 1979. Arsenic speciation in sea water and interstitial waters: the influence of biological–chemical interactions on the chemistry of a trace element. *Limnol. Oceanogr.*, **24**, 440–452.

Andreae, M.O. 1986. Organoarsenic compounds in the environment. In P.J. Craig (Ed.), *Organometallic Compounds in the Environment, Principles and Reactions*, pp. 198–228. John Wiley, New York.

Bates, M.N., Smith, A.H. and Hopenhayn-Rich, C. 1992. Arsenic ingestion and internal cancers: a review. *Am. J. Epidemiol.*, **135**, 462–476.

Braman, R.S. and Foreback, C.C. 1973. Methylated forms of arsenic in the environment. *Science*, **182**, 1247–1249.

Buchet, J.P. and Lauwerys, R. 1994. Inorganic arsenic metabolism in humans. In W.R. Chappell, et al. (Ed.), *Arsenic Exposure and Health*, pp. 181–189. Science and Technology Letters, Northwood.

Byrne, A.R. et al. 1995. Arsenobetaine and other arsenic species in mushrooms. *Appl. Organomet. Chem.*, **9**, 305–313.

Chappell, W.R. et al. 1997. Inorganic arsenic: A need and an opportunity to improve risk assessment. *Environ. Health Perspect.*, **105**, 1060–1067.

Chatterjee, A. et al. 1995. Arsenic in ground water in six districts of West Bengal, India: the biggest arsenic calamity in the world. Part 1. Arsenic species in drinking water and urine of the affected people. *Analyst*, **120**, 643–650.

Chen, C.-J., Chuang, Y.-C., Lin, T.-M. and Wu, H.-Y. 1985. Malignant neoplasms among residents of a Blackfoot disease-endemic area in Taiwan: High-arsenic artesian well water and cancers. *Cancer Res.*, **45**, 5895–5899.

Chen, C.J., Chen, C.W., Wu, M.M. and Kuo, T.L. 1992. Cancer potential in liver, lung, bladder, and kidney due to ingested inorganic arsenic in drinking water. *Br. J. Cancer*, **66**, 888–892.

Chowdhury, T.R. et al. 1997. Arsenic in ground water in six districts of West Bengal, India: The biggest arsenic calamity in the world; The status report up to August, 1995. In C.O. Abernathy, et al. (Ed.), *Arsenic Exposure and Health Effects*, pp. 93–111. Chapman & Hall, London.

Crecelius, E.A. 1977. Changes in the chemical speciation of arsenic following ingestion by man. *Environ. Health Perspect.*, **19**, 147–150.

Cullen, W.R. and Dodd, M. 1989. Arsenic speciation in clams of British Columbia. *Appl. Organomet. Chem.*, **3**, 79–88.

Cullen, W.R. and Reimer, K.J. 1989. Arsenic speciation in the environment. *Chem. Rev.*, **89**, 713–764.

Edmonds, J.S. and Francesconi, K.A. 1981. Arseno-sugars from brown kelp (*Ecklonia radiata*) as intermediates in cycling of arsenic in a marine ecosystem. *Nature*, **289**, 602–604.

Edmonds, J.S. and Francesconi, K.A. 1983. Arsenic-containing ribofuranosides: isolation from brown kelp *Ecklonia radiata* and N.M.R. spectra. *J. Chem Soc., Perkin Trans.*, **1**, 2375–2382.

Edmonds, J.S. et al. 1977. Isolation, crystal structure and synthesis of arsenobetaine, the arsenical constituent of the western rock lobster Panulirus longipes cygnus George. *Tetrahedron Lett.*, **18**, 1543–1546.

Edmonds, J.S., Morita, M. and Shibata, Y. 1987. Isolation and identification of arsenic-containing ribofuranosides and inorganic arsenic from Japanese edible seaweed *Hizikia fusiforme*. *J. Chem Soc., Perkin Transactions 1*, 577–580.

Edmonds, J.S., Shibata, Y., Francesconi, K.A., Rippingale, R.J. and Morita, M. 1997. Arsenic transformations in short marine food chains studied by HPLC-ICP MS. *Appl. Organomet. Chem.*, **11**, 281–287.

Francesconi, K.A. and Edmonds, J.S. 1993. Arsenic in the sea. *Oceanogr. Mar. Biol. Annu. Rev.*, **31**, 111–151.

Francesconi, K.A. and Edmonds, J.S. 1997. Arsenic and marine organisms. *Adv. Inorg. Chem.*, **44**, 147–189.

Geiszinger, A., Goessler, W., Kuehnelt, D., Francesconi, K. and Kosmus, W. 1998. Determination of arsenic compounds in earthworms. *Environ. Sci. Technol.*, **32**, 2238–2243.

Goessler, W., Rudorfer, A., Mackey, E.A., Becker, P.R. and Irgolic, K.J. 1998. Determination of arsenic compounds in marine mammals with high-performance liquid chromatography and an inductively coupled plasma mass spectrometer as element-specific detector. *Appl. Organomet. Chem.*, **12**, 491–501.

Hasegawa, H. 1997. The behavior of trivalent and pentavalent methyl arsenicals in Lake Biwa. *Appl. Organomet. Chem.*, **11**, 305–311.

Hopenhayn-Rich, C. et al. 1996. Bladder cancer mortality associated with arsenic in drinking water in Argentina. *Epidemiology*, **7**, 117–124.

Koch, I., Wang, L., Cullen, W.R. and Reimer, K.J. 1998. The speciation of arsenic in Yellowknife biota. "The 81st Canadian Society for Chemistry Conference and Exhibition." *Whistler, B.C., Canada*, 427.

Kuehnelt, D., Goessler, W. and Irgolic, K.J. 1997a. Arsenic compounds in terrestrial organisms I: *Collybia maculata*, *Collybia butyracea* and *Amanita muscaria* from arsenic smelter sites in Austria. *Appl. Organomet. Chem.*, **11**, 289–297.

Kuehnelt, D., Goessler, W. and Irgolic, K.J. 1997b. Arsenic compounds in terrestrial organisms II: Arsenocholine in the mushroom *Amanita muscaria*. *Appl. Organomet. Chem.*, **11**, 459–470.

Kuehnelt, D., Goessler, W., Schlagenhaufen, C. and Irgolic, K.J. 1997c. Arsenic compounds in terrestrial organisms III: Arsenic compounds in *Formica* sp. from an old arsenic smelter site. *Appl. Organomet. Chem.*, **11**, 859–867.

Lai, V.W., Cullen, W.R., Harrington, C.F. and Reimer, K.J. 1997. The characterization of arsenosugars in commercially available algal products including one of terrestrial origin, *Nostoc* sp. *Appl. Organomet. Chem.*, **11**, 797–803.

Larsen, E.H. 1995. Speciation of dimethylarsinyl-riboside derivatives (arsenosugars) in marine reference materials by HPLC-ICP-MS. *Fresenius J. Anal. Chem.*, **352**, 582–588.

Larsen, E.H., Pritzl, G. and Hansen, S.H. 1993. Arsenic speciation in seafood samples with emphasis on minor constituents: An investigation using high-performance liquid chromatography with detection by inductively coupled plasma mass spectrometry. *J. Anal. Atom. Spectrom.*, **8**, 1075–1084.

Le, X.C., Cullen, W.R. and Reimer, K.J. 1993. Determination of urinary arsenic and impact of dietary arsenic intake. *Talanta*, **40**, 185–193.

Le, X.C., Cullen, W.R. and Reimer, K.J. 1994a. Human urinary arsenic excretion after one-time ingestion of seaweed, crab, and shrimp. *Clin. Chem.*, **40**, 617–624.

Le, X.C., Cullen, W.R. and Reimer, K.J. 1994b. Speciation of arsenic compounds in some marine organisms. *Environ. Sci. Technol.*, **28**, 1598–1604.

Le, X.C. and Ma, M. 1997. Speciation of arsenic compounds by using ion-pair chromatography with atomic spectrometry and mass spectrometry detection. *J. Chromatogr. A*, **764**, 55–64.

Le, X.C. and Ma, M. 1998. Short-column liquid chromatography with hydride generation atomic fluorescence detection for the speciation of arsenic. *Anal. Chem.*, **70**, 1926–1933.

Le, X.C., Ma, M. and Wong, N.A. 1996. Speciation of arsenic compounds using high-performance liquid chromatography at elevated temperature and selective hydride generation atomic fluorescence detection. *Anal. Chem.*, **68**, 4501–4506.

Luo, Z.D. et al. 1997. Chronic arsenicism and cancer in Inner Mongolia—consequences of well-water arsenic levels greater than 50 μg/L. In C.O. Abernathy, et al. (Ed.), *Arsenic Exposure and Health Effects*, pp. 55–68. Chapman & Hall, London.

Ma, M. and Le, X.C. 1998. Effect of arsenosugar ingestion on urinary arsenic speciation. *Clin. Chem.*, **44**, 539–550.

NRC 1999. *Arsenic in Drinking Water*. National Academy Press, Washington, DC.

Pergantis, S.A., Francesconi, K.A., Goessler, W. and Thomas-Oates, J.E. 1997. Characterization of arseno-sugars of biological origin using fast atom bombardment tandem mass spectrometry. *Anal. Chem.*, **69**, 4931–4937.

Shibata, Y., Jin, K. and Morita, M. 1990. Arsenic compounds in the edible red alga, Porphyra tenera, and in *nori* and *yakinori*, food items produced from red algae. *Appl. Organomet. Chem.*, **4**, 255–260.

Shibata, Y. and Morita, M. 1988. A novel, trimethylated arseno-sugar isolated from the brown alga *Sargassum thunbergii*. *Agric. Biol. Chem.*, **52**, 1087–1089.

Shibata, Y. and Morita, M. 1992. Characterization of organic arsenic compounds in bivalves. *Appl. Organomet. Chem.*, **6**, 343–349.

Shibata, Y., Morita, M. and Fuwa, K. 1992. Selenium and arsenic in biology: Their chemical forms and biological functions. *Adv. Biophys.*, **28**, 31–80.

Shiomi, K., Kakehashi, Y., Yamanaka, H. and Kikuchi, T. 1987. Identification of arsenobetaine and a tetra-methylarsonium salt in the clam *Meretrix lusoria*. *Appl. Organomet. Chem.*, **1**, 177–183.

Shiomi, K., Sugiyama, Y., Shimakura, K. and Nagashima, Y. 1995. Arsenobetaine as the major arsenic compound in the muscle of two species of freshwater fish. *Appl. Organomet. Chem.*, **9**, 105–109.

Stoeppler, M. and Vahter, M. 1994. Arsenic. In R.F.M. Herber and M. Stoeppler (Ed.), *Trace Element Analysis in Biological Specimens*, pp. 291–320. Elsevier, Amsterdam.

Tseng, W.P. 1977. Effects and dose–response relationships of skin cancer and Blackfoot disease with arsenic. *Environ. Health Perspect.*, **19**, 109–119.

Vahter, M.E. 1994. Species differences in the metabolism of arsenic compounds. *Appl. Organomet. Chem.*, **8**, 175–182.

WHO 1981. *Environmental Health Criteria 18: Arsenic*. Environmental Health, WHO, Geneva.

Whyte, J.N.C. and Englar, J.R. 1983. Analysis of inorganic and organic-bound arsenic in marine brown algae. *Botanica Marina*, **26**, 159–164.

Yamauchi, H., Takahashi, K., Mashiko, M., Saitoh, J. and Yamamura, Y. 1992. Intake of different chemical species of dietary arsenic by the Japanese, and their blood and urinary arsenic levels. *Appl. Organomet. Chem.*, **6**, 383–388.

Arsenic Exposure and Health Effects
W.R. Chappell, C.O. Abernathy and R.L. Calderon (Editors)
© 1999 Elsevier Science B.V. All rights reserved.

Dietary Exposure to Inorganic Arsenic

R.A. Schoof, J. Eickhoff, L.J. Yost, E.A. Crecelius, D.W. Cragin,
D.M. Meacher, D.B. Menzel

ABSTRACT

Background exposures to inorganic arsenic have been estimated to be in the range of 16–19 μg/day with diet being the primary source. However, data on concentrations of inorganic arsenic in food are scanty. Existing literature includes data for a limited number of foods, and suggests that daily dietary inorganic arsenic intake for various age groups ranges from 8 to 14 μg/day in the United States and from 5 to 13 μg/day in Canada. The current study was conducted to obtain data for a larger number and variety of foods, and to more precisely estimate background dietary exposure ranges in the United States. This analysis presents results of a market basket survey of 40 commodities selected to represent samples of most major food types and to account for the majority of dietary exposure to inorganic arsenic. Samples were analyzed for total arsenic using NaOH digestion and inductively coupled plasma-mass spectrometry. Separate aliquots were analyzed for As^{3+}, As^{5+}, MMA, and DMA using an HCl digestion and hydride atomic absorption. The highest inorganic arsenic concentrations were found in grains and produce. Total arsenic concentrations were highest in seafood, and were consistent with results from previous studies; however, average inorganic arsenic concentrations in seafood were lower than previously reported. Data from the foods surveyed were coupled with a United States food consumption database to estimate the distribution of the dietary inorganic arsenic intake among adults. For most adults, dietary inorganic arsenic intake is predicted to fall within the range of 1–20 μg/day.

Keywords: inorganic arsenic, dietary exposures, arsenic in food

INTRODUCTION

In evaluating the significance of exposures to inorganic arsenic, it is important to consider the range of background exposures. Because inorganic arsenic occurs naturally in the environment and is present in most foods, arsenic exposure is a typical part of everyday life. For adults, typical background exposures to inorganic arsenic have been estimated to range from 16–19 μg/day (Valberg et al., 1997). An illustration of the magnitude of exposure (based on absorbed daily doses) associated with the primary background sources of arsenic exposure for adults is shown in Figure 1. For drinking water and soil, natural arsenic concentrations are highly variable, so exposures associated with both typical and high exposures are illustrated. Drinking water and diet are generally the most significant sources of background exposure.

Arsenic intake from water in the United States has been estimated to average about 5 μg/day for adults, based on surveys of arsenic in public drinking water supplies (ATSDR, 1993).[1] However, more recent surveys suggest average intake may be lower (Menzel et al., 1998). The range of exposures to inorganic arsenic from drinking water is controlled largely by the range of arsenic concentrations in drinking water supplies. In North America, concentrations of arsenic occurring naturally in groundwater range from below 1 μg/L to well above 50 μg/L (Borum and Abernathy, 1994).

Based on data from the U.S. Food and Drug Administration's Total Diet Studies, total daily intake of all forms of arsenic from the typical diet in the United States is approximately 50 μg for an adult (Borum and Abernathy, 1994). Some of the dietary arsenic is present in organic forms; however, these forms generally are not thought to be a health concern. Using estimates of the fraction of total arsenic that is likely to be inorganic in different food categories, Borum and Abernathy (1994) calculated that a typical adult diet contains 11–14 μg/day of inorganic arsenic. Our recent analysis (Yost et al., 1998) using revised estimates of the fraction of total arsenic that is inorganic suggested a similar intake (i.e., 14 μg/day of inorganic arsenic for American adults).

Although there is general agreement regarding typical amounts of arsenic intake from the diet, little information is available regarding the expected range of intakes. Dabeka et al. (1993) conducted a survey of total arsenic in food samples collected from six Canadian cities, and determined that dietary intake estimates for total arsenic based on samples collected in each city varied by as much as 53 percent among the cities. This result might be explained in part by the wide variability of total arsenic content that was observed both among samples within food groups (e.g., less than 1.3 to as much as 536 ng/g in individual samples of meat and poultry) and in mean concentrations between food groups (e.g., seafood concentrations are typically much higher than those in other food groups). Since the organic forms of arsenic predominate in seafood and the total arsenic concentration in these foods is several orders of magnitude higher than the concentrations in other food groups, some indication of the variation in intake of inorganic arsenic may be obtained by considering the range of arsenic intake from food other than seafood. In the Dabeka et al. (1993) study, the mean total arsenic content for 10 food groups, excluding fish and shellfish, ranged from 3.0 to 24.5 ng/g.

Mohri et al. (1990) conducted duplicate diet studies in Japan that offer some insights into the variation in inorganic arsenic intake in individuals. For six men and six women followed for three days, daily inorganic arsenic intake ranged from 1.2 to 31.7 μg, with a mean of 13.7 μg. Total arsenic intake ranged from 31 to 682 μg/day, with an average of 202 μg/day. In another group of two men and two women followed for seven days, dietary inorganic arsenic ranged from 1.8 to 22.6 μg/day, with a mean of 10.3 μg/day; total arsenic ranged from 27 to 376 μg/day, with a mean of 182 μg/day. The diets of the study participants consisted mainly of rice with cooked or raw seafood accompanied by soup and vegetables, including seaweed.

---

1   Essentially all of the arsenic in drinking water is thought to be inorganic.

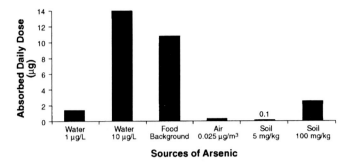

Sources of Arsenic

Fig. 1. Typical adult exposures to inorganic arsenic. The expected absorbed doses for arsenic from water supplies having 1 µg/L or 10 µg/L were estimated assuming a water consumption rate of 1.4 L/day and complete absorption. For food the estimated dose of 11 µg arsenic/day is based on a daily dietary intake of 13.5 µg of bioaccessible inorganic arsenic, assuming that 80 percent of the bioaccessible inorganic arsenic in food is absorbed into the body. Uthus (1994b) suggests that absorption of arsenic from food ranges from 80 to 100 percent; thus, the assumption that 80 percent of the acid-extractable inorganic arsenic is absorbed into the body represents a conservative estimate (i.e., absorption might be up to 100 percent). For air, the assumed concentration was 0.025 µg/m³ (mean urban concentration) with an inhalation rate of 13.3 m³/day. No adjustment was made for fractional absorption. For soil, absorbed doses are estimated for two concentrations assuming that an adult ingests 0.05 g of soil per day, and that arsenic in soil is 50 percent as bioavailable as arsenic in water.

The present study was conducted to obtain data for a larger number and variety of foods, and to estimate the range of background dietary exposures to total and inorganic arsenic in the United States.

METHODOLOGY

Sample Collection and Preparation

A market basket survey was conducted of 40 commodities selected to include representative samples of most major food types and to account for more than 90 percent of dietary exposure to inorganic arsenic. Commodities sampled included cereals (flour, corn meal, and rice); vegetables (ten kinds); fruits (six kinds, plus three kinds of juice); meat, poultry, fish, eggs, peanuts (peanut butter); milk; fats (soybean oil, butter); and salt, beer, and tap water. A detailed description of the selection of the commodities and survey design was provided previously (Schoof et al., 1999). Four samples of each commodity were collected from large supermarkets in two towns in Texas, and were prepared according to the protocols for preparation and cooking used for the U.S. Food and Drug Administration's Total Diet Study (Pennington, 1992).

Sample Analyses

Samples were analyzed for total arsenic by inductively coupled plasma-mass spectrometry (ICP-MS) following digestion in a 2N NaOH solution for 16 hrs at 80°C (method detection limit = 3.6 ng/g wet weight). Separate aliquots were analyzed for As^{3+}, As^{5+}, monomethyl-arsonic acid (MMA), and dimethylarsinic acid (DMA) (method detection limits = 1–2 ng/g wet weight) by hydride atomic absorption (AA) spectroscopy according to U.S. Environmental Protection Agency (EPA) Method 1632 (U.S. EPA, 1996) following digestion in 2N HCl at 80°C for 16 hrs. A detailed description of analytical methods and quality control procedures has been provided elsewhere (Schoof et al., 1999).

Dietary Exposure Estimates

The inorganic arsenic concentration data from the foods surveyed were then coupled with food consumption data from the U.S. Department of Agriculture's (USDA's) Continuing Surveys of Food Intake by Individuals (CSFII) for 1989–1992 to estimate the distribution of inorganic arsenic intake among adults (USDA, 1992; 1993; 1994b). The list of foods was expanded from the 40 commodities sampled to 156 foods by applying inorganic arsenic concentrations found in surveyed foods to similar foods that were not surveyed, or by applying standard concentration factors to processed forms of commodities that were surveyed.

Title 40 of the *Code of Federal Regulations,* Part 180.41 (40 CFR 180.41) was the reference source used to determine appropriate extrapolations from representative commodities to other commodities in a crop group. These crop groupings were developed by EPA's pesticide office for use by pesticide registrants when planning pesticide field trials. When residue data to establish pesticide residue tolerance levels are provided for representative crops, tolerances may also be established for other commodities listed in the crop group based on the residue levels observed in the representative crops. Extrapolations made based on crop groupings and from surveyed foods to other related foods are shown in Table 1.

The inorganic arsenic concentration for the analyzed survey food was applied to all forms of the food. For example, the food "corn meal" was surveyed because it was the best match for "corn/grain-endosperm"; however, the inorganic arsenic concentration for corn meal was also applied to other forms of field corn (i.e., corn/grain-bran and popcorn). Likewise, the concentration in whole wheat flour was applied to whole wheat, wheat bran, and wheat germ.

For tomatoes, the concentration in raw tomatoes was extrapolated to processed tomato products (sauce, purée, juice, paste, catsup) using processing factors based on standard water evaporation or removal during processing. This principle was applied to dried forms of foods, as well. Processing factors have also been established for juices and juice concentrates. In the case of orange juice, the juice concentrate was analyzed for inorganic arsenic; therefore, the concentrate value was adjusted to reflect appropriate dilution when used to represent single-strength juice. For apple and grape juice, single-strength juice was analyzed in the survey; thus, the single-strength juice values were adjusted to reflect appropriate concentration for concentrated forms of the juice. The processing factors were

TABLE 1

Extrapolations from surveyed foods to other foods

Surveyed Food	Associated Crop Grouping
Extrapolations Based on Crop Groupings:	
Carrots and potatoes	Root and tuber vegetables—180.41 (c)(1)
Onions	Bulb vegetables—180.41 (c)(3)
Spinach	Leafy vegetables—180.41 (c)(4) (a separate analysis was conducted for lettuce; therefore, the spinach concentration was not used for lettuce)
Watermelon (pulp)	Melons—180.41 (c)(9)
Cucumbers	Squash/Cucurbit vegetables—180.41 (c)(9)
Oranges	Citrus fruit—180.41 (c)(10)
Apples	Pome Fruit—180.41 (c)(11)
Peaches	Stone Fruit (except cherries)—180.41 (c)(12)
Wheat flour	Cereal Grains (except corn and rice)—180.41 (c)(15)
Additional Extrapolations for Associated Foods:	
Bananas	Plantains
Green beans	Yellow wax beans
Green peas	Snowpeas
Corn sugar	Corn sugar molasses
Beef	Other ruminants (goats, sheep, veal)
Chicken	Turkey and other poultry
Saltwater finfish	Unspecified finfish
Whole milk	Milk sugar and milk-based water

developed for ENVIRON's[2] Exposure® series software (TAS-ENVIRON, 1995) and are based on information provided in USDA's *Composition of Foods: Raw, Processed, Prepared, Agricultural Handbook No. 8* (USDA, 1994a).

The use of these processing factors is supported by available total arsenic monitoring data (FDA's Total Diet Study [FDA, 1997] and the Canadian study by Dabeka et al., 1993). Arsenic concentrations in juices or processed foods (e.g., concentrated tomato products) were generally higher than concentrations in the raw commodity, suggesting that arsenic is not necessarily removed with water during processing.

Using the commodity and food arsenic values, a detailed distributional dietary exposure analysis was developed by applying USDA CSFII food consumption data. Daily exposure was estimated for each surveyed individual (aged 18–59 years) by multiplying the inorganic arsenic concentration for each food by the consumption of that food and summing the exposure for all foods consumed by the individual that day. The distribution of exposure among individuals was then determined by ranking each individual's exposure to determine the percentiles of exposure. Fourteen distribution percentiles for arsenic intake ranging from the 10th percentile to the 100th percentile were calculated for the total United States population aged 18–59 years.

RESULTS

Mean inorganic arsenic concentrations for the 40 commodities tested ranged from 0.4 ng/g for beef to 73.7 ng/g for uncooked rice, while total arsenic concentrations ranged from 1.4 ng/g for lettuce to 2,360 ng/g for saltwater finfish. Seventeen of the 40 commodities had mean inorganic arsenic concentrations greater than the method detection limit of 2 ng/g for inorganic arsenic, while 28 commodities had total arsenic concentrations greater than the method detection limit of 3.6 ng/g. The highest inorganic arsenic concentrations were found in grains and produce, including raw rice, flour, corn meal, grape juice, watermelon, grapes, spinach, peas, carrots, and onions (Figure 2). Total arsenic concentrations were highest in seafood, but inorganic arsenic, MMA, and DMA were generally low in these foods (Table 2). Shrimp and rice were the only foods with substantial concentrations of DMA (means of 34 and 91 ng/g, respectively). Detailed results for all of the commodities tested have been presented previously (Schoof et al., 1999).

The estimated distribution of inorganic arsenic intake for adults based on the food survey data is shown in Table 3. The inorganic arsenic concentration data from the foods surveyed were coupled with a database for food consumption in the United States as described in the

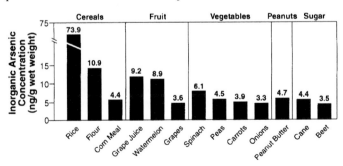

Fig. 2. Food groups and commodities with the highest inorganic arsenic concentrations among the commodities tested.

2   Formerly TAS-ENVIRON.

TABLE 2

Arsenic concentrations in seafood (ng/g wet weight $n=4$)

	Total Arsenic	Inorganic Arsenic	DMA	MMA
Saltwater Finfish	466–6080	<1	<1–5	<0.5
Canned Tuna	156–770	<1	1–6	<0.5
Shrimp	473–2820	1–3	21–57	<0.5
Freshwater Finfish	25–555	<1–2	<1–4	<0.5

TABLE 3

Estimated daily dietary intake of inorganic arsenic among adults (U.S. Population 18–59 years, all regions)

Percentile of Exposure	Exposure (μg/kg BW-day)	Intake (μg/person-day)*
10	0.014	1.0
20	0.019	1.3
30	0.024	1.7
40	0.029	2.0
50	0.034	2.4
60	0.040	2.8
70	0.049	3.4
80	0.063	4.4
90	0.095	6.7
95	0.133	9.3
97.5	0.179	12.5
99	0.228	16.0
99.5	0.279	19.5
Mean	0.046	
Standard deviation	0.048	

*Estimated based on the assumption of 70 kg average body weight.

methods section. Table 3 presents μg/kg BW-day estimates of dietary exposure to inorganic arsenic for adults 18–59 years of age. Estimated intake (μg/person-day) for a 70-kg adult is also shown. For most adults, dietary inorganic arsenic intake is predicted to fall in the range of 1–20 μg/day, with a mean intake of 3.2 μg/day.

DISCUSSION AND CONCLUSIONS

Rice has been shown to have consistently high concentrations of inorganic arsenic, with average concentrations for uncooked polished rice ranging from the 74 ng/g reported in this study to as high as 110 ng/g in other studies (Schoof et al., 1998; Yost et al., 1998). The greater uptake of arsenic by rice compared to other produce tested likely relates to substantial differences in the chemistry of the soil in which rice is grown in comparison with that in which other plants are grown. Specifically, rice is the only one of the terrestrial plants evaluated that is grown with the roots under water for most of the growing period. The flooded soils create an anaerobic environment that facilitates the release of arsenic from soil (Hess and Blanchar, 1977). The finding of higher inorganic arsenic in rice is also consistent with the relatively high inorganic arsenic concentrations observed in marine plants (Sanders, 1979).

Inorganic arsenic concentrations in the seafood surveyed in the present study were generally lower than concentrations previously reported (Table 4). These differences could have several causes, including regional differences in arsenic concentrations in seafood, differences in methods of preparation and sample handling, or differences in analytical methods. All of the samples in the present study were purchased at large supermarkets in the United States. In previous studies, samples were purchased in markets in Japan (Mohri et al.,

TABLE 4

Comparison of seafood arsenic concentrations (ng/g wet weight *n*=4)

	Total Arsenic	Inorganic Arsenic
Saltwater finfish		
Cod, halibut, orange roughy (U.S., *N*=4)[1]	466–6080	<1
Sole (Canada, *N*=1)[2]	4000	22
Salmon, red seabream (Japan, *N*=2)[3]	728–1684	5
Cod, seabream, plaice, sole (Belgium, *N*=32)[4]	1500–28000	10–28[a]
Canned Tuna		
Canned Tuna (U.S., *N*=4)[1]	156–770	<1
Canned Tuna (Canada, *N*=1)[2]	1100	25
Shrimp		
Shrimp (U.S., *N*=4)[1]	473–2820	1–3
Shrimp (Japan, *N*=1)[3]	1170	37
Shrimp (Canada, canned *N*=1)[2]	650	100
Freshwater finfish		
Catfish, rainbow trout (U.S., *N*=4)[1]	25–555	<1–2
Pickerel (Canada, *N*=1)[2]		22
	140	

[a] Acid hydrolysis.
[1] Schoof et al. (1999); [2] Yost et al. (1998); [3] Mohri et al. (1990); [4] Buchet et al. (1994).

1990), Belgium (Buchet et al., 1994), and Canada (Yost et al., 1998). In the present study, finfish was baked, while shrimp was microwaved. In the Japanese study, seafood was boiled or grilled, while there was no reported indication that the Belgian or Canadian samples were cooked. Both the present study and the Belgian study used HCl digestion with hydride AA analysis for inorganic arsenic, while the Japanese study used NaOH digestion followed by hydride AA analysis. At present no obvious reason for the differences among studies is apparent from this information.

The range of predicted inorganic arsenic intakes of 1–20 μg/day based on the food survey data is clearly consistent with previous studies. As described earlier, duplicate diet studies of Mohri et al. (1990) found a range of approximately 1–32 μg/day in Japanese adults. As might be expected in a population with very high intakes of rice, the upper end of this range exceeds that predicted for the general United States population. A distribution similar to the distribution for Japanese adults might be expected in subpopulations within the United States that have high rice intakes. In general, the food survey data suggest that individuals who have a high intake of grains and produce (versus protein and fat-rich foods) may be expected to have the highest intake of inorganic arsenic.

This information on relative concentrations of inorganic arsenic in different food groupings should be particularly useful in the design of studies of the nutritional role of inorganic arsenic. It has been proposed that arsenic is an essential nutrient (Nielsen, 1998; Uthus, 1994a,b; Uthus and Seaborn, 1996); however, studies of arsenic essentiality have not clearly distinguished among the forms of arsenic in the diet. Future studies should focus on the role of inorganic arsenic, and should consider the relatively high inorganic arsenic concentrations found in grains when attempting to create arsenic deficient diets for laboratory animal studies. If inorganic arsenic is confirmed to be a required nutrient, the low intakes predicted for much of the United States population could indicate the presence of deficiencies in fairly large segments of the population.

ACKNOWLEDGEMENTS

The authors would like to thank the following individuals for their assistance with this study: Darin Waylett of ENVIRON for assistance with study design and sample collection; Chuck

Apts, Linda Bingler, and Peg O'Neill of Battelle Marine Sciences for chemical analyses; Allyson Templeton of Exponent for spreadsheets and calculations; Laura Jones of Exponent for review of analytical methods; and Professor M. B. Dillencourt, Professor L. Bic, and Edward Lee of the University of California at Irvine for preliminary designs and excellent constructive criticism.

REFERENCES

ATSDR. 1993. *Toxicological Profile for Arsenic*. Agency for Toxic Substances and Disease Registry, U.S. Public Health Service, Atlanta, GA.

Borum, D.R., Abernathy, C.O. 1994. Human oral exposure to inorganic arsenic. In: Chappell, W.R., Abernathy, C.O., and Cothern, C.R. (eds.), *Arsenic Exposure and Health*. Science and Technology Letters, Northwood, UK.

Buchet, J.P., Pauwels, J., Lauwerys, R. 1994. Assessment of exposure to inorganic arsenic following ingestion of marine organisms by volunteers. *Environ. Res.*, 66, 44–51.

Dabeka, R.W., McKenzie, A.D., Lacroix, G.M.A., Cleroux, C., Bowe, S., Graham, R.A., Conacher, H.B.S., Verdier, P. 1993. Survey of arsenic in total diet food composites and estimation of the dietary intake of arsenic by Canadian adults and children. *J. AOAC Int.*, 76, 14–25.

FDA. 1997. Total Diet Study monitoring data, 1985 to 1994. Dataset. U.S. Food and Drug Administration, Washington, DC.

Hess, R.E., Blanchar, R.W. 1977. Dissolution of arsenic from waterlogged and aerated soil. *Soil Sci. Soc. Am. J.*, 41, 861–865.

Menzel, D.B., Dillencourt, M.D., Meacher, D.M., Bic, L.F., Schoof, R.A., Yost, L.J., Cragin, D.W., Farr, C.H. 1998. Estimation of multimedia inorganic arsenic exposure in the U.S. population. Paper presented at the Third International Conference on Arsenic Exposure and Health Effects, July 12–15, 1998, San Diego, CA.

Mohri, T., Hisanaga, A., Ishinishi, N. 1990. Arsenic intake and excretion by Japanese adults: a 7-day duplicate diet study. *Food Chem. Toxicol.*, 28, 521–529.

Nielsen, F.H. 1998. Ultratrace elements in nutrition: current knowledge and speculation. *J. Trace Elements Exper. Med.*, 11, 251–274.

Pennington, J.A.T. 1992. Preparation and cooking instructions for total diet study foods. Appendix B. In: *Appendices for the 1990 Revision of the Food and Drug Administration's Total Diet Food List and Diets*. Center for Food Safety and Applied Nutrition (CFSAN), U.S. Food and Drug Administration (FDA). NTIS # PB92-176239.

Sanders, J.G. 1979. The concentration and speciation of arsenic in marine micro-algae. *Estu. Coastal Mar. Sci.*, 9, 95–99.

Schoof, R.A., Yost, L.J., Eickhoff, J., Crecelius, E.A., Cragin, D.W., Meacher, D.M., Menzel, D.B. 1999. A market basket survey of inorganic arsenic in food. *Food Chem. Toxicol.*, 37, 837–844.

Schoof, R.A., Yost, L.J., Crecelius, E., Irgolic, K., Goessler, W., Guo, H.-R., and Greene, H.L. 1998. Dietary arsenic intake in Taiwanese districts with elevated arsenic in drinking water. *Human Ecol. Risk Assess.*, 4, 117–135.

TAS-ENVIRON. 1995. Exposure 1® chronic dietary exposure analysis software utilizing 1989–1992 U.S. food consumption patterns. TAS-ENVIRON, Arlington, VA.

USDA (U.S. Department of Agriculture). 1992. Nationwide food consumption survey: continuing survey of food intakes by individuals 1989–90. Human Nutrition Information Service (HNIS). Dataset.

USDA (U.S. Department of Agriculture). 1993. Nationwide food consumption survey: continuing survey of food intakes by individuals 1990–91. Human Nutrition Information Service (HNIS). Dataset.

USDA (U.S. Department of Agriculture). 1994a. Composition of foods: raw, processed, prepared, agricultural handbook No. 8. Human Nutrition Information Service (HNIS). Dataset.

USDA (U.S. Department of Agriculture). 1994b. Nationwide food consumption survey: continuing survey of food intakes by individuals 1991–92. Human Nutrition Information Service (HNIS). Dataset.

U.S. EPA. 1996. *Method 1632, Inorganic Arsenic in Water by Hydride Generation Quartz Furnace Atomic Absorption*. EPA/821/R-95/028. U.S. Environmental Protection Agency, Office of Water, Office of Science and Technology, Engineering and Analysis Division, Washington, DC.

Uthus, E.O. 1994a. Arsenic essentiality studies and factors affecting its importance. In: Chappell, W.R., Abernathy, C.O., and Cothern, C.R. (eds.), *Arsenic Exposure and Health*. pp. 199–208. Science and Technology Letters, Northwood, U.K.

Uthus, E.O. 1994b. Estimation of safe and adequate daily intake for arsenic. In: Hertz, L.O., Abernathy, C.O., and Olin, S.S. (eds.), *Risk Assessment of Essential Elements*. pp. 273–282. ILSI Press, Washington, DC.

Uthus, E.O., and Seaborn, C.D. 1996. RDA workshop: new approaches, endpoints and paradigms for RDAs for mineral elements. *J. Nutr.*, 126:9 Suppl. (Sep. 1996) pp. S2452–S2459.

Valberg, P.A., Beck, B.D. Bowers, T.S., Keating, J.L., Bergstrom, P.D., and Boardman, P.D. 1997. Issues in setting health-based cleanup levels for arsenic in soil. *Reg. Toxicol. Pharm.*, 26, 219–229.

Yost, L.J., Schoof, R.A., and Aucoin, R. 1998. Intake of inorganic arsenic in the North American diet. *Human Ecol. Risk Assess.*, 4, 137–152.

Arsenic Exposure and Health Effects
W.R. Chappell, C.O. Abernathy and R.L. Calderon (Editors)
1999 Elsevier Science B.V.

Exposure to Inorganic Arsenic from Fish and Shellfish

Joyce Morrissey Donohue, Charles O. Abernathy

ABSTRACT

Humans are exposed to arsenic (As) from many sources such as food, water, air and soil. Most foods contain both organic and inorganic forms of As and the inorganic compounds are generally considered to be more toxic. Although fish and shellfish are major contributors to dietary As among seafood consumers, over 90% of the As in seafood is generally organic rather than inorganic. Thus, it is important to know the relative levels of various As species in fish and shellfish when estimating risks from seafood consumption. Data were collected from published and unpublished literature on the concentrations of total, inorganic and organic As present in fish and shellfish. Distributions were skewed with median concentrations, in this instance, a better representation of central tendency than mean concentrations. The data were used to estimate total exposure to inorganic As from consumption of fish/shellfish for several exposure scenarios applicable to seafood-consuming populations, including subsistence groups. Data on fish and shellfish consumption patterns were derived from the 1989–1991 U.S. Department of Agriculture Continuing Survey of Food Intake by Individuals. Organic As in ocean and estuarine fish and shellfish is primarily present as arsenobetaine (AsB) with smaller amounts as arsenocholine (AsC) or other organic compounds. Less is known about the identity of the organic arsenic in freshwater fish. Data on the toxicokinetics of AsB and AsC demonstrate that the As in these compounds is apparently not bioavailable for interaction with other biological molecules.

Keywords: arsenic, arsenobetaine, arsenocholine, fish, shellfish

The opinions expressed in this manuscript are those of the authors and do not necessarily represent the opinions or policies of the US Environmental Protection Agency

INTRODUCTION

Quantitative data on arsenic (As) concentrations and the speciation of As compounds in fish and shellfish are limited but are generally consistent with the hypothesis that most of the arsenic is organic rather than inorganic. This is an important point because organic arsenicals are generally considered to be less toxic than inorganic arsenic (Sabbioni et al., 1991; Buchet at al., 1996). Thus, when evaluating exposure to As from drinking water and the diet, the total As levels in foods should not be treated as equivalent to the inorganic As. This difference needs to be considered when developing exposure scenarios for dietary arsenic intake.

In ambient water, inorganic As is generally present as arsenite and/or arsenate ions. Depending on the oxidation-reduction status of the environment, arsenite (+3) and arsenate (+5) can be interconverted. In addition, physiological conversions of these two ions may also be possible (Marafante and Vahter, 1987). The arsenite anion is considered to be the more toxic of the inorganic species and is believed to be responsible for the carcinogenic activity of inorganic As (Rossman, 1981; Wang and Rossman, 1996). In most surface waters, As is generally found as arsenate, while ground waters contain arsenite and/or arsenate. There is usually very little organic arsenic in ambient water (Irgolic, 1994).

There are a number of organic As compounds that have been identified in foods. These compounds include arsenobetaine (AsB), arsenocholine (AsC), arsenosugars, monomethyl arsonic acid (MMA), dimethyl arsinic acid (DMA) and trimethyl arsine oxide (Chew, 1996). These As compounds are found in a variety of foods, however, seafood is generally the richest source of dietary As.

The predominant organic As compound in marine fish and shellfish is AsB (Ballin et al., 1994; Lawrence et al., 1986). Arsenocholine is also present both free and in tissue phospholipids (Ballin et al., 1994; Lawrence et al., 1986). The amount of AsB exceeds the AsC (Ballin et al., 1994; Chew, 1996; Lawrence et al., 1986). In both compounds, As has replaced the nitrogen of the natural metabolite (choline; betaine). Arsenic is incorporated into the betaine and choline molecules by microorganisms, phytoplankton, zooplankton and algae (Ballin et al., 1994). The fish obtain AsB and AsC from their food supply.

In freshwater fish, some data indicate that the principal organic arsenical may not be either AsB or AsC. A single compound was isolated by Lawrence et al. (1986) from all freshwater fish samples analyzed. This compound accounted for at least 70 to 85% of the As recovered. The compound was not identified, but it appeared to be more hydrophilic than AsB. In contrast, Shiomi et al., (1995) identified AsB as the major organoarsenical in two species of freshwater fish.

Small amounts of MMA and DMA have been found in fish and shellfish (Buchet et al., 1994; Caroli et al., 1994; Chew 1996; Velez et al., 1995, 1996). Concentrations were generally ≤1% for MMA and ≤2% for DMA, although higher amounts were occasionally reported. Buchet et al. (1994) found that the recovery of MMA and DMA varied with the extraction technique and between samples from the same fish. For this reason, the data for these As species must be viewed with caution. Arsenosugars are also found in seafoods (Larsen, 1995); they are apparently synthesized by marine algae, especially seaweeds and become incorporated into fish and shellfish tissues through their diet.

EVALUATION OF ANALYTICAL DATA

In order to estimate the amounts and types of As in fish and shellfish, quantitative analytical data from published and unpublished reports were compiled. Separate databases were established for marine fish, shellfish and freshwater fish. The fish/shellfish species, total As concentration, inorganic As concentration, percent inorganic As, and reference were entered into the database for each sample analyzed. The data were sorted to array the concentration data in increasing numeric order. After sorting, the mean, median, and 95th percentile values were determined for each data set. The data on the organic As compounds found in fish and shellfish were far more limited than the data for total and inorganic As. Accordingly the organic data were not entered into a database, but were merely grouped and tabulated.

The data compiled in this report have several limitations. About 50% of the data came from a review paper by Chew (1996) which did not record analytical methods or limits of detection for individual assays. In the Chew (1996) report, samples that were below the limit of detection were reported as having no As. The remainder of the data came from primary reference materials and did contain method and detection limit information, but many of these had little information on sample holding conditions or preparation methods for fish tissues. Some studies evaluated total and organic As rather than total and inorganic As. In those cases, the inorganic As was determined by the difference between the total and the organic concentrations. In some studies, recovery of AsB and AsC from spiked samples was not complete. In such cases, the inorganic As values determined by difference may overestimate the actual amount present. However, we believe that it is important to compile and catalog the existing data base as it provides a reference point for subsequent exposure evaluations.

TOTAL ARSENIC

The data on the total As in fish and shellfish are summarized in Table 1. Data were available for 77 marine fish samples, 57 marine/estuarine shellfish samples and 24 samples of freshwater fish. The data indicate that there is considerable variability in the total As present in marine fish and shellfish; variability is apparent between and within species. For example, the data for five samples of cod range from 0.6 to 7 mg/kg and the data for five samples of mackerel range from 0.3 to 2.0 mk/kg (Ballin et al, 1994; Chew, 1996; Lawrence et al., 1986). Data for 8 samples of herring were more consistent with six samples containing 1 mg/kg when rounded to one significant figure, one having 2 mg/kg and the last having 4 mg/kg (Ballin et al, 1994; Chew, 1996; Lawrence et al., 1986). Two plaice samples had 9 and 32 mg/kg total As, respectively (Ballin et al., 1994). In all situations where the database included data for more than one sample of a marine fish species, the values did not differ by more than an order of magnitude.

The data for shellfish were more variable than the marine fish data. The analytical results from twenty mollusc samples ranged from 1 to 126 mg/kg, the values for four clam samples ranged from 1 to 67 mg/kg. As was the case for marine fish species, there was little consistency within a species (Ballin et al., 1994, Chew, 1996; Lawrence et al., 1994).

The analytical data on freshwater fish species, while more limited than those for marine fish, displayed an order of magnitude difference for different samples of the same species. The As concentrations from four samples of perch ranged from 0.02 to 2 mg/kg while four samples of carp ranged from 0.05 to 0.5 mg/kg (Ballin et al., 1994; Lawrence et al., 1986). Smelt

TABLE 1

Arsenic in marine fish, shellfish and freshwater fish tissues

Total Arsenic	Marine Fish (mg/kg)	Shellfish (mg/kg)	Freshwater Fish (mg/kg)
Number of samples	77	57	24
Range	0.19–65	0.2–125.9	0.007–1.46
Mean	5.1	15.9	0.3
Median	2.1	4.2	0.06
95%	17	57	NA

NA = Not applicable.

and trout were the only species from the 24 samples analyzed that had more than 1 mg/kg total arsenic (Ballin et al., 1994; Shomi et al., 1995)

It is readily apparent from the range, median, and mean values in Table 1, that shellfish have the highest concentrations of total As. The lowest amounts of total As were found in freshwater fish. The mean concentration of As in shellfish was three times that for marine fish and about 50 times that found in freshwater fish. The median concentration for shellfish was twice that for marine fish and 60 times that for freshwater fish. In all cases, the mean concentration was greater than twice the median concentration showing that the distribution is skewed towards the high end of the range and the majority of the samples fall at the low end of the range. Thus, in this instance, the median value is a better measure of central tendency than the mean.

INORGANIC ARSENIC

The data for inorganic As in seafoods are summarized in Table 2. There were only three values available for the inorganic As in freshwater fish. Thus, mean, median and 95% values were not determined for this data set. Inorganic As concentrations in marine fish and shellfish were less variable than the total As concentrations. In both situations, there were a number of samples that were reported as having no inorganic As or amounts that were less than the limit of detection. Similar to the situation for total As, the mean concentration in shellfish is greater than that for marine fish. In this instance the mean is four times greater while the median concentration is three times greater.

Borum and Abernathy (1994) reported that the amount of inorganic As in marine fish and shellfish is about 10% of the total. Among the data collected for this project, there were only three marine fish specimens that exceeded 4% inorganic arsenic. One anchovy sample had 5% inorganic arsenic; the amount in a second sample was less than the detection limit. A sample of sturgeon had 7% inorganic As and one of shark had 10%. Among the shellfish data collected, there were only 3 that exceeded 3% inorganic As. One cockle sample had 8% and a mussel sample 11% (Lopez et al., 1994). Chew (1996) reported values of less than 3% for other samples of these same species. A sample of *Barnea dilatata* was reported to have 95% of its As in the inorganic form (Chew, 1996). However only one sample was analyzed and that sample was at the low end of the total arsenic distribution with 0.2 mg/kg total As.

ORGANIC ARSENIC COMPOUNDS

Table 3 summarizes the data collected for the organic As compounds in fish and shellfish. As mentioned above, the data are somewhat limited, however, they indicate that, in most cases, more than 85% of the As is present as organic rather than inorganic compounds (Alberti et al., 1995; Ballin et al., 1994; Lawrence et al., 1986; Velez et al., 1995, 1996). The most prevalent organic species, especially in marine fish, is AsB (Ballin et al., 1994; Lawrence et al., 1986). The

TABLE 2

Inorganic arsenic in marine fish, shellfish and freshwater fish tissues

Inorganic Arsenic	Marine Fish (mg/kg)	Shellfish (mg/kg)	Freshwater Fish (mg/kg)
Number of samples	43	53	3
Range	0–0.16	0–0.57	0.001–0.04
Mean	0.02	0.08	ND
Median	0.01	0.03	ND
95%	0.1	0.3	ND

ND = Not determined.

TABLE 3

Organic arsenic compounds in marine fish and shellfish

Organoarsenic Species	Concentration Range	Median	Number of Samples
Arsenobetaine	9.3–100%	81%	38
Arsenocholine	ND–15%	ND	16
Arsenophospholipids	0.17–15%	1.1%	10
Arsenosugars	18%	NA	2
Monomethylarsonic acid	<0.0006–6.7%	<0.0008%	9
Dimethylarsinic acid	ND–12.9%	1.9%	15
Hydrophilic unidentified compound (freshwater fish, salmon)	41–85%	73%	10

ND = Not detected; NA = not applicable.

compound present in the next highest concentration is either AsC or DMA. The AsC was estimated by Edmonds and Francesconi, (1993) to be less than 1% of the total As. The data surveyed for this report are similar except in one case, where AsC accounted for 15% of the total As (Lawrence et al., 1986).

Arsenobetaine, AsC and the unidentified organic As compound in freshwater fish are hydrophilic and have little tendency to bioaccumulate in edible fish tissues. These compounds are unlikely to be present in fish oils or adipose tissues due to their hydrophilic nature although some AsC may be present in membrane phospholipids. Based on the metabolism of betaine and choline, it is expected that most of the AsC from fish will be converted to AsB and excreted. A small portion of AsC may become incorporated in phospholipids and retained; another small amount may be converted to trimethyl arsine oxide (Chew, 1996).

RELATIVE TOXICITY

Organic arsenicals, especially AsB and AsC are significantly less acutely toxic than most inorganic arsenic species (Table 4). Although there have been no chronic exposure studies involving AsB or AsC, available evidence indicates that they would have little chronic toxicity. For example, AsB is metabolically inert in mammalian systems. Almost all of the radiolabeled arsenic in AsB administered orally or intravenously to rats, mice or guinea pigs was excreted in three days (Vahter et al., 1983; Yamauchi et al., 1986). In rats and mice, more than 99% of the excreted label was found in the urine as AsB. In comparable studies using AsC, there was greater label retention with 70–80% excreted in three days (Marafante et al., 1984). Since extracts from mouse urine showed that more than 90% of the water-soluble As from AsC was excreted as AsB, it is reasonable to assume that some of the retained AsC is incorporated in membranes as phosphatidyl choline compounds or in lipoprotein complexes and, thus, will have little tendency to bioaccumulate as inorganic As. Each of these factors diminishes human health concerns related to exposure to organic As compounds in fish and shellfish. Additional data suggesting that the organic As compounds from fish and shellfish do not bioconcentrate is provided by data showing that samples of human milk from 88 mothers from the Faroe Islands did not show elevated As in their transition milk despite consumption of diets rich in seafoods (Grandjean et al., 1995). Moreover, in people consuming seafood in their diet, the amount of inorganic As released during the digestive process is minimal (Buchet et al., 1996).

There are experimental laboratory data that also suggest that AsB possesses little toxicity. It is not cytotoxic to the BALB 3T3 cell line as compared to sodium arsenite. It also is not retained by these cells nor does it have neoplastic transforming ability in this cell line

TABLE 4

Acute LD_{50}s of Various Arsenicals in Mice and Rats*

Arsenical	Oral LD_{50} (mg/kg)
Sodium Arsenite	15–42
Arsenic Trioxide	34–34
Calcium Arsenate	20–800
MMA	700–1800
DMA	1200–2600
Arsenobetaine	$\geq 10 \times 10^3$
Arsenocholine	6.5×10^3

*Data compiled from Done and Peart 1971; Cannon et al. 1983; Fielder et al., 1986, Kaise and Fukui, 1992; RTECS, 1985–1986.

(Sabbioni et al., 1991). Cannon et al. (1983) reported that AsB had no mutagenic activity in the Ames *Salmonella typhimurium* assay, with or without metabolic activation. The available evidence, taken as a whole, indicates that the major organic arsenicals in seafood possess little or no toxic potential.

EXPOSURE ESTIMATES

The data on As in seafoods summarized above support several hypotheses regarding concentrations of inorganic As in fish and shellfish. It appears that the inorganic As is usually only 3 to 4% of the total arsenic. However, there is a wide distribution in the amounts of both total and inorganic As in fish and shellfish tissues and the distribution is skewed towards the high side of the distribution. This makes the median concentrations from this data set a better measure of central tendency for both total and inorganic As than the mean concentrations. Differences in inorganic As between samples may reflect differences in the As concentrations of the source waters.

In marine fish and shellfish, the organic As is mostly AsB with lesser amounts of the other organic As compounds. There appears to be more of the other organoarsenicals in shellfish than marine fish. An unidentified hydrophilic, organic arsenical compound was the predominant organic compound in freshwater fish rather than AsB in work completed by Lawrence et al. (1986). This compound accounted for most of the organic As in nine samples from six species. However, Shiomi et al. (1995) identified AsB as the major As compound in a freshwater smelt and trout . Freshwater species contain far less As than marine species; this may reflect an inability of the freshwater food chain organisms to incorporate inorganic As into biological molecules.

The data on the amounts of inorganic As in fish and shellfish can be used to estimate inorganic As exposure from the diet for individuals consuming diets rich in fish and shellfish. For this analysis, the mean, median, and 95 percentile inorganic As values were combined with several fish/shellfish intake values to estimate exposure to inorganic As through seafood. The reference diet for the exposure calculation was a diet where the seafood component was 90% marine fish and 10% shellfish.

Using the concentration data discussed above, the inorganic As concentration for the seafood exposure estimate was calculated using the following equation:

0.9 (Marine Fish Inorganic As Conc) + 0.1 (Shellfish Inorganic As Conc.)

= Seafood Inorganic As Conc.

Estimates of the seafood inorganic As were derived for the mean, median and 95% values from Table 2.

The seafood As concentrations were combined with data on fish and shellfish consumption from the US Department of Agriculture (USDA) to estimate exposure. For the seafood consuming portion of the population, the seafood consumption values were derived from the 1989–1991 USDA Continuing Survey of Food Intake by Individuals (CSFII; USEPA, 1997). The USDA survey collects data on three consecutive days of food intake from participants. Data for one day is provided through a 24-hour recall interview and data for two days through food intake records kept by the respondent.

The USEPA derived exposure data for seafood consumers by segregating the CSFII respondents that consumed seafoods at least once during the three day survey period and looking at their fish and shellfish consumption. The resultant intake estimates apply only to the seafood-consuming population and represent a rather skewed distribution. However, for the purpose of defining an exposure on the high end of the distribution curve the data are appropriate and useful.

The maximum seafood intake for seafood-consuming adults (461 g/day) from the CSFII data was used to simulate the eating habits of subsistence fishers such as Eskimos and other Native American tribes that consume a diet that is very high in fish and shellfish. This was the highest intake value reported to CSFII. The maximum reported value was about four times the 95th percentile value in the 1989–1991 CSFII survey (USEPA, 1997). The 50th percentile value from the USDA data (32 g/day) was used to estimate the As exposure of the average seafood consumer (USEPA, 1997). The seafood exposure scenarios described above apply only to those individuals who routinely consume fish and/or shellfish as a dietary protein source. Most of the general population consumes fish and shellfish only occasionally, and some individuals never eat fish or shellfish. Thus, the general population has a lower exposure averaged over time. The EPA has previously used a daily fish intake of 6.5 g/day to represent these individuals. This value was derived from the 1973–1974 data from the National Purchase Dairy Survey (USEPA, 1998). USEPA is now recommending a daily intake value of 18.1 g/day (all fish species) for adults in the general population based on the 1989 to 1991 CSFII data. Refer to USEPA (1998) for specifics on the determination of this value.

The seafood ingestion value for the general population is a normalized value which recognizes that, on the days that fish and/or shellfish are consumed, the intake will be higher than 18.1 grams but there will also be many days in the course of a year when there is no consumption of either fish or shellfish. As a worst case, the 95% inorganic As concentration was used for the general population arsenic exposure calculation. Individuals who consume fish or shellfish only occasionally tend to have a few species they favor (e.g., tuna, shrimp) and the species of preference may be among the higher As species.

The exposure to inorganic As from the seafood in the diet was calculated by multiplying the 95%, mean and median inorganic As concentrations in dietary seafoods by the average daily seafood consumption. The results of these calculations are summarized in Table 5.

The data in Table 5 demonstrate that exposure to inorganic As from fish and shellfish is minimal for most segments of the population, even those who eat fish routinely. The only groups with significant inorganic As exposure are subsistence fishers who consistently consume species that are high in inorganic As such as herring, sardines and conger eel. Since subsistence fishermen are dependent on the species available to them, it is expected that the diet will have at least seasonal variety. Even among subsistence groups, the inorganic As exposure decreased by more than 75% when the mean inorganic As was used for the calculation in place of the 95% value.

For an average fish consumer, the inorganic As intake is 4 μg/day when species with high inorganic As concentration are consistently consumed. It drops by an order of magnitude if the median inorganic As value is used for the calculation. A member of the population who consumes fish only occasionally has an estimated normalized exposure of 2 μg/day if that person selects fish or shellfish with high concentrations of inorganic As. Tuna, salmon, cod,

TABLE 5

Inorganic arsenic exposure from fish and shellfish consumption

Consumer Category	Inorganic Arsenic μg/kg	Fish/Shellfish Consumption g/day	Inorganic Arsenic Exposure** μg/day
High Fish–High (95%) Arsenic	120	461*	60
High Fish–Mean Arsenic	26	461*	10
High Fish–Median Arsenic	12	461*	6
Average Fish–High (95%) Arsenic	120	32*	4
Average Fish–Mean Arsenic	26	32*	1
Average Fish–Median Arsenic	12	32*	0.4
General Population	120	18.1*	2

*1989–1991 data from the USDA Continuing Survey of Food Intake (USEPA, 1997).
**Values reported to one significant figure.

haddock, sole and shrimp, which are popular among occasional fish consumers, tend to have inorganic As values closer to the mean concentration than the 95% value. Ingestion of these seafoods would drop the normalized exposure to about 0.5 μg/day.

The estimates for exposure to inorganic As through seafoods presented in Table 5 were based on a diet that was a mix of 90% marine fish and 10% shellfish. Since the concentrations of inorganic As are greater in shellfish than fish, increasing shellfish consumption will increase exposure. Consumption of freshwater fish in place of marine fish will tend to decrease exposure since freshwater species appear to contain far less inorganic As than their marine counterparts.

The arsenic exposure estimates presented in this report mitigate some concerns regarding inorganic As exposure from fish in the diet. However, since inorganic As is only a small portion (4%) of the total As, seafoods do result in a significant exposure to total As. To the extent that organic As compounds are trimethylated species such as AsB, AsC and trimethyl arsine oxide, toxicokinetic data support the conclusion that there is little, if any interaction of the arsenic metabolite with other biomolecules, and, thus, the potential for toxicity is low. However, in cases where DMA is found in fish/shellfish species, low toxicity cannot be assumed because there are some data that suggest that DMA may be a tumor promoter (Yamamoto et al., 1995). If DMA is a promoter, its presence could pose a risk factor for carcinogenicity. The lack of data on the nature of the organoarsenical(s) present in freshwater fish precludes making an estimate of the risk presented from freshwater fish species.

CONCLUSIONS

Data from 99 samples on inorganic As in fish and shellfish support a conclusion that concentrations seldom exceed 4%. Maximum, mean and median inorganic As concentrations for shellfish are greater than those for marine fish. Within the data set evaluated, the distribution of inorganic As is skewed to the high side of the range. Accordingly, the median concentration is a better measure of central tendency than the mean. There were few data on the concentrations of inorganic As in freshwater species. Where there were data for multiple samples of the same species, variability between samples rarely exceed an order of magnitude.

The principal organic As compound in marine fish and shellfish was AsB which accounted for 80% or more of the arsenic in over half of the samples. Much smaller amounts of AsC, DMA, MMA and arsenosugars were present. For freshwater fish there is less data and it is not consistent. Lawrence et al. (1986) reported that the principal organic As species was an unidentified hydrophilic compound while Shiomi et al. (1995) identified it as AsB. In mammalian species, AsB is rapidly absorbed and excreted as the parent compound, decreasing the toxicological concern. The available database, although limited, suggests that AsB and AsC are less toxic than the inorganic As.

Estimates of exposure to inorganic As from seafoods were developed for a diet that was 90% marine fish and 10% shellfish. As expected, subsistence seafood consumers reliant on species with the highest inorganic As concentrations had the highest exposure (~60 µg/day). Substitution of the mean inorganic As concentration in the calculation lowered the exposure to ~10 µg/day and use of the median value lowered it to ~6 µg/day. Average seafood consumers had a maximum exposure of ~4 µg/day and those consumers who do not routinely consume seafoods had a maximum exposure of ~2 µg/day. The data presented in this report illustrate the importance of considering speciation of the arsenic when estimating dietary exposure.

REFERENCES

Alberti, J., Rubio, R., Rauret. G. 1995. Extraction method for arsenic speciation in marine organisms. *J. Anal. Chem.*, **351**, 420–425.

Ballin, U., Kruse, R., Russel, H.S. 1994. Determination of total arsenic and speciation of arseno-betaine in marine fish by means of reaction-headspace gas chromatography utilizing flame-ionization detection and element specific spectrometric detection. *Fresenius J. Anal. Chem.*, **350**, 54–61.

Borum, D.R., Abernathy, C.O. 1994. Human oral exposure to inorganic arsenic. In: W.R. Chappell, C.O. Abernathy and C.R. Cothern (eds.), *Arsenic Exposure and Health*, pp. 21–30. Science and Technology Letters, Northwood, England.

Buchet, J.P., Lison, D., Ruggeri, M., Foa, V., Elia, G. 1996. Assessment of exposure to inorganic arsenic, a human carcinogen, due to the consumption of seafood. *Arch. Toxicol.*, **70**, 773–778.

Buchet, J.P., Pauwels, J., Lauwerys, R. 1994. Assessment of exposure to inorganic arsenic following ingestion of marine organisms by volunteers. *Environ. Res.*, **66**, 44–51.

Cannon, J.R., Saunders, J.B., Tola, R.F. 1983. Isolation and preliminary toxicological evaluation of arseno-betaine—the water-soluble arsenical constituent from the hepatopancreas of the western rock lobster. *Sci. Total Environ.*, **31**, 181–185.

Caroli, S., La Torre, F., Petrucci, F., Violante, N. 1994. On-line speciation of arsenical compounds in fish and mussel extracts by HPLC-ICP-MS. Environ. *Sci. Pollut. Res.*, **1**, 205–208.

Chew, C. 1996. Toxicity and exposure concerns related to arsenic in seafood: An arsenic literature review for risk assessments. Prepared for USEPA Region 10 ESAT, Seattle Washington as Technical Instruction Document 10-9601-815 under Work Unit document 4038.

Done, A.K., Peart, A.J. 1971. Acute toxicities of arsenical herbicides. *Clin. Toxicol.*, **4**, 343–355.

Edmonds, J.S.D., Francesconi, K.A. 1993. Arsenic in seafood: Human health aspects and regulations. *Marine Pollut. Bull.*, **26**, 665–674.

Fielder, R.J., Dale, E.A., Williams, S.D. 1986. *Toxicity Review 16. Inorganic Arsenic Compounds*. HMSO Publications, London.

Grandjean, P., Weihe, P., Needham, L.L., Burse, V.W., Patterson, D.G., Sampson, E.J., Jorgesen, P.J., Vahter, M. 1995. Relation of a seafood diet to mercury, selenium, arsenic, and polychlorinated biphenyl and other organochlorine concentrations in human milk. *Environ. Res.*, **71**, 29–38.

Irgolic, K. 1994. Determination of total arsenic and arsenic compounds in drinking water. In: W.R. Chappell, C.O. Abernathy and C.R. Cothern (eds.), *Arsenic Exposure and Health*, pp. 51–60. Sicence and Technology Letters, Northwood, England.

Kaise, T., Fukui, S. 1992. The chemical form and acute toxicity of arsenic compounds in marine organisms. *Appl. Organomet. Chem.*, **6**, 155–160.

Larsen, E. 1995. Speciation of dimethyl-riboside derivatives (arsenosugars) in marine reference materials by HPLC-ICP-MS. *J. Anal. Chem.*, **352**, 582–588.

Lawrence, J.F., Michalik, P., Tam, G., Conacher, H.B.S. 1986. Identification of arsenobetaine and arseno-choline in Canadian fish and shellfish by high-performance liquid chromatography with atomic absorption detection and confirmation by fast atom bombardment mass spectrometry. *J. Agric. Food Chem.*, **34**, 315–319.

Lopez. J.C., Montoro, R., Cervera, M.L., de la Guardia, M. 1994. Determination of inorganic arsenic in sea food products by microwave-assisted distillation and atomic absorption spectrometry. *J. Anal. Atom. Spectros.*, **9**, 651–656.

Marafante, E., Vahter, M., Dencker, L. 1984. Metabolism of arsenocholine in mice, rats, and rabbits. *Sci. Total Environ.*, **34**, 223–240.

Marafante, E., Vahter, M. 1987. Solubility, retention and metabolism of intratracheally and orally administered inorganic arsenic compounds in the hamster. *Environ. Res.*, **42**, 72–87.

Montgomery, R., Conway, T.W., Spector, A.A. 1990. *Biochemistry: A Case Oriented Approach*. The C.V. Mosby
Company. St. Louis, MO.

Rossman, T.G. 1981. Enhancement of UV-mutagenesis by low concentrations of arsenite in *E. coli. Mutat. Res.*,
91, 207–211.

RTECS. Registry of Toxic Effects of Chemical Substances, 1985–1986.

Sabbioni, E., Fischbach, M., Pozzi, G., Pietra, R., Gallorini, M., Piette, J.L. 1991. Cellular retention, toxicity and
carcinogenic potential of seafood arsenic. I. Lack of cytotoxicity and transforming activity of arsenobetaine
in the BALB/3T3 cell line. *Carcinogenesis*, **12**, 1287–1291.

Shiomi, K., Sugiyama, Y., Shimakura, K., Nagashima, Y. 1995. Arsenobetaine as the major arsenic compund in
the muscle of two species of freshwater fish. *Appl. Organomet. Chem.*, **9**, 105–109. US EPA. 1989. Exposure
Factors Handbook. United States Environmental Protection Agency. Office of Health and Environmental
Assessment. EPA 600 8-89-043.

US EPA. 1997 Draft. Mercury Study Report to Congress, Volume IV: An Assessment of Exposure to Mercury
in the United States. United States Environmental Protection Agency, Office of Air Quality Planning and
Standards; Office of Research and Development EPA 4521/R-97-006.

US EPA. 1998. AWQC Derivation Methodology Human Health Technical support Document. Office of Water. EPA 822-B-98-005

Vahter, M., Marafante, E., Dencker, L. 1983. Metabolism of arsenobetaine in mice, rats and rabbits. *Sci. Total
Environ.*, **30**, 197–211.

Velez, D., Ybanez, N., Montoro, R. 1996. Monomethyarsonic and dimethyarsinic acid contents in seafood. *J.
Agric. Food Chem.*, **44**, 959–864.

Velez, D., Ybanez, N., Montoro, R. 1995. Percentages of total arsenic represented by arsenobetaine levels in
manufactured seafood products. *J. Agric. Food Chem.*, **43**, 1289–1294.

Wang, Z., Rossman, T.G. 1996. The carcinogenicity of arsenic. In: L.W. Chang, L. Magos and T. Suzuki (eds.),
Toxicology of Metals, pp. 221–229. CRC Press Lewis Publishers, Boca Raton, FL.

Yamamoto, S., Konishi, Y., Matsuda, T., Shibata, M.A., Matsui-Yausa, I., Otani, S., Kuroda, K., Endo, G.,
Fukushima, S. 1995. Cancer induction by an organic arsenic compound, dimethylarsinic acid (cacodylic
acid) in F344/DuCrj rats after pretreatment with five carcinogens. *Cancer Res.*, **55**, 1271–1276.

Yamauchi, Y., Kaise, T. and Yamamura, Y. 1986. Metabolism and excretion of orally administered arsenobetaine in the hamster. *Bull. Environ. Contam. Toxicol.*, **36**, 350–355.

Arsenic Exposure and Health Effects
W.R. Chappell, C.O. Abernathy and R.L. Calderon (Editors)

Application of the Risk Assessment Approaches in the USEPA Proposed Cancer Guidelines to Inorganic Arsenic

Harvey J. Clewell, Annette M. Shipp, Melvin E. Andersen,
Janice W. Yager, Kenny S. Crump

ABSTRACT

There is convincing evidence from a number of epidemiological studies that exposure to inorganic arsenic in air or in drinking water has been associated with an increased incidence of cancer in exposed populations. However, there is less agreement on two critical risk assessment issues: (1) quantification of the dose–response for these exposed populations, especially in the low-dose region, and (2) extrapolation of that dose–response relationship to current exposures of the public in air and drinking water. Recent evaluations of the biological basis for inorganic arsenic carcinogenicity have suggested that the mode of action is nonlinear and may even have an effective threshold. Therefore, under the new USEPA proposed cancer guidelines, inorganic arsenic might better be evaluated using the margin of exposure (MOE) approach rather than linear extrapolation as is now used. The purpose of this investigation was to explore the application of the MOE approach to inorganic arsenic using epidemiological data for both oral and inhalation routes of exposure. Unfortunately, while qualitative data support a nonlinear mode of action for the carcinogenicity of inorganic arsenic, quantitative data are inadequate to support a determination of the exposure levels at which nonlinearity might occur. On the basis of benchmark analyses of recently published epidemiological data it would appear that the risk of cancer from lifetime consumption of inorganic arsenic in drinking water at the current Maximum Contaminant Level (MCL) may be significant. In contrast, current environmental arsenic inhalation exposures appear unlikely to entail significant risks of cancer.

Keywords: arsenic, cancer, risk assessment, benchmark dose

INTRODUCTION

The recently proposed USEPA (1996) guidelines for carcinogen risk assessment provide a major departure from the prior approach in which low-dose linearity and a lack of a threshold were assumed for any potential carcinogenic agent. In the new guidelines, both mode of action and pharmacokinetic data are explicitly considered in the hazard assessment and are factored into a weight-of-evidence evaluation of the conditions under which the chemical should be considered to be a human carcinogen. In the dose–response assessment step, a mathematical dose–response model is applied to estimate the 95% lower bound on the dose associated with a specific increase in lifetime risk, e.g., one in ten (LED_{10}). This dose level then serves as the point of departure for extrapolation. In the case of a carcinogen that is presumed to act by a linear mechanism, the dose or intake at an acceptable level of risk, which would be defined by the risk manager, is estimated by linear extrapolation from the point of departure to the origin. However, if the mode of action is presumed to be nonlinear, then the point of departure is simply compared with anticipated environmental exposure levels; the resulting ratio is termed the Margin of Exposure (MOE); the adequacy of the MOE is evaluated by the risk manager.

Inorganic arsenic is classified as a human carcinogen by both the oral and inhalation routes of exposure. There is convincing evidence from a number of epidemiological studies that exposure to inorganic arsenic in air or in drinking water has been associated with an increased incidence of cancer in exposed populations. Recent evaluations of the biological basis for inorganic arsenic carcinogenicity have suggested that the mode of action is nonlinear and may even have an effective threshold. If that is the case, then under the new cancer guidelines, inorganic arsenic might better be evaluated using the MOE approach rather than linear extrapolation as is now used. The purpose of this investigation was to explore the application of the new USEPA (1996) guidelines to inorganic arsenic.

APPROACH

The investigation focused on several of the most recently published studies providing evidence of an association between inorganic arsenic exposure in air or drinking water and the incidence of cancer. Skin tumors were not considered. The approach and steps taken are those outlined in USEPA (1996).

Critical Study Selected — Inhalation Exposure

Enterline et al. 1995: The mortality experience of 2802 men exposed to inorganic arsenic at a copper smelter for at least a year during the period 1940–64 was evaluated, with follow-up through 1986. Exposure to inorganic arsenic was estimated from air samples taken starting in 1938 and from measurements of urinary arsenic measured in workers starting in 1948. Significant excesses in mortality from all malignant neoplasms, cancer of the respiratory system, cancer of the large intestine, and bone cancer were reported.

Critical Studies Selected — Oral Exposure

Bates et al. 1995: Data from the Utah respondents to the National Bladder Cancer Study conducted in 1978 were analyzed. Drinking water concentrations were in the range 0.0005–0.16 mg/L. No statistically significant association between inorganic arsenic exposure and bladder cancer was detected in the original analysis of the data. However, using the benchmark dose methodology described in this paper, it was possible to obtain a quantitative estimate of the dose–response relationship between bladder cancer and inorganic arsenic exposure.

Chen et al. 1992: An association between inorganic arsenic exposure and cancer mortality for the liver, lung, bladder, and kidney was demonstrated for residents in southwestern Taiwan. Inorganic arsenic levels in drinking water ranged from 0.01 to 1.752 mg/L.

Chiou et al. 1995: Internal cancer incidence for individuals with blackfoot disease in southwestern Taiwan was found to be increased compared to healthy residents of the same area in a 7-year prospective study.

Guo et al. 1994: An association between inorganic arsenic well-water concentrations and the incidence of urinary bladder and kidney cancer was demonstrated using data on 243 townships in Taiwan.

Tsuda et al. 1995: An association between inorganic arsenic exposure and cancer mortality for lung and urinary tract was demonstrated for Japanese residents exposed from 1955 to 1959 to inorganic arsenic drinking water concentrations of up to 3 mg/L, with follow-up through 1992.

Evaluation of Other Key Data

A review of the literature, as summarized in Clewell et al. (1999) concluded the following:

- Saturation of metabolism in the dose-range associated with tumors does not appear to be adequate to produce a major impact on the dose–response;
- The overall body of literature is most consistent with a central role for arsenite acting as a co-mutagen;
- The activity of arsenite appears to result from an interaction with proteins responsible for DNA repair and replication leading to increased likelihood of DNA strand breaks and chromosomal aberrations, thereby resulting in genomic instability;
- A co-mutagenicity mode of action is proposed in which inorganic arsenic acts primarily on intermediate cells deficient in cell cycle control at a late stage in a pre-existing carcinogenic process. This interaction enhances genomic fragility and accelerates conversion of pre-malignant lesions to more aggressive, clinically observable tumors.

Extrapolation Approach

The data strongly suggest that the mode of action for inorganic arsenic is nonlinear and may even have an effective threshold, in which case extrapolation should be based on the Margin of Exposure approach. However, at the present time there is no indication at what dose this threshold may exist, and indeed the dose–response curve may be linear in the dose range of interest. Therefore, both linear and MOE approaches were applied using the LED_{10} as the point of departure.

The benchmark dose, or ED_{10}, was defined as the concentration of inorganic arsenic in air ($\mu g/m^3$) or water (mg/L) from which a person exposed constantly over a lifetime would have a 10% increased risk of dying of the cancer endpoint being modeled. The LED_{10} is the 95% statistical lower bound on the ED_{10}. For each study and endpoint, a linear relative risk model was used to describe cancer incidence or mortality in terms of cumulative inorganic arsenic exposure.

RESULTS

Inhalation

The highest relative risk in the Enterline et al. 1995 study was observed for respiratory cancer mortality. The ED_{10} and LED_{10} obtained for this endpoint were 0.354 mg/m^3 and 0.168 mg/m^3, respectively.

Oral

The results of the benchmark modeling for the drinking water studies are summarized in Table 1. Although models assume both a zero lag and a 15-year lag (assuming exposures in the most recent 15 years do not affect risk of mortality), these different lags gave very similar results in all cases. Consequently, only results for zero lag are presented in Table 1.

TABLE 1

EDs and LEDs corresponding to 10% additional risk from lifetime exposure to inorganic arsenic in air or drinking water

Cancer	Study	ED_{10} (mg/L)	LED_{10} (mg/L)
All	Chiou et al. (1995)	0.251	N/A*
	Tsuda et al. (1995)	0.0259	0.0163
Lung	Chen et al. (1992), females	0.654	0.510
	Chen et al. (1992), males	0.370	0.293
	Chiou et al. (1995)	0.399	N/A
	Tsuda et al. (1995)	0.0174	0.0101
Urinary	Tsuda et al. (1995)	0.0621	0.0267
Bladder	Bates et al. (1995)	0.131	N/A
	Chen et al. (1992), females	4.91	2.35
	Chen et al. (1992), males	1.72	0.698
	Chiou et al. (1995)	4.95	N/A
	Guo et al. (1994), females	6.6	N/A
	Guo et al. (1994), males	4.7	N/A
Kidney	Chen et al. (1992), females	0.190	0.144
	Chen et al. (1992), males	4.96	1.16
	Guo et al. (1994), females	24.6	N/A
	Guo et al. (1994), males	22.7	N/A
Liver	Chen et al. (1992), females	1.38	0.492
	Chen et al. (1992), males	0.351	0.247
	Tsuda et al. (1995)	0.8266	0.2762
Uterine	Tsuda et al. (1995)	0.2977	0.1059

*Could not be calculated due to inadequate presentation of data

In all of the analyses reported in Table 1, an acceptable fit was obtained using a linear model. Oral ED_{10}s ranged from 0.02 to 25 mg/L, and LED_{10}s ranged from 0.01 to 2.35 mg/L. The lowest ED_{10} and LED_{10} (highest potency) were obtained for lung cancer mortality in the Japanese cohort which had been exposed for only 5 years (Tsuda et al., 1995). Potencies roughly an order of magnitude lower were obtained for kidney cancer mortality in Taiwanese females (Chen et al., 1992) and for bladder cancer in Utah (Bates et al., 1995).

The results of these benchmark calculations are compared with the current drinking water standard and USEPA unit risks in Table 2.

TABLE 2

Comparison of risk estimates for inorganic arsenic exposure

Route	LED_{10}	MCL[a]	LED_{10}–Linear Unit Risk	Current USEPA Unit Risk
Oral	10 μg/L	50 μg/L	$0.01\ (\mu g/L)^{-1}$	$5 \times 10^{-5}\ (\mu g/L)^{-1}$ [b]
	144 μg/L		$7 \times 10^{-4}\ (\mu g/L)^{-1}$	
Inhalation	168 μg/m³	NA	$6 \times 10^{-4} (\mu g/m^3)^{-1}$	$4.3 \times 10^{-3} (\mu g/m^3)^{-1}$

(a) Maximum Contaminant Level for drinking water promulgated by the USEPA.
(b) Based on skin tumor incidence (all others based on internal cancers).

DISCUSSION

In the case of inorganic arsenic, compelling qualitative data suggest that the carcinogenic mode of action is fundamentally nonlinear (Clewell et al., 1999). However, departure from the linear default requires quantitative evidence for a nonlinear cancer dose–response between the exposure levels associated with tumors (i.e., the $LED_{10}s$) and the exposure levels currently of concern (e.g., community air or drinking water concentrations). No such data are available for inorganic arsenic.

The proposed USEPA (1996) guidelines provide essentially no guidance for evaluating the adequacy of an MOE for the case of human cancer mortality data. For inhalation exposures to inorganic arsenic in ambient air, it is likely that MOEs well above 1000 would generally be obtained. In the case of drinking water exposures, the lowest LED_{10}, 0.01 mg/L, is from a study (Tsuda et al., 1995) in which the exposure was actually to much higher concentrations (around 1 mg/L) for a short period (about 5 years). If the dose–response for the carcinogenicity of inorganic arsenic is indeed strongly nonlinear between 1 and 0.01 mg/L, the tumor response for a given cumulative inorganic arsenic exposure in this short-term study may greatly overestimate the risk for a lifetime exposure at a lower concentration which would be associated with the same cumulative inorganic arsenic dose.

Nevertheless, if the lowest LED_{10} for chronic exposure, 0.144 mg/L is used, the MOE for the current MCL is only a factor of 3. Use of this LED_{10}, which was obtained for kidney cancer mortality in Taiwanese females (Chen et al., 1992), would also be consistent with the ED_{10} of 0.131 mg/L obtained for bladder cancer in Utah (Bates et al., 1995). Although the latter study failed to achieve a statistically significant association between bladder cancer and inorganic arsenic exposure, a positive trend was identified in the case of smokers.

CONCLUSIONS

1. Although qualitative data support a nonlinear mode of action for the carcinogenicity of inorganic arsenic, quantitative data are inadequate to support a determination of the exposure levels at which nonlinearity might occur.

2. On the basis of currently available epidemiological data it would appear that the risk of cancer from lifetime consumption of inorganic arsenic in drinking water at the current Maximum Contaminant Level (MCL) may be significant. The lowest LED_{10} for lung cancer from the ingestion of inorganic arsenic in drinking water over a lifetime is a factor of 5 *lower* than the current MCL. This estimate is based on a relatively short (5-year) drinking water exposure in Japan (Tsuda et al., 1995), and could overestimate the potency for lifetime exposure. However, $ED_{10}s$ and $LED_{10}s$ from chronically-exposed populations in the U.S. and Taiwan (Bates et al., 1995; Chen et al., 1992) are less than a factor of 3 higher than the MCL.

3. In contrast with the results for drinking water, the LED_{10} of 168 $\mu g/m^3$ obtained for inhalation suggests that current environmental inorganic arsenic exposures are associated with MOEs of greater than 1000, and are unlikely to entail significant risks of cancer.

ACKNOWLEDGMENT

This work was supported by funds provided by the Electric Power Research Institute.

REFERENCES

Bates, M.N., Smith, A.H., Cantor, K.P. 1995. Case-control study of bladder cancer and arsenic in drinking water. *Am. J. Epidemiol.*, **141**, 523–530.

Chen, C.-J., Chen, C., Wu, M.-M., Kuo, T.-L. 1992. Cancer potential in liver, lung, bladder and kidney due to ingested inorganic arsenic in drinking water. *Br. J. Cancer*, **66**, 888–892.

Chiou, H.-Y., Msueh, Y.-M., Liaw, K.-.F, Horng, S.-F., Chiang, M.-H., Pu, Y.-S., Lin, J., Huang, C.-H., Chen, C.-J. 1995. Incidence of internal cancers and ingested inorganic arsenic: a seven-year follow-up study in Taiwan. *Cancer Res.*, **55**, 1296–1300.

Clewell, H.J., Gentry, P.R., Barton, H.A., Shipp, A.M., Yager, J.W., Andersen, M.E. 1999. Requirements for a biologically-realistic cancer risk assessment for inorganic arsenic. *Int. J. Toxicol.*, **18**, 131–147.

Enterline, P., Day, R., Marsh, G. 1995. Cancers related to exposure to arsenic at a copper smelter. *Occup. Environ. Med.*, **52**, 28–32.

Guo, H.-R., Chiang, H.-S., Hu, H., Lipsitz, S.R., Monson, R.R. 1994. Arsenic in drinking water and urinary cancers: a preliminary report. In: W.R. Chappell, C.O. Abernathy, and C.R. Cothern (eds). *Arsenic. Exposure and Health.* pp. 119–124. Science and Technology Letters, Northwood, U.K.

Tsuda, T., Babazono, A., Yamamoto, E., Kurumatani, N., Mino, Y., Ogawa, T., Kishi, Y., Aoyama, H. 1995. Ingested arsenic and internal cancer: a historical cohort study followed for 33 years. *Am. J. Epidemiol.*, **141**, 198–209.

USEPA. 1996. *Proposed Guidelines for Carcinogen Risk Assessment.* EPA/600/P-92/003C. Office of Research and Development, Washington, DC.

APPENDIX—DOCUMENTATION OF ANALYTICAL APPROACH

Parameters in a dose–response model were estimated by maximizing the likelihood of the data. The solver routine in Excel was used for most calculations.

The ED_{10} for a particular cancer category was defined as the concentration of inorganic arsenic in air ($\mu g/m^3$) or water (mg/L) to which a person would be exposed to constantly over a lifetime that would increase their risk of dying of that type of cancer by 10%, i.e., as the concentration, e, that satisfies

$$LR(e) - LR(0) = 0.1$$

where $LR(e)$ is the probability of dying of the cancer of interest when exposed throughout life to e $\mu g/m^3$ inorganic arsenic in air or e mg/L inorganic arsenic in drinking water. In most cases, $LR(e)$ was calculated using a life table approach. In this approach, age was divided into 18 intervals: 1. [1 ≤ age < 5]; 2. [5 ≤ age < 10]; 3. [10 ≤ age < 15]; ... 17. [80 ≤ age < 85]; 18. [85 ≤ age]. The lifetime risk, $LR(e)$, was calculated as

$$LR(e) = \sum_{\kappa} P_{\kappa} f_{\kappa}(e)$$

where the sum is over the 18 age intervals, P_i is the probability of surviving until the beginning of the ith age interval, computed recursively by $P_1 \equiv 1$,

$$P_i = P_{i-1}[1 - \Lambda_{i-1}b_{i-1}], i = 2, \ldots 18,$$

where b_i is the death rate for all causes per year for the ith age interval, $f_i(e)$ is the probability of dying of the cancer of interest during the ith age interval, conditional on surviving until the beginning of the interval, and Λ_i is the width of the i-th age interval ($\Lambda_1 = 4$, $\Lambda_i = 5$, $i = 2, \ldots,$ 18). $f_i(0)$ is given by $\Lambda_i a_i$, where a_i is the background death rate per year from the cancer of interest during the i-th age interval. For $e > 0$, $f_i(e)$ is the modification of $\Lambda_i a_i$ determined by the dose–response model used to model the arsenic effect. In the calculations herein, the b_i and a_i are U.S. death rates for 1989, either for males, females or both. Although P_i depends slightly on e (The probability of surviving the i-th age interval becomes smaller as the probability of dying of the cancer of interest increases.), this effect will be small as long as the increased risk is no greater than 0.1. In order to simplify computations, this dependence of P_i on e was not modeled.

To determine the statistical lower bound (the LED), the equation, $LR(e) - LR(0) = 0.1$, was solved explicitly for e and the LED was computed as the solution to the following optimization problem: Minimize e by modifying the parameters in the dose–response model subject to the constraint $2(L_{\text{Max}} - L) \le 2.7055$, where L_{Max} is the maximum value of the

TABLE A-1

Lung cancer mortality data vs. Inorganic arsenic exposure (Enterline et al., 1995)

Mean Inorganic Arsenic Exposure (mg-yr/m³)	Lung Cancer Deaths			ED_{10} ($\mu g/m_3$)	LED_{10} ($\mu g/m_3$)
	Observed	Expected	Predicted under model	354	168
0.405	22	14.3	26.1		
1.31	30	17.1	31.9		
2.93	36	17.2	33.1		
5.71	36	17.0	34.6		
12.3	39	15.5	35.4		
28.3	20	7.04	20.5		
59.0	5	1.58	6.5		

(Chi-square goodness of fit p-value (5 d.f.) = 0.51).

log-likelihood corresponding to the maximum likelihood estimates of the model parameters, L is the log-likelihood as a function of the parameters of the dose–response model, and 2.7055 is the 90% percentage point of the chi-square distribution with one degree of freedom.

In application of this methodology to different arsenic data sets, different analytic expressions were used for $f_i(e)$ to account for the different forms of the data present in different studies, which required that different modeling approaches be applied. Specific expressions for $f_i(e)$ used in specific studies are described below.

Enterline et al.

Enterline *et al.* (1995), in their Table 2, present mean exposure ($\mu g/m^3$-years inorganic arsenic in air) and observed and expected cases (based on rates for Washington State white males) of lung cancer for the entire cohort for seven groups of person-years categorized by cumulative arsenic exposure (Table A-1). The observed number of cases of lung cancer in a particular cumulative exposure group was modeled as a Poisson exposure variable with expected value

$$c * EXP_0 * (1 + \beta X),$$

where EXP_0 is the expected number of lung cancers based on the comparison population (Washington State), X is the average cumulative exposure to inorganic arsenic ($\mu g/m^3$-years) for the group, and c and β are parameters estimated from the data by maximum likelihood. Table A-1 displays the data from Enterline et al. used in the modeling, the numbers of lung cancers predicted in exposure group by this model and the p-value from a standard chi-square goodness of fit of the model to the data. This model provided a good fit to these data ($p = 0.51$).

The ED_{10} and LED_{10} were calculated using the life table approach described above. The probability, $f_i(e)$, of dying of lung cancer during the i-th age interval, given survival to the beginning of that interval, corresponding to the model used to fit the data is

$$f_i(e) = \Lambda_i * a_i * (1 + 3 * \beta * e * t_i),$$

where t_i is the mid-point of the i-th age interval. The factor of 3 converts from occupational to continuous exposure. (Assuming a worker works for 240 days per year and breathes half as much during a shift as person normally breath in a 24-hour period, $2*365/240 = 3.0$). The resulting ED_{10} and LED_{10} are listed in Table A-1.

Tsuda et al.

Tsuda et al. (1995), in their Table 3, report observed and expected cases for all cancers, lung cancer, urinary cancer, liver cancer, uterine cancer and colon cancer among 443 persons followed for 33 years (from 1959 through 1992) following five years of drinking contaminated well water. Expected incidence was based on sex–age-specific rates between 1960 and 1989 for either Japan (urinary cancer, which included cancer of the bladder and renal pelvis) or Niigata Prefecture (all other cancers). These data were classified by the average arsenic concentration in a person's well water: less than 0.05 mg/L, 0.05 mg/L to 0.99 mg/L and greater than or equal to 1 mg/L. The average concentrations in these categories were assumed to be 0 mg/L, 0.5 mg/L, and 2 mg/L, respectively. (The maximum concentration in any well was 3 mg/L.)

The observed number of cases of a particular category of cancer was modeled as a Poisson variable with expected value

$$EXP_0*(1 + \beta X),$$

where EXP_0 is the expected number of cancers based on rates for Japan or Niigata Prefecture, and X is the estimated average cumulative exposure to inorganic arsenic in mg/L-years (computed by multiplying the average well water concentration by 5 years). Table A-2 displays the data from Tsuda et al. for each cancer category, the number of cancers predicted

TABLE A-2

Model fits to data in Tsuda et al. (1995) and corresponding ED_{10} and LED_{10}

	Arsenic conc. (mg/L)	Average cumulative exposure (mg/L yrs)	Observed	Expected	Expected under model	No Lag ED_{10} (mg/L)	No Lag LED_{10} (mg/L)	15 Year Lag ED_{10} (mg/L)	15 Year Lag LED_{10} (mg/L)
All Cancers	< 0.05	0	11	14.1	14.1	0.0259	0.0163	0.0328	0.0207
	0.05–0.99	2.5	5	3.86	6.26				
	1.0+	10	18	4.96	17.3				
Chi-square goodness of fit p-value (1 d.f.)					0.33				
Lung Cancer	< 0.05	0	0	1.55	1.6	0.0174	0.0101	0.0221	0.0129
	0.05–0.99	2.5	1	0.43	1.86				
	1.0+	10	8	0.51	7.3				
Chi-square goodness of fit p-value (1 d.f.)					0.16				
Urinary Cancer	< 0.05	0	0	0.3	0.300	0.0621	0.0267	0.0780	0.0336
	0.05–0.99	2.5	0	0.08	0.560				
	1.0+	10	3	0.1	2.50				
Chi-square goodness of fit p-value (1 d.f.)					0.33				
Liver Cancer	< 0.05	0	0	0.85	0.850	0.8266	0.2762	1.0476	0.3500
	0.05–0.99	2.5	0	0.25	0.552				
	1.0+	10	2	0.28	1.64				
Chi-square goodness of fit p-value (1 d.f.)					0.22				
Uterine Cancer	< 0.05	0	0	0.47	0.470	0.2977	0.1059	0.3760	0.1338
	0.05–0.99	2.5	0	0.1	0.361				
	1.0+	10	2	0.15	1.71				
Chi-square goodness of fit p-value (1 d.f.)					0.35				
Colon Cancer	< 0.05	0	2	0.67	0.670	Infinite	0.1185	Infinite	0.1486
	0.05–0.99	2.5	0	0.17	0.170				
	1.0+	10	0	0.22	0.220				
Chi-square goodness of fit p-value (1 d.f.)					0.082				

by the above model (expected number under model) and the *p*-value from a standard chi-square test of the model fit to the data. The model provided an acceptable fit ($p > 0.05$) in each case.

The ED_{10} and LED_{10} were calculated using the life table approach described above. In implementing this approach, $f_i(e)$, the probability of dying of the cancer of interest during the *i*-th age interval, was assumed to be of the form

$$f_i = \Lambda_i{}^*a_i{}^*[1 + \beta^*e^*Max(t_i - Lag,0)],$$

where t_i is the mid-point of the *i*-th age interval. Two values for Lag were considered: Lag = 0 and Lag = 15 years. The use of the value, Lag = 15, is consistent with the assumption that inorganic arsenic exposure during the most recent 15 years of life does not affect the mortality rate. Since no lag was incorporated into the modeling of the mortality data (due to the fact that the data were not presented in a suitable form), use of a 15-year lag carries the implicit assumption that no excess cancers were observed during the first 15 years following cessation of exposure. The resulting ED and LED are listed in Table A-2.

Chen et al.

Chen et al. (1992) studied the relationship of liver cancer, lung cancer, bladder cancer, and kidney cancer to inorganic arsenic in drinking water. For each cancer, the number of deaths during a 13-year period were recorded, cross-classified by sex, age (<30, 30–49, 50–69, 70+) and inorganic arsenic level (<0.10, 0.10–0.29, 0.30–0.59, 0.60+ mg/L). The numbers of person-years of observation were similarly cross-classified. The assumed average ages in the four age categories were 25, 40, 60, and 80, respectively, and the assumed average inorganic arsenic water concentrations in the four water categories were 0.02, 0.2, 0.45 and 1.0 mg/L, respectively.

The data for males and females were modeled separately. The relative risk of dying of the cancer of interest was assumed to depend linearly upon cumulative exposure lagged either zero (Lag = 0) or 15 years (Lag = 15). Specifically, for a given gender, the number of cancer deaths in the *i*-th age and *j*-th exposure category was modeled as a Poisson variable with expected value

$$PY_{ij}{}^*a_i[1 + \beta^*(t_i - Lag)^*e_j],$$

where PY_{ij} is the number of person-years of observation in *ij*-th cell, t_i is the mid-point of the *i*-th age interval, e_j is the assumed average inorganic arsenic water concentration in the *j*-th exposure category. The use of the value, Lag = 15, is consistent with the assumption that inorganic arsenic exposure during the most recent 15 years of life does not affect the mortality rate. Five parameters were estimated from the data: a_i, the background death rate per person-year from the cancer of interest in the *i*-th age category, for $i = 1,2,3,4$, and β, the parameter that gauges the carcinogenic potency of inorganic arsenic.

Table A-3 displays, for both Lag = 0 and Lag = 15 years, the observed number of cancers in each cell, the number of cancers predicted by the above model from Chen et al. for each cancer category and each gender, the number of cancers predicted by the above model (expected number under model), and the *p*-value from a standard chi-square test of the model fit to the data. The model provided an acceptable fit ($p > 0.05$) in each of these eight cases.

The EDs and LEDs were calculated using the life table approach described above. In implementing this approach, f_i, the probability of dying of the cancer of interest during the *i*-th age interval, given survival to the beginning of the interval was modeled in exactly the same manner as described earlier for the Tsuda et al. data. Since male and female data were modeled separately, sex-specific background rates were used in the life table calculation. The resulting ED_{10} and LED_{10} are listed in Table A-3.

TABLE A-3a

Model fits to data for males in Chen et al. (1992) and corresponding ED_{10} and LED_{10}

Arsenic conc. (mg/L)	Average cumulative exposure (mg/L yrs)	Observed for Age:				No lag — Expected Under Model for:				ED_{10} (mg/L)	LED_{10} (mg/L)	15 Year lag — Expected Under Model for:				ED_{10} (mg/L)	LED_{10} (mg/L)
		<30	30-49	50-69	70+	<30	30-49	50-69	70+			<30	30-49	50-69	70+		
Liver Cancer < 0.1	0.05	1	13	20	4	1.53	12.4	21.7	4.24	0.351	0.247	1.64	12.8	21.5	4.12	0.256	0.184
0.1 - 0.29	0.2	0	7	15	4	0.874	7.11	13.1	2.64			0.901	7.19	1.31	2.62		
0.3 - 0.59	0.4	3	11	26	6	1.54	12.9	25.8	5.55			1.52	12.8	25.8	5.61		
0.6 +	1	1	11	17	1	1.05	9.60	17.4	2.56			0.937	9.23	17.6	2.65		
Death Rate (a_j) * 100,000						1.34	34.4	94.3	85.8			1.45	35.6	93.5	82.6		
Chi-square goodness of fit *p*-value (11 d.f.)						0.93						0.91					
Lung Cancer < 0.1	0.05	1	3	20	14	0.830	5.13	23.2	11.2	0.370	0.293	0.929	5.41	23.4	10.9	0.359	0.293
0.1 - 0.29	0.2	0	2	13	11	0.503	3.21	15.8	7.93			0.527	3.28	15.8	7.89		
0.3 - 0.59	0.4	1	11	40	16	0.945	6.31	33.9	18.4			0.924	6.25	33.9	18.5		
0.6 +	1	1	4	26	6	0.723	5.34	26.1	9.56			0.620	5.06	26.0	9.74		
Death Rate (a_j) * 100,000						0.706	13.7	95.8	210			0.812	14.7	96.4	203		
Chi-square goodness of fit *p*-value (11 d.f.)						0.41						0.35					
Bladder Cancer < 0.1	0.05	0	1	8	10	0	1.85	9.03	5.34	1.72	0.698	0	1.99	9.03	5.17	1.61	0.719
0.1 - 0.29	0.2	0	0	6	2	0	1.45	7.96	5.04			0	1.48	7.96	5.01		
0.3 - 0.59	0.4	0	5	22	11	0	3.32	20.0	13.5			0	3.28	20/0	13.6		
0.6 +	1	0	4	19	9	0	3.38	18.0	8.10			0	3.24	18.0	8.22		
Death Rate (a_j) * 100,000						0	4.39	31.5	81.6			0	4.88	31.5	77.0		
Chi-square goodness of fit *p*-value (11 d.f.)						0.52						0.46					
Kidney Cancer < 0.1	0.05	0	0	4	3	0	0.493	4.04	2.26	4.96	1.16	0	0.527	4.10	2.23	4.94	1.15
0.1 - 0.29	0.2	0	0	2	1	0	0.317	2.83	1.67			0	0.324	2.84	166		
0.3 - 0.59	0.4	0	2	9	2	0	0.636	6.22	3.96			0	0.628	6.21	3.97		
0.6 +	1	0	0	3	4	0	0.554	4.91	2.11			0	0.521	4.86	2.13		
Death Rate (a_j) * 100,000						0	1.30	16.3	41.8			0	1.42	16.7	40.9		
Chi-square goodness of fit *p*-value (11 d.f.)						0.56						0.56					

TABLE A-3b

Model fits to data for females in Chen et al. (1992) and corresponding ED_{10} and LED_{10}

Arsenic conc. (mg/L)	Average cumulative exposure (mg/L yrs)	Observed for Age:				No lag						15 Year lag					
						Expected Under Model for:				ED_{10} (mg/L)	LED_{10} (mg/L)	Expected Under Model for:				ED_{10} (mg/L)	LED_{10} (mg/L)
		<30	30-49	50-69	70+	<30	30-49	50-69	70+			<30	30-49	50-69	70+		
Liver Cancer																	
<0.1	0.05	0	3	9	4	0.311	4.73	9.23	3.09	1.38	0.492	0.333	4.91	9.28	3.05	1.06	0.386
0.1-0.29	0.2	0	2	8	2	0.174	2.65	5.85	2.28			0.179	2.69	5.86	2.27		
0.3-0.59	0.4	1	7	10	3	0.305	4.88	11.3	4.04			0.300	4.85	11.3	4.07		
0.6+		0	4	7	2	0.210	3.74	7.64	1.58			0.187	3.56	7.59	1.61		
Death Rate (a_i) * 100,000						0.307	13.9	39.5	54.9			0.334	14.5	39.8	53.8		
Chi-square goodness of fit p-value (11 d.f.)						0.89						0.88					
Lung Cancer																	
<0.1	0.05	1	3	20	14	0.539	6.08	18.5	4.89	0.654	0.510	0.605	6.35	18.2	4.69	0.601	0.304
0.1-0.29	0.2	0	2	13	11	0.328	3.84	13.7	4.33			0.344	3.90	13.6	4.30		
0.3-0.59	0.4	1	11	40	16	0.626	7.90	29.9	8.76			0.612	7.84	30.0	8.86		
0.6+	1	1	4	26	6	0.507	7.18	24.0	4.02			0.438	6.91	24.2	4.15		
Death Rate (a_i) * 100,000						0.514	16.9	73.3	78.7			0.594	17.9	71.8	73.7		
Chi-square goodness of fit p-value (11 d.f.)						0.72						0.80					
Bladder Cancer																	
<0.1	0.05	0	3	11	9	0	1.01	11.9	5.84	4.91	2.35	0	1.08	11.9	5.67	4.58	2.16
0.1-0.29	0.2	0	0	8	2	0	0.721	10.2	6.08			0	0.738	10.2	6.05		
0.3-0.59	0.4	0	0	22	15	0	1.62	24.4	13.4			0	1.61	24.4	13.5		
0.6+	1	0	2	27	6	0	1.65	21.5	6.69			0	1.57	21.5	6.80		
Death Rate (a_i) * 100,000						0	2.64	43.0	83.8			0	2.90	43.1	79.6		
Chi-square goodness of fit p-value (11 d.f.)						0.28						0.30					
Kidney Cancer																	
<0.1	0.05	0	0	1	1	0	0.473	1.13	0.802	0.190	0.144	0	0.452	1.08	0.772	0.108	0.0998
0.1-0.29	0.2	0	1	3	1	0	0.798	2.14	1.72			0	0.792	2.13	1.72		
0.3-0.59	0.4	0	2	7	5	0	2.46	6.61	4.71			0	2.47	6.62	4.73		
0.6+	1	0	4	6	3	0	3.27	7.14	2.76			0	3.29	7.18	2.78		
Death Rate (a_i) * 100,000						0	0.277	0.702	1.64			0	0.200	0.423	0.911		
Chi-square goodness of fit p-value (11 d.f.)						1.0						1.0					

TABLE A-4

Model fits to data in Chiou et al. (1995) and corresponding ED_{10}

	Cumulative arsenic exposure (mg/L yrs)	Relative risk	95% Lower bound	95% Upper bound	Ln(RR)	Ln(LB)	Ln(UB)	Standard error	ED_{10} (mg/L)
All Sites	0	1							0.251
	10	1.6	1	2.7	0.470	0.0	0.993	0.253	
	30	2	1.2	3.6	0.693	0.182	1.28	0.280	
Lung Cancer	0	1							0.399
	10	3.1	0.8	12.2	1.13	-0.223	2.50	0.695	
	30	4.7	1.2	18.9	1.55	0.182	2.94	0.703	
Bladder Cancer	0	1							4.95
	10	2.1	0.6	7.2	0.742	-0.511	1.97	0.634	
	30	5.1	1.5	17.3	1.63	0.405	2.85	0.624	

Chiou et al.

In their Table 1, Chiou et al. (1995) presented relative risks and corresponding 95% CI (obtained by Cox regression) for cancer at all sites, lung cancer and bladder cancer, adjusted for age, sex, and cigarette smoking, and classified by cumulative exposure to inorganic arsenic (0, 0.1–19.9, and 20+ mg/L-years). The average inorganic arsenic exposures in these groups were assumed to be 0, 10 and 30 mg/L-years, respectively. The relative risks and 95% lower and upper bounds were log-transformed. The standard error associated with each log-transformed relative risk was estimated as the width of the 95% log-transformed confidence interval divided by 2 and by 1.96, the normal variate corresponding to 97.5% probability. A summary of these calculations is provided in Table A-4.

The log-transformed relative risks were then fit in a weighted regression with the intercept fixed at zero, and the weights being the inverses of the estimated standard errors. The slope, β, from this regression was used to compute the ED using the same life table approach as described with the analysis of the Tsuda et al. data. These calculations include the approximation $e^{\beta X} = 1 + \beta X$, which should be reasonably accurate at the level of increased risk of interest (0.1). Due to limitations in the data provided by Chiou et al., LED were not determined from these data.

Bates et al.

In their Table 3, Bates et al. presented odds ratios and 90% confidence intervals for bladder cancer, adjusted for sex, age, smoking, years of exposure to chlorinated surface water, history of bladder infection, urbanization and ever employed in a high-risk occupation, and classified by cumulative exposure to inorganic arsenic in water (less than 19 mg, 19 to 33 mg, 33 to 53 mg, and greater than or equal to 53 mg). The average exposure in each category was assumed to be 0, 26, 43 and 80 mg, respectively. In order to make the exposures commensurate with those available from other studies, they were converted to mg/L-years by assuming subjects consumed 1.5 liters of water per day. From this point on, the analysis proceeded exactly as the analysis applied to the data from Chiou et al., with the odds ratios being treated as relative risks. A summary of these calculations is provided in Table A-5.

Guo et al.

In their "Model 1", Guo et al. (1994) modeled the standardized incidence rate (SIR) as

$$SIR = \alpha + \beta X + \gamma U + \delta T,$$

TABLE A-5

Model fits to bladder cancer data in Bates et al. (1995) and corresponding ED_{10}

Exposure (mg/L yrs)	Relative risk	95% Lower bound	95% Upper bound	Ln(RR)	Ln(LB)	Ln(UB)	Standard error	ED_{10} (mg/L)
0	1							0.131
0.0475	1.56	0.8	3.2	0.445	–0.22 3	1.16	0.421	
0.0785	0.95	0.4	2	–0.0513	–0.916	0.693	0.489	
0.146	1.41	0.7	2.9	0.344	–0.35 7	1.06	0.432	

where *SIR* is the standardized incidence rate (standardized to the world standard population in 1976) per 100,000 over the eight year study period for a township, *X* is the mean inorganic arsenic level (mg/L) of wells in the township, *U* is the urbanization index for the township, and *T* is the average number of cigarettes sold per day in the township. They determined the inorganic arsenic coefficient, β, for three types of urinary cancer. The values of β for bladder cancer were 225.41 (males) and 162.11 (females), and for transitional cell kidney cancer they were 46.97 (males) and 43.36 (females). All of these values were statistically different from zero. The β for renal cell kidney cancer were negative and not statistically different from zero.

According to this model the additional lifetime risk from *X* mg/L inorganic arsenic in drinking water is

$$\beta X/100,000/(8 \text{ years}) * (75 \text{ years})$$

and the corresponding ED_{10} water concentration is

$$X = 0.1/[\beta/100,000/(8 \text{ years}) * (75 \text{ years})].$$

The following ED_{10}s were obtained:

Cancer type	Males	Females
Bladder cancer	4.7 mg/L	6.6 mg/L
Transitional cell kidney cancer	22.7 mg/L	24.6 mg/L

Since no *p*-values or standard errors were provided, it was not possible to calculate LEDs for this study.

Arsenic Exposure and Health Effects
W.R. Chappell, C.O. Abernathy and R.L. Calderon (Editors)
© 1999 Elsevier Science B.V. All rights reserved.

Emerging Epidemics of Arseniasis in Asia

Chien-Jen Chen, Lin-I Hsu, Chin-Hsiao Tseng, Yu-Mei Hsueh,
Hung-Yi Chiou

ABSTRACT

Both acute and chronic health effects of inorganic arsenic on human beings involve several organ systems. In recent decades, emerging epidemics of arseniasis resulting from environmental and occupational exposures have been reported in Asia. There are four sources of arsenic exposure including well water, coal burning, mining and smelting, and geothermal drilling. The endemic areas of arseniasis through drinking high-arsenic well water are located in Bangladesh, Xingjiang, Inner Mongolia and Shaanxi of China, West Bengal of India, and southwestern and northeastern Taiwan. The endemic area of coal burning-related arseniasis is located in Guizhou of China, while those of industrial arseniasis are located in Yunnan of China, Toroku, Matsuo and Nakajo of Japan, Mindanao of Phillipines, and Ronpibool of Thailand. Clinical manifestations of arseniasis include skin hyperpigmentation and depigmentation, palmoplantal hyperkeratosis, dermatitis, gastroenteritis, bronchitis, peripheral polyneuritis and polyneuropathy, hepatopathy, conjunctivitis, lens opacity, diabetes mellitus, mental retardation, ischemic heart disease, electrocardiographic abnormality, cerebrovascular accident, peripheral vascular disease and limb gangrene, microcirculation abnormality, hypertension, and cancers of the skin and various internal organs. The clinical manifestations vary with the source and duration of arsenic exposure, while nutritional status and arsenic methylation capability may be involved in the determination of individual susceptibility to arseniasis.

Keywords: epidemic, arseniasis, Asia

INTRODUCTION

Arsenic is a ubiquitous element present in various compounds throughout the earth's crust. It is widely distributed and mainly transported in the environment by water. Arsenic is used in pigments and dyes, in preservatives of animal hides, and in the manufacture of semiconductors, glass, agricultural pesticides and various pharmaceutical substances. Humans are exposed to low levels of arsenic in water, air, food and beverages. For most people, food constitutes the largest source of arsenic intake, with smaller amounts coming from drinking water and air. Above-average levels of exposure are usually observed among people who live in areas where drinking water has an elevated level of inorganic arsenic because of natural mineral deposits or contamination from human activities, among workers involved in arsenic-related industries, and among patients treated with arsenic-containing drugs (World Health Organization, 1981).

Arsenic exposure from environmental, occupational or medicinal sources enters the human body mainly through ingestion and inhalation, and rarely through skin absorption. Most ingested and inhaled arsenic is absorbed rapidly via the gastrointestinal tract and lungs into the bloodstream and spreads to a large number of organs, including the lungs, liver, kidneys, and skin (Vahter and Norin, 1980). The biotransformation of inorganic arsenic in the human body is very complicated. A substantial fraction of absorbed arsenate (AsV) is reduced in the blood to arsenite (AsIII) (Bertolero et al., 1981; McBride et al.,1978), which is then taken up by hepatocytes and methylated. Inorganic arsenic is methylated to become monomethylarsonic acid (MMA) and dimethylarsinic acid (DMA) in humans. AsIII and AsV are detoxified in humans by methylation to MMA and then further methylated to DMA, which are excreted in the urine together with unchanged AsIII and AsV (Braman and Foreback, 1973; Buchet et al., 1981). The methylation may be considered a detoxification mechanism, because the methylated metabolites, in comparison with inorganic arsenic, are less reactive with tissue constituents (Tam et al., 1978), less toxic, and more readily excreted in the urine (Vahter, 1986, 1988; Le et al., 1994).

Inorganic arsenic may affect many organ systems, including the dermal, respiratory, gastrointestinal, hepatic, cardiovascular, nervous, renal, hematopoietic, and ophthalmic systems (Chen and Lin, 1994; Chen et al., 1997a). Arsenic is not carcinogenic in animal species, but it is documented to cause cancers of the skin, lung, liver, kidney, bladder, and prostate in humans (Chen and Lin, 1994; Chen et al., 1997b). Emerging epidemics of arseniasis have been recently reported from several Asian countries including Bangladesh, China, India, Japan, Philippines, Taiwan, and Thailand. The arsenic exposure and clinical manifestations of arseniasis in these countries are reviewed and compared.

ARSENIC EXPOSURE AND POPULATION AT RISK

Table 1 shows the source of arsenic exposure and exposed populations of arseniasis-endemic areas in Asia. There are four sources of arsenic exposure: well-water, coal burning, mining and smelting, and geothermal drilling. The population exposed to arsenic through industry-related contamination is much smaller than the population exposed to arsenic through drinking well-water containing high levels of arsenic from geological sources or through burning arsenic-containing coal for cooking, heating and drying crops. Endemic areas of arseniasis due to drinking high-arsenic well-water are located in Bangladesh (Dhar et al., 1997), China (Cao, 1996), India (Chowdhury et al., 1997) and Taiwan (Tseng, 1968; Chiou et al., 1997). More than 52 million people have been exposed to arsenic at a level higher than the World Health Organization maximum contamination guideline ($10\,\mu g/l$) in drinking water. The exposure to high-arsenic well-water started in the early 1920s in southern Taiwan (Tseng et al., 1968), early 1950s in Inner Mongolia (Luo, 1996), late 1950s in northeastern Taiwan (Chiou et al., 1997 and 1996), early 1960s in Xinjiang, China (Wang, 1996) and West

TABLE 1

Area, source of arsenic exposure and population of emerging epidemics of arseniasis in Asia

Area	Source	Population	Reference
Bangladesh	Well-water	50,000,000	Dhar et al. (1997)
China			
Guizhou	Coal burning	200,000	Cao (1996)
Inner Mongolia	Well-water	600,000	Cao (1996)
Shaanxi	Well-water	1,000,000	Cao (1996)
Xinjiang	Well-water	100,000	Cao (1996)
Yunnan	Metal smelting	100,000	Niu (1996)
India			
West Bengal	Well-water	1,000,000	Chowdhury et al. (1997)
Japan			
Toroku/Matsuo	Metal smelting	217 patients	Hotta (1989)
Nakajo	Contaminated water	44 patients	Tsuda et al. (1989)
Philippines			
Mindanoa	Geothermal drilling	?	Hironaka (1995)
Taiwan			
Southwest coast	Well-water	100,000	Tseng (1968)
Northeast coast	Well-water	100,000	Chiou et al. (1997)
Thailand			
Ronpibool	Mining	1,000 patients	Choprapawon and Rodcline (1997)

Bengal, India (Bagla and Kaiser, 1996), 1970s in Bangladesh (Quamruzzaman, 1997) and Shanxi, China (Cao, 1996). Arsenic exposure through drinking well-water stopped in the late 1970s in southwestern Taiwan and in the early 1980s in Xinjiang. Residents in Bangladesh, West Bengal (India), Inner Mongolia and Shanxi (China), and northeastern Taiwan still use the high-arsenic well-water for cooking and drinking.

High-arsenic coal has been used for cooking, heating and drying crops for decades in Guizhou, China (Zhen et al., 1996). More than 200,000 people are exposed to arsenic in air, water and food in the endemic area in Guizhou. There have been several reports on industry-related exposure to arsenic through metal smelting in Yunnan, China (Niu, 1996) and Toroku and Matsuo, Japan (Yokoi, 1995), through drinking water contamination from factory waste in Nakajo, Japan (Tsuda et al., 1989) and geothermal power plant in Mindanao in the Philippines (Hironaka, 1995), and through tin mining in Ronpibool, Thailand (Choprapawon and Rodcline, 1997). There are 100,000 people living in the arseniasis-endemic area in Yunnan, China. The number of patients diagnosed with arseniasis has been certified to be 217 in Toroku and Matsuo as well as 44 in Nakajo. More than 1,000 arseniasis cases have been identified in Ronpibool, Thailand.

NON-CANCER NON-VASCULAR MANIFESTATIONS OF ARSENIASIS

The clinical manifestations of arseniasis among exposed populations in Shaanxi and Yunan (China) and Mindanao (Philippines) have not been well studied and reported. Table 2 shows dermal, gastrointestinal, respiratory and neural manifestations of arseniasis in Asia. Hyper-pigmentation (melanosis) and hyperkeratosis are characteristic skin lesions of arseniasis. They are observed in all endemic areas of arseniasis. However, the prevalence and severity of these skin lesions vary from area to area, even in areas where arsenic exposure levels are similar. Individual susceptibility, both hereditary and acquired, may play roles in the modification of risk of arsenic-induced hyperpigmentation and hyperkeratosis. Dermatitis is also a skin lesion of arseniasis prevalent in Bangladesh (Dhar et al., 1997), West Bengal, India

TABLE 2

Comparison of dermal, gastrointestinal, respiratory and neural manifestations of arseniasis in Asia

Area	Melanosis/ keratosis	Dermatitis	Gastroenteritis	Bronchitis	Polyneuritis/ polyneuropathy
Bangladesh	+/+	+	+	+	
China					
Guizhou	+/+		+		+
Inner Mongolia	+/+		+		+
Xinjiang	+/+		+		+
India					
West Bengal	+/+	+	+	+	+
Japan					
Toroku	+/+	+	+	+	+
Nakajo	+/+		+	+	+
Taiwan					
Southwest coast	+/+				
Northeast coast	+/+				
Thailand					
Ronpibool	+/+				

(Chowdhury et al., 1997) and Toroku, Japan (Hotta, 1989). Gastroenteritis is prevalent among residents in arseniasis-endemic areas in Bangladesh (Chowdhury et al., 1997), Guizhou (Zhen et al., 1996), Inner Mongolia (Ma et al., 1996), Xinjiang (Wang, 1996), West Bengal (Chowdhury et al., 1997), Toroku (Hotta, 1989) and Nakajo (Tsuda et al., 1989); but not in Taiwan and Ranpibool. An increased prevalence of bronchitis has been observed among arsenic-exposed populations in Bangladesh (Dhar et al., 1997), West Bengal (Chowdhury et al., 1997), Toroku (Hotta, 1989) and Nakajo (Tsuda et al., 1989); but not in China, Taiwan and Thailand. Peripheral polyneuritis and polyneuropathy are prevalent in arseniasis-endemic areas in Guizhou (Zhen et al., 1996), Inner Mongolia (Ma et al., 1996), Xinjiang (Wang, 1996), West Bengal (Chowdhury et al., 1997), Toroku (Hotta, 1989) and Nakajo (Tsuda et al., 1989); but not in Bangladesh, Taiwan and Thailand.

TABLE 3

Comparison of hepatic, ophthalmic, metabolic and mental manifestations of arseniasis in Asia

Area	Hepatopathy	Conjunctivities	Lens opacity	Diabetes mellitus	Mental retardation
Bangladesh	+	+	+	+	
China					
Guizhou	+	+			
Inner Mongolia	+				+
Xinjiang	+				
India					
West Bengal	+	+			
Japan					
Toroku	+		+		
Nakajo	+	+			
Taiwan					
Southwest coast			+	+	
Northeast coast					
Thailand					
Ronpibool					+

Table 3 compares hepatic, ophthalmic, metabolic and mental manifestations of arseniasis in Asia. Hepatopathy is a prevalent manifestation of arseniasis in Bangladesh (Dhar et al., 1997), Guizhou (Zhen et al., 1996), Inner Mongolia (Ma et al., 1996), Xinjiang (Wang, 1996), West Bengal (Chowdhury et al., 1997), Toroku (Hotta, 1989) and Nakajo (Tsuda et al., 1989). Conjunctivitis is prevalent among arsenic-exposed populations in Bangladesh (Dhar et al., 1997), Guizhou (Zhen et al., 1996), West Bengal (Chowdhury et al., 1997), and Nakajo (Tsuda et al., 1989). An increased prevalence of lens opacity has been reported in Toroku (Hotta, 1989) and southwestern Taiwan (See, 1997), and there is a biological gradient between lens opacity and arsenic exposure in southwestern Taiwan (See, 1997). The prevalence of diabetes mellitus in the arseniasis-endemic area is higher than that in the non-endemic area in Taiwan (Lai et al., 1994), and a dose–response relationship between prevalence of diabetes mellitus and arsenic exposure level has been observed in Bangladesh (Rahman et al., 1998) and Taiwan (Lai et al., 1994). A elevated prevalence of mental retardation was reported among children in arseniasis-endemic areas in Inner Mongolia (Ma et al., 1996) and Ronpibool (Choprapawon, 1996).

VASCULAR EFFECTS

Table 4 shows circulatory manifestations of arseniasis in Asia. An increased prevalence of ischemic heart disease has been reported among residents in arseniasis-endemic areas in Inner Mongolia (Ma et al., 1996), Xinjiang (Wang, 1996), Toroku (Hotta, 1989) and Nakajo (Tsuda et al., 1989). A dose–response relationship between ischemic heart disease and arsenic exposure has been observed in southwestern Taiwan (Chen et al., 1996a). Abnormal electro-cardiographic findings have been found to be prevalent among arsenic-exposed populations in Inner Mongolia (Ma et al., 1996) and Xinjiang (Wang, 1996). An elevated risk of cerebrovascular accident has been reported in Toroku (Hotta, 1989) and Nakajo (Tsuda et al., 1989). Arsenic exposure is associated with cerebrovascular accident in a dose–response relationship in southwestern (Chen et al., 1996b) and northeastern (Chiou et al., 1997) Taiwan.

Peripheral vascular disease and limb gangrene have been well documented as a chara-cteristic vascular effect of arsenic exposure. They are reported in Bangladesh (Dhar et al.,

TABLE 4

Comparison of circulatory manifestations of arseniasis in Asia

Area	ISHD	EKG anomaly	CVA	PVD/Limb gangrene	Microcirculation abnormality	Hypertension
Bangladesh				+		+
China						
Guizhou						
Inner Mongolia	+	+		+	+	+
Xinjiang	+	+		+		
India						
West Bengal				+		
Japan						
Toroku	+		+	+		
Nakajo	+		+			
Taiwan						
Southwest coast	+		+	+	+	+
Northeast coast			+			
Thailand						
Ronpibool						

ISHD: ischemic heart disease; EKG: electrocardiograph; CVA: cerebrovascular accident.

1997), Inner Mongolia (Ma et al., 1996), Xinjiang (Wang, 1996), West Bengal (Chowdhury et al., 1997) and Toroku (Hotta, 1989). A biological gradient between arsenic exposure and peripheral vascular disease, clinical or subclinical, has been reported in southwestern Taiwan (Tseng, 1977; Tseng et al., 1995a). In addition to long-term exposure to high-arsenic well-water, under-nourishment also plays a role in the development of peripheral vascular disease known as blackfoot disease in southwestern Taiwan (Chen et al., 1988). Patients affected with blackfoot disease have an increased risk of dying from ischemic heart disease and various cancers (Chen et al., 1988). Abnormal microcirculation has been found to be prevalent among residents in arseniasis-endemic areas in Inner Mongolia (Ma et al., 1996) and southwestern Taiwan (Tseng et al., 1995b). The prevalence of hypertension is higher in the arseniasis-endemic area than in the non-endemic area in southwestern Taiwan (Chen et al., 1995) and Inner Mongolia (Ma et al., 1996); and a dose–response relationship between arsenic exposure and hypertension prevalence has been found in southwestern Taiwan (Chen et al., 1995) and Bangladesh (Rhaman et al., 1999).

CARCINOGENICITY

Inorganic arsenic has been well documented as a human carcinogen of the skin and lung (World Health Organization, 1980). Significantly high prevalence of skin cancer has been observed in all arseniasis-endemic areas in Asia. The dose–response relationship between arsenic exposure and skin cancer risk has been reported in southwestern Taiwan (Tseng et al., 1968). The risk of arsenic-induced skin cancer is elevated among those who have poor nutritional status and/or liver dysfunction (Hsueh et al., 1995). Low serum β-carotene level and poor arsenic methylation capability have been found to modulate the risk of arsenic-induced skin cancer (Hsueh et al., 1997). Patients affected with arsenic-induced skin cancer have an increased frequency of spontaneous and arsenic-induced sister chromatid exchanges and delayed proliferation of cultured peripheral lymphocytes (Hsu et al., 1997).

Increased risk of malignant neoplasms of all sites combined has been observed in Inner Mongolia (Ma et al., 1996), Xinjiang (Wang 1996) , Toroku (Hotta, 1989) and Nakajo (Tsuda et al., 1989). There is a biological gradient between arsenic exposure and cancers of all sites combined in Taiwan (Chen et al., 1988). In a nested case-control study carried out in the arseniasis-endemic area in southwestern Taiwan, there is an increased risk of developing malignant neoplasms among those who have an increased frequency of chromosome-type chromosome aberration (Liu et al., 1999). Long-term exposure to arsenic has been associated with an increased risk of several internal cancers in southwestern Taiwan (Chen et al., 1985). Increased risk of lung cancer has been reported among residents in arseniasis-endemic areas in Inner Mongolia (Ma et al., 1996), Xinjiang (Wang, 1996), Toroku (Hotta, 1989) and Nakajo (Tsuda et al., 1989). A dose–response relationship between arsenic exposure and lung cancer risk is found in southwestern and northeastern Taiwan (Chen et al., 1988, 1992; Chiou et al., 1999).

Arsenic-exposed populations in Guizhou (Zhen et al., 1996), Inner Mongolia (Ma et al., 1996) and Nakajo (Tsuda et al., 1989) have an increased risk of liver cancer. There is a biological gradient between arsenic exposure and liver cancer risk among residents in arseniasis-endemic areas in southwestern and northeastern Taiwan (Chen et al., 1988, 1992; Chiou et al., 1999). Risk of cancers of the urinary bladder and kidney is increased among arsenic-exposed populations in Inner Mongolia (Ma et al., 1996), Toroku (Hotta, 1989) and Nakajo (Tsuda et al., 1989). A dose–response relation between arsenic exposure and cancers of the urinary bladder and kidney, mainly transitional cell carcinoma, has been observed in southwestern and northeastern Taiwan (Chen et al., 1988, 1992; Chiou et al., 1999). Other cancers have also been associated with the long-term exposure to arsenic. They include nasal cavity cancer (Chen and Wang, 1990) and prostate cancer (Chen et al., 1988; Chen and Wang,

TABLE 5

Comparison of cancers of the skin and internal organs related to arseniasis in Asia

Area	All sites	Skin	Nasal cavity	Lung	Liver	Urinary bladder	Kidney	Prostate	Other
Bangladesh		+							
China									
Guizhou		+			+				
Inner Mongolia	+	+		+	+	+	+		esophagus
Xinjiang	+	+		+					
India									
West Bengal		+							
Japan									
Toroku	+	+		+		+	+		
Nakajo	+			+	+	+	+		cervix uteri
Taiwan									
Southwest coast	+	+	+	+	+	+	+	+	
Northeast coast	+	+	+	+	+	+	+		stomach
Thailand									
Ronpibool	+								

1990) in Taiwan, esophagus cancer in Inner Mongolia (Ma et al., 1996), and cervical cancer in Nakajo (Tsuda et al., 1989).

CONTROL OF ARSENIASIS IN ASIA

The control of arseniasis in Asia needs an international and multidisciplinary effort. This effort should include: (1) a comprehensive survey and monitoring of the arsenic level in the environment and foodstuffs, (2) identification and surveillance of health hazards due to arsenic exposure through ingestion or inhalation, (3) provision of safe drinking water for residents in endemic area where drinking water has a high level of arsenic, (4) replacement of high-arsenic coal for cooking and heating, (5) improvement of arsenic-related industries to reduce environmental contamination, (6) promotion of health education to prevent or reduce the exposure to arsenic, (7) establishment and strengthening of the infrastructure of governmental and non-governmental agencies to cope with the problems, and (8) facilitation of international collaboration in the exchange of knowledge and experiences in the control and treatment of arseniasis.

REFERENCES

Bagla, P., Kaiser, J. 1996. India's spreading health crisis draws global arsenic experts. *Science*, 274, 174–175.

Bertolero, F., Marafante, E., Edel Rade, J., Pietro, R., Sabbioni, E. 1981. Biotransformation and intracellular binding of arsenic in tissues of rabbits after intraperitoneal administration of As-74 labeled arsenite. *Toxicology*, 20, 35–44.

Braman, R.S., Foreback, C.C. 1973. Methylated forms of arsenic in the environment. *Science*, 182, 1247–1249.

Buchet, J.P., Lauwerys, R., Roels, H. 1981. Urinary excretion of inorganic arsenic and its metabolites after repeated ingestion of sodium metaarsenite by volunteers. *Int. Arch. Occup. Environ. Health*, 48, 111–118.

Cao, S. 1996. Current status of inorganic arsenic pollution in mainland China. *Chinese J. Public Health*, 15, s6–10.

Chen, C.J., Chuang, Y.C., Lin, T.M., Wu, H.Y. 1985. Malignant neoplasms among residents of a blackfoot disease-endemic area in Taiwan: High-arsenic artesian well water and cancers. *Cancer Res.*, 45, 5895–5899.

Chen, C.J., Wu, M.M., Lee, S.S., Wang, J.D., Cheng, S.H., Wu, H.Y. 1988a. Atherogenicity and carcinogenicity of high-arsenic artesian well water: Multiple risk factors and related malignant neoplasms of blackfoot disease. *Arteriosclerosis*, 8, 452–460.

Chen, C.J., Kuo, T.L., Wu, M.M. 1988b. Arsenic and cancers. *Lancet*, 2, 414–415.

Chen, C.J., Wang, C.J. 1990.Ecological correlation between arsenic level in well water and age-adjusted mortality from malignant neoplasms. *Cancer Res.*, 50, 5470–5474.

Chen, C.J., Chen, C.W., Wu, M.M., Kuo, T.L. 1992. Cancer potential in liver, lung, bladder, and kidney due to ingested inorganic arsenic in drinking water. *Br. J. Cancer*, 66, 888–892.

Chen C.J., Lin, L.J. 1994. Human carcinogenicity and atherogenicity induced by chronic exposure to inorganic arsenic. In: J.O. Nriagu (ed.), *Arsenic in the Environment, Part II: Human Health and Ecosystem Effects*, pp. 109–132. John Wiley & Sons, Inc., New York.

Chen, C.J., Hsueh, Y.M., Lai, M.S., Shyu, M.P., Chen, S.Y., Wu, M.M., Kuo, T.L., Tai, T.Y. 1995. Increased prevalence of hypertension and long-term arsenic exposure. *Hypertension*, 25, 53–60.

Chen, C.J., Chiou, H.Y., Chiang, M.H., Lin, L.J., Tai, T.Y. 1996a. Dose–response relationship between ischemic heart disease mortality and long-term arsenic exposure. *Arterioscl. Thromb. Vasc. Biol.*, 16, 504–510.

Chen, C.J., Chiou, H.Y., Hsueh, Y.M., Huang W.I., Hsu, Y.H., Lin L.J., Chu, T.H., Wei, M.L., Chen, H.C., Hsu, L.I., Hsieh F.I. 1996b. Epidemiological studies on long-term arsenic exposure and cardiovascular disease in Taiwan: An review. *Chinese J. Public Health*, 15, s59–67.

Chen, C.J., Chiou, H.Y., Huang, W.I., Chen, S.Y., Hsueh, Y.M., Tseng, C.H., Lin, L.J., Shyu, M.P., Lai, M.S. 1997a. Systemic non-carcinogenic effects and developmental toxicity of inorganic arsenic. In: C.O. Abernathy, R.L. Calderon and W.R. Chappell (eds.), *Arsenic Exposure and Health Effects*, pp. 124–134. Chapman & Hall, London.

Chen, C.J., Hsueh, Y.M., Chiou, H.Y., Hsu, Y.H., Chen, S.Y., Horng, S.F., Liaw. K.F., Wu, M.M. 1997b. Human carcinogenicity of inorganic arsenic. In: C.O. Abernathy, R.L. Calderon and W.R. Chappell (eds.), *Arsenic Exposure and Health Effects*, pp. 232–242. Chapman & Hall, London.

Chiou, H.Y., Huang, W.I., Su, C.L., Chang, S.F., Hsu, Y.H., Chen, C.J. 1997. Dose-response relationship between prevalence of cerebrovascular disease and ingested inorganic arsenic. *Stroke*, 28, 1717–1723.

Chiou, H.Y., Hsu, Y.H., Wei, M.L., Tseng, C.H., Chen, C.J. 1999. Incidence of transitional cell carcinoma due to arsenic in drinking water: A follow-up study of 8102 residents in an arseniasis-endemic area in northeastern Taiwan. *Am. J. Epidemiol.* (submitted).

Choprapawon, C. 1996. Does arsenic affect children's intelligence? *Asia Arsenic Network Newsletter*, 1, 6–7.

Choprapawon, C., Rodcline, A. 1997. Chronic arsenic poisoning in Ronpibool Nakhon Sri Thammarat, the Southern Province of Thailand. In: C.O. Abernathy, R.L. Calderon and W.R. Chappell (eds.), *Arsenic Exposure and Health Effects*, pp. 69–77. Chapman & Hall, London.

Chowdhury, T.R., Mandal, B.Kr., Samanta, G., Basu, G.Kr., Chowdhury, P.P., Chanda, C.R., Karan, N.Kr., Lodh, D., Dhar, R.Kr., Das, D., Saha, K.C., Chakraborti, D. 1997. Arsenic in groundwater in six districts of West Bengal, India: The biggest arsenic calamity in the world: The status report up to August, 1995. In: C.O. Abernathy, R.L. Calderon and W.R. Chappell (eds.), *Arsenic Exposure and Health Effects*, pp. 93–111. Chapman & Hall, London.

Dahr R.Kr., Biswas B.Kr., Samanta, G., Mandal B.Kr., Chakraborti, D., Roy, S., Jafar, A., Islam, A., Ara, G., Kabir, S., Khan, A.W., Ahmed, S.A., Hadi, S.A. 1997. Groundwater arsenic calamity in Bangladesh. *Current Sci.*, 73, 48–59.

Horinaka, H. 1995. Mindanao Island, the Philippines: Arsenic found near geothermal power plant. Asia Arsenic Network, Miyazaki, Japan.

Hotta, N. 1989. Clinical aspects of chronic arsenic poisoning due to environmental and occupational pollution in and around a small refining spot. *Jpn. J. Constit. Med.*, 53, 49–70.

Hsu, Y.H., Li, S.Y., Chiou, H.Y., Yeh, P.M., Liou, J.C., Hsueh, Y.M., Chang, S.H., Chen, C.J. 1997. Spontaneous and induced sister chromatid exchanges and delayed cell proliferation in peripheral lymphocytes of Bowen's disease patients and matched controls of arseniasis-hyperendemic villages in Taiwan. *Mutation Res.*, 386, 241–251.

Hsueh, Y.M., Cheng, G.S., Wu, M.M., Yu, H.S., Kuo, T.L., Chen, C.J. 1995. Multiple risk factors associated with arsenic-induced skin cancer: Effects of chronic liver disease and malnutritional status. *Br. J. Cancer*, 71, 109–114.

Hsueh, Y.M., Chiou, H.Y., Huang, Y.L., Wu, W.L., Huang, C.C., Yang, M.H., Lue, L.C., Chen, G.S., Chen, C.J. 1997. Serum β-carotene level, arsenic methylation capability, and incidence of skin cancer. *Cancer Epidemiol. Biomark. Prev.*, 6, 589–596.

Lai, M.S., Hsueh, Y.M., Chen, C.J., Shyu, M.P., Chen, S.Y., Kuo, T.L., Wu, M.M., Tai, T.Y. 1994. Ingested inorganic arsenic and prevalence of diabetes mellitus. *Am. J. Epidemiol.*, 139, 484–492.

Le, X.C., Cullen, W.R., Reimer, K.J. 1994. Human urinary arsenic excretion after one-time ingestion of seaweed crab and shrimp. *Clin. Chem.*, 40, 617–624.

Liu, S.H., Lung, J.C., Chen, Y.H., Yang, T., Hsieh, L.L., Chen, C.J., Wu, T.N. 1999. Increased chromosome-type chromosome aberration frequencies as biomarkers of cancer risk in a blackfoot endemic area. *Cancer Res.*, 59, (in press)

Luo, Z.D. 1996. Current status of inorganic arsenic pollution in Inner Mongolia. *Chinese J. Public Health*, 15, s11–14.

Ma, L., Luo, Z.D., Zhang, Y.M., Zhang, G.Y., Dai, X., Liang, X.F., Ren, X.Y., Zhang, M.Y. 1996. Current status of endemic arsenicism in Inner Mongolia. *Chinese J. Public Health*, 15, s15–39.

McBride, B.C., Merilees, H., Cullen, W.R., Pickett, W. 1978. Arsenic. In: F.E. Brinckman and J.M. Bellama (eds.), *Organometals and Organometalloids. ACS Symposium Series 82*, pp. 94–115. American Chemical Society, Washington, DC.

Niu, S. 1996. Current status of arsenic poisoning and its control in mainland China. *Chinese J. Public Health*, **15**, s1–5.

Quamruzzaman, Q. 1997. *Arsenic pollution in groundwater of Bangladesh.* Dhaka Community Hospital Trust, Dhaka, Bangladesh.

Rhaman, M., Tondel M., Ahmad S.A., Axelson, O. 1998. Diabetes mellitus associated with arsenic exposure in Bangladesh. *Am. J. Epidemiol.*, **148**, 198–203.

Rhaman, M. and Axelson, O. 1999. Hypertension and arsenic exposure in Bangladesh. *Hypertension*, **33**, 74–78.

See. L.C. 1997. Epidemiological study on senile cataract in Taiwan. PhD thesis, National Taiwan University.

Tam, K.H., Charbonneau, S.M., Bryce, F., Lacroix, G. 1978. Separation of arsenic metabolites in dog plasma and urine following intravenous injection of 74 As. *Anal. Biochem.*, **86**, 505–511

Tseng, C.H., Chong, C.K., Chen, C.J.. Tai, T.Y. 1995a. Dose–response relationship between peripheral vascular disease and ingested inorganic arsenic among residents in blackfoot disease endemic villages in Taiwan. *Atherosclerosis*, **120**, 125–133.

Tseng, C.H., Chong, C.K., Chen, C.J. 1995b. Abnormal peripheral microcirculation in seemingly normal subjects living in blackfoot disease-hyperendemic villages in Taiwan. *Int. J. Microcirc.*, **15**, 21–27.

Tseng, W.P., Chu, C.M., How, S.W., Fong, J.M., Lin, C.S.. Yeh, S. 1968. Prevalence of skin cancer in an endemic area of chronic arsenicism in Taiwan. *J. Natl. Cancer Inst.*, **40**, 453–463

Tseng, W.P. 1977. Effects and dose–response relationships of skin cancer and blackfoot disease with arsenic. *Environ. Health Perspect.*, **19**, 109–119.

Tsuda, T. 1989. Malignant neoplasms among residents who drank well water contaminated by arsenic from a king's yellow factory. *J. UOEH*, **11**, 289–301.

Vahter, M., Norin, H. 1980. Metabolism of 74 As-labeled trivalent and pentavalent inorganic arsenic in mice. *Environ. Res.*, **21**, 446–457.

Vahter, M. 1986. Environmental and occupational exposure to inorganic arsenic. *Acta Pharmacol. Toxicol.*, **59**, 31–34.

Vahter, M. 1988. Arsenic. In: Clarkson, T.W., Friberg, L., Nordberg, G.F. and Sager, P.R. (eds.), *Biological Monitoring of Toxic Metals*, pp. 303–321. Plenum, New York.

Wang, L.F. 1996. Current status of studies and control of arsenic health hazards in Xinjiang. *Chinese J. Public Health*, **15**, 53–58.

World Health Organization.1980. IARC monographs on the evaluation of the carcinogenic risk of chemical to humans: Some metal and metallic compounds. pp. 39–141. IARC Scientific Publ. No. 23. International Agency Research on Cancer, Lyon, France.

World Health Organization. 1981. *Environmental Health Criteria 18: Arsenic.* World Health Organization, Geneva.

Yokoi, H. 1995. Toroku and Matsuo, Japan: Villages poisoned for half a century. Asia Arsenic Network, Miyazaki, Japan.

Zhen, B.S., Zhang, G., Yu, X.Y., Long, G.P., Zhou, D.X., Liou, D.N. 1996. An environmental geochemical study on arsenic poisoning induced by high-arsenic coal. *Chinese J. Public Health*, **15**, 44–48.

Arsenic Exposure and Health Effects
W.R. Chappell, C.O. Abernathy and R.L. Calderon (Editors)
© 1999 Elsevier Science B.V. All rights reserved.

The Present Situation of Chronic Arsenism and Research in China

G.F. Sun, G.J. Dai, F.J. Li, H. Yamauchi, T. Yoshida, H. Aikawa

ABSTRACT

Since chronic endemic arsenism was found in Taiwan in 1968, the disease has been reported in the continent of China only in recent years. Although its history is short, the situation and spread of the disease are very serious. Up to now, the disease has been found in Xinjiang, Inner Mongolia, Guizhou and Shanxi provinces. The population exposed has been over 2 million and diagnosed arsenism patients number up to 20,000. Arsenism found in Shanxi province was only reported in 1996 and the population involved was up to one million. Now the epidemic area is expanding. Chronic arsenic poisoning in China can be divided into three types according to the exposure sources. One is caused by drinking water, in which the As concentration is very high. Xinjiang, Inner Mongolia, Shanxi and Taiwan belong to the drinking water type. Most of the people drink water from a pump well more than 20 m deep and the As concentration ranges from 0.15 to 2.0 mg/L according to different areas, the highest being up to 4.4 mg/L which was 88 times the 0.05 mg/L China standard. The second is caused by coal burning. People use coal containing high levels of As as fuel for cooking and drying grain in the kitchen and so inhale high levels of As concentration in the air. In Guizhou the As content in coal was up to 9600 mg/kg and As in kitchen air was 0.003–0.11 mg/m³. The third type is caused by industry exposure, mainly copper smelting and arsenic mining. There are about 30,000 workers in the industry exposed to As. Besides the skin changes in all three types of arsenic poisoning, lung cancer incidence was high in coal burning and industrial exposure, especially in As mines. The workers in a tin mine in Yunnan province were exposed to high levels of As and the cancer incidence was $716.9/10^5$ which was 82 times that of controls. Average As content in the lungs of the cancer patients was 43.33 mg/kg. Another characteristic of endemic arsenism in China is that the fluoride content was also high in both drinking water and burned coal with high As concentration. The "Fluoride and Arsenic Society of China (FASC)" was established in 1996 and is now organizing research into the field of fluoride and arsenic throughout the country.

Keywords: chronic arsenic poisoning, drinking water, lung cancer

In China, chronic endemic arsenism was first found in Jiayi and Tainan, Taiwan province in 1968. The area of the disease was about 300 km² and the population of the disease was about 150 thousand. The number of patients with "Black foot" disease amounted to 1636 by the end of 1985 (Zheng et al., 1995).

However, the disease on the continent of China has been reported only in recent years. Although its history is short, the situation and the spread of the disease are very serious. In the mainland, the disease was first found in Kuitun, Xinjiang province in 1980. There were more than 2000 patients out of a population of about 100,000 (Wang et al., 1983). In 1989, a large-scale survey was conducted in five cities and 11 counties of Inner Mongolia Autonomous Region. There were 1774 patients and the population involved reached 600 thousand. In a seriously affected village of this area, the prevalence rate was 74.3% (Sun et al., 1995). By the end of 1994, 2600 patients were found in a population of 200,000 in six cities or counties of Guizhou province (Zheng et al., 1994). It was not until 1996 that arsenism was found in Datong Basin, Shanxi province. There were 3000 patients and the exposed population was up to one million.

So far it is estimated that the total population exposed has been over two million and the number of diagnosed patients has reached 20,000. Furthermore, the epidemic area is still expanding.

Chronic endemic arsenic poisoning in China can be divided into three types according to the exposure sources. The first is caused by drinking water, in which As concentration is very high. Xinjiang, Inner Mongolia, Shanxi and Taiwan belong to this type. Most of the people drink water from pump wells deeper than 20 m, and the As concentration ranges between 0.15 and 2.0 mg/l, which is 88 times that of the 0.05 mg/l China standard. The second is caused by coal burning. People use coals containing high levels of As as fuel for cooking, keeping warm and drying grain in the kitchen, so that As is absorbed by the respiratory tract, skin and digestive tract. It is a type of exposure unique to China. Guizhou belongs to this type, where the As content in coal is as high as 9600 mg/Kg and As in kitchen air was 0.003–0.11 mg/m³ (Zheng et al., 1994). The third type is caused by industry exposure, mainly copper smelting and arsenic mining. There are now about 30,000 workers in the industry who are exposed to As through their occupation. There is a tin mine in Yunnan province where the workers were exposed to high levels of As and the cancer incidence was $716.9/10^5$, which was 82 times that of controls. Average As content in the lungs of the cancer patients was 43.33 mg/kg. Another characteristic of endemic arsenism in China was caused by absorbing excess fluoride and arsenic simultaneously. This type was first reported in 1985. In some areas of China, fluoride and As contents were all very high in drinking water and burning coal, which leads to combined poisoning with fluoride and arsenic.

Chronic endemic arsenism in China also shows the following characteristics:
• The disease attacks people aging from 3 to 80 years old;
• No significant difference was found between males and females (Sun et al., 1995; Wang et al., 1994);
• It is common for more than one members of a family to be affected (Wang et al., 1994);
• The prevalence rate rises with increasing age and extended length of residence. In addition, if the patients leave the disease areas and emigrate to non-disease areas, their condition will gradually improve, even returning to normal states. Conversely, immigrants to the disease areas or residents living for long periods of time are inclined to suffer from the disease, and their condition will deteriorate sharply (Liu et al., 1988).

Arsenic exposure results in a wide range of health effects. One of the most typical clinical manifestations is skin lesion, which includes hyperkeratosis, hyperpigmentation, depigmentation and skin cancer (Wang et al., 1994). The second distinguishing feature is nerve system injury, consisting of central nervous dysfunction, peripheral neuritis and polyneuropathy. It is not uncommon to see neurasthenia or eyesight impediment among the patients. In addition to these characteristic symptoms, other arsenic-induced health effects of cardiovascular diseases, digestive system and respiratory system diseases are all prevalent in the epidemic areas, such as Raynaud syndrome, vasculitis, gastroenteritis, liver and spleen swelling, liver cirrhosis, abnormal function of liver and bronchitis. There are also reports suggesting an increased risk of various internal cancers including lung cancer, bladder cancer, breast carcinoma and so on.

REFERENCES

Liu, H.D. et al. 1988. *Foreign Medical Sciences Geographical Fascicle,* 2, 49–52.
Sun, T.Z. et al. 1995. Endemic Arsenism Special Issue, *Chinese J. Endemiol.,* 1–3.
Wang, L.F. et al. 1994. *Internal Mongolia Endemiology Preventive and Cure Research,* **19** (suppl.), 37–40.
Wang, L.F. et al. 1983. *Chinese J. Endemiol.,* **2** (2), cover 3.
Zheng, B.S. et al. 1994. *Internal Mongolia Endemiology Preventive and Cure Research,* **19** (suppl.), 41–43.
Zheng, B.S. et al. 1995. *Internal Mongolia Endemiology Preventive and Cure Research,* **20** (1), 40–42.

Arsenic Exposure and Health Effects
W.R. Chappell, C.O. Abernathy and R.L. Calderon (Editors)
1999 Elsevier Science B.V.

127

Human Exposure to Arsenic and Health Effects in Bayingnormen, Inner Mongolia

Heng Z. Ma, Ya J. Xia, Ke G. Wu, Tian Z. Sun, Judy L. Mumford

ABSTRACT

Arsenic is naturally occurring in ground water in large areas of Inner Mongolia with concentrations ranging from $<50~\mu g/l$ to 1.8 mg/l. Bayingnormen (Ba Men; 543 villages with arsenic concentrations $>50~\mu g/l$) in the west and Tumet (with 81 villages $>50~\mu g/l$) in central Inner Mongolia are the two major endemic areas with high concentrations of arsenic in ground water. Drinking water is the only significant source of arsenic contamination. This paper reports arsenic exposure and health effects associated with arsenic exposure in Ba Men. Up to 1995, a total of 1,447 cases (81% of all cases in Inner Mongolia) of arsenicism (mainly skin hyperkeratosis, depigmentation, hyperpigmentation and skin cancer) were confirmed in Ba Men and more cases have been reported since then. The prevalence of arsenicism shows a dose–response relationship with concentration of arsenic in drinking water. Patients with arsenicism range from 5 to 80 years old with peak prevalence among the 40–49 age group. Other clinical symptoms among the people exposed to arsenic in Ba Men are (1) central and peripheral neuro effects, including peripheral neuritis, Raynaud's Syndrome, (2) gastro-enteritis, hypertrophy and abnormal functions in liver, (3) peripheral and cardiovascular effects, including myocardial ischemia, arrhythmia, (4) pulmonary effects, (5) hematological effects showing morphological changes and membrane damage in red blood cells and (6) Bowen's Disease and skin tumors. These studies showed that arsenic exposure resulted in a wide range of health effects in Inner Mongolia.

Keywords: arsenic, arsenicism, skin hyperkeratosis, hyperpigmentation

INTRODUCTION

Inner Mongolia autonomous region is located in northern China. The Northern boundary of Inner Mongolia is formed by Mongolia and Russia. In Inner Mongolia, there are four cities and eight regions (leagues). Huhhot is the capital city of Inner Mongolia and Bayingnormen (Ba Men) is one of the eight regions in Inner Mongolia. Arsenic is naturally occurring in ground water in large areas of Inner Mongolia with concentrations ranging from $<50\,\mu g/L$ to 1.8 mg/L. It is estimated that a total of 300,000 residents live in the areas with arsenic concentrations greater than $>50\,\mu g/L$ in Inner Mongolia. Two major areas are Bayingnormen (Ba Men; 543 villages with arsenic concentrations $>50\,\mu g/L$) in the west and Tumet (with 81 villages $>50\,\mu g/l$) in central Inner Mongolia. Keqi (with two villages $>50\,\mu g/L$) located in the northeast is the minor site of arsenicism. Luo et al. previously reported chronic arsenicism in skin in eastern Tumet (Luo et al., 1997). This paper reports health effects associated with arsenic exposure in Ba Men.

In March 1990, a severe case of skin hyperkeratosis and skin cancer was first reported in Ba Men. This prompted the investigation of health problems related to arsenic exposure in this region. The Bayingnormen Region is situated in western Inner Mongolia (see Figure 1). Its neighbor on the east side is the city of Baotou and on the west is the Alashan Region. The Region is separated from the Yikezhao Region by the Yellow River (Huang He) in the south and on the north is linked with Mongolia. The Region's land covers 64 thousand square kilometers with a total population of 1.7 million. Ba Men, including six counties and one city, is an agricultural region producing wheat, corn, potatoes, sunflowers, and fruit. In Ba Men,

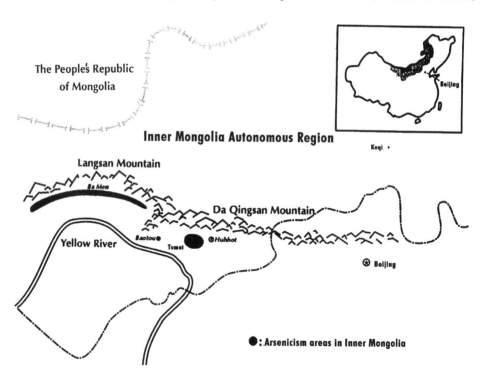

Fig. 1. Map of Inner Mongolia Autonomous Region. In the top right-hand corner, the map of China shows Inner Mongolia Autonomous Region as the lightly shaded region and Ba Men as the darker shaded area. The main map shows Ba Men and Tumet Arsenicism Endemic Areas in Inner Mongolia.

the area containing the ground water with elevated levels of arsenic used to be a sunken lake basin located in the northwest of Baotao (see Figure 1). Its arsenic concentrations, ranging from 50 μg/L to 1.8 mg/L, is due to naturally occurring geological formations (Yu et al., 1995). In the north, there are the Langsan mountains which contain rocks high in arsenic and in the south there is the Yellow River which has a high water table owing to its high demand for irrigation in the region. In Ba Men, elevated levels of arsenic were found in the underground water in the belt-shaped lake basin area with an area of 300 km by 20 km as shown in Figure 1. The three major sites with elevated levels of arsenic in ground water are Hanginhougi County, Wuyan County and Linhe. The studies reported here were from the results of arsenic investigation in these three sites (Sun et al., 1995, Ma et al., 1995).

METHOD

Arsenic concentrations in well-water were determined by a standard colorimetric method using silver diethyldithiocarbamate (Sun et al., 1995, Fan et al., 1993). This method which was validated with the analysis results using an atomic absorption method has been approved by the Chinese government for water quality analysis. Hair and urine sample collection and analysis were conducted as described previously (Chou et al., 1995). Arsenic-related health effects, including cases of arsenicism and other clinical symptoms, were investigated in residents living in the Ba Men arsenic belt region with ground water higher than 50 μg/L and compared with the results from the residents exposed to <50μg/L. Arsenicism was defined as subjects exposed to arsenic concentrations >50 μg/L for at least 6 months with the following symptoms: (1) skin hyperkeratosis, hyper- and de-pigmentation, tumors or other lesions and/or (2) peripheral neuritis and urinary arsenic >90 μg/L or hair arsenic >0.6 μg/g.

RESULTS AND DISCUSSION

Arsenic Exposure

Prior to 1978, the Ba Men residents in a village shared wells, usually shallow wells (3–5 m deep), which contain arsenic mostly lower than 50μg/l. After 1978, each family had their own well usually 10–30 m deep and elevated levels of arsenic were frequently found in the wells with a depth of 15–35 m (Yu et al., 1995). See Table 1. In the arsenic belt region, it is estimated that 30% of the wells used by the residents showed arsenic concentrations higher than 50 μg/L. Drinking water is the only significant source of arsenic contamination. The analysis of the samples from soil, food, and surface water from Ba Men showed that arsenic concentrations in these samples are all below China national standards (Sun et al., 1995). In this region, no arsenic pesticides were used. Most of the industries in Ba Men are agriculture-related and there are no arsenic emissions or discharge from industries. Analysis of dissolved oxygen in well-water showed that the well-water from Ba Men is in a reducing environment so the arsenic in the wells contains As^{+3}. After storage undergoing oxidation at room temperature, most arsenic (>80%) is oxidized to arsenic^{+5}. In Inner Mongolia, it is known that some well-water showed high concentrations of fluoride, especially in Tumet endemic area. In Ba Men, fluoride concentrations in well-water vary greatly with a mean concentration of 1.8±1.2 mg/L (N=170). The Chinese standard for fluoride concentration in drinking water is 1 mg/L.

Arsenic Analysis in Hair and Urine

Urinary and hair arsenic data of the residents of the arsenic belt region in Ba Men substantiate the exposure to arsenic in these subjects. Table 2 gives urinary and hair arsenic from Ba Men residents exposed to various concentrations of arsenic (Chou et al., 1995). These data showed that there is a correlation between arsenic concentrations in hair and in nails and the arsenic concentrations in well-water.

TABLE 1

Well depth and arsenic concentration

Well depth (m)	Total no. wells	>50 μg/L	% Wells with arsenic concentrations >50 μg/L.
1-4	195	14	7.2
5-9	329	22	6.7
10-14	2387	168	7.0
15-19	1752	442	25.2
20-24	846	290	34.3
25-29	197	88	44.7
30-34	146	58	39.7
>35	40	6	15.0

Source: Yu et al. (1995).

TABLE 2

Arsenic in urine and hair

Arsenic concentration in drinking water	Urine (μg/L)	Hair (μg/g)
<50 μg/L	23±2.1 (N=68)	0.096±0.017 (N=18)
50–299 μg/L	270±20.7 (N=113)	1.53±0.45 (N=14)
300-599 μg/L	315±32.5 (N=86)	4.75±0.61 (N=28)
>600 μg/L	562±43.2 (N=90)	5.56±0.46 (N=32)

Source: Chou et al. (1995).

Health Effects

Clinical examinations showed the presence of skin hyperkeratosis, hyperpigmentation, and depigmentation and other skin lesions. Up to 1995, a total of 1447 cases (81% of all cases in Inner Mongolia) of arsenicism were confirmed in Ba Men and more cases have been reported since then. Table 3 shows the prevalence of arsenicism in Ba Men residents exposed to different concentrations of arsenic in drinking water. These data showed a dose–response relationship between chronic arsenicism and arsenic concentrations in drinking water. Patients with arsenicism range from 5 to 80 years old with peak prevalence among the 40–49 age group. In Ba Men, the arsenic exposure is about 18–20 years and cases of Bowen's disease and skin cancer have been reported. Other clinical symptoms among the people exposed to arsenic in Ba Men are (1) central and peripheral neuro effects, including peripheral neuritis, Raynaud's phenomenon, (2) gastroenteritis, hypertrophy and abnormal functions in liver, (3) peripheral and cardiovascular effects, including myocardial ischemia, arrhythmia, (4) pulmonary effects, and (5) hematological effects showing morphological changes and membrane damage in red blood cells (Ma et al., 1995). Clinical examination of 530 individuals from the endemic site in Hanginhougi in Ba Men showed 55% with arsenicism, and these

TABLE 3

Prevalence of chronic arsenicism

Arsenic concentration (μg/L)	N	Cases	Prevalence (%)
<50	1624	0	0
50–199	641	29	4.5
200–399	321	39	12.1
400–649	1021	278	23.1
>650	1179	581	49.3

Source: Yu et al. (1995).

cases also showed elevated rates of cardiovascular (41% with abnormal ECG) and peripheral vascular diseases, peripheral (27% in peripheral neuritis) and central nervous dysfunction (abnormal EEG 40%).

CONCLUSION

Ba Men residents in endemic areas are exposed to a wide range of arsenic levels mainly via drinking water. Arsenic exposure was confirmed by the elevated levels of arsenic in hair and urine which showed a dose–response relationship. The clinical investigations showed that arsenic exposure resulted in a wide range of health effects in Ba Men.

ACKNOWLEDGMENT

The research described here has been reviewed by the National Environmental and Health Effects Research Laboratory, U.S. Environmental Protection Agency, and approved for publication. Approval does not signify that the contents necessarily reflect the views and policies of the Agency nor does mention of trade names or commercial products constitute endorsement or recommendation for use.

REFERENCES

Chou, H.J., Ma, H.Z., Yu, K.J., Xian Y.J., Lee, Y.H., and Kuo X. J. 1995. Evaluation of using urinary and hair arsenic for diagnosis of endemic arsenicism. *J. Chinese Endemic Diseases*, Special Suppl., 36–38.

Fan, C.W., Naren, G., Zhang Y.M. et al. 1993. Analysis for arsenical water and approach for reason of rich arsenic in Western Huhhot Basin. *Environ. Health*, **10** (2), 56–58.

Luo, Z.D., Zhang, Y.M., Ma, L., Zhang, He, X., Wilson, R.. Byrd, D.M., Griffiths, J.F., Lai, S. He, L., Grumski, K., Lamm, S.H. 1995. Chronic arsenicism and cancer in Inner Mongolia—consequences of well water arsenic levels greater than 50 μg/l. In: C.O. Abernathy, R.L. Calderon, W.R. Chapell (ed), *Arsenic Exposure and Health Effects*, pp. 55–68. Chapman & Hall, London.

Ma, H.Z., Guo, X.J., Yu, G.J., Wu, K.G., Xia, Y. J., Dang, Y.H., Li, Y.H., Zheng, Z., Zhou, H.J., Wang, F.Z., Li, Z.Y., Li, Z.Z., Wu, R.N. 1995. Clinical features of arsenicism in endemic area (Inner Mongolia) with arsenic contamination in drinking water. *J. Chinese Endemic Diseases*, Special suppl. 17–24.

Sun, T.Z., Wu, K.G., Xing, C.M., Ma, H.Z., Yu, G.J., Li, Y.H., Xia, Y.J., Guo, X.J., Zheng, Z. 1995. Epidemiological Investigation of Arsenicism of Endemic area in Inner Mongolia. *J. Chinese Endemic Diseases*, Special Suppl., 1–4.

Yu, G.J., Ma, H.Z., Wu, K.G., Xia, Y.J., Li, Y.H., Zheng, Z., Qin, Y.X., Zhou, J.H. 1995. Investigation of Environmental Arsenic in Arsenicism Endemic Area of Inner Mongolia. *J. Chinese Endemic Disease*, Special Suppl., 10–14.

Arsenic Exposure and Health Effects
W.R. Chappell, C.O. Abernathy and R.L. Calderon (Editors)
1999 Elsevier Science B.V.

Drinking Water Arsenic: The Millard County, Utah Mortality Study

Denise Riedel Lewis

ABSTRACT

The EPA risk assessment for drinking water arsenic is based on skin cancer and wide-ranging exposures of between 10 and 1,820 μg/L drinking water arsenic from studies conducted in Taiwan in the 1960s. This investigation in a cohort of 4,058 residents from Millard County, Utah, represents one of the larger studies in a U.S. population. Exposure was based on drinking water arsenic concentrations performed by the Utah Health Laboratory using EPA approved test methods. Median arsenic concentrations in drinking water ranged from 14 μg/L to 166 μg/L. The cohort was established based on historic membership records of the Church of Jesus Christ of Latter-Day Saints (LDS). Current vital status was determined by the LDS. Death certificates of the deceased were collected and reviewed for cause of death. Cause of death was coded using the International Cause of Death (ICD) version 9 coding rubric. Results from the standard mortality ratio (SMR) analysis include statistically significant associations for hypertensive heart disease (SMR=2.20), nephritis and nephrosis (SMR=1.72), and prostate cancer (SMR=1.45) among males, and hypertensive heart disease (SMR=1.73) and all other heart disease including pulmonary heart disease and diseases of the pericardium (SMR=1.43) among females. These results indicate that cancer and other health effects may be important at low exposure levels of less than 200 μg/L. A Cox proportional hazards analysis using an exposure matrix to include the number of years in residence and the median arsenic level for the town of residence is underway.

Keywords: arsenic, drinking water, mortality, cancer, cardiovascular effects, Utah

The views expressed in this report are those of the author and do not necessarily reflect the policies or opinions of the US EPA.

INTRODUCTION

The U.S. National Interim Primary Drinking Water Regulation was set at 50 μg/L in 1975 (US EPA, 1976). Cross-sectional studies that were completed in the 1960's from Taiwan (Tseng et al., 1968; Tseng, 1977) that reported associations with blackfoot disease, a vaso-occlusive disorder that has never been observed in U.S. populations, and skin cancer have been used in a risk assessment of skin cancer (US EPA, 1988). Previous studies of arsenic in drinking water in the U.S. have evaluated non-melanoma skin cancer (Harrington et al., 1978; Kreiss et al., 1983; Morton et al., 1976; Valentine, 1994; Wong et al., 1992), bladder cancer (Bates et al., 1995), vascular disease (Engel & Smith, 1994), reproductive effects (Aschengrau et al., 1989; Zierler et al., 1988), and toxic effects (Feinglass et al., 1973; Southwick et al., 1982; Warner et al., 1994). The results from these studies have been largely negative.

In the late 1970s, EPA conducted a small study of acute health effects in Millard County, Utah on a population from several small towns exposed to drinking water with a mean arsenic concentration of at least 150 μg/L (range 53 to 750 μg/L). In order to conduct a mortality study, a cohort of Millard County residents was established based on the 1970s studies. The intent of the current study is to examine the health effects of chronic consumption of arsenic-contaminated drinking water in a U.S. population. This paper describes the results of an analysis of drinking water arsenic exposures of less than 200 μg/L and cancer and non-cancer health effects in a U.S. population. Results for cancer and non-cancer causes of death are presented with drinking water arsenic exposure concentrations that consider residence time in the geographic study area.

METHODOLOGY

Cohort Assembly

The cohort was assembled from historical "ward" membership records of the Church of Jesus Christ of Latter Day Saints (LDS) also known as the Mormons. All members who ever lived in a ward during a specific time period were registered in the records. The records were compiled by ward members. In this study, the boundaries of the LDS Church wards are closely aligned with the respective town boundaries. The wards and dates from the historical membership books, which were used in constructing the cohort for the towns included were Delta 1921–1924 (original ward), 1927–1941 (first ward), 1939–1941 (second ward), and 1918–1941 (third ward); Hinckley (1932–1941); Deseret (1933–1945); Oasis (1900–1945); and Abraham (1900–1944). Information abstracted from the historic ward records included individual characteristics such as church ward, family relationships, birth date, death date, location of death, and when the person moved into or out of that church ward. Additional information was collected from other sources including death certificates, and drinking water arsenic concentrations from historic records from the State of Utah.

Water Samples

Potential drinking water arsenic exposure for cohort members was determined by historical records of arsenic measurements in drinking water maintained by the State of Utah dating back to 1964. An overview of arsenic concentrations in drinking water and sources of exposure information for the study area was presented in a previous feasibility assessment (Lewis et al., 1998). In this study, arsenic exposure levels for the communities were based on measurements that were performed by the Utah State Health Laboratory, which participated in EPA's quality assurance program and water quality proficiency testing. The samples must have originated from a clearly located water source used for culinary or potable purposes only (not for agricultural or irrigation purposes). The analysis date had to have been 1976 or later, when the sample collection method involved acidification of the collection containers.

TABLE 1

Distribution of arsenic drinking water concentrations from historical and recent arsenic measurement data for communities in the study area (Lewis et al., 1999)

Town	Number	Median	Mean	Min. arsenic concentration (ppb)	Max. arsenic concentration (ppb)	Standard deviation
Hinckley	21	166	164.4	80	285	48.1
Deseret	37	160	190.7	30	620	106.6
Abraham	15	116	134.2	5.5	310	67.2
Sugarville	6	92	94.5	79	120	15.3
Oasis	7	71	91.3	34	205	57.8
Sutherland	19	21	33.9	8.2	135	31.8
Delta	46	14	18.1	3.5	125	17.7

In all, 151 samples of drinking water were used in assessing the potential exposure of cohort members to arsenic in drinking water. The distribution of the concentrations of arsenic in drinking water in the study communities is provided in Table 1 in order of highest to lowest median concentration. The Delta water samples came from the Delta public water system, and samples from Abraham, Deseret, Oasis, Sugarville, and Sutherland were taken from private drinking water wells. No additional water samples were taken for Hinckley because the original wells were abandoned in 1981 when a new, low arsenic source (less than 50 μg/L) of public drinking water was provided to Hinckley residents.

Arsenic Exposure Index

An arsenic exposure index score was calculated for each individual in the cohort. The exposure index was derived from the number of years of residence in the community and the median arsenic concentration of drinking water arsenic in the community (Lewis et al., 1999). Residence was determined by the members' entry into historical LDS Church censuses that were conducted roughly every five years by the LDS between 1914 and 1962 to determine where individual members lived throughout the world. Census years were 1914, 1920, 1925, 1930, 1935, 1940, (1945 skipped), 1950, 1955, 1960, and 1962. Data extracted from the censuses included date of census and residence at the time of the census.

Based on the exposure index values, three exposure categories were used in the Cox proportional hazards analysis (see description below). The referent group comprised those with exposures of less than 1,000 μg/L*years. These individuals were believed to be at low risk for outcomes with long latencies of 20 years or so if their exposures were at 50 μg/L on average, hence 50 μg/L drinking water arsenic × 20 years = 1,000 μg/L*years. The other exposure groups were 1,000 to 3,999 μg/L*years; 4,000 to 5,999 μg/L*years; and greater than or equal to 6,000 μg/L*years.

Statistical Analysis

The cohort data analysis uses standardized mortality ratios (SMRs) as the measure of association (Rothman and Greenland, 1998). The OCMAP program (Marsh et al., 1989), adapted to a non-occupational cohort, was used to compare the observed number of deaths with the expected number of deaths generated from death rates from the white male and white female general population of Utah within a given underlying cause of death category. Death rates for the State of Utah were available for the years 1960 through 1992 for diseases other than cancer, and from 1950 to 1992 for cancers. The death rates were applied in 5-year increments, with the exception of the 1990–1992 period. For those who died of causes other than cancer before 1960, the 1960–1964 death rates for causes other than cancer were applied.

Similarly, for those who died of cancer before 1950, the 1950–1954 cancer death rates were applied. For those who died after 1992, the 1990–1992 death rates for the cancer or non-cancer cause of death were applied.

A Cox proportional hazards analysis using internal comparison groups (exposure to arsenic of greater than or equal to 1,000 μg/L*years versus low exposure to arsenic of less than 1,000 μg/L*years) to calculate relative risks as the epidemiologic measure of association between mortality and exposure is underway. This analysis utilizes SAS software (SAS Version 6.12, 1996) and the procedure PHREG for proportional hazards (Allison, 1995). The Mantel–Haenszel test for trend was computed to examine whether risk of mortality from selected outcomes is associated with the drinking water arsenic exposure categories in a dose–response relationship (Mantel, 1963).

RESULTS

Cohort members were enrolled from historical LDS ward registries: 1,191 (29.4%) from Delta; 1,192 (29.4%) from Hinckley, and the remaining 1,675 (41.2%) were enrolled from historical ward registries from the surrounding areas of Deseret, Abraham, and Oasis. In all, 2,092 (51.6%) were male and 1,966 (48.5%) were female. At the end of cohort assembly in November 1996, 1,551 (38.2%) were alive, 2,203 (54.3%) were deceased, and 300 (7.4%) were lost-to-follow-up. Four individuals were less than 1 year of age and were not included in further analysis.

Non-Cancer SMR Results

Selected results for non-cancer causes of death are presented in Table 2. Death from hypertensive heart disease was significantly increased for both females (SMR=1.73) and males (SMR=2.20). Nephritis and nephrosis was significantly increased for males (SMR= 1.72) and non-significantly elevated for females (SMR=1.21). All other heart disease, a category that excludes major causes like arteriosclerosis and cerebrovascular disease, was significantly increased for females only (SMR=1.43). Non-cancer causes of death that were elevated, but not statistically significant included benign neoplasms for females (SMR=1.96) and males (SMR=1.05) and arteriosclerosis for both females (SMR=1.18) and males (SMR= 1.24). Diabetes mellitus was elevated for females only (SMR=1.23). Non-cancer causes of death that were observed significantly less in the study population than in the general population of Utah included ischemic heart disease for females and males, and cerebro-vascular disease and non-malignant respiratory causes for males.

Cancer SMR Results

Among cancer outcomes, only prostate cancer (males only) was significantly elevated (SMR=1.45) (Table 2). Prostate cancer was also the only endpoint that appeared to follow a dose–response type of relationship whereby the SMRs increased with the exposure (data not shown, Lewis et al., 1999). Other cancer causes of death were non-significantly increased in the study population and included kidney cancer for females (SMR=1.60) and males (SMR=1.75). Melanoma of the skin (SMR=1.82) and cancer of the biliary passages (SMR= 1.42) was non-significantly elevated among females. Cancer outcomes that were observed significantly less often in the study population than in the general population of the state of Utah included cancer of the digestive organs and peritoneum in females and males, respiratory system cancer in females and males, breast and pancreatic cancer in females, and cancer of the large intestine in males. Other outcomes not listed in Table 2 include cancer of the stomach, biliary passages and liver, uterine cancer, other female cancers, bladder cancer, central nervous system cancer, and cancer of the lymphatic and hematopoietic tissue. The results for these outcomes were decreased associations that were nonsignificant.

TABLE 2

Selected standard mortality ratios for non-cancer and cancer causes of death

	Females		Males	
	SMR	(95%CI)	SMR	(95%CI)
Non-Cancer Outcome				
Significantly Increased:				
Hypertensive Heart Disease	1.73	(1.11–2.58)	2.20	(1.36–3.36)
Nephritis and Nephrosis	see below		1.72	(1.13–2.50)
All Other Heart Disease	1.43	(1.11–1.80)	(0.94)	non-significant
Elevated, Not Significantly Increased:				
Benign Neoplasms	1.96	(0.85–3.86)	1.05	(0.29–2.69)
Arteriosclerosis	1.18	(0.68–1.88)	1.24	(0.69–2.04)
Diabetes Mellitus	1.23	(0.86–1.71)	(0.79)	non-significant
Nephritis and Nephrosis	1.21	(0.66–2.03)	see above	

Significantly Decreased Non-Cancer (SMR < 1.0): Ischemic Heart Disease (females and males); Cerebrovascular Disease (males); Respiratory Causes (males); All Heart Disease (females and males); Bronchitis, Emphysema and Asthma (males).

	Females		Males	
Cancer Outcome				
Significantly Increased:				
Prostate Cancer	—		1.45	(1.07–1.91)
Elevated, Not Significantly Increased:				
Kidney Cancer	1.60	(0.44–4.11)	1.75	(0.80–3.32)
Melanoma of the Skin	1.82	(0.50–4.66)	(0.83)	non-significant
Biliary Passages	1.42	(0.57–2.93)	(0.85)	non-significant
All Other Malignant Neoplasms	1.34	(0.84–2.03)	(0.96)	non-significant

Significantly Decreased Cancer Outcome (SMR < 1.0): Digestive Organs and Peritoneum (females and males); Respiratory System (females and males); Breast (females); Pancreas (females); Large Intestine (males); All Malignant Cancers (females and males).

Despite recent reports of bladder cancer associated with increased intake of drinking water arsenic in Argentina (Hopenhayn-Rich et al., 1996), there were only 5 bladder cancer deaths observed in the Utah mortality cohort. Bladder cancer incidence in Utah is relatively low with bladder cancer accounting for 5.2% of all incident cancers among males (incidence rate = 16.2 per 100,000) and 1.4% of all incident cancers among females (incidence rate = 3.3 per 100,000) (Parkin et al., 1997). The average annual age-adjusted bladder cancer mortality rate for Utah from 1987 to 1991 was 4.1 per 100,000 for males and 1.0 per 100,000 for females (Ries et al., 1994). This proportionately low rate of incidence and low mortality for bladder cancer combined with the fact that bladder cancer was associated with slightly higher concentrations of arsenic in Argentina may partially explain the lack of an association with bladder cancer in this study. In the Cox proportional hazards analysis (see the next section), there is a non-significant association with bladder cancer but only in the highest exposure group with very low numbers.

Preliminary Cox Proportional Hazards Results

To evaluate drinking water arsenic exposure using internal comparisons, Cox proportional hazards models were calculated. The outcome variable was the specific cause of death. The exposure variable was based on median arsenic exposure times the number of years in residence at the median exposure (see section on *Arsenic Exposure Index*). The exposure categories included less than 1,000 μg/L*years; 1,000 to 3,999 μg/L*years; 4,000 to 5,999 μg/

TABLE 3

Utah mortality cohort: selected Cox proportional hazards analysis[1] preliminary relative risk results

	1000–3999 μg/L*yrs[2]	4000–5999 μg/L*yrs[2]	≥6,000 μg/L*yrs[2]	Trend[3]
	RR	RR	RR	
Non-Cancer Outcome				
All Causes	1.06	1.33**	1.52**	$p=0.001$
Diabetes Mellitus	1.73	0.84	1.31	n.s.
Cerebrovascular Dis.	1.00	0.99	1.54*	n.s.
All Heart Disease	0.95	1.39**	1.51**	n.s.
Ischemic Heart Dis.	0.88	1.66**	1.41*	n.s.
Hypertension w/ Ht. Dis.	1.32	2.46	4.14**	n.s.
Arteriosclerosis	0.75	0.68	0.70	n.s.
All Other Heart Dis.	1.08	0.56	1.16	$p < 0.05$
Nephritis & Nephrosis	1.34	0.31	2.00	$p=0.06$
Cancer Outcome				
Malignant Neoplasms	1.09	1.33	2.04*	$p < 0.05$
Digestive Org. & Perit.	0.88	1.14	2.05*	n.s.
Stomach	1.71	1.60	4.71*	n.s.
Pancreas	8.96*	11.65*	9.43	n.s.
Prostate (males only)	1.85	1.57	2.60*	$p < 0.05$
Bladder	0.0	0.0	7.67	n.s.

[1] Model adjusted for birth year (before 1900, 1900–1920, 1920–1950), gender, and cumulative exposure to arsenic (years of residence).
[2] Exposure categories based on median level of arsenic exposure (μg/L) times the number of years of residence. Estimates are relative to less than 1,000 μg/L*yrs.
[3] Mantel–Haenszel chi-square test for trend. **Significant at $p < 0.01$; *significant at $p < 0.05$; n.s. non-significant at $p=0.05$.

L*years; and greater than or equal to 6,000 μg/L*years. Covariates included in the models are birth cohort (born before 1900, 1900–1920, 1920–1950), gender, and number of years in the cohort. Relative risk results for selected mortality outcomes are presented in Table 3 for noncancers and cancers. Also shown are the results of the Mantel–Haenszel test for trend. Evidence of an increase in risk with an increase in exposure can be observed for all causes of death, cerebrovascular disease, all heart disease, hypertension with heart disease, all other heart disease, and nephritis and nephrosis, however not all of these are significant according to the Mantel–Haenszel test. For cancer mortality, significant trends for all malignant neoplasia and prostate cancer were observed. Non-significant trends were observed for cancer of the digestive organs and peritoneum, and stomach cancer. Analysis of the trend associations is continuing. There were only five deaths from bladder cancer, two in the lowest exposure group at <1,000 μg/L*years, and three in the highest group at ≥6,000 μg/ L*years (RR=7.67, $p > 0.05$).

DISCUSSION

The results of this mortality study indicate that epidemiologic studies of drinking water arsenic and health outcomes in U.S. populations are possible. To improve exposure assessment, drinking water arsenic concentration data may be collected from several different sources including water quality and water rights offices which not only can provide information about the concentration, but often the exact location and intended use, i.e., whether the source provides water for human consumption, irrigation, or livestock use. In addition, residence history data should be collected whenever possible so that location of residence can be paired with drinking water arsenic concentration for use in an exposure matrix similar to the one described in this paper or elsewhere (Chiou et al., 1995).

The epidemiologic measures of association that have been presented, specifically the SMR and the relative risk (RR), indicate that lower exposures to drinking water arsenic of 200 µg/L or less for median concentration of exposure appear to be important contributors to certain types of cardiovascular disease and cancer causes of death. Since this paper was presented in August 1998, a full report of the SMR results has been published (Lewis et al., 1999). The Cox proportional hazards (RR) analysis is ongoing and is expected to be complete by mid to late 1999.

CONCLUSIONS

While the existing reports on drinking water arsenic and health effects often share similar findings, the types of exposures incurred by the study populations are often very different. In the United States alone, a considerable amount of the naturally occurring drinking water arsenic in the west is believed to be due to previous volcanic activity that has resulted in arsenic uptake in the water. However, other areas of the United States and other countries may have different geologic explanations for the occurrence of arsenic in the water. While the results of this study are from a population where previous volcanic activity is believed to be a major factor for increased drinking water arsenic concentrations, additional studies are necessary to describe the various exposure scenarios from geologically diverse locations and to examine whether the type and magnitude of the associated health effects are similar. Future plans for additional studies of drinking water arsenic in the United States include convening a workshop of selected states to discuss potential exposure to drinking water arsenic and to discuss sources of data on health effects to plan additional epidemiologic studies.

ACKNOWLEDGMENTS

The author wishes to acknowledge the following individuals. Jerry Rench (currently with RTI, Inc., Rockville, Maryland), J. Wanless Southwick, Rita Ouellet-Hellstrom, Ron O'Day and Linda Dudley at SRA Technologies, Falls Church, Virginia. Larry Scanlan, Becky Hylland, and Jerry Olds with the State of Utah. Neil Forster, Director of Public Works, Delta, Utah. Ward Petersen of Hinckley, Utah and Rawlin Dalley with Hinckley Public Works, Hinckley, Utah.

REFERENCES

Allison, P.D. 1995. Chapter 5: Estimating Cox regression models with Proc PHREG. In: Allison, P.D., *Survival Analysis Using the SAS System. A Practical Guide*, pp. 111–184. SAS Institute, Inc., Cary, North Carolina, USA.

Aschengrau, A., Zierler, S., Cohen, A. 1989. Quality of community drinking water and the occurrence of spontaneous abortion. *Arch. Environ. Health*, **44(5)**, 283–289.

Bates, M.N., Smith, A.H., Cantor, K.P. 1995. Case-control study of bladder cancer and arsenic in drinking water. *Am. J. Epidemiol.*, **141**, 523–530.

Chiou, H-Y., Hsueh, Y-M., Liaw, K-F., Horng, S-F., Chiang, M-H., Pu, Y-S., Lin, J. S-N., Huang, C-H., Chen, C-J. 1995. Incidence of internal cancers and ingested inorganic arsenic: a seven-year follow-up study in Taiwan. *Cancer Res.*, **55**, 1296–1300.

Engel, R.R., Smith, A.H. 1994. Arsenic in drinking water and mortality from vascular disease: an ecologic analysis in 30 counties in the United States. *Arch. Environ. Health*, **49**, 418–427.

Feinglass, E.J. 1973. Arsenic intoxication from well water in the United States. *N. Engl. J. Med.*, **288**, 828–830.

Harrington, J.M., Middaugh, J.P., Morse, D.L., Housworth, J. 1978. A survey of a population exposed to high concentrations of arsenic in well water in Fairbanks, Alaska. *Am. J. Epidemiol.*, **108**, 377–385.

Hopenhayn-Rich, C., Biggs, M.L., Fuchs, A., Bergoglio, R., Tello, E.E., Nicolli, H., Smith, A.H. 1996. Bladder cancer mortality associated with arsenic in drinking water in Argentina. *Epidemiology*, **7**, 117–124.

Kreiss, K., Zack, M.M., Feldman, R.G., Niles, C.A., Chirico-Post, J., Sax, D.S. 1983. Neurologic evaluation of a population exposed to arsenic in Alaskan well water. *Arch. Environ. Health*, **38**, 166–121.

Lewis, D.R., Southwick, J.W., Ouellet-Hellstrom, R., Rench, J., Calderon, R. 1999. Drinking water arsenic in Utah: A cohort mortality study. *Environ. Health Perspect.*, **107**, 359–365.

Lewis, D.R., Southwick, J.W., Scanlan, L.P., Rench, J., Calderon, R. 1998. The feasibility of conducting epidemiologic studies of waterborne arsenic. A mortality study in Millard County, Utah. *J. Environ. Health*, **60**, 14–19.

Mantel, N. 1963. Chi-square tests with one degree of freedom: extension of the Mantel–Haenszel procedure. *J. Am. Stat. Assoc.*, **58**, 690–700.

Marsh, G.M., Preininger, M., Ehland, J., Caplan, R., Bearden, A., CoChien, H., Paik, M. 1989. OCMAP, OCMAP/PC. Mainframe and Microcomputer Version 2.0. University of Pittsburgh, Pittsburgh, PA.

Morton, W., Starr, G., Pohl, D., Stoner, J., Wagner, S., Weswig, P. 1976. Skin cancer and water arsenic in Lane County, Oregon. *Cancer*, **37**, 2523–2532.

Parkin, D.M., Whelan, S.L., Ferlay, J., Raymond, L., Young, J. (eds.). 1997. *Cancer Incidence in Five Continents, Vol. VII*. International Agency for Research on Cancer, Lyon, France.

Ries, L.A.G., Miller, B.A., Hankey B.F., Kosary C.L., Harras, A., Edwards, B.K. (eds.). 1994. SEER Cancer Statistics Review, 1973–1991: Tables and Graphs, National Cancer Institute. NIH Pub. No. 94–2789. Bethesda, Maryland, USA.

Rothman, K.J., Greenland, S. 1998. Introduction to Categorical Statistics. In: K.J. Rothman and S. Greenland, (eds.). *Modern Epidemiology, 2nd ed.*, pp. 234–236. Lippincott-Raven Publishers, Philadelphia, Pennsylvania, USA.

SAS Version 6.12. 1996. SAS Institute, Inc., Cary, North Carolina, USA.

Southwick, J.W., Western, A.E., Beck, M.M., Whitley, J., Isaacs, R. 1982. *Community health associated with arsenic in drinking water in Millard County, Utah*. U.S. Environmental Protection Agency. Pub. No. EPA-600/S1-81-064. EPA Health Effects Research Laboratory, Cincinnati, Ohio, USA.

Tseng, W.P. 1977. Effects and dose–response relationships of skin cancer and blackfoot disease with arsenic. *Environ. Health Perspect.*, **19**, 109–119.

Tseng, W-P., Chu, H-M., How, S-W., Fong, J-M., Lin, C-S. and Yeh, S. 1968. Prevalence of skin cancer in an endemic area of chronic arsenicism in Taiwan. *J. Natl. Cancer Inst.*, **40**, 453–463.

US EPA (1976). National Interim Primary Drinking Water Regulations. Washington, DC, US Environmental Protection Agency (EPA/570/9-76-003).

US EPA (1988) Special report on ingested arsenic: skin cancer and nutritional essentiality. Washington, DC, US Environmental Protection Agency Risk Assessment Forum (EPA/625/3-87/013).

Valentine, J.L. 1994. Review of health assessments for US/Canada populations exposed to arsenic in drinking water. In: Chappell, W.R., Abernathy, C.O. and Cothern, C.R. (eds.). *Arsenic Exposure and Health* (a special issue of Environmental Geochemistry and Health Volume 16), pp. 139–152. Laws and Stimson Associates, Surrey, UK.

Warner, M.L., Moore, L.E., Smith, M.T., Kalman, D.A., Fanning, E., Smith, A.H. 1994. Increased micronuclei in exfoliated bladder cells of individuals who chronically ingest arsenic-contaminated water in Nevada. *Cancer Epidemiol. Biomarkers Prevent.*, **3**, 583–590.

Wong, O., Whorton, M.D., Foliari, D.E., Lowengart, R. 1992. An ecologic study of skin cancer and environmental arsenic exposure. *Int. Arch. Occup. Environ. Health*, **64**, 235–241.

Zierler, S., Theodore, M., Cohen, A., Rothman, K. 1988. Chemical quality of maternal drinking water and congenital heart disease. *Int. J. Epidemiol.*, **17**, 589–94.

Association between Chronic Arsenic Exposure and Children's Intelligence in Thailand

Unchalee Siripitayakunkit, Pongsakdi Visudhiphan, Mandhana
Pradipasen, Thavatchai Vorapongsathron

ABSTRACT

Previous studies have reported high arsenic level in hair of children at Ronpiboon subdistrict. It is possible that the accumulation of arsenic in their bodies may adversely affect intelligence. This study aims to explore the relationship between arsenic level in hair and intelligence of children. We measured the arsenic level in hair using atomic absorption spectrophotometry method as the indicator of chronic arsenic exposure and IQ with Wechsler Intelligence Scale Test for Children. Potential confounders were collected at the same time period of this cross-sectional study between 16 January and 5 March, 1995. To explore the association, multiple classification analysis was conducted with data from 529 children aged 6–9 years who had lived in Ronpiboon district since birth. This study found an association between arsenic hair levels and children's intelligence. After adjusting for confounders, we observed a statistically significant relationship that arsenic could explain 14% of variance in children's IQ. This result revealed that chronic arsenic exposure as shown by hair samples was related to retardation of intelligence in children. Prevention of further arsenic exposure and health status monitoring of children with arsenic accumulation should be implemented.

Keywords: association, arsenic exposure, arsenic level in hair, IQ, WISC, AAS

INTRODUCTION

Arsenic in water supplies has occasionally caused poisoning, usually of a chronic rather than acute nature (Ferguson and Gavis, 1972). Chronic environmental exposure to well water naturally high in arsenic has been described in Chile, Taiwan, Japan, and other parts of the world, including instances where keratosis, and possibly skin cancer, have resulted from such exposures (Armstrong et al., 1984). In Thailand, environmental arsenic exposure has received attention since 1987 because of skin manifestations or "Black Fever" resulting from ingestion of water containing arsenic (Division of Environmental Health, 1992; Choprapawan, 1994). Most of the inhabitants of Ronpiboon subdistrict prefer the daily drinking and use of well-water which has been contaminated with arsenic for a long time. They have habitually ingested this well-water for drinking and cooking because it is sweet, delicious and full (Ajjimangkul, 1992; Division of Environmental Health, 1992; Choprapawan, 1994). Chronic arsenic poisoning will increase in severity day after day unless inhabitants stop ingesting well-water and the drinking water supply is made safe for them. Neither the routine monitoring of arsenic contamination of well-water nor the monitoring of public drinking water for safety is presently mandated in Ronpiboon subdistrict. The population living in this area is still at risk of gradual chronic arsenic exposure.

Examinations of skin lesions from chronic arsenic poisoning in October 1987 found that children aged less than 10 years had proportionally more severe skin lesions than other age groups. Nevertheless, children aged 12–15 years (44%) had high arsenic levels in their hair and in their nails (78%) (Piampongsant and Udomnitikul, 1989; Choprapawan, 1994). A recent report indicated 89.8% of children aged 0–9 years were found with an arsenic level of ≥ 0.2 ppm in their hair (Rodklai, 1994). Many children had arsenic accumulated in their bodies though they had no skin changes. What are the health effects of chronic arsenic poisoning to these children? Is it possible that the high arsenic concentrations in their bodies will affect intellectual development? This question is interesting, yet there is no clear answer. Prior to the study, the pilot results showed that eight pupils who had accumulated high arsenic levels in their hair ($>1\,\mu g/g$) had IQs between 68 and 103, and two of them were classified as mental defective. The other four pupils with normal hair arsenic levels had IQs between 83 and 122, two of them had above average intelligence. The results were interesting and suggested a further need to explore the effect of chronic arsenic exposure on the development of intelligence in young children. The children in Ronpiboon subdistrict constitute a critical group for chronic low-level arsenic exposure. It is worthwhile to undertake epidemiological investigation to determine the effect of arsenic exposure in this child group, in order to obtain information on the possible effect on the intelligence of children. The present study aimed to test the association between chronic arsenic exposure, indicated by the arsenic levels in hair, and children's intelligence among children living in Ronpiboon district since birth. It is a pioneer study conducted in humans, to try to determine the exposure factor related to the developmental defect of children with high arsenic concentrations in their hair.

METHODOLOGY

Study Population

The children who were born between 1986 and 1989 in Ronpiboon and Soa Thong subdistricts were the study population. To prevent a distorted association of results, the following criteria were employed: (1) the subjects must have lived in the study area since birth; (2) the subjects' parents must be a married couple, live together and look after the children; (3) the subjects were six to nine years old at the time of the IQ test. The subjects' selection was done using simple random sampling. We selected fifteen schools from 21 schools (71.4%) and selected 529 subjects from 838 children who met the criteria (63.1%).

Data Collection

Data were collected between 16 January and 5 March, 1995. The collection of data was done for two days in each school as follows: On the first day: (1) Each subject was interviewed using the child interview form. This form was constructed to collect variables: gender, parent arguing, child rearing and food intake. (2) The subject was measured for visual acuity and hearing with Snellen's chart and Impact Audiometer Model 1001, respectively. (3) Child's hair was cut approximately 2–3 cm from the scalp and about 1 g of each. (4) Then, each subject got the father's questionnaire for completion by his father at home. The questionnaire was developed to assess father's variables: education, history of slow learning, occupation and income. On the second day: (1) We received the fathers' questionnaire from the subjects. (2) Each subject was administered with the Wechsler Intelligence Scale Test for Children (WISC) (Wechsler, 1949) for IQ determination by the psychologist. (3) Subjects' mothers were interviewed for data according to the maternal interview form. This form was constructed to collect variables: prenatal factors, perinatal factors, postnatal factors, birth order, birth weight, breast feeding, illness history, food intake, family size, child rearing, mother's education, occupation, and income. Then, they were tested for intelligence with the Progressive Matrices slides in groups of 10–15 persons.

Arsenic Analysis

The hair samples were sent to the Faculty of Pharmaceutical Science, Prince of Songkla University for arsenic analysis with atomic absorption spectrophotometry (AAS) method using a GBC 906 automatic multi-element atomic absorption spectrophotometer with the GBC HG 3000 hydride generator (Chapple and Danby, 1990).

Statistical Analysis

Descriptive statistics of frequency and percent distribution were calculated to depict basic characteristics of the sample children as well as for arsenic level in hair and IQ. The chi-square test was performed to examine the difference of potential confounders distributed among subjects. We explained the association between arsenic level in hair and intelligence of children by simultaneously adjusting for confounders in multiple classification analysis. The data was analysed by the statistical package of SPSS/PC+ version 4 (Norusis, 1990).

RESULTS

Characteristics of the Subjects

The total study subjects numbered 529 persons: 353 from Ronpiboon subdistrict and 176 from Soa Thong subdistrict. The subjects consisted of children from kindergarten, grades I and II, 30.8%, 36.5%, 32.7%, respectively. The male to female ratio was 1.08:1. The percentage of children of the first and the second birth order was 35.2% and 28.5%, respectively. Eighty-two percent of the subjects were from families of 3 to 6 persons. Half of the subjects were from low income families ($\leq$ 5,000 baht/month).

Distribution of Arsenic Level in Hair and IQ

The mean hair arsenic concentration for all subjects was 3.52 μg/g (SD = 3.58), the median hair arsenic was 2.42 μg/g. The range of arsenic levels in hair was 0.48 to 26.94 μg/g. Around half of the children (55.4%) had arsenic levels between 1.01 and 3 μg/g. Only 44 of 529 (8.3%) children had normal arsenic levels in hair ($\leq$1 μg/g) as shown in Table 1. The mean IQ of the study subjects was 90.44 points. The mean of Verbal IQ, Performance IQ, and each WISC subtest scores are presented in Table 2.

TABLE 1

Number and percentage of arsenic levels in hair of children aged 6–9 years in Ronpiboon district, 1995

Arsenic level (μg/g)	Number	%
≤1	44	8.3
1.01–2	146	27.6
2.01–3	147	27.8
3.01–4	60	11.3
4.01–5	37	7.0
5.01–10	71	13.4
>10	24	4.6
Total	529	100.0

TABLE 2

Mean scaled score of WISC subtests and IQ of children aged 6–9 years, Ronpiboon district, Nakorn Si Thammarat province, 1995

Variables	Mean	SD	Range
Information	8.42	2.71	3-18
Comprehension	8.02	2.55	3-18
Arithmetic	10.51	2.89	3-20
Similarities	8.74	2.85	2-18
Digit Span	10.27	3.10	2-20
Picture Completion	7.40	2.42	2-18
Picture Arrangement	7.36	2.38	2-16
Block Design	8.98	2.31	2-18
Object Assembly	6.96	2.31	1-16
Coding	10.30	2.84	4-19
Verbal IQ	94.70	11.94	63-130
Performance IQ	87.47	11.67	55-128
IQ	90.44	11.45	54-123

Most of the children's IQs were classified in the average and the dull normal groups at 45.7% and 31.6%, respectively. Around half of children (48.4%) had below average IQ with 13.8% and 3%, respectively, in the borderline and mental defective groups as shown in Figure 1. The lowest IQ of total subjects was 54, the highest was 123 and no subject scored at the very superior level.

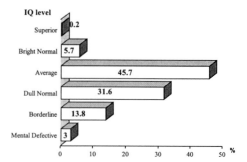

Fig. 1. Level and percentage of IQ score of children aged 6–9 years, Ronpiboon district, Nakorn Si Thammarat province, 1995.

TABLE 3

Percentage of IQ scores in children aged 6–9 years by arsenic level, Ronpiboon district, Nakorn Si Thammarat province, 1995

IQ level	Arsenic level (μg/g)			
	≤1 (%)	1.01–2 (%)	2.01–5 (%)	>5 (%)
Mental Defective (≤69)	0	2.1	2.9	6.3
Borderline (70–79)	11.4	12.3	16.0	11.6
Dull Normal (80–89)	22.7	27.4	33.2	37.9
Average (90–109)	56.8	50.7	43.0	40.0
Bright Normal (110–119)	9.1	6.8	4.9	4.2
Superior (120–129)	0	0.7	0	0
Total	100.0	100.0	100.0	100.0
	$n = 44$	$n = 146$	$n = 244$	$n = 95$

Association between Arsenic Level and IQ

Results of IQ level by arsenic group are presented in Table 3. The percentage of children in the average IQ group decreased remarkably from 56.8 to 40.0 as the arsenic level increased. The percentage in the above average IQ group decreased from 9.1 to 4.2. On the other hand, the percentage at the dull normal level increased as the arsenic level increased from 22.7% to 37.9%, and percentage at the mental defective level increased from 0 to 6.3%.

The findings of arsenic levels in hair showed that few of the total subjects (8.3%) had normal hair arsenic (≤1 ppm) and this group was only in Soa Thong subdistrict. Thus it was not possible to analyse IQ comparing normal with high arsenic level groups. The student t-test and chi-square test showed no significant differences between normal (≤1 ppm) and slightly elevated (1.01–2 ppm) arsenic groups for all variables. Comparison of IQ between the arsenic subjects ≤2 ppm and arsenic subjects >2 ppm was performed. Multiple classifications were applied to selecting the parsimonious model for explanation of children's IQ variation. This model consisted of children's age and four significant variables associated with children's IQ as follows: arsenic level, father's occupation, mother's intelligence score and family income. Arsenic could explain about 14% of the variance in IQ after controlling for other risk factors ($p=0.002$). Mean IQ in the low arsenic group was two points above the grand mean while mean IQ in the high arsenic group was 0.75 points below the grand mean and that in the very high arsenic group was two points below the grand mean (Table 4). The mean of children's IQs in different hair arsenic groups is presented in Table 5.

DISCUSSION

Since this is the first analysis of association between arsenic exposure and intellectual ability, we are unable to compare this study with other studies for a consistency phenomenon of this relationship. The findings will be mainly discussed within the context of this study.

Arsenic Level and Intelligence

Arsenic can damage the central nervous system, chronic encephalopathy symptoms include diminished recent memory and organic cognitive impairment (Morton and Dunnette, 1994). However, CNS impairment has been less frequently observed. A follow-up study in the Morinaga powdered-milk poisoning case in 1969–1971, reported the victims had a lower IQ than their siblings (Ohira and Aoyama, 1973). The other study presented proportion of victims (21%) with an IQ of less than 85 which exceeded average numbers (Yamashita et al, 1972). The WISC was administered for the assessment of children's IQ in this study. The

TABLE 4

Multiple classification analysis of IQ in children aged 6–9 years, Ronpiboon district, Nakorn Si Thammarat province, 1995 (grand mean = 90.44)

	n	Unadjusted		Adjusted for independents + covariates		p-value
		Deviation	Eta	Deviation	Beta	
Arsenic (ppm)			0.13		0.14	0.002***
Low (≤2)	190	1.98		1.97		
High (2.01–5)	244	−0.77		−0.75		
Very high (>5)	95	−1.97		−2.02		
Father's occupation			0.24		0.17	0.000***
Farmer	211	−2.65		−2.23		
Labourer and miner	248	0.54		0.95		
Trader and government employee	70	6.08		3.33		
Maternal Intelligence score			0.28		0.24	0.000***
Very low (≤ 10)	39	−4.19		−3.67		
Low (11–20)	255	−1.70		−1.55		
Medium (21–30)	137	−0.23		0.08		
High (>30)	98	6.42		5.38		
Family income (baht)			0.21		0.10	0.059
≤ 5,000	262	−0.76		−0.29		
5,001–10,000	224	−0.65		−0.43		
>10,000	43	8.02		4.02		
Multiple R^2					0.16	
Multiple R					0.40	

Level of significance: ***$p < 0.001$.

TABLE 5

Mean of children's IQs in different arsenic groups after controlling for confounders

Arsenic groups (ppm)	Mean (point)	95% CI
Low (≤ 2)	92.41	91.70–93.12
High (2.01–5)	89.69	88.98–90.40
Very high (> 5)	88.42	87.71–89.13

findings showed the percentage of IQ less than 85 was increased with increasing concentrations of arsenic in hair. When the intelligence classification was applied, it showed no mental defect in the normal arsenic group. The percent mental defective increased with increasing hair arsenic levels as shown in Table 3.

The WISC is commonly used in Thai children's IQ assessment. It has been employed to determine IQ of Thai children for more than 40 years, but there is no study of Thai standard IQ by the WISC at present. The interpretation of IQ in this study can be compared to data of the standardization of the WISC (Vogel, 1986). When compared with WISC standard, the below average IQ level in this study (48.4%) was much higher than those in the standard (26.2%). Moreover, the IQ of children in Ronpiboon district among the different arsenic levels showed an increasing proportion of the below average IQ level (<90) with increasing concentrations of arsenic found in hair. This finding suggests a possible association between hair arsenic concentrations and children's IQ.

The findings after controlling for other risk factors showed a significant inverse relationship between arsenic levels and IQ ($p=0.002$). Arsenic levels could explain 14% of the variation in IQ. Other variables were significant explanatory variables of IQ; namely, maternal intelligence, occupation of father, children's age ($p \leq 0.002$), and family income ($p=0.06$).

The brain growth spurt runs from mid-pregnancy to about 3 to 4 years in humans. This period is a focus of early central nervous system (CNS) development, it is characterized by glial proliferation, and subsequently myelination. Exposure to adverse environmental circumstances and abnormal hormonal influences is dangerous at this time. It can lead to permanent impairment or alteration of CNS function that cannot be reversed (Brook, 1982; Meyer-Bahlberg, 1978). In humans, effects of metallic compounds on the developing CNS indicate that arsenic should have an adverse effect (Nordberg, 1988). The children living in the arsenic contamination area had been exposed continuously to low-dose arsenic since conception. Arsenic accumulation can be detected in hair which reflects recent chronic arsenic exposure. These children who had high arsenic concentrations in hair and lower IQ development suggested that chronic low-dose arsenic exposure might adversely affect the CNS.

Neither the arsenic concentrations in biological samples at birth nor the intelligence records of children were available in Ronpiboon district. A prospective study for exploring the chronic arsenic effects in children could not be performed in the limit time. To clarify this association, further research should be conducted in a follow-up study. However, the finding of an inverse association between arsenic concentrations in hair and the IQ of the children living in Ronpiboon district is sensitive to public concern. It demands immediate proper management from the appropriate organizations.

Limitations of the Study Design

The study was a cross-sectional design. It could not evaluate cause and effect because the measurements of chronic arsenic exposure and intelligence were made at the same point in time. Thus, it is impossible to determine a causal relationship. A better study design would be to undertake a prospective study of a birth cohort to relate a longitudinal assessment of arsenic exposure from birth to subsequent measures of intellectual development.

The rationale of cross-sectional design in this study was as follows: (1) There was no clear evidence of a relationship between human arsenic level and intelligence in the literature. The pioneer study should be cross-sectional to explore the possibility of this relationship. (2) No arsenic biomarkers exist which reflect arsenic exposure in children since birth. The prevalence survey of arsenic level in hair of schoolchildren and in the drinking water is a first step in establishing a database. (3) The time period of the cross-sectional study was appropriate to indicate the trend of the association of hair arsenic level and intelligence. The results are more timely for use in the planning and management of problem solving. The prospective results would not be available for many years.

However, the weak point of this study was the difficulty in evaluating a causal relationship. It could not establish the time precedence of arsenic exposure, as the level of arsenic was measured at the same time as IQ. The study suggests that the IQ deficit in children is the result of excess arsenic exposure. This association can be explained by the biological plausibility and dose–response trend. The association is more plausible as study subjects were born in 1986–1989, in a period of chronic arsenic poisoning problem in Ronpiboon subdistrict. At that time, a high proportion of drinking water was contaminated with arsenic above 0.05 ppm. This birth cohort has been continuously exposed to arsenic since birth because of their non-mobility.

Another weak point of this study was the small number of children with low hair arsenic (≤ 1 ppm) concentrations, which therefore could not be compared to the high arsenic (>1

ppm) children. If there were enough subjects with low arsenic exposure, the strength of association might have been greater. Therefore, future studies should screen arsenic exposure by biomarker prior to the outcome measurement. To confirm arsenic exposure, two biological samples for arsenic analysis (i.e., hair and urine) should be collected. Consequently, the ability to separate inorganic and organic arsenic in urine, needs to be developed in Thailand. Total urinary arsenic can be misleading because seafood consumption elevates urine arsenic. It is difficult to have Ronpiboon district's residents cease consuming seafood since this area is located in the southern part of Thailand and these residents habitually consume seafood.

The association found in this study should be non-spurious because of significant risk factor control. Thus, the reasonable and reliable findings of the study are compelling. However, several methodological difficulties were overcome in the conduct of this study. They were (1) selecting an appropriate and reliable biological marker of chronic arsenic exposure, the laboratory for hair arsenic analysis by AAS method; (2) measuring intelligence with an instrument of adequate sensitivity, WISC (the WISC was administered by experienced clinical psychologists); (3) identifying, measuring, and controlling for factors that might confounder modify the arsenic effect. These factors were identified from literature reviews, and some of them were suggested by mentors, researchers, and experts; (4) calculating and recruiting a sample large enough to provide adequate statistical power to detect a small effect; and (5) avoiding bias in sample selection by simple random sampling from the children who met the inclusion criteria.

CONCLUSIONS

This study concludes that most children aged 6–9 years of Ronpiboon and Soa Thong subdistricts had elevated arsenic concentrations in hair (>1 ppm). The mean arsenic level in hair was 3.52 ppm (SD=3.58), and the median was 2.42 ppm. The range of hair arsenic was 0.48 to 26.94 ppm. The below average IQ level (<90), and the average level (90–109) represented 48.4%, and 45.7%, respectively, of total children. After controlling for risk factors, arsenic levels in hair were inversely associated with IQ ($p=0.002$). Arsenic levels could explain why the mean IQs were different in the groups of children with varying arsenic exposure.

The management of risk to children should be the following: (1) The health personnel of public health offices in Ronpiboon district and Nakorn Si Thammarat province should monitor the health status of children for early detection of abnormal IQ findings. Especially, children who have high arsenic hair concentrations need to have frequent physical examinations. (2) The curriculum at primary education level needs to be revised for the children with poor IQ by the Ministry of Education and the Ministry of Public Health. (3) The children with IQs below average should receive special teaching in school.

The important implementation is the limitation of chronic arsenic exposure from ingestion as follows: (1) The inhabitants in Ronpiboon and Soa Thong subdistricts need to discontinue using well-water for drinking and food preparation. (2) The organizations responsible should inform the community leaders, inhabitants, and students in Ronpiboon district about the severity of chronic arsenic poisoning that might retard the intelligence of children. This information is important to parents, and it is hopefully the trigger for awareness of contaminated arsenic ingestion. (3) Sufficient distribution of safe drinking water needs to be provided for residents of Ronpiboon and Sao Thong subdistricts. (4) The monitoring of drinking water in these critical areas should be established.

In order to clarify the association between arsenic and intelligence, studies should be designed and performed for confirmation in the form of a longitudinal study beginning at birth. These children should be followed up for evaluation of prolonged or permanent intellectual impairment.

ACKNOWLEDGMENTS

This study was supported by Thailand Health Research Institute, National Health Foundation and Division of Epidemiology, Ministry of Public Health.

The authors gratefully acknowledge the assistance of individuals and organizations who made this study possible: Drs. Yupin Songpaisan, Somchai Supanvanich of Department of Epidemiology, Faculty of Public Health, Mahidol University for valuable suggestions; Sudaruk Lue and the personnel in Faculty of Pharmaceutical Science, Prince of Songkla University for arsenic analysis; the team of psychologists for evaluation of children's IQ; Amara Thonghong and health personnel from Division of Epidemiology, Ronpiboon Hospital, Health Stations and Public Health Office at Ronpiboon district for assistance in data collection; Dr. Stephen L. Hamann for grammatical correction; Dr. Chanpen Choprapawan for valuable help; and subjects, with their parents, for their cooperation.

REFERENCES

Ajjimangkul, S. 1992. *Arsenic poisoning prevention behavior in under five children group.*

Armstrong, C.W., Stroube, R.B., Robio, T., Siudyla, E.A., Miller, G.B. 1984. Outbreak of fatal arsenic poisoning caused by contaminated drinking water. *Arch. Environ. Health.*, **39** (4), 276–279.

Brook, C.G.D. 1982. *Growth assessment in childhood and adolescence.* Blackwell Scientific Publications, Oxford.

Chapple, G., Danby, R. 1990. *The determination of arsenic, selenium, and mercury levels in U.S. EPA quality control samples using the GBS HG 3000 continuous-flow hydride generator.* GBC Scientific Equipment Pty. Ltd., Victoria.

Choprapawan, C. 1994. *Arsenic poisoning problem at Ronpiboon district, Nakorn Si Thammarat province.*

Division of Environmental Health, Ministry of Public Health. 1992. *Conclusion of arsenic poisoning situation at Ronpiboon district, Nakorn Si Thammarat province.*

Ferguson, J.F., Gavis, J. 1972. A review of the arsenic cycle in natural waters. *Water Res.*, **6**, 1259–1274.

Meyer-Bahlberg, H.F.L., Feinman, J.A., MacGillivray, M.H., Aceto, T. 1978. Growth hormone deficiency, brain development, and intelligence. *Am. J. Dis. Child.*, **132**, 565–572.

Morton, W.E., Dunnette, D.A. 1994. Health effects of environmental arsenic. In: Nriagu, J.O. (ed.), *Arsenic in the Environment Part II: Human Health and Ecosystem Effects*, pp. 17–34. John Wiley & sons, Inc., New York.

Nordberg, G.F. 1988. Current concepts in the assessment of effects of metals in chronic low-level exposures-considerations of experimental and epidemiological evidence. *Sci. Total. Environ.*, **71**, 243–252.

Norusis, M.J. 1990. SPSS/PC+ Statistics™ 4.0 for the IBM PC/XT/AT and PS/2. SPSS Inc., Chicago.

Ohira, M., Aoyama, H. 1973. Epidemiological studies on the Morinaga powdered milk poisoning incident. *Jpn. J. Hyg.*, **27**, 500–531.

Piamphongsant, T., Udomnitikul, P. 1989. Arsenic levels in hair and nail samples of normal adolescents in Amphoe Ronpiboon. *Bull. Dept. Med. Services*, **14**, 225–229.

Rodklai, A. 1994. *Prevalence survey of chronic arsenic poisoning and health status of population in Ronpiboon district, Nakorn Si Thammarat province.*

Vogel, J.L. 1986. *Thinking about Psychology.* Nelson-Hall, Inc., Chicago.

Wechsler, D. 1949. *Wechsler Intelligence Scale for Children Manual.* The Psychological Corporation, New York.

Yamashita, N., Doi, M., Nishio, M., Hojo, H., Tanaka, M. 1972. Recent observations of Kyoto children poisoned by arsenic tainted "Morinaga Dry Milk". *Jpn. J. Hyg.*, **27**, 364–399.

Arsenic Exposure and Health Effects
W.R. Chappell, C.O. Abernathy and R.L. Calderon (Editors)
© 1999 Elsevier Science B.V. All rights reserved.

Reproductive and Developmental Effects Associated with Chronic Arsenic Exposure

C. Hopenhayn-Rich, I. Hertz-Picciotto, S. Browning, C. Ferreccio,
C. Peralta

ABSTRACT

Chronic exposure to inorganic arsenic is known to cause cancer and non-cancer health effects in humans. The evidence from animal studies clearly shows that arsenic is teratogenic, and the findings of limited human studies suggest that inorganic arsenic may be associated with several reproductive/developmental outcomes, including increased rates of spontaneous abortion, low birth weight, congenital malformations, pre-eclampsia and infant mortality. The city of Antofagasta, located in northern Chile, has a history of high arsenic exposure in drinking water. Due to changes in the sources of water, there were considerably high arsenic levels in the public drinking water supply from 1958 to 1970 (over 800 μg/L), which decreased gradually to the current concentrations close to 50 μg/L. A number of studies have reported various health effects associated with the high exposure period, including skin alterations typically linked to arsenic exposure and increases in bladder and lung cancer. We conducted an ecologic study of infant mortality rates in Chile from 1950 to 1996, comparing Antofagasta to low arsenic exposure areas. Temporal and cross-regional comparisons showed a general steady decline over time in late fetal, neonatal and post-neonatal mortality rates for all locations, consistent with improvements in standard of living and health care. However, comparatively high rates were observed in Antofagasta for the three outcomes studied during the 12-year period of highest arsenic exposure, compared to Santiago and Valparaíso, two locations used as reference groups. While not definitive, these findings support a role for arsenic in the observed increases in mortality rates. Given the worldwide public health concern for arsenic effects, more population studies are needed in the area of human reproductive and developmental effects.

Keywords: arsenic, reproduction, developmental effects, Chile, drinking water, infant mortality, environment

INTRODUCTION

Arsenic is a naturally occurring element present throughout the earth's crust. Both organic and inorganic arsenic are present in the environment, but the inorganic forms are considered much more toxic (henceforth referred to simply as "arsenic", unless otherwise noted). Elevated human exposure to arsenic occurs mainly from mining and smelting of metals, pesticide production and application, medicinal treatments, and ingestion of arsenic-rich water, usually from natural contamination (IARC, 1980; ATSDR, 1993). A number of populations worldwide have been and/or are currently exposed to high arsenic levels in drinking water, and in recent years a growing number of exposed groups have been identified, including populations living in regions of India, Bangladesh, Thailand, Mexico, Chile, Argentina, China, Hungary and Finland. In the United States, an estimated 350,000 people currently receive drinking water containing more than 50 μg/L of arsenic, the current standard set by the U.S. Environmental Protection Agency (EPA, 1987). Although higher exposures are more common in western states of the United States, in recent years concern has grown in other states in areas where private well use is common and arsenic has been more extensively measured and often found to be near the EPA standard or the World Health Organization recommended lower standard of 10 μg/L (e.g. Minnesota, New Hampshire and Michigan) (Small-Johnson et al., 1998; Karagas et al., 1998; Michigan Department of Public Health, 1982).

Chronic arsenic exposure at high doses has neurologic, dermatologic, vascular and carcinogenic effects (IARC, 1980; ATSDR, 1993). Exposure to arsenic from drinking water increases the risks of skin, lung and bladder cancer and possibly other target sites (Wu et al., 1989; Chen et al., 1992; Hopenhayn-Rich et al., 1998; Smith et al., 1998), and also appears to increase the risk of diabetes (Shibata et al., 1994; Rahman et al., 1998).

Despite extensive research on the health risks of arsenic exposure, the potential impact of arsenic on human reproduction has been given minimal attention. However, there is sufficient data from animal studies, supported by *in vitro* and mechanistic information, and limited data from human studies, to indicate that arsenic may be associated with adverse reproductive effects in humans. The purpose of this paper is two-fold: first, to give an overview of existing evidence from the perspective of plausible effects in humans, and second, to present the descriptive results of an ecologic study of early infant mortality and its relationship to arsenic exposure from drinking water.

REVIEW OF THE EVIDENCE FOR REPRODUCTIVE TOXICITY

Human Studies of Arsenic and Pregnancy Outcomes

Several studies from the Ronnskar copper smelter in Sweden reported reproductive effects among female employees and nearby residents (Nordstrom et al., 1978a; Nordstrom et al., 1978b; Nordstrom et al., 1979a; Nordstrom et al., 1979b). For 662 births occurring between 1930 and 1959, the average birth weight of babies born to women employees (3,394 g) was significantly lower than those born to residents from an unexposed town distant from the smelter (3,460 g) ($p < 0.05$), and was lowest for those whose mothers worked in the most highly exposed jobs (e.g. smelting and cleaning operations) (3,087 g) (Nordstrom et al., 1979a). An increasing trend in the rates of spontaneous abortion was observed comparing all women employed at the smelter during pregnancy (14%), those who worked and whose husbands also worked at the smelter (19%) and those in high exposure jobs (28%) (Nordstrom et al., 1979a). A larger epidemiologic study of pregnancy outcomes for all women (*N*=4427) born after 1929 who lived in four areas of increasing distances from the smelter found a dose–response increase for the risk of spontaneous abortion with residential proximity to the smelter (Nordstrom et al., 1978b). Congenital malformations occurred in 6%

of babies born to female employees who worked during pregnancy, compared to 2% among employees who did not work while pregnant ($p < 0.005$) (Nordstrom et al., 1979b).

Although arsenic exposures in and around the Ronnskar smelter were high, confounding from lead or copper could not be excluded. Moreover, no adjustments were made for the effects of other potential confounding risk factors, such as maternal age, which is known to have a strong relationship with spontaneous abortion and congenital anomalies.

In southeast Hungary, rates of spontaneous abortions and stillbirths for the period 1980–87 were examined in an area with drinking water arsenic levels over 100 μg/L and a control area with low arsenic levels (Borzsonyi et al., 1992). Spontaneous abortions were 1.4-fold ($p < 0.02$) and stillbirths 2.8-fold ($p < 0.02$) higher in the exposed region, but the effect of potential confounders was not assessed. An increased incidence of spontaneous abortions and perinatal death in areas of Argentina with high arsenic in water has been reported, but no further detail was provided (Castro, 1982). In Bulgaria, the incidence of toxemia of pregnancy and mortality from congenital malformations in an area close to a smelter with environmental contamination from various metals were significantly higher than the national rates (Zelikoff et al., 1995).

Three U.S. studies reported adverse reproductive effects associated with relatively low water arsenic levels. An ecologic study examined mortality from cardiovascular-related causes in 30 counties from 11 states which had recorded arsenic measurements in public drinking water supplies greater than 5 μg/L (mean range: 5.4–91.5). This investigation found increases in mortality from congenital anomalies of the heart and other anomalies of the circulatory system over a 16-year period (1968–74) for two Nevada counties classified in the highest exposure group (29 and 46 μg/L average arsenic concentration in water) (Engel and Smith, 1994). In a case-control study conducted in Massachusetts, the adjusted prevalence odds ratio (POR) for congenital heart disease, comparing those whose water supply had arsenic levels above the detection limit (reported as 0.8 μg/L) with those below the limit, was not elevated (Zierler et al., 1988), but when stratified by type of heart disease, the POR for coarctation of the aorta was 3.4 (1.3–8.9). In a study of spontaneous abortions in the same area, Aschengrau et al. reported an adjusted odds ratio of 1.5 for the group with the highest arsenic concentrations: 1.4-1.9 μg/L (Aschengrau et al., 1989). We note that in these studies even the "exposed" areas had arsenic levels that are extremely low and therefore the results are difficult to interpret.

A hospital case-control study investigated the occurrence of stillbirths in relation to residential proximity to an arsenical pesticide production plant in Texas (Ihrig et al., 1998). Exposure was categorized in three groups according to arsenic air levels. An increasing trend in the risk of stillbirths was observed, significant for the high exposure group. When stratified by ethnicity, however, the findings remained significant for Hispanics only. It should be noted that this study was quite small, and more so when stratified by exposure and ethnic sub-groups; additionally, other exposures from the chemical plant were possible and were not measured in the study. Therefore, the results should be considered suggestive but not conclusive.

Overall, the epidemiologic evidence on adverse reproductive outcomes, although it suffers from methodological limitations, suggests positive associations with arsenic exposures.

Animal Teratogenicity

The teratogenicity of arsenic in animals is well-documented (Willhite and Ferm, 1984; Golub, 1994, Golub et al., 1998; DeSesso et al., 1998). Arsenic-induced defects include anophthalmia, exencephaly, and malformations of the genito-urinary, skeletal and cardiovascular systems. In general, there is consistent evidence from numerous animal studies showing that arsenic causes neural tube defects, and mechanistic hypotheses have been proposed (Shalat et al.,

1996). In addition, arsenic induces other forms of developmental toxicity, including death and growth retardation in the fetuses of exposed pregnant laboratory animals. Findings are consistent across studies, with effects found to be dependent on dose, route and timing of administration (Golub et al., 1998).

However, in a recent article, DeSesso et al. (1998) concluded that all previous positive findings of congenital malformations could be dismissed based on weaknesses in study design or interpretation of results. One of their criticisms of earlier work was that the high doses administered to pregnant animals generally caused maternal toxicity. Although many of the studies did find developmental effects at doses high enough to induce maternal toxicity, the existing evidence supports an increase in fetal abnormalities that is not secondary to maternal effects (Golub, 1994). Given the extensive number of peer-reviewed, published papers totaling over 50 studies describing positive, consistent results across different laboratories, using several animal species and modes of exposure, we should consider the results and interpretation provided by DeSesso et al. cautiously. It is also important to point out that arsenic appears to be unique in relation to its effects on humans: while an adequate animal model has not yet been found for the carcinogenicity of arsenic, there is now ample evidence of high cancer risks for several target sites in humans (skin, bladder and lung, and possibly others). It seems that humans have higher susceptibility to the toxic effects of arsenic compared to most commonly used laboratory animals. Therefore, human developmental effects may occur at lower doses than those determined from animal models, and as with cancer and other health outcomes, they will only be identified through epidemiological studies of chronically exposed populations.

Concentration in Tissues and Placental Transfer

Human autopsy studies have found high levels of arsenic accumulation in tissues of cancer target organs associated with arsenic exposure (Dang et al., 1983; Gerhardsson et al., 1988). However, these studies did not report concentrations in reproductive organs. In animal studies, rodents (Danielsson et al., 1984; Calvin and Turner, 1982), rabbits (Vahter and Marafante, 1983) and hamsters (Marafante and Vahter, 1987) have been shown to accumulate arsenic in the testis and epididymis. In female rats, arsenic concentrations in the ovaries were as high as in the liver (Ramos et al., 1995). Sheep fed a mixture of metals from emissions from a copper and zinc plant showed preferential accumulation in different organs by different metals with arsenic showing higher concentrations in ovaries than in kidneys or liver (Bires et al., 1995).

Placental transfer of inorganic arsenic is known to occur in both animals and humans (Ferm, 1977; Squibb and Fowler, 1983; Willhite and Ferm, 1984; Hood et al., 1987; Nicholson et al., 1982; Tabacova et al., 1994). A recent study in an area of Argentina with high arsenic in drinking water (250 μg/L), found a close relation between placental and cord blood arsenic levels, indicating considerable placental transfer of arsenic to the developing fetus during pregnancy (Concha et al., 1998).

Mechanisms

Several possible biological mechanisms can be postulated to support the evidence of arsenic-induced reproductive/developmental effects. Shalat et al. (1996) proposed plausible mechanistic roles for arsenic in the pathogenesis of neural tube defects, based on inhibition and disruption of cell proliferation, cell metabolism and placental/embryonal vascularization. All these processes can be affected by arsenic and in turn can affect neural tube formation.

In humans, methylation of inorganic arsenic appears to be the main detoxification pathway. Glutathione (GSH) and associated enzymes have been found to play a role in several steps of this methylation process (Thompson, 1993; Chiou et al., 1997). Since GSH is involved in many biological pathways, it has been proposed that high arsenic exposures may decrease the availability of GSH, and perhaps even more in the presence of other contaminants that also use GSH for detoxification (Hopenhayn-Rich et al., 1996). The potential oxidative damage due to depletion of GSH provides possible pathways for reproductive toxicity, such as congenital malformations, pre-eclampsia and abnormal sperm development. Explanted mouse embryos and yolk sacs treated with arsenic showed malformative syndromes in parallel with decreasing levels of GSH, the teratogenicity was dose- and timing-dependent, and it was further enhanced by inhibition of GSH synthesis (Zelikoff et al., 1995). High levels of placental arsenic in women living near a smelter were associated with a lower percentage of placental GSH, and a concomitant increase in lipid peroxides capable of causing oxidative damage to the developing embryo (Tabacova et al., 1994). Arsenic concentrations were also associated with increased lipid peroxidation and lower levels of GSH in female treated rats (Ramos et al., 1995).

Arsenic may also cause male-mediated reproductive effects. GSH is normally found in high levels in rat testis and is believed to play an important role during meiosis in normal spermatogenesis (Calvin and Turner, 1982). In addition, membranes of mature sperm are susceptible to oxidative damage, and GSH forms part of the antioxidant defense systems in sperm (Lai et al., 1994). Therefore, decreased availability of GSH could lead to oxidative stress and lipid peroxidation of the sperm membrane, and to defective sperm function (Grieveau, 1995). Given the strong evidence of arsenic-induced clastogenicity and aneuploidy in other human cells (Nordenson et al., 1981; Moore et al., 1997; Gonsebatt et al., 1997), together with the accumulation of arsenic in the testis where sperm are formed, arsenic may also have a direct effect on human sperm chromosomes. Chromosomal aberrations, aneuploidy or other cytogenetic damage to sperm during spermatogenesis could lead to transgenerational effects including spontaneous abortion or aneuploid offspring.

Summary of the Evidence

It is well established that arsenic is genotoxic and causes cancer at various target sites in humans. It has also been proposed that arsenic is a likely reproductive toxicant. The human evidence, although limited, is consistent with animal, and *in vitro* laboratory results, and is suggestive of effects at various endpoints. Therefore, further studies of arsenic effects on human reproduction are warranted.

ECOLOGIC STUDY OF INFANT MORTALITY

INTRODUCTION

A retrospective ecologic study design was employed to investigate trends in infant mortality during the period 1950 to 1996 for a region of Chile with naturally elevated arsenic exposure in comparison to other regions with low background arsenic levels in the water. In addition, infant mortality rates were examined over time within the presumed high arsenic exposure area with respect to the temporal variation of arsenic levels in the water supply.

The population under study resides in Chile's Region II. This is generally a very dry area, with the Atacama Desert occupying a vast proportion of its territory. The surface water that supplies most of the region comes from rivers originating in the Andes mountains. Antofagasta, the largest city in the region with a current population of around 250,000 people, has a well-documented history of arsenic exposure. In 1958, due to insufficient water supply to serve the growing population and the decrease in water availability, water from the Toconce River was introduced as the main new water source. This river contains naturally occurring arsenic and at the time that inhabitants first used it, arsenic levels in this public water supply were approximately 800 μg/L. An increase in specific health effects in this population began shortly after the change in water supply. The first reported cases of chronic arsenic poisoning appeared at the regional hospital in Antofagasta in 1962 (Zaldivar, 1980). A number of publications document the effects of exposure to arsenic in Antofagasta during that time (Zaldivar et al., 1981; Borgoño et al., 1977; Arroyo-Meneses, 1991; Smith et al., 1998). After 12 years of exposure to the high concentrations of the Toconce River, an arsenic removal plant was installed at the public water supply company.

METHODS

Vital Statistics Data

We obtained vital statistics data from the Instituto Nacional de Estadísticas (INE) in Chile, which centralizes local and national annual information on population vital statistics and census data. For this study, we obtained natality and mortality data from the period 1950 to 1996 from the yearly published books housed at INE's central office. The information in these reports has been collated from birth and death certificates obtained from civil registration and health department offices from each local government responsible for routinely collecting this data. The INE infant mortality data are classified separately as late fetal deaths (over 28 weeks of gestation, mainly consisting of stillbirths) and infant deaths, the latter being further divided into those under 28 days of age and those 28 days to one year of life. We will refer to these groups as fetal, neonatal and post-neonatal, respectively, according to standard definitions.

Exposure Data

Historical water measurement data from public water supplies for Antofagasta, obtained from a compilation of existing sources (water company records and the regional health service), are given in Table 1. As mentioned previously, water from the Toconce River was introduced as the primary new water source for the entire city in 1958, serving all the population using piped water. Based upon the data collected, the mean arsenic level during the period 1958–1970 was estimated to be 860 $\mu g/L$. Before that, arsenic levels in the water averaged 90 $\mu g/L$. In 1970, the city installed an arsenic removal plant for this water supply. Subsequent measurements established a decline in the arsenic levels over the next 26 years to the current levels close to 50 $\mu g/L$. The primary, extremely high exposure period for Antofagasta was during the period 1958–1970.

To enhance validity, we examined several comparison groups and present results from the following in this report: all of Chile, the metropolitan region of Santiago, which currently concentrates about 40% of the country's population, and the county of Valparaíso. Santiago includes the capital and surrounding areas. Valparaíso, like Antofagasta, is a coastal town and they share similar size and sociodemographic characteristics. Neither Santiago nor Valparaíso have historical evidence of high arsenic contamination, and recent water surveys

TABLE 1

Average arsenic levels in Antofagasta. Data represents an average of existing arsenic water measurements (Pedreros, 1994).

Year	Concentration ($\mu g/L$)
1950–1957	90
1958–1970	860
1971–1979	110
1980–1987	70
1988–1996	40

conducted by the National Environmental Commission (FONDEF, 1997) and by our group show arsenic levels to be low (<5 μg/L). In general, aside from Region II, no major populations in Chile have had such high exposures to arsenic from drinking water (Smith et al., 1998).

ANALYSIS

For this study, we retrieved yearly vital statistics information from 1950 to 1996, to examine infant mortality rates in the county of Antofagasta over time, and to compare Antofagasta rates to those in low arsenic areas. Although we used the county as the unit of analysis, for Antofagasta and Valparaíso the majority of their population lives in cities bearing the same names (99% and 97%, respectively).

Changes over time in the number and geographic boundaries of geopolitical localities had to be considered in developing a common geographic unit for the analysis of the data. Birth and mortality data are reported by different aggregate geographic regions. Since 1976, Chile has been stratified into regions, numbered 1 through 12 from north to south (plus the metropolitan area of Santiago which is considered separately). Regions have been further divided into provinces, and then into comunas (equivalent to counties). Prior to 1976, Chile was divided into 25 provinces, and throughout the years, several changes occurred in the number and jurisdiction of geographic delimitations. One of us (C.F.) undertook an in-depth reclassification of geographic areas to achieve compatibility across the entire period. The comuna was the smallest unit of analysis with identifiable vital statistics information, standardized across time.

Further, births and deaths are reported by both place of occurrence and place of maternal residence. Since we were interested in the relationship between infant mortality and arsenic exposure from drinking water (either through maternal or infant ingestion) we used location of maternal residence for coding the birth and death data.

Infant mortality rates were calculated by dividing the number of deaths by the number of live births per location and multiplying by 1000. For late fetal mortality, we divided the number of fatalities by the number of live births + fetal deaths, as is standard practice, to obtain the death rate per total births. After calculating yearly mortality rates for each group, considerable variation was noted from year to year, given the relatively small number of events. Therefore, we grouped the rates into 4-year periods (except for the last period for which we only had 3 years of data), to achieve higher stability and still maintain a distinct period of highest arsenic exposure in Antofagasta.

Since the patterns of mortality in the three outcomes studied did not vary greatly between Santiago and Valparaíso, the two reference groups, we concentrated further analysis in comparing Antofagasta to Valparaíso, given the greater homogeneity found within a county (as opposed to the large Santiago metropolitan area) and the similarities shared by these two counties in particular. For each mortality outcome, we calculated rate differences for each 4-year time period.

RESULTS

The period studied is one of general decrease worldwide in fetal and infant mortality due mainly to improvements in living conditions and health care. Figure 1 shows mortality rates in Chile as a whole for the 3 outcomes studied. Figures 2, 3 and 4 show the 4-year fetal, neonatal and post-neonatal mortality rates, respectively, for Antofagasta, Santiago and Valparaíso. For Antofagasta, the rates were generally higher at the beginning of the study period (1950–57), but a clear pattern of elevated rates in comparison to Santiago and Valparaíso is observed during the high arsenic years (1958–69). After 1970, mortality rates

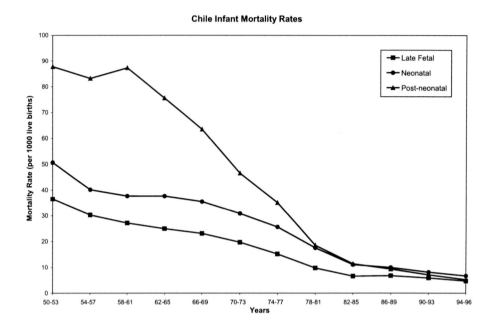

Fig. 1. Late fetal, neonatal and post-neonatal mortality rates for all of Chile, 1950–1996.

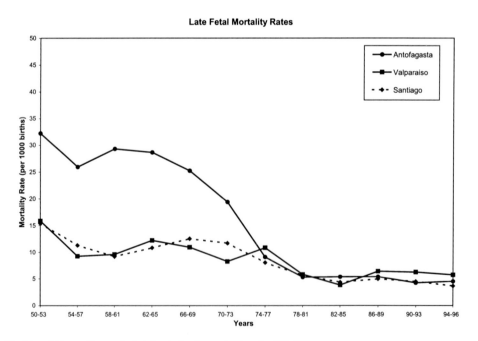

Fig. 2. Late fetal mortality rates for Antofagasta, Santiago and Valparaíso, 1950–1996.

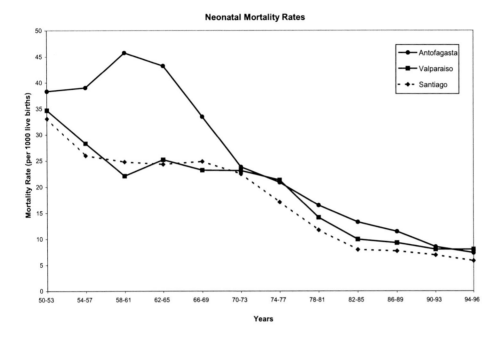

Fig. 3. Neonatal mortality rates for Antofagasta, Santiago and Valparaíso, 1950–1996.

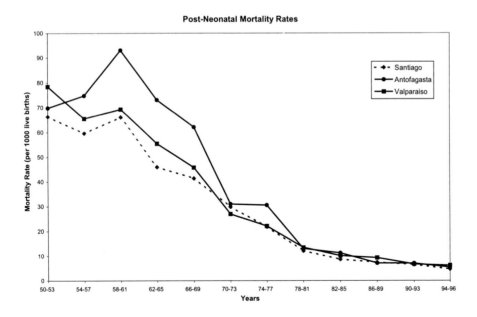

Fig. 4. Post-neonatal mortality rates for Antofagasta, Santiago and Valparaíso, 1950–1996.

TABLE 2

Mortality rates and rate differences for Antofagasta and Valparaiso. Mortality rates were calculated per 1000 live births for neonatal and postneonatal period, and per 1000 total births (live + late fetal) for the late fetal period.

Year	Late Fetal				Neonatal				Post-neonatal			
	ANT.	VALP.	R.D.	95% C.I.	ANT.	VALP.	R.D.	95% C.I.	ANT.	VALP.	R.D.	95% C.I.
1950–53	32.2	15.9	16.3	15.4–17.3	38.4	34.7	3.7	3.0–4.4	69.8	78.4	–8.6	(7.8–9.4)
1954–57	25.9	9.3	16.7	15.8–17.4	39.0	28.3	10.7	9.9–11.5	74.8	65.5	9.2	8.5–10.1
1958–61	29.3	9.6	19.7	18.9–20.5	45.7	22.1	23.6	22.7–24.5	93.0	69.3	23.7	22.8–24.6
1962–65	28.6	12.3	16.4	15.6–17.0	43.2	25.2	18.0	17.2–18.8	73.1	55.4	17.6	16.8–18.6
1966–69	25.2	11.0	14.3	13.5–14.9	33.5	23.2	10.3	9.6–11.0	62.1	45.8	16.3	15.4–17.2
1970–73	19.4	8.3	11.1	10.5–11.7	23.8	23.1	0.7	0.2–1.3	31.0	26.9	4.0	3.5–4.7
1974–77	9.1	10.9	–1.7	(1.3–2.3)	20.9	21.3	–0.5	(0.1–0.9)	30.5	22.2	8.4	7.6–9.0
1978–81	5.3	5.9	–0.5	(0.3–0.9)	16.5	14.1	2.3	1.9–2.9	13.0	13.5	–0.5	(0.03–1.0)
1982–85	5.4	3.8	1.6	1.3–1.9	13.2	9.9	3.3	2.8–3.8	11.2	10.1	1.1	0.7–1.5
1986–89	5.4	6.5	–1.0	(0.7–1.5)	11.4	9.3	2.2	1.7–2.5	7.1	9.3	–2.2	(1.8–2.6)
1990–93	4.3	6.3	–2.1	(1.6–2.4)	8.5	8.0	0.5	0.1–0.9	7.0	6.7	0.3	0.04–0.6
1994–96	4.5	5.8	–1.2	(0.9–1.7)	7.3	8.0	–0.7	(0.3–1.2)	5.4	6.1	–0.8	(0.3–1.1)

Abbreviations: ANT.= Antofagasta; VALP= Valparaíso; R.D.= Rate Difference; C.I.= Confidence Interval.

level off and become very similar to those in the two comparison areas. The increases vary by outcome, but are generally more pronounced in the first 4-year period (1958–61). Table 2 shows the comparison of rates between Antofagasta and Valparaíso. The rate differences, which were reflected in the graphs, peak in the 1958–61 period for all three outcomes.

DISCUSSION

The results of this study show decreasing trends in fetal, neonatal and post-neonatal mortality over time in all of Chile, as well as in Santiago and Valparaíso. These are consistent with observations in many countries in the latter half of this century. The main factors reported to affect infant mortality in Chile are improvement in the standard of living, development of programs for maternal and infant care (prenatal, nutritional supple-mentation, health education, etc.) and the decrease in birth rate, most noticeable since the 1960s (INE, 1994). The same tendency is observed in Antofagasta, except for increases observed in 1958–61 which decline thereafter but remain elevated relative to Santiago and Valparaíso until 1974–77 for fetal mortality, and until 1970–73 for neonatal and post-neonatal mortality. These results show a generally close temporal relationship with the considerably high arsenic levels present in the city's water supply from 1958 to March of 1970. As in the rest of the country, mortality rates in Antofagasta did decline gradually during this time period probably due to the time trend effect, that is, other, non-arsenic related factors, such as improvement in health care and standard of living, which were improving in Antofagasta alongside the rest of Chile. Moreover, starting in the 1960s, Region II experienced a surge of economic growth due mostly to the great expansion associated with copper mining (mainly in the high desert Chuquicamata mine). The rate of growth in the region as a whole greatly surpassed that of the rest of the country. A study of the development of production by regions showed that adjusted annual per capita income in Region II had the greater increase, by far, than any other region (CIEPLAN, 1994). This regional expansion brought improve-ments in health care as well, which likely explain, at least in part, the fact that in the period before the arsenic highest years, Antofagasta had higher mortality rates for the three outcomes examined, while after the arsenic removal plant was installed, rates declined to levels similar to those of the comparison regions (as indicated by the rate differences).

Ecologic studies, in which both exposures and outcomes are measured at the group level and not at the individual level, are subject to the bias known as the ecologic fallacy. This problem arises when the exposure does not apply to all persons in the group, such that those experiencing the adverse outcomes might not be the same individuals who are experiencing the exposure. However, since water is consumed by virtually all residents, and in this case the public water supply came from one main source, this bias seems unlikely to have influenced the results of this study. The distinct temporal pattern of Antofagasta rates relative to other regions also argues against this bias. As with other epidemiologic study designs, ecologic studies can be subject to confounding and effect modification. Although it is possible that other factors not accounted for in this analysis could have been present concurrently with the high arsenic that could explain or exacerbate the observed increase in infant mortality, we could not identify any other factors during that time period that would so closely relate to the timing of the arsenic exposure. The change in water supply was an indisputable event, and no other contaminant has been described to account for the many documented health effects suffered by the population of Antofagasta due to arsenic exposure during that time period.

With regards to the validity of these data, the accuracy of birth and death certification is hard to assess. However, there is evidence of under-ascertainment, which has improved over time. In Chile, children whose births are not registered by March of the year after they were born are counted in a separate, late registration category. A study of birth registrations from 1955–1988 showed variations ranging from 11.4% to 4.2% in late birth registrations (INE/CELADE, 1990). Variations were also observed between regions, with the northern-most ones (which include Antofagasta, Valparaíso and Santiago) having the lowest late registration rates. The increase in births that occur in hospitals (42.3% in 1953 versus 99.7% in 1994) and births assisted by medical professionals (57.5% in 1953 vs. 100% in 1994) also impacts completeness of birth registration. Regional statistics show Antofagasta, Santiago and Valparaíso were three of the four provinces with the highest percentage of births assisted by physicians in 1953 relative to the rest of Chile (INE, 1953).

Death registration is also characterized by omissions, especially for newborns that died in the first hours or days after birth. A study conducted in Santiago in 1968–69 showed that over half of the babies born in hospitals who died were not registered (Legarreta et al., 1973). Another study in Santiago found 13% of the deaths of children under the age of five (5) were not registered, and most of this under-registration occurred for deaths in the neonatal period. In Chile, the certification of infant deaths by a physician increased from 55% in 1952 (INE, 1953) to 96% in 1994 (INE, 1994).

The omissions in death reporting and the late birth registrations described above will affect mortality rates, which will be biased up or down depending whether birth or death omissions were higher. However, given the much greater number of births than deaths, under-reporting of deaths will have a much greater effect on the rates. There is no evidence that under-reporting of deaths was significantly different between Antofagasta, Valparaíso and Santiago. Moreover, if the increases in fetal and infant mortality observed in Antofagasta were at least, in part, attributable to the high arsenic water levels during those years, the omissions would underestimate rather than over-estimate the effect.

In this paper, we have reviewed the literature for reproductive effects of arsenic and have found suggestive evidence for arsenic-related human developmental toxicity. Furthermore, we have conducted a study of infant mortality rates in Chile comparing high and low areas of arsenic exposure. The period of highest exposure was accompanied by higher rates in Antofagasta, as compared to the unexposed areas of Santiago and Valparaíso. While not definitive, these findings support a role for arsenic in the increased mortality. If arsenic reproductive toxicity is specific to certain causes of death, then the actual magnitude of the effect would probably be larger than that observed here. Further studies are warranted to investigate the risks of specific causes of late fetal and infant death in very highly exposed populations, as well as of other less drastic but nevertheless adverse developmental effects at moderately elevated levels.

ACKNOWLEDGMENTS

This work was partially funded by a grant from the Andrew W. Mellon Foundation to the Carolina Population Center, and by the National Center for Environmental Assessment of the U.S. Environmental Protection Agency. This chapter was reviewed and approved for publication by the US EPA. The views communicated in this manuscript are solely the opinions of the authors and should not be inferred to represent those of the US EPA. We wish to acknowledge the help of Dan Remley, Bin Huang, Alex Dmitrienko and Kjell Johnson and the support of the Center for Health Services Management and Research, University of Kentucky.

REFERENCES

Arroyo-Meneses, A. 1991. Hidroarsenicismo cronico en la region de Antofagasta. *Primera Jornada Sobre Arsenicismo Laboral y Ambiental II Region*, 35–47.

Aschengrau, A., Zierler, S., Cohen, A. 1989. Quality of Community Drinking Water and the Occurrence of Spontaneous Abortion. *Arch. Environ. Health*, **44**(5), 283–290.

ATSDR, 1993 Toxicological Profile for Arsenic, Update, Report No. TP-92/102. Agency for Toxic Substances and Disease Registry, U.S. Dept. of Health and Human Services, Atlanta, GA.

Bires, J., Maracek, I., Bartko, P., Biresova, M., Weissova, T. 1995. Accumulation of trace elements in sheep and the effects upon quantitative and qualitative ovarian changes. *Vet. Human Toxicology*, **37** (4), 349–356.

Borgoño, J.M., Vicent, P., Venturino, H., Infante, A. 1977. Arsenic in the drinking water of the city of Antofagasta: epidemiological and clinical study before and after the installation of a treatment plant. *Environ. Health Perspect.*, **19**, 103–105.

Borzsonyi, M., Bereczky, A., Rudnai, P., Csanady, M., Horvath, A. 1992. Epidemiological studies on human subjects exposed to arsenic in drinking water in Southwest Hungary. *Arch. Toxicol.*, **66**, 77–78.

Calvin, H.I., Turner, S.I. 1982. High levels of glutathione attained during postnatal development of rat testis. *J. Exp. Zool.*, **219**, 389–393.

Castro, J.A. 1982. Efectos carcinógenicos, mutagénicos y teratogénicos del arsénico. *Acta Bioquim. Clin. Latinoamericana*, **16**(1), 3–17.

Chiou, H-Y., Hsueh, Y-M., Hsieh, L-L., Hsu, L-I., Hsu, Y-H., Hsieh, F-I., Wei, M-L., Chen, H-C., Yang, H-T., Leu, L-C., Chu, T-H., Chen-Wu, C., Yang, M-H., Chen, C-J. 1997. Arsenic methylation capacity, body retention, and null genotypes of glutathione S-Transferase M1 and T1 among current arsenic-exposed residents in Taiwan. *Mutat. Res.*, **386**, 197–207.

CIEPLAN Corporación de Investigaciones Económicas para Latinoamérica - Ministerio del Interior, Subsecretaría de Desarrollo Regional y Administrativo. 1994. *Evolución del Producto por Regiones 1960–1992*, Santiago, Chile.

Concha, G., Vogler, G., Lexcano, D., Nermell, B., Vahter, M. 1998. Exposure to Inorganic Arsenic Metabolites during Early Human Development. *Toxicol. Sci.*, **44**, 185–190.

Dang, H.S., Jaiswal, D.D., Somasundaram, S. 1983. Distribution of arsenic in human tissues and milk. *Sci. Total Environ.*, **29**, 171–175.

Danielsson, B.R., Dencker, L., Lindgren, A., Tjalve, H. 1984. Accumulation of toxic metals in male reproduction organs. *Arch. Toxicol.*, **Suppl. 7**, 177–180.

DeSesso, J.M., Jacobson, C.F., Scialli, A.R., Farr, C.H., Holson, J.F. 1998. An assessment of the developmental toxicity of inorganic arsenic. *Reproduct. Toxicity*, **12**(4), 385–433.

Engel, R.E., Smith, A.H. 1994. Arsenic in drinking water and mortality from vascular disease: an ecological analysis in 30 counties in the United States. *Arch. Environ. Health*, **49**(5), 418–127.

EPA (Environmental Health Protection Agency) 1987. Estimated national occurrence of arsenic in public drinking water supplies. EPA contract No. 68-01-7166.

Ferm, V.H. 1977. Arsenic as a tetagenic agent. *Environ. Health Perspect.*, **19**, 215–217.

FONDEF—Facultad de Ciencias Fisicas y Matemáticas de la Universidad de Chile. 1997. Determinación de línea base de arsénico ambiental, Santiago, Chile.

Gerhardsson, L., Brune, D., Nordberg, G.F., Wester, P.O. 1988. Multielemental assay of tissues of deceased smelter workers and controls. *Sci. Total Environ.*, **74**, 97–110.

Golub, M.S. 1994. Maternal toxicity and the identification of inorganic arsenic as a developmental toxicant. In: *Reproductive Toxicology*. pp. 283–295. Elsevier, Amsterdam and New York.

Golub, M.S., Macintosh, M.S., Baumrind, N. 1998. Developmental and reproductive toxicity of inorganic arsenic: animal studies and human concerns. *J. Toxicol. Environ. Health*, Part B, 199–241.

Gonsebatt, M.E., Vega, L., Salazar, A.M., Montero, R., Guzmán, P., Blas, J., Del Razo, L.M., García-Vargas, G., Albores, A., Cebrián, M.E., Kelsh, M., Ostrosky-Wegman, P. 1997 Cytogenetic effects in human exposure to arsenic. *Mutat Res.*, **386**, 219–228.

Grieveau, J.F., Dumont, E., Renard, P., Callegari, J.P., Le Lannou, D. 1995. Reactive oxygen species, lipid peroxidation and enzymatic defense systems in human spermatozoa. *J. Reproduct. Fertil.*, **103**, 17–26.

Hood, R.D., Vedel-Macrander, G.C., Zaworotko, M.J., Tatum, F.M., Meeks, R.G. 1987. Distribution, metabolism, and fetal uptake of pentavalent arsenic in pregnant mice following oral or intraperitoneal administration. *Teratology*, **35**, 19–25.

Hopenhayn-Rich, C., Biggs, M.L., Smith, A.H., Kalman, D.A., Moore, L.E. 1996. Methylation study in a population environmentally exposed to high arsenic drinking water. *Environ. Health Perspect.*, **104**, 620–628.

Hopenhayn-Rich, C., Biggs, M.L., Smith, A.H. 1998. Lung and kidney cancer mortality associated with arsenic in drinking water in Córdoba, Argentina. *Int. J. Epidemiol.*, **27**, 561–569.

IARC. World Health Organization. 1980. *Monographs on the Evaluation of the Carcinogenic Risk of Chemicals to Humans: Some Metals and Metallic Compounds*. Lyon, France.

Ihrig, M.M., Shalat, S.L., Baynes, C. 1998. A hospital-based case-control study of stillbirths and environmental exposure to arsenic using an atmospheric dispersion model linked to a geographical information system. *Epidemiology*, **9**(3), 290–294.

INE (Instituto Nacional de Estadísticas - Ministerio de Salud. Servicio Registro Civil e Identificación) 1994. *Anuario de Demografía*, Santiago, Chile.

INE (Instituto Nacional de Estadísticas - Ministerio de Salud. Servicio Registro Civil e Identificación) 1953. *Movimiento Demográfico*, Santiago, Chile.

INE/CELADE (Instituto Nacional de Estadísticas y Centro Latinoamericano de Demografía - Ministerio de Economía, Fomento y Reconstrucción) 1990. *Chile, Estimación de la Oportunidad de Inscripción de los Nacimientos—Total País y Regiones 1955–1988*, Santiago, Chile.

Karagas, M.R., Tosteson, T.D., Blum, J., Morris, J.S., Baron, J.A., Klaue, B. 1998. Design of an Epidemiologic Study of Drinking Water Arsenic Exposure and Skin and Bladder Cancer Risk in a U.S. Population. *Environ. Health Perspect.*, **106**(4), 1047–1050.

Lai, M-S., Hsueh, Y-M., Chen, C-J., Shyu, M-P., Chen, S-Y., Kuo, T-L., Wu, M-M., Tai, T-Y. 1994. Ingested inorganic arsenic and prevalence of diabetes mellitus. *Am. J. Epidemiol.*, **139**(5), 484–492.

Legarreta, A., Aldea, A., López, L. 1973. Omisión del registro de defunciones de niños ocurridas en maternidades, Santiago, Chile. *Bol. Sanit. Panam.*, **75**(4), 308–314.

Marafante, E., Vahter, M. 1987. Solubility, retention, and metabolism of intratracheally and orally administered inorganic arsenic compounds in the hamster. *Environ Res.*, **42**, 72–82.

Michigan Department of Public Health. 1982. Arsenic in Drinking Water—A Study of Exposure and a Clinical Survey.

Moore, L.E., Smith, A.H., Hopenhayn-Rich, C., Biggs, M.L., Kalman, D.A., Smith, M.T. 1997. Micronuclei in Exfoliated Bladder Cells among Individuals Chronically Exposed to Arsenic in Drinking Water. *Cancer Epidemiol. Biomarkers Prev.*, **6**, 31–36.

Nicholson, W.J., Perkel, G., Selikoff, I.J. 1982. Occupational exposure to asbestos: population at risk and projected mortality— 1980–2030. *Am. J. Ind. Med.*, **3**, 259–311.

Nordenson, I., Sweins, A., Beckman, L. 1981. Chromosome aberrations in cultured human lymphocytes exposed to trivalent and pentavalent arsenic. *Scand. J. Work Environ. Health*, **7**, 277–281.

Nordstrom, S., Beckman, L., Nordenson, I. 1978a. Occupational and environmental risk in and around a smelter in northern Sweden—I. Variations in birth weight. *Hereditas*, **88**, 43–46.

Nordstrom, S., Beckman, L., Nordenson, I. 1978b. Occupational and environmental risk in and around a smelter in northern Sweden—III. frequencies of spontaneous abortion. *Hereditas*, **88**, 51–54.

Nordstrom, S., Beckman, L., Nordenson, I. 1979a. Occupational and environmental risk in and around a smelter in northern Sweden—V. Spontaneous abortion among female employees and decrease birth weight in their offspring. *Hereditas*, **90**, 291–296.

Nordstrom, S., Beckman, L., Nordenson, I. 1979b. Occupational and environmental risks in and around a smelter in northern Sweden—VI. congenital malformations. *Hereditas*, **90**, 297–302.

Pedreros, R. 1994. Presencia de Arsénico en II Región y Estudio Retrospectivo de Cohortes Expuestas. Presented at Segundas Jornadas de Arsenicismo Laboral y Ambiental, Servicio Nacional de Salud, Antofagasta, Chile.

Rahman, M., Tondel, M., Ahmad, S.A., Axelson, O. 1998. Diabetes Mellitus Associated with Arsenic Exposure in Bangladesh. *Am. J. Epidemiol.*, **148**(2), 198–203.

Ramos, O., Carrizales, L., Yañez, L., Mejía, J., Batres, L., Ortíz, D., Diaz-Barriga, F. 1995 Arsenic Increased Lipid Peroxidation in Rat Tissues by a Mechanism Independent of Glutathione Levels. *Environ. Health Perspect.*, **103** (Suppl.1), 85–88.

Shalat, S.L., Walker, D.B., Finnell, R.H. 1996. Role of arsenic as a reproductive toxin with particular attention to neural tube defects. *J. Toxicol. Environ. Health*, **48**, 253–272.

Shibata, A., Ohneseit, P.F., Tsai, Y.C., Spruck, C.H., Nichols, P.W., Chiang, H., Lai, M., Jones, P.A. 1994. Mutational spectrum in the p53 gene in bladder tumors from the endemic area of black foot disease in Taiwan. *Carcinogenesis*, **15**(6), 1085–1087.

Small-Johnson, J., Soule, R., Durkin, D., and Minnesota Dept. of Health. Minnesota Arsenic Research Study (MARS): Geochemical Parameters, Human Exposures and Health Effects Associated with Low-Level Arsenic in Drinking Water. Presented at the International Society of Environmental Epidemiology Conference, Boston, Massachusetts, August 1998.

Smith, A.H., Goycolea, M., Haque, R., Biggs, M.L. 1998. Marked increase in bladder and lung cancer mortality in a region of northern Chile due to arsenic in drinking water. *Am. J. Epidemiol.*, **147**(7), 660–669.

Squibb, K.S., Fowler, B.A. 1983. The toxicity of arsenic and its compounds. *Biolog. Environ. Effects Arsenic*, 233–69.

Tabacova, S., Baird, D.D., Balabaeva, L., Lolova, D., Petrov, I. 1994. Placental arsenic and cadmium in relation to lipid peroxides and gluthathione levels in maternal-infant pairs from copper smelter area. *Placenta*, **15**(8), 873–881.

Thompson, D.J. 1993. A chemical hypothesis for arsenic methylation in mammals. *Chem. Biol. Interact.*, **88**, 89–114.

Vahter, M., Marafante, E. 1983. Intracellular interaction and metabolic fate of arsenite and arsenate in mice and rabbits. *Chem. Biol. Interact.*, **47**, 29–44.

Willhite, C.C., Ferm, V.H. 1984. Prenatal and development of toxicology of arsenicals. *Nutritional and Toxicological Aspects of Food Safety*, Ch. 9, pp. 205–228.

Wu, M-M., Kuo, T-L., Hwang, Y-H., Chen, C-J. 1989. Dose–response relation between arsenic concentration in well water and mortality from cancers and vascular diseases. *Am. J. Epidemiol.*, **130** (6), 1123–1132.

Zaldivar, R. 1980. A morbid condition involving cardiovascular, bronchopulmonary, digestive and neural lesions in children and young adults after dietary arsenic exposure. *Zbl. Bakt. I Abt. Orig. B*, **179**, 44–56.

Zaldivar, R., Prumes, L., Ghai, G.L. 1981. Arsenic dose in patients with cutaneous carcinomata and hepatic haemangioendothelioma after environmental and occupational exposure. *Arch. Toxicol.*, **47**, 145–154.

Zelikoff, J.T., Bertin, J.E., Burbacher, T.M., Hunter, E.S., Miller, R.K., Silbergeld, E.K., Tabacova, S., Rogers, J.M. 1995. Health Risks Associated with Prenatal Metal Exposure. *Fundam. Appl. Toxicol.*, **25**, 161–170.

Zierler, S., Theodore, M., Cohen, A., Rothman, K.J. 1988. Chemical quality of maternal drinking water and congenital heart disease. *Int. J. Epidemiol.*, **17**(3), 589–594.

Arsenic Exposure and Health Effects
W.R. Chappell, C.O. Abernathy and R.L. Calderon (Editors)

Groundwater Arsenic Contamination and Suffering of People in Bangladesh

Uttam K. Chowdhury, Bhajan K. Biswas, Ratan K. Dhar,
Gautam Samanta, Badal K. Mandal, Tarit Roy Chowdhury,
Dipankar Chakraborti, Saiful Kabir, Sibtosh Roy

ABSTRACT

The total area and population of Bangladesh are 148,393 km^2 and 120 million respectively. We have been working in Bangladesh for about 3 years. To date, we have analysed 9089 water samples collected from 60 districts and found arsenic concentrations in 41 districts to be above 50 $\mu g/l$. The area and population of these 41 districts are 89,186 km^2 and 76.9 million respectively. This does not mean the total population in these 41 districts are drinking contaminated water and suffering from arsenicosis, but no doubt they are at risk. About 3000 each of hair and nail samples from people living in arsenic-affected villages (including patients) have so far been analysed and 97% of the hair samples contain arsenic above the toxic level and 95% of the nail samples contain above the normal level. Out of the 41 districts where arsenic has been found above 50 $\mu g/l$ we have, so far, surveyed 22 districts for arsenicosis patients, and in 21 districts we have identified people suffering from arsenic-induced skin lesions. We have examined, at random, in these affected villages 6973 people including children, and out of them, 33.1% are found to have arsenical skin lesions. The School of Environmental Studies has worked for the last 10 years in West Bengal. It appears that the arsenic calamity in Bangladesh may be more severe than in West Bengal. To combat the situation, Bangladesh needs a proper utilization of its vast surface and rain water resources. Proper water resource management is a proposed solution.

Keywords: groundwater, arsenic contamination, Bangladesh, arsenic in water, hair, nail, urine, skin-scales; arsenical skin lesions; arsenic body burden

INTRODUCTION

The world's four major arsenic calamities are in Asia. The countries affected are Bangladesh (Dhar et al., 1997; Biswas et al., 1998), West Bengal, India (Das et al., 1994; Chatterjee et al., 1995; Das et al., 1995; Das et al.,1996; Mandal et al., 1996; Roy Chowdhury et al., 1997; Mandal et al., 1997; Guha Mazumder et al., 1997; Mandal et al., 1998), Inner Mongolia, P.R. China (Xiao, 1997; Gao, 1997), Xinjiang, P.R. China (Lian and Jian, 1994) and Taiwan (Tseng et al., 1961; Yeh, 1962). Arsenic contamination is the single biggest threat to Bangladesh's ground-water resources. The problem is gradually assuming alarming proportions. Out of the total 64 districts of Bangladesh, so far we have analysed water from 9089 tubewells in 60 districts. The four districts that have not yet been surveyed are Khagrachari, Rangamati, Banderban and Cox's Bazar. In 52 districts we have found arsenic in groundwater above the WHO recommended value of arsenic in drinking water (10 μg/l) (WHO, 1993). So far from our preliminary survey, groundwater of eight districts (i.e., Panchagarh, Thakurgaon, Nilphamari, Dinajpur, Gaibanda, Naogaon, Moulabi Bazar, Patuakhali) is found to be safe with respect to the WHO recommended value in drinking water. So far, we have identified 41 districts where groundwater contains arsenic more than 50 μg/l, the WHO maximum permissible limit. The area and population of these 41 districts are 89,186 km^2 and 76.9 million, respectively. This does not, however, mean that all the people have been drinking contaminated water and are affected by arsenic poisoning. What it means is that they run the risk of being affected, since they are living in areas where elevated arsenic in drinking water has been found.

A preliminary survey was conducted for arsenicosis patients in 98 villages in 22 districts where we have found arsenic in ground water more than 50 μg/l and we have identified arsenicosis patients in 95 villages in 21 districts. Out of the 6973 people we examined and recorded at random from affected villages, 33.6% showed arsenical skin lesions. The reason for such a high percentage is that our survey was carried out in highly arsenic-contaminated villages. However, to obtain a more realistic picture an extensive survey is necessary. Other than the dermatological symptoms, we also analysed at random approximately 3000 each of hair and nail samples from subjects in the affected villages. Arsenic concentration of above 95% of these biological samples are above normal level or toxic level. Thus, many may be subclinically affected. Normally, children under 11 years do not show skin lesions. Exceptions are observed when arsenic in drinking water is very high or poor nutrition status is present combined with moderately high arsenic exposures. During our survey over the last 10 years in highly arsenic-affected areas of West Bengal, India, we have found around 2–3% of the children under 11 years show arsenical skin lesions. However, we have found 18.7% in highly arsenic-affected areas in Bangladesh. We feel children are more affected and are at higher risk in Bangladesh than in West Bengal.

The present paper presents a status report of the arsenic problem of Bangladesh and the resulting suffering of people on the basis of analytical reports of water, biological specimens, dermatological features and the survey carried out in the last three years.

METHODOLOGY

Flow injection-hydride generation atomic-absorption spectrometry (FI-HG-AAS) was used for analysis of the water, urine and digested biological samples. The flow injection system was assembled from commercially available instruments and accessories in our laboratory. A Perkin-Elmer Model 3100 atomic absorption spectrometer (USA) equipped with Perkin-Elmer EDL system-2, arsenic lamp (lamp current 400 mA) were used. The detailed de-scription of the instrumentation and FI-HG-AAS procedure have been described in our earlier publications (Chatterjee et al., 1995; Das et al., 1995; Samanta and Chakraborti, 1997). For high concentrations of arsenic in water, our modified spectrophotometric method using Ag-DDTC in CHCl$_3$ with hexamethylenetetramine (Chakraborti et al., 1982) was used.

Reagents and Glassware

All reagents are of analytical grade. A solution of 1.5% $NaBH_4$ (Merck, Germany) in 0.5% NaOH (E. Merck, India) and 5.0 M solution of HCl (E.Merck, India) were used for flow injection analysis. Details of the reagents and glassware are given elsewhere (Chatterjee et al., 1995; Das et al., 1995).

Sample Collection and Digestion

Water samples were collected from tubewells (borewells). The mode of collection and details of the borewells are as described earlier (Chatterjee et al., 1993; Chatterjee et al., 1995). Samples of hair, nail, skin-scale and urine were collected from arsenic victims, from people who have no skin lesions but drink arsenic-contaminated water and also from those who live in a contaminated village but their tubewells are safe to drink. The procedure for cleaning the sample and the mode of digestion have been previously described (Chatterjee et al., 1995; Das et al., 1995; Mandal et al., 1996).

Procedure for Determination of Arsenic in Water, Hair, Nails, Skin-scale and urine

Detailed procedures for determination of arsenic in water, hair, nail, skin-scale and urine have been described elsewhere (Chatterjee et al., 1995; Das et al., 1995; Roy Chowdhury et al., 1997; Biswas et al., 1998). In this study we have determined total arsenic in water both by FI-HG-AAS and spectrophotometric methods. By using our modified spectrophotometric method, water samples containing high arsenic (above 40 $\mu g/l$) can be determined with 95% confidence whereas by FI-HG-AAS we can determine arsenic concentrations of 3 $\mu g/l$ with 95% confidence. Arsenic in hair, nail, skin-scale and urine (inorganic arsenic and its metabolites) was measured by FI-HG-AAS.

RESULTS AND DISCUSSIONS

Concentration of Arsenic in Drinking Water

Arsenic in groundwater has been found in West Bengal, India in the area of sediment of Younger Deltaic Deposition (YDD). It appears the same deposition extended eastward, covering the arsenic affected districts of Bangladesh (Figure 1). During the last 3 years, 9089 water samples were collected covering 60 districts out of the total 64 districts of Bangladesh. So far, from the water analysis report, 52 districts have been identified where arsenic in groundwater is more than WHO recommended value (10 $\mu g/l$) in drinking water and in 41 districts arsenic concentration above the WHO maximum permissible limit (50 $\mu g/l$). The total number of water samples analysed from these 41 districts was 7816 and out of that 52.75% have arsenic above 50 $\mu g/l$ with a maximum concentration 2500 $\mu g/l$. Table 1 shows the detailed analytical report of 41 districts where we have found arsenic in groundwater more than 50 $\mu g/l$.

Table 2 gives an overview of arsenic groundwater contamination in districts of Bangladesh. Figure 2 shows the percentage of 9089 water samples in 60 districts at different arsenic concentration ranges and a comparative study with 42,225 water samples of West Bengal. We have not yet surveyed 4 districts and in 8 districts, so far, it appears groundwater is safe to drink according to WHO recommended value of arsenic in drinking water. A detailed survey is necessary to know the actual groundwater contamination. The position of different districts and their arsenic situation is presented in Figure 3.

Although this is a preliminary survey and only 9089 water samples have been tested, it appears from Table 1 that in many districts a higher concentration of arsenic exists. We have so far analysed water from 42,225 tubewells in arsenic-affected areas of West Bengal. A comparative study with analysed samples of Bangladesh with that of West Bengal shows (Table 3) that highly elevated concentrations of arsenic in groundwater are more frequently

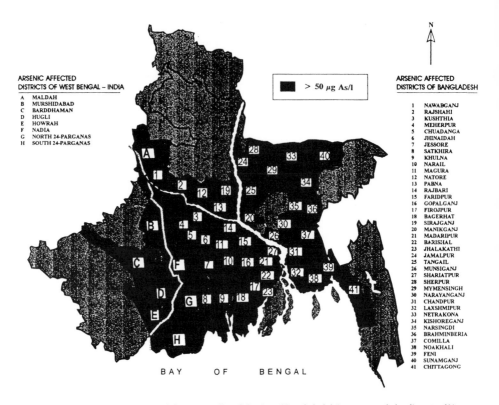

Fig. 1. Map showing most of the arsenic affected districts of Bangladesh lying on extended sediments of Younger Deltaic Deposition (YDD) of West Bengal.

encountered in Bangladesh than in West Bengal. Out of 9089 samples, 189 have arsenic more than 1000 μg/l. From our experience of the last 10 years we expect that those drinking contaminated water above 1000 μg/l of arsenic will develop arsenical skin lesions, including children.

So far in Bangladesh we have no estimation about the population drinking arsenic contaminated water above 50 μg/l. Our detailed study on Samta village of Sharsa Thana of Jessore District (Biswas et al., 1998) indicates that out of 265 tubewells of the village, 242 contain arsenic above 50 μg/l and 91% of the population (total population 4841) were drinking contaminated water above 50 μg/l (our survey shows 100% people in that area use tubewell water for drinking and cooking). Samta village is not an isolated example in Bangladesh. From our analytical report we find Chandipur, Shibpur, Augankhil, villages of the Laxshmipur district, Hatkhopa village in the Narayanganj district, Rajarampur village in the Nawabganj district, Lakuriakandi village in the Noakhali district and Sahapur in the Chandpur district also show the same trend. However, a detailed survey is necessary to ascertain the total population drinking arsenic-contaminated water above 50 μg/l. In one study, we made an estimation of the population drinking arsenic-contaminated water from only Bazars and Hats (*Bazars* and *Hats* are places where the villagers come to sell or buy). During our survey in arsenic-affected districts (Figure 3) we collected water from 83 tubewells from 83 Bazars/Hats in 13 districts (Figure 4) and found 77% of these tube-wells have arsenic above 50 μg/l (range 50 μg/l to 690 μg/l). We obtained information from the

TABLE 1

Analytical report of 41 districts of Bangladesh where arsenic (As) found in groundwater $>50\,\mu g/l$

District	No. of Thanas*	No. of Thanas surveyed	No. of Thanas where As		No. of samples	Distribution of total samples in different As concentration ($\mu g/l$) range							
			$>10\,\mu g/l$	$>50\,\mu g/l$		<10	10–49	50–99	100–299	300–499	500–699	700–1000	>1000
Nawabganj	5	5	4	4	622	184	99	48	173	53	32	9	24
Rajshahi	13	5	3	2	315	249	55	8	3				
Pabna	9	9	8	5	354	204	61	19	36	15	6	5	8
Kushtia	6	5	5	5	292	128	101	22	14	9	4	4	10
Meherpur	2	1	1	1	79	7	12	19	32	8	1		
Chuadanga	4	4	3	2	28	5	4	2	4	6	6	1	
Jhinaidah	6	1	1	1	26	10	9	3	2	1	–	1	
Jessore	8	5	5	5	665	113	75	219	162	44	31	19	2
Satkhira	7	4	4	4	156	28	53	23	41	7	3	1	
Khulna	14	7	7	6	862	484	185	66	93	25	7	2	
Bagerhat	9	3	3	3	188	18	24	19	57	47	13	10	
Narayanganj	5	2	2	2	188	48	5	2	54	28	24	17	10
Faridpur	8	3	3	3	324	64	97	36	79	21	14	5	8
Rajbari	4	3	3	2	70	35	27	3	4	–		1	
Magura	4	2	2	2	34	14	13	2	2		1	1	1
Chandpur	7	4	4	4	211	3	6	14	54	106	23	3	2
Noakhali	6	4	4	4	198	3	4	11	53	27	30	16	54
Laxshmipur	4	4	4	4	1326	45	109	173	405	223	175	127	69
Munsiganj	6	4	4	4	123	10	6	11	61	29	6		
Madaripur	4	3	3	3	80	2	14	16	23	15	7	3	
Shariatpur	6	4	4	3	70	30	19	8	6	6	1		
Narail	3	2	2	2	32	19	10	2	1				
Barisal	10	4	4	4	116	24	12	15	48	11	3	3	
Pirojpur	6	3	3	3	65	35	15	3	10	2			
Jhalakathi	4	3	3	2	28	17	6	2	2	1			
Gopalganj	5	3	3	3	98	18	23	9	20	13	7	8	
Natore	6	3	3	1	109	91	14	3	1				
Comilla	12	4	3	3	135	14	1	1	24	54	33	8	
Manikganj	7	3	3	3	116	43	22	26	25				
Feni	5	2	2	1	30	5	18	4	3				
Narsingdi	6	3	2	2	163	87	10	9	23	23	8	2	1
Chittagong	20	13	6	2	283	243	20	8	12	–	–	–	–
Sherpur	5	2	2	2	52	39	7	2	4				
Netrakona	10	2	2	1	29	24	2	3					
Mymensingh	12	3	3	2	49	30	17	2					
Jamalpur	7	2	1	1	39	23	6	4	6				
Tangail	11	2	1	1	10	9		1					
Kishoreganj	13	2	2	2	78	31	28	12	7				
Sunamganj	10	2	2	2	34	4	14	12	4				
Sirajganj	9	4	3	2	42	18	21	3					
Brahminberia	7	2	2	2	47	12	9	9	17				
Total	305	146	129	110	7766	2470	1233	854	1565	774	435	246	189

*Each district consists of several thana.

Bazars/Hats on the approximate population drinking that water (on average 1000 people drink water from each contaminated tubewell). A simple calculation shows that about 0.048 million people from 13 districts we surveyed were drinking contaminated water only from Bazars and Hats. An extrapolation of the data to Bazar/Hats of 41 arsenic-affected districts indicated that approx. 3.3 million people are drinking contaminated water only from Bazars

TABLE 2

An overview of arsenic (As) concentration of groundwater of Bangladesh

Total area (km²)	Total population (million)	Total no. of districts in Bangladesh	No. of districts not yet surveyed	No. of districts where As <10 μg/l	No. of districts where As 10–50 μg/l	No. of districts where As >50 μg/l	Total area of the districts where As >50 μg/l (km²)	Total population of districts where As >50 μg/l (million)
148,39 3	120	64	4	8	11	41	89,186	76.9

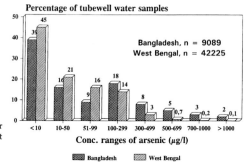

Fig. 2. Comparative study of groundwater arsenic situation in Bangladesh and West Bengal, India.

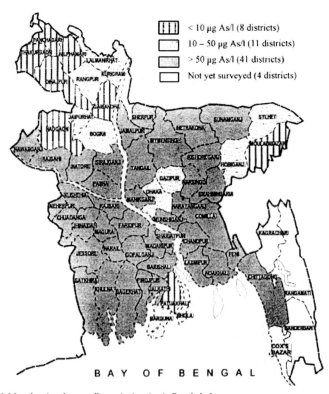

Fig. 3. Map showing the overall arsenic situation in Bangladesh.

TABLE 3

Distribution of arsenic in tubewell water in 60 districts of Bangladesh and 8 districts of West Bengal, India as of May, 1998

Country	Total no. tubewell water samples analysed	Distribution of no. of tubewell samples in different concentration range (μg/l) of arsenic			
		<10	10–50	>50	>1000
Bangladesh	9089	3507 (38.59%)	1459 (16.05%)	4123 (45.36%)	189 (2.08%)
West Bengal, India					
	42225	19001 (44.9%)	8867 (20.9%)	14357 (34%)	36 (0.09%)

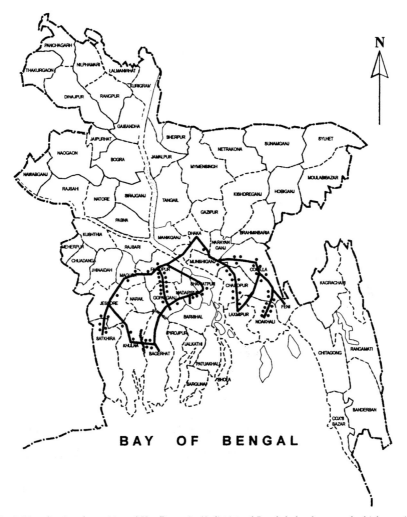

Fig. 4. Map showing the position of Hats/Bazars in 13 districts of Bangladesh where people drink arsenic contaminated water (>50 μg/l).

and Hats (information shows 4356 Bazar/Hats are in 41 districts where we found arsenic in groundwater more than 50 μg/l). Although, at this moment, we have no estimation of the total population drinking arsenic-contaminated water above 50 μg/l, we suspect on the basis of our field survey and 9089 water analyses that more people in Bangladesh are drinking arsenic-contaminated water than in West Bengal.

People with Arsenical Skin Lesions in 21 Districts of Bangladesh

Out of the 41 districts where we have found arsenic in groundwater more than 50 μg/l, so far we have surveyed 22 districts for arsenicosis patients and in 21 districts we have identified people having arsenical skin manifestations. We surveyed only 98 villages in these districts and in 95 villages we have identified patients. Figure 5 shows the districts where arsenicosis patients have been identified. So far from these 95 villages, we had surveyed at random 6973 people and out of that 2309 (33.1%) have been found with arsenical skin lesions. Table 4 shows the details of the districts/Thanas/Villages and populations including male, female and children that we have surveyed in these districts. We have information from many other villages in these 21 districts. Further, we have information from four more districts—Barisal, Jamalpur, Narail, and Shariatpur—where people have arsenical skin lesions. Yet, we do not know how many districts out of the 41 have arsenicosis patients. A true picture will come when a detailed survey is made in these 41 districts. From our field survey experience we feel we have identified a small portion of the total affected areas. We have identified all possible arsenical symptoms (except blackfoot disease) available in the literature from arsenic-affected villages. Figure 6 shows the abundance of the arsenic symptoms among 2110 adults who have arsenical skin lesions. Other than these common symptoms, we have also

TABLE 4

Detail study report of 21 districts of Bangladesh where arsenic patients identified

District	Area (km²)	Total no. Thanas	Population (million)	Total no. Thanas surveyed	Total no. Thanas where patients identified	Total no. villages surveyed	Total no. villages where patients identified	Total population surveyed	Total patients identified	Total male patients	Total female patients	Total child patients
Nawabganj	1702	5	1,232,000	1	1	4	4	236	143	85	58	–
Kushtia	1621	6	1,563,000	3	3	8	8	404	176	72	68	36
Rajshahi	2407	13	1,988,000	2	2	2	2	76	46	28	18	–
Meherpur	716	2	511,000	1	1	3	3	210	79	53	20	6
Pabna	2371	9	2,016,000	4	4	6	6	572	165	76	75	14
Chuadanga	1158	4	844,000	1	1	2	2	204	101	41	43	17
Jessore	2567	8	219,200	3	3	6	6	1445	424	164	191	69
Khulna	4395	14	2,130,000	2	2	5	5	304	102	59	35	8
Gopalganj	1490	5	1,097,000	2	2	3	3	31	22	8	10	4
Madaripur	1145	4	1,106,000	1	1	3	3	79	18	3	14	1
Satkhira	3858	7	1,660,000	1	1	2	1	138	54	31	23	–
Bagerhat	3959	9	1,489,000	2	2	9	9	438	196	132	59	5
Magura	1049	4	752,000	3	3	3	3	100	38	14	20	4
Faridpur	2073	8	1,558,000	1	1	6	6	288	96	45	42	9
Noakhali	3601	6	2,347,000	3	3	9	9	425	99	44	49	6
Laxshmipur	1456	4	1,391,000	2	2	14	13	1540	422	251	156	15
Narsingdi	1141	6	1,710,000	1	1	1	1	33	12	8	3	1
Rajbari	1119	4	865,000	1	1	2	2	16	8	4	4	–
Narayanganj	759	5	1,819,000	1	1	4	3	209	50	16	32	2
Chandpur	1704	7	2,149,000	3	3	3	3	150	42	23	17	2
Comilla	3085	12	4,263,000	3	3	3	3	75	16	14	2	–
Total	43,376	142	32,709,200	41	41	98	95	6973	2309	1171	939	199

Fig. 5. Map showing the districts where arsenic patients identified.

identified patients who have undergone amputation due to gangrene and also have squamous cell carcinoma. It is a common enquiry as to how many people have died so far from arsenic toxicity in the affected villages. We have no available statistics on this; however, when we survey in villages, we hear from arsenic-affected villagers that in many arsenic-affected families one or two died and in some cases more than two. There is no available documentation that they died of arsenic toxicity but what affected villagers say is, all those who died had as severe arsenical skin lesions as they have.

At present there is almost no available medicine for chronic arsenic patients. We have observed that by drinking safe water, eating nutritious food and doing some physical exercise, those who have very preliminary arsenic symptoms may get better. During our last 10 years experience in West Bengal and 3 years in Bangladesh we have noticed that those having diffuse melanosis and light spotted melanosis with the above prescription have a good prognosis. But when the spotted melanosis becomes appreciably visible, use of safe

Fig. 6. Distribution of arsenical skin lesions among the adults of 21 districts of Bangladesh. SM-P = Spotted melanosis on palm, DM-P = Diffuse melanosis on palm, SM-T = Spotted melanosis on trunk, DM-T = Diffuse melanosis on trunk, SK-P = Spotted keratosis on palm, DK-P = Diffuse Keratosis on palm, SK-S = Spotted keratosis on sole, DK-S = Diffuse keratosis on sole, LEUCO = Leuco melanosis, WB-M = Whole body melanosis.

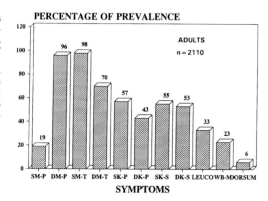

PERCENTAGE OF PREVALENCE

ADULTS
n = 2110

SM-P DM-P SM-T DM-T SK-P DK-P SK-S DK-S LEUCO WB-M DORSUM

SYMPTOMS

water does not substantially diminish the spotted melanosis and some of these are converted to leucomelanosis. This is again true for early keratosis. Mild keratosis may decrease, but not severe keratosis. During the last 10 years we have studied two groups (Chakraborti et al., 1998 and Mandal et al., 1998) and have found that arsenical skin lesions even remained the same for those having appreciably visible skin lesions after eight and two years of drinking safe water. During our field survey in Bangladesh on 19th July, 1998 we met a young man (age 33) in Augankhil village, Thana: Ramganj, Dist: Laxshmipur who left for Dubai in 1994 and had come back a few months before. He was examined by our dermatologist who found spotted melanosis (+ +), leucomelanosis (+), whole body melanosis (+). He had no keratosis when he left for Dubai. He told us that even after drinking water from Dubai for the last four years, his skin lesions remained almost unchanged; however, he is feeling better and the black shade of his skin has disappeared but more white spots are now visible on the skin.

The dermatological features of one patient from each district where we have identified patients with arsenical skin lesions is presented in Table 5. Figures 7–11 show photographs of

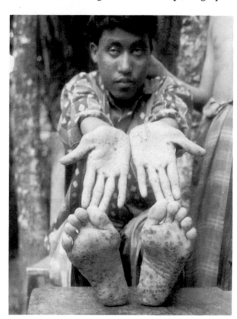

Fig. 7. Diffuse and spotted keratosis on palm and sole. Village: Lakuriakandi, District: Noakhali, Bangladesh.

TABLE 5

Dermatological features of 21 patients of 21 districts of Bangladesh

Sl. No.	District	Sex & Age	Melanosis						Keratosis					Non Petting Oedema	Conjunc- tional congestion	Bowans	Carcinoma
			Palm		Trunk		Leuco	Whole Body	Palm		Sole		Dorsum				
			SP	DIF	SP	DIF			SP	DIF	SP	DIF					
P-1	Nawabganj	M,46	+	+	+	++	+	–	+	+	+	++	–	–	–	–	–
P-2	Kushtia	F,11	++	++	+	+	–	++	+	++	++	++	–	–	+	–	–
P-3	Meherpur	M,7	–	+	++	+	–	–	+	–	+	+	–	–	–	–	–
P-4	Chuadanga	F,35	–	++	+++	++	+	++	++	++	+++	++	+	+	–	–	–
P-5	Jessore	F,30	+	+	+++	++	+	+++	+++	+++	+++	+++	+	+	+	–	–
P-6	Satkhira	M,25	–	–	+++	++	++	–	+	++	++	+++	+	–	–	–	–
P-7	Khulna	M,35	–	+	+++	+	+	++	++	++	+++	++	–	–	–	–	Gangrene
P-8	Magura	F,26	–	–	++	+	+	+	+++	+	+++	+	–	–	+	–	–
P-9	Pabna	M,14	–	++	++	+++	+	++	++	+++	++	+++	+	–	+	–	–
P-10	Rajbari	M,28	–	–	++	++	+	++	++	+++	++	+++	–	–	–	–	–
P-11	Faridpur	F,15	+	+	+++	++	+	+	+	–	++	++	–	–	–	+	–
P-12	Gopalganj	M,55	+	+	+++	+	–	+	–	–	+	+	–	–	–	–	–
P-13	Bagerhat	M,45	+++	–	+++	++	–	+	+++	++	+++	+++	++	–	+++	–	–
P-14	Madaripur	F,30	+	+	++	++	+	+	++	+	++	+++	+	–	–	–	–
P-15	Narayanganj	F,25	–	+	++	+	+	+	+	+	+	+	–	–	–	–	–
P-16	Chandpur	M,55	–	+	++	+	+	+	++	+	+	++	–	–	–	–	–
P-17	Laxshmipur	M,22	–	+	+++	+	+	+	+++	++	+++	++	++	–	–	–	–
P-18	Narsingdi	M,42	+++	+	+	+	–	–	+++	++	+++	++	–	–	–	–	–
P-19	Comilla	F,15	–	–	+++	–	++	++	++	+	+++	+	–	++	–	–	–
P-20	Noakhali	M,2 0	–	–	+++	+	+	++	+++	+	+++	+	–	–	–	–	Mucus membrane melanosis
P-21	Rajshahi	F,16	+	+	++	++	+	–	+	++	+	++	–	–	–	–	–

SP = spotted. DIF = diffuse; + = mild, ++ = moderate, +++ = severe.

Right: Fig. 8. Diffuse, spotted leuco melanosis and diffuse and spotted keratosis on palm. Village: Krishnakati, District: Satkhira, Bangladesh.

Below: Fig. 9. Diffuse and spotted keratosis on palm. Village: Bailtali, District: Bagerhat, Bangladesh.

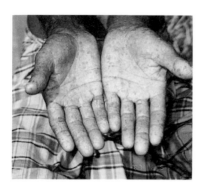

five patients from Table 5. Table 6 shows arsenic in hair, nail, and skin-scale of those five patients. Arsenic in urine is a direct indication that subjects are presently drinking contaminated water. In our study area in many cases we have not found high arsenic in urine as the people are aware of their contaminated tubewell and are consuming other sources of water. However, arsenic in hair, nail, skin-scale does not decrease as quickly as urinary arsenic. Even those drinking safe water may have some elevated level of arsenic in urine because drinking water is not the only source of arsenic (Mandal et al., 1998) in contaminated villages.

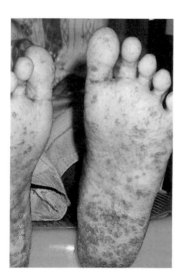

Fig. 11. Spotted keratosis on sole. Village: Orain, District: Comilla, Bangladesh.

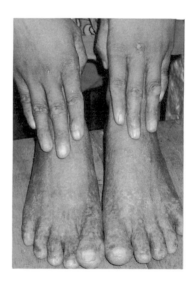

Fig. 10. Dorsum on feet and hands. Village: Shibpur, District: Laxshmipur, Bangladesh.

TABLE 6

Arsenic concentration in hair, nail and skin-scales of five arsenic patients

Patient	District	Village	Hair (µg/kg)	Nail (µg/kg)	Skin-scale (µg/kg)
P-6	Satkhira	Krishnakati	6,990	14,000	11,250
P-13	Bagerhat	Bailtali	2,050	3,150	2,140
P-17	Laxshmipur	Shibpur	19,510	40,100	9,260
P-19	Comilla	Orain	3,170	4,290	–
P-20	Noakhali	Lakuriakandi	5,580	10,000	5,080

Analyses of Hair, Nails, Skin-scale and Urine for Arsenic from Villagers in Affected Districts

Hair, nail, skin-scale and urine samples were collected from villagers in arsenic-affected villages during our field survey. Out of these samples about 78% are from patients with arsenical skin lesions. The analytical results of hair, nail, skin-scale and urine analysis are presented in Table 7.

Although 78% of the biological specimens (hair, nails) are from arsenicosis patients, the results in Table 7 show that about 97% of hair samples contain arsenic at toxic levels and about 95% of nail samples contain arsenic above normal level. To better understand the arsenic concentration in urine, hair and nails, samples of the control population, we collected water and biological samples from Chittagong, where we could not find arsenic above 50 µg/l out of 110 tubewells from the Thana Patiya and Boalkhali area. Table 8 shows the results. While 1–2% of the control population show arsenic above the normal level, the explanation for this might be that they had stayed occasionally in a contaminated area.

During our field survey we noticed that in a family, all adult members may drink the same contaminated water, but not necessarily all will have arsenical skin manifestations. We observed that about 40% of the family members may not show the arsenical skin lesions although all the members have high arsenic in hair and nails. It is possible that a group may not show skin lesions but they may have internal damage and may be subclinically affected. Figures 12 and 13 show the average and maximum arsenic content in hair and nails from villagers of the 21 districts where we have identified arsenicosis patients.

TABLE 7

Status of biological samples collected from the people of villages of Bangladesh where tubewells water are arsenic-contaminated

Parameters	Arsenic content in hair* (µg/kg)	Arsenic content in nail** (µg/kg)	Arsenic content in urine*** (µg/l)	Arsenic content in skin-scale**** (µg/kg)
No. of valid observation	2,942	2,940	1,043	349
Mean	4,050	9,250	495	5,730
Minimum	280	260	24	600
Maximum	28,060	79,490	3,086	53,390
Standard deviation	4,040	8,730	493	9,790
% of samples having arsenic above normal/toxic level[#]	97.3[#]	95.58	99.3	—

*Normal level of arsenic in hair 80–250 µg/kg with 1000 µg/kg being the indication of toxicity (Arnold et al., 1990).
**Normal level of arsenic nail 430–1080 µg/kg (Ioanid et al., 1961).
***Normal excretion of arsenic in urine range from 5–40 µg/day (Ferman and Johnson, 1990).
****There is no normal value for skin-scale in literature.
[#]Above toxic level

TABLE 8

Parametric presentation of arsenic in urine, hair and nails of control population of Chittagong, Bangladesh where arsenic in tubewell water <50 µg/l

Parameters	Mean	Range	Standard deviation	No. of samples
Urine (µg/l)	31	6–94	20	62
Hair (µg/kg)	410	120–850	180	62
Nail (µg/kg)	830	90–1580	680	62

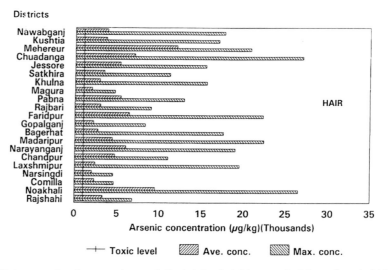

Fig. 12. Average and maximum arsenic concentration in hair collected from arsenic victims and people drinking arsenic contaminated water from 21 districts of Bangladesh.

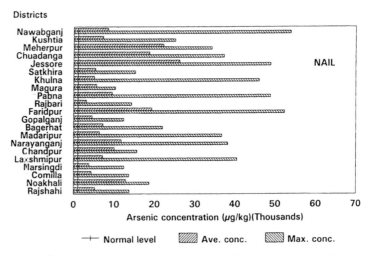

Fig. 13. Average and maximum arsenic concentration in nails collected from arsenic victims and people drinking arsenic contaminated water from 21 districts of Bangladesh.

Fig. 14. Distribution of common arsenical skin lesion among the children of 21 districts of Bangladesh. SM-P = Spotted melanosis on palm, DM-P = Diffuse melanosis on palm, SM-T = Spotted melanosis on trunk, DM-T = Diffuse melanosis on trunk, SK-P = Spotted keratosis on palm, DK-P = Diffuse Keratosis on palm, SK-S = Spotted keratosis on sole, DK-S = Diffuse keratosis on sole, WB-M = Whole body melanosis

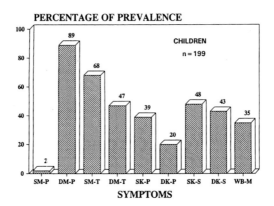

PERCENTAGE OF PREVALENCE

CHILDREN
n = 199

SYMPTOMS

Arsenic-Affected Children in 21 Districts

Studies over the years have shown that the children are at higher risk of arsenic exposure (Zaldivar, 1974; Borgono and Grebier, 1972; Jensen et al., 1991; Zaldivar and Guiller, 1977). Our West Bengal experience shows that normally children under 11 years of age do not show arsenical skin manifestations. However, we have observed a few exceptions when (1) the arsenic content in water consumed by children is very high—around 700 $\mu g/l$ and above— and (2) even with water containing arsenic around 50 $\mu g/l$ or above with poor nutrition status (Das et al., 1995; Mandal et al., 1998; Roy Chowdhury et al., 1997). We have observed that children of affected villages of Bangladesh are more affected than in West Bengal. In Bangladesh, from arsenic-affected villages of 21 districts, we have examined 1063 children and out of that 199 (18.7%) have been found with arsenical skin lesions. In some affected villages the percentage of affected children is even higher. Some examples are Benagari (district: Chuadanga), Samta (district: Jessore), and Ramkrishnapur (district: Kushtia). The total number of children affected will only be known after a thorough survey. Normally the arsenical symptoms we observe in children are diffuse melanosis and spotted melanosis. Keratosis on palm and sole are not so common in children in West Bengal but in Bangladesh spotted keratosis and diffuse keratosis are quite common. Figure 14 shows the abundance of arsenical skin lesions in 199 children under 11 years whom we had examined in 21 districts. Figure 15 shows a group of children and Table 9 shows skin lesions and arsenic in hair and

Fig. 15. A group of affected children having arsenical melanosis and keratosis. Village: Ramkrishnapur, District: Kushtia, Bangladesh.

TABLE 9

Skin lesions and arsenic in hair and nails of a group of children at Ramkrishnapur village, Veramara thana, Kushtia district, Bangladesh

Subject	Sex & Age	Melanosis					Keratosis				Arsenic concentration	
		Palm		Trunk		Whole Body	Palm		Sole		Hair (μg/kg)	Nail (μg/kg)
		SP	DIF	SP	DIF		SP	DIF	SP	DIF		
1	M,10	–	+	+	+	–	–	–	+	+	1,470	7,620
2	F,4	–	+	+	–	+	++	+	+	++	12,300	17,550
3	F,10	–	+	+	+	+	+	–	+	–	4,940	13,920
4	F,5	–	+	–	+	–	–	–	–	–	–	25,060
5	M,9	–	+	+	+	–	+	+	++	+	6,580	18,570
6	M,7	–	++	++	+	++	++	+	++	+	–	8,410
7	M,8	–	+	+	–	–	+	–	–	–	–	8,370
8	M,7	–	+	–	–	–	–	–	–	–	1,130	7,510

SP = Spotted, DIF = diffuse.
Leuco-melanosis: all negative.

nail samples from children in Ramkrishnapur village of Veramara Thana of Kushtia district. They used to drink arsenic-contaminated water (1500 μg/l). This is, however, one of the villages where children are seriously affected.

CONCLUSION

When we started our survey during October, 1995 we had information on only 8 districts and 12 villages. At present we have found 41 districts where tubewell water have arsenic more than 50 μg/l. We have found arsenic in groundwater more than 50 μg/l in 317 villages. Out of those 41 districts, we have surveyed 22 districts so far for arsenicosis patients. In 21 of these districts we have found 2309 patients with arsenical skin lesions. So far, patients have been identified with arsenical skin lesions in 95 villages. When the magnitude of the problem is compared with arsenic-contaminated areas of West Bengal, Bangladesh appears to be much more affected. In Noakhali district (Thana Senbagh), we have found the village Chiladi has 21 tubewells with 17 containing more than 1000 μg/l of arsenic that villagers are using for drinking and cooking.

A survey of the affected districts of Bangladesh is badly needed to ascertain the actual magnitude of the calamity. After the international arsenic conference in Bangladesh from 8–12 February 1998, jointly organised by the School of Environmental Studies, Jadavpur University and Dhaka Community Hospital, Bangladesh, two reports were published by the London *Guardian*, a daily newspaper (Pearce, 1998). The articles contained interviews with international organisations and arsenic experts. In that published report the World Bank's local chief Pierre Mills, said, "Tens of millions of people are at risk." The World Bank further mentions that 43,000 villages are presently at risk or could be in the future. A report by the World Health Organisation (WHO) predicts that, within a few years, one in 10 adult deaths across southern Bangladesh could be from cancers triggered by arsenic.

During February 1960, when groundwater was first withdrawn in the Nadia district of West Bengal, panicky villagers ran from the spot, crying, "Devil's water is coming, Devil's water is coming". They were reluctant to use the groundwater. But government and international funding agencies convinced them that the use of underground water would not only cure their water-borne disease but also bring economic wealth and happiness by increasing crop production. Slowly, the villagers got accustomed to using groundwater. The villagers no longer call it "Devil's water". If one asks a villager what he desires most, he will say a "tubewell of his own". We have made the green revolution at the cost of underground water and have neglected available surface water. Villagers no longer pray to the Rain God for water. Whenever they need water, they start the pumps and get more water than they need. Previously, we had only one crop a year, but now four crops are very common. At any time of the year, you will find the paddy fields green with crops. Were we not aware of the dangers of underground water use? Yes, we were. The first patient of the crippling fluorosis disease, which is caused by the drinking of fluoride-contaminated water, was identified in the Prakasam district of Andhra Pradesh as far back as 1937. Slowly, information about people suffering from fluorosis started coming from more and more states. During 1986, nine states were found to be affected. At present (1998), thousands are suffering from fluorosis in 16 states, including West Bengal. UNICEF, in one of its reports, wrote: "the number of people affected by fluorosis, which causes dental problems, gastroenteritis and crippling is estimated at an appalling 25 million" (UNICEF, WATSAN India 2000, 1995).

Over-exploitation of underground water for agriculture is becoming very common in developing countries e.g. India, Bangladesh and China. Millions are exposed to arsenic and fluoride contamination in India and China alone, and a huge population is suffering from arsenicosis and fluorosis because of drinking contaminated groundwater. Any country where exploitation of water goes similarly unchecked could be making itself vulnerable to a similar calamity. We cannot go on exploiting our natural resources recklessly. The problem of arsenic-contaminated groundwater in Bangladesh and West Bengal will take a very serious turn if alternative water sources are not explored. Both these affected areas have huge wetlands, high rainfall (more than 2000 mm/year), flooded river basin, ox-bow lakes and we need to use these sources for our sustainable development. To combat the arsenic calamity we badly need proper water resource management.

REFERENCES

Arnold, H.L., Oden, R.B., James, W.D. 1990. *Diseases of the Skin. Clinical Dermatology.* W.B. Saunder Company, Philadelphia, USA, 8th ed. pp. 121–122.

Biswas, B.K., Dhar, R.K., Samanta, G., Mandal, B.K., Chakraborti,D., Faruk,I., Islam, K.S., Chowdhury, M.M., Islam, A., Roy, S. 1998. Detailed study report of Samta. One of the arsenic-affected villages of Jessore District, Bangladesh. *Curr. Sci.,* **74(2)**, 134–145.

Borgono, J.M., Greiber, R. 1972. Epidemiological study of arsenicism in the city of Antofagasta. In: Hemphill D.D. (ed.) *Trace Substances in Environmental Health–V. 5*, pp. 13–24. Columbia, University of Missouri Press.

Chatterjee, A., Das, D., Mandal, B.K., Roy Chowdhury, T., Samanta, G., Chakraborti D. 1995. Arsenic in groundwater in six districts of West Bengal, India: the biggest arsenic calamity in the world. Part I: Arsenic species in drinking water and urine of the affected people. *The Analyst,* **120,** 643–650.

Chakraborti, D., Valentova, M., Sucha, L. 1982. Decomposition of materials containing traces of arsenic and its spectrophotometric determination. Prague Institute of Chemical Technology of Czechoslovakia. *Anal. Chem.,* **H-17,** 31–42.

Chatterjee, A., Das, D., Chakraborti, D. 1993. A study of ground water contamination by arsenic in the residential area of Behala, Calcutta due to industrial pollution. *Environ. Poll.,* **80(1),** 57–65.

Chakraborti, D., Samanta, G., Mandal, B.K., Roy Chowdhury, T., Chanda, C.R., Biswas, B.K., Dhar, R.K., Basu, G.K., Saha, K.C. 1998. Calcutta's industrial pollution: Groundwater arsenic contamination in a residential area and sufferings of people due to industrial effluent discharge — An eight-year study report. *Curr. Sci.,* **74(4),** 346–355.

Dhar, R.K., Biswas, B.K., Samanta, G., Mandal, B.K., Chakraborti, D., Roy, S., Jafar, A., Islam, A., Ara, G., Kabir, S., Khan A. W., Ahmed, S.A., Hadi, A. 1997. Groundwater arsenic calamity in Bangladesh. *Curr. Sci.,* **73(1),** 48–59.

Das, D., Chatterjee, A., Samanta, G., Mandal, B., Roy Chowdhury, T., Samanta, G., Chowdhury, P.P., Chanda, C., Basu, G., Lodh, D., Nandi, S., Chakraborti, T., Mandal, S., Bhattacharya, S.M., Chakraborti, D. 1994. Arsenic contamination in groundwater in six districts of West Bengal, India: the biggest arsenic calamity in the world. *The Analyst,* **119,** 168N–170N.

Das, D., Chatterjee, A., Mandal, B.K., Samanta, G., Chakraborti, D., Chanda, B. 1995. Arsenic in groundwater in six districts of West Bengal, India: the biggest arsenic calamity in the world. Part II: Arsenic concentration in drinking water, hair, nail, urine, skin-scale and liver tissue (Biopsy) of the affected people. *The Analyst*, **120**, 917–924.

Das, D., Samanta, G., Mandal, B.K., Roy Chowdhury, T., Chanda, C.R., Chowdhury, P.P., Basu, B.K., Chakraborti, D. 1996. Arsenic in groundwater in six districts of West Bengal, India. *Environ. Geochem. Health*, **18(1)**, 5–15.

Ferman G., Johnson, L.R. 1990. Assessment of occupational exposure to inorganic arsenic based on urinary concentration and speciation of arsenic. *Br. J. Indust. Med.* **47**, 342–348.

Guha Mazumder, D.N., Das Gupta, J., Santra, A., Pal, A., Ghosh, A., Sarkar, S., Chattopadhyay, N., Chakraborti, D. 1997a. Non-cancer effects of chronic arsenicosis with special reference to liver damage. In: Abernathy, C.O., Calderon, R.L. and Chappell, W.R. (eds.) *Arsenic Exposure and Health Effects*. 10, pp. 112–113. Chapman and Hall, New York.

Gao, C. 1997. The geological setting of arsenic contamination in Inner Mongolia, China. *The Association for the Geological Collection in Japan (Chigaku Dantai Kenkyukai)*, **11**, 28–33

Ioanid, N.B., Bors, G., Popa, I. 1961. Beitrage Zur Kenntnis des normalen Arsengehaltes Von Nageln und des Gehaltes in den Fallen Von Arsenpolyneuritis. *Dtsch. Gesamte Gerichtt. Med.* **52**, 90–94.

Jensen, G.E., Christensen, J.M., Poulsen, O.M. 1991. Occupational and environmental exposure to arsenic increased urinary arsenic level in children. *Sci. Total Environ.*, **107**, 169–177.

Lian, F.W., Jian, Z.H. 1994. Chronic arsenisim from drinking water in some areas of Xinjiang, China. In: Nriagu. J.O. (ed.), *Arsenic in the Environment. Part II: Human Health and Ecosystem Effects*, pp. 159–172. John Wiley and Sons, New York.

Mandal, B.K., Roy Chowdhury, T., Samanta, G., Basu, G.K., Chowdhury, P.P., Chanda, C.R., Lodh, D., Karan, N.K., Dhar, R.K., Tamili, D.K., Das, D., Saha, K.C., Chakraborti, D. 1996. Arsenic in groundwater in seven districts of West Bengal, India—the biggest arsenic calamity in the world. *Curr. Sci.*, **70(2)**, 976–986.

Mandal, B.K., Roy Chowdhury, T., Samanta, G., Basu, G.K., Chowdhury, P.P., Chanda, C.R., Lodh, D., Karan, N.K., Dhar, R.K., Tamili, D.K., Das, D., Saha, K.C., Chakraborti, D. 1997. Chronic arsenic toxicity in West Bengal. *Curr. Sci.*, **72(2)**, 114–117.

Mandal, B.K., Roy Chowdhury, T., Samanta, G., Mukherjee, D.P., Chanda, C.R., Saha, K.C., Chakraborti, D. 1998. Impact of safe water for drinking and cooking on five arsenic affected families for 2 years in West Bengal, India. *Sci. Total Environ.*, **218**, 185–201.

Pearce, F. 1998. Arsenic in groundwater. *The Guardian*, London daily newspaper, 19th February, pp 1–3.

Roy Chowdhury, T., Mandal, B.K., Samanta, G., Basu, G.K., Chowdhury, P.P., Chanda, C.R., Karan, N.K., Dhar, R.K., Lodh, D., Das, D., Saha, K.C., Chakraborti, D. 1997. Arsenic in groundwater in six districts of West Bengal, India: The biggest arsenic calamity in the world. The status report upto August, 1995. In: Abernathy, C.O., Calderon, R.L., Chappell, W.R. (eds.), *Arsenic Exposure and Health Effects*, 9, pp. 91–111. Chapman and Hall, New York.

Samanta, G., Chakraborti, D. 1997. Flow-Injection Atomic Absorption Spectrometry for the Standardization of arsenic, lead and mercury in environmental and biological standard reference materials. *Fresenius J. Anal. Chem.*, **357**, 827–832.

Tseng, W.P., Chen, W.Y., Sung, J.L., Chen, J.S. 1961. A clinical study of black foot disease in Taiwan: An endemic peripheral vascular disease. *Mem. Coll. Med. Natl. Taiwan Univ.*, **7**, 1–18.

UNICEF, 1995. *WATSAN India 2000*, pp. 42.

WHO, 1993. Guideline for drinking water quality, recommendations (2nd ed), World Health Organisation, Geneva, 1, 41.

Xiao, J.G. 1997. Report from Inner Mongolia, China. *Asia Arsenic Network — Newsletter, Japan*, **2**, 7–9.

Yeh, S. 1962. Relative incidence of skin cancer in Chinese in Taiwan: with special reference to arsenical cancer. *The Conference on Biology of Cutaneous Cancer*, April 6–11, pp. 81–102. Philadelphia, PA.

Zaldivar, R. 1974. Arsenic concentration of drinking water and foodstuffs causing endemic chronic poisoning. *Beitr. Pathol.*, **151**, 384–400.

Zaldivar, R., Guillier, A. 1977. Environmental and clinical investigation on epidemic chronic arsenic poisoning in infants and children. *Zentralbl. Bakteriol. Parasitenkd. Infektionskr. Hyg. At. 1: Orig. Reihe B.* **165**, 226–243.

Arsenic Exposure and Health Effects
W.R. Chappell, C.O. Abernathy and R.L. Calderon (Editors)
© 1999 Elsevier Science B.V. All rights reserved.

Inorganic Arsenic and Prenatal Development: A Comprehensive Evaluation for Human Risk Assessment

Joseph F. Holson, John M. DeSesso, Anthony R. Scialli, Craig H. Farr

ABSTRACT

Several publications have suggested an association between inorganic arsenic and cranial neural tube defects. A comprehensive program was developed to assess all aspects of this putative association. The program began with an exhaustive critical review of experimental and human literature, followed by a series of new laboratory studies. Analysis of the experimental literature revealed that extant literature is inadequate for estimating human risk. Most of the studies were not designed for risk assessment and did not use routes of administration relevant to potential human exposures. The available experimental literature indicated, however, that while intravenous and intraperitoneal injections of arsenic could produce structural malformations, oral and inhalatory exposures to arsenic could not. Studies of arsenic and human pregnancy are limited in number and are based on ecologic designs assessing drinking water, airborne dusts, and areas near smelters or other industrial activity. These studies do not support an association between arsenic exposure and adverse pregnancy outcome, because they did not measure arsenic exposure during pregnancy and failed to control for recognized confounders. A series of new laboratory studies was conducted in which rats were exposed to arsenic trioxide by oral or inhalatory routes. Exposure began two weeks prior to mating and continued throughout gestation. Even at arsenic doses/exposures toxic to the dam, fetal survival, growth, and development were not affected. This comprehensive evaluation of arsenic— the literature evaluation and the new laboratory studies—indicates that humans do not seem to be at risk for birth defects at environmentally relevant levels and routes of exposure. New, well-designed epidemiology studies would be useful in evaluating this conclusion.

Key Words: arsenic, developmental toxicity, teratogenicity, neural tube defects, human, animal

INTRODUCTION

Inorganic arsenic is widely distributed in all environmental media. Consequently, most people experience a measurable arsenic intake each day. For most, food is the major source of arsenic (Gunderson, 1995). Much work has been done over the past 30 years to investigate the developmental toxicity of inorganic arsenic in animals. Intraperitoneal (IP) or intravenous (IV) injections of sodium arsenate (As^{+5}) or arsenite (As^{+3}) to pregnant rats (e.g., Beaudoin, 1974; Umpierre, 1981), hamsters (e.g., Ferm and Carpenter, 1968; Hood and Harrison, 1982), and mice (e.g., Hood and Bishop, 1972; Hood, 1972) result in embryolethality, increased resorptions, and fetal malformations, especially cranial neural tube defects. These studies are inadequate for estimating human health risks, however, in particular because the injection routes of administration are not relevant to human exposures.

Our goal was the evaluation of the potential for environmentally relevant exposures to inorganic arsenic to cause structural malformations in humans. To do that, a four-part comprehensive assessment program was designed:

1. Conduct an exhaustive analysis of the experimental and human literature;
2. Confirm the earlier literature using high-dose, single-injection protocols;
3. Assess the maternal and developmental toxicity of repeated oral exposure to arsenic;
4. Assess the maternal and developmental toxicity of repeated inhalatory exposure to arsenic.

ANALYSIS OF THE EXPERIMENTAL AND HUMAN LITERATURE

Over 300 publications were reviewed in the course of this analysis. The results were presented at the 1998 Teratology Society meeting (DeSesso et al., 1998a) and have been published elsewhere (DeSesso et al., 1998b). We refer the reader to that publication for a full presentation of the data from the cited experimental and epidemiological studies.

Whole Animal Studies

Most of the whole animal studies reviewed were not designed to assess the risk of arsenic to human pregnancy. Instead, they investigated the pathogenesis of maldevelopment, most often that of cranial neural tube closure defects. These studies usually used single, high doses of arsenic administered by IV or IP injection at the critical period for neural tube closure. This study design is fine for its stated purpose, but results from these studies have been inappropriately used to support the notion that inorganic arsenic is likely a human teratogen.

Many of the studies suffered from other shortcomings. Most used small numbers of animals, lacked a dose–response design, and were not conducted in accordance with U. S. and international guidelines for assessing developmental toxicity. In addition, the methods and results were not clearly reported in many of the publications.

In spite of these difficulties, however, several conclusions are readily apparent. High doses of inorganic arsenic injected IV or IP do produce cranial neural tube closure, eye, kidney, and skeletal defects in rodents. These defects are elicited only at doses that are lethal (or nearly so) to many of the pregnant animals. In addition, post-implantation loss is increased, and fetal weight is decreased in survivors (Ferm and Carpenter, 1968; Holmberg and Ferm, 1969; Ferm et al., 1971; Hood, 1972; Hood and Bishop, 1972; Hood and Pike, 1972; Beaudoin, 1974; Burk and Beaudoin, 1977; Hood et al., 1978; Umpierre, 1981; Willhite, 1981; Hood and Harrison, 1982; Hood and Vedel-Macrander, 1984; Carpenter, 1987; Morrissey and Mottet, 1983; Mason et al., 1989).

In contrast to the above findings, when arsenic is administered by routes that are more relevant to potential human exposures (oral gavage, diet, drinking water, or inhalation), offspring do not display neural tube or other defects, even at maternally lethal doses. Extreme dosages by these routes may cause increased post-implantation loss and decreased fetal weights, though effects varied among the studies (James et al., 1966; Schroeder and Mitchener, 1971; Matsumoto et al., 1973; Kojima, 1974; Hood et al., 1977; Hood et al., 1978; Baxley et al., 1981; Hood and Harrison, 1982; Nagymajtényi et al., 1985; Seidenberg et al., 1986; Morrissey et al., 1990). Only two repeated-dose studies conducted under EPA guidelines were found in the literature base; in these studies, repeated oral administration of inorganic arsenic (as arsenic acid) throughout organogenesis did not produce structural malformations in mice or rabbits. In addition, post-implantation loss and decreased fetal weight occurred only at maternally toxic/lethal dose levels (Nemec et al., 1998).

Thus, one can conclude from the extant literature that

1. IV or IP injection of maternally toxic doses of inorganic arsenicals can cause cranial neural tube and other defects in rodents;
2. Administration of inorganic arsenicals to laboratory animals by routes relevant to humans exposures does not result in malformations.

In Vitro *and* In Ovo *Studies*

Sustained, high concentrations of inorganic arsenic in whole embryo culture (Müller et al., 1986; Chaineau et al., 1990; Mirkes and Cornel, 1992; Tabacova et al., 1996) or the chick egg (Ancel and Lallemand, 1941; Ancel, 1946–1947; Ridgway and Karnofsky, 1952; Birge and Roberts, 1976; Peterková and Puzanová, 1976; Gilani and Alibhai, 1990) can decrease embryonic growth, produce malformations, and cause embryolethality. However, these results are of questionable value in estimating human risk, most notably because the exposure concentrations and durations are not relevant to the *in vivo* situation. In addition, these systems lack the maternal-placental unit which normally modifies embryonal exposure by metabolizing and excreting xenobiotics.

Epidemiological Studies

Investigations of the effects of arsenic on human development have focused on exposures in smelter environs (Nordström et al., 1978a; 1978b; 1979a; 1979b; Beckman and Nordström, 1982; Tabacova et al., 1994a; 1994b), from drinking water (Zierler et al., 1988; Aschengrau et al., 1989; Börzsönyi et al., 1992), and from nearby industrial activity (Ihrig, 1997; Ihrig et al., 1998). Some of these studies have concluded that prenatal arsenic exposure is associated with increased rates of spontaneous abortion and stillbirth and decreased birthweight. However, all of the studies are of an ecologic design. Maternal arsenic exposure was not quantified in any study, and confounding exposures to other agents were present in many of the studies. The interpretation of most studies is further complicated by the fact that the authors did not control for maternal age and health status, prenatal care, smoking and alcohol use, and socioeconomic status. The only study that investigated arsenic and neural tube defects did not demonstrate a relationship between the two (Texas Department of Health, 1992); however, this study was limited by its small size and its attempt to estimate arsenic exposure months during the first trimester of pregnancy from urinary data collected months after delivery.

The preceding studies are weak. As such, they do not support an association between prenatal exposure to inorganic arsenic and adverse pregnancy outcomes in humans.

Conclusion

With the exception of the repeat-dose studies conducted in mice and rabbits under EPA guidelines (Nemec et al., 1998), the available experimental and human data were not adequate for use in determining whether prenatal exposure to inorganic arsenic has the potential to cause structural malformations in humans. In order to develop a database suitable for use in estimating risk, new laboratory animal studies were designed and conducted.

CONFIRMATION OF EARLIER LITERATURE USING A HIGH-DOSE, SINGLE-INJECTION PROTOCOL

The first task of the experimental phase of this comprehensive program was to repeat the single-exposure study design used frequently by earlier investigators. The results of these experiments have been presented (Stump et al., 1998a), and the data are being published in full (Stump et al., in press).

Sodium arsenate (As^{+5}; 5, 10, 20, or 35 mg/kg) was administered to Crl:CD®(SD)BR rats by a single IP injection; arsenic trioxide (As^{+3}) was given by a single IP injection (1, 5, 10, or 15 mg/kg) or by oral gavage (5, 10, 20, or 30 mg/kg). Controls received deionized water. Administration was on gestational day (GD) 9, the day most critical for neural tube closure in the rat.

Intraperitoneal injection of sodium arsenate (35 mg/kg, GD 9) and arsenic trioxide (10 mg/kg, GD 9) caused decreases in maternal food consumption, body weight, and body weight gain. Among offspring, these maternally toxic doses increased post-implantation loss, decreased fetal weights, and increased incidences of neural tube closure and craniofacial defects. At lower doses which were not toxic to the dams (5, 10, and 20 mg/kg of sodium arsenate, and 1 and 5 mg/kg arsenic trioxide), developmental toxicity was not observed. Based on the results of this study, the rat was considered to be an appropriately responsive model for studying the potential developmental effects of inorganic arsenic.

Oral gavage of arsenic trioxide (30 mg/kg, GD 9) was lethal to 7 of 25 dams and caused decreases in maternal food consumption, body weight, and body weight gain among the survivors. This maternally toxic dose caused an increase in post-implantation loss but did not affect fetal weight or induce malformations. At lower doses which were not toxic to the dams (5, 10, and 20 mg/kg), developmental toxicity was not observed.

Thus, the results of earlier investigations were confirmed: single, high-dose IP injections of inorganic arsenic during a critical time in development result in neural tube closure defects and other malformations (e.g., Hood and Bishop, 1972; Hood, 1972), while single oral exposure to even higher doses does not (e.g., Hood et al., 1978; Baxley et al., 1981).

ASSESSMENT OF THE MATERNAL AND DEVELOPMENTAL TOXICITY OF REPEATED ORAL EXPOSURE TO ARSENIC

The second task of the experimental phase of this comprehensive program was to assess the developmental effects of arsenic after repeated oral exposure. The results of these experiments have been presented (Stump et al., 1998b), and the data are being published in full (Holson et al., in press a).

Arsenic trioxide (As^{+3}; 1, 2.5, 5, or 10 mg/kg/d) was administered to Crl:CD®(SD)BR rats by daily oral gavage from 14 days prior to mating through GD 19. Because rats sequester arsenic in their erythrocytes (Hunter et al., 1942; Vahter, 1994), pre-mating exposure was used in an attempt to establish a steady state and to ensure that arsenic administered during gestation would be available to the conceptus.

Repeated oral exposure to arsenic trioxide at 10 mg/kg/d decreased maternal food consumption and net body weight gain. This dose level also produced stomach lesions (adhesions and eroded areas) evident at necropsy. The maternal LOAEL was set at 5 mg/kg/d based on transiently decreased maternal food consumption during pre-mating exposure, and 2.5 mg/kg/d was determined to be the maternal NOAEL.

Arsenic trioxide at 10 mg/kg/d decreased fetal weight but did not increase post-implantation loss or fetal malformations. No developmental effects were seen at lower doses, and the developmental NOAEL was thus set at 5 mg/kg/d. It was concluded that repeated oral exposure to arsenic trioxide, at doses as high as the maternally toxic dose of 5 mg/kg/d, did not adversely affect the developing rat.

ASSESSMENT OF THE MATERNAL AND DEVELOPMENTAL TOXICITY OF REPEATED INHALATORY EXPOSURE TO ARSENIC

The third task of the experimental phase of this comprehensive program was to assess the developmental effects of arsenic after repeated inhalational exposure. The results of these experiments have been presented (Stump et al., 1998c), and the data are being published in full (Holson et al., in press b).

Arsenic trioxide (As^{+3}; 0.3, 3, or 10 mg/m^3) was administered to Crl:CD®(SD)BR rats by daily whole-body inhalational exposure, 6 hours/day, from 14 days prior to mating through GD 19. Again, pre-mating exposure was used in an attempt to establish a steady state of arsenic within the dam and to ensure that arsenic administered during gestation would be available to the conceptus. In order to minimize oral intake of arsenic by grooming, following exposure, each animal was wiped with a disposable paper towel moistened with a mild soap solution.

Particle sizes, expressed as median mass aerodynamic diameter (mean ± SD), were determined to be 2.1 ± 0.13, 1.9 ± 0.29, and 2.2 ± 0.13 μm at the three exposure levels; the mean geometric standard deviations were 1.74, 1.94, and 1.87, respectively. These analyses indicate that the test material atmosphere was generated consistently and that the dust was respirable; in fact, the particle size and size distribution of arsenic trioxide used in this study were near the optimum for alveolar deposition in the rat (Lewis et al., 1989; Schlesinger, 1995; Ménache et al., 1996).

Inhalatory exposure to arsenic trioxide at 10 mg/m^3 decreased maternal food intake, decreased maternal net body weight gain, and produced signs of respiratory distress. Because effects were not seen at lower exposures, the maternal NOAEL was determined to be 3.0 mg/m^3.

At exposure levels as high as 10 mg/m^3, arsenic trioxide did not affect post-implantation loss, fetal weight, or the incidence of malformations. It was concluded, therefore, that inhalatory exposure to arsenic trioxide, at concentrations as high as the maternally toxic level of 10 mg/m^3, did not adversely affect the developing rat, and that the developmental NOAEL is ≥10 mg/m^3.

SAFE EXPOSURE LEVELS

Experimental maternal and developmental NOAELs from oral studies with As^{+3} and As^{+5} in mice, rats, and rabbits (Nemec et al., 1998; Holson et al., in press b) can be integrated with a reasonably high soil arsenic level (100 mg As/kg soil) and an estimation of the oral bioavailability of arsenic from soil (20%; Davis et al., 1992; Freeman et al., 1993, 1995). Extrapolating based on body weight, pregnant women could ingest *3–30 pounds of arsenic-containing soil per day* without exceeding the established maternal and developmental NOAELs for laboratory animals. Obviously, such an exposure scenario is highly unrealistic.

Likewise, these NOAELs can be integrated with extremely high arsenic water levels (2.5 mg As/L; 50 times the U. S. maximum contaminant level [EPA, 1998]) and an assumption of nearly complete oral bioavailability of arsenic from water (Cohen et al., 1998): Extrapolating based on body weight, women could drink *10–100 liters of such water per day* without exceeding the established maternal and developmental NOAELs for laboratory animals. Again, this is an unrealistic exposure scenario.

CONCLUSIONS

Based upon our extensive analysis of the literature and our laboratory results, we conclude that:

1. The frequently reported association between *in utero* exposure to inorganic arsenicals and the production of neural tube defects in offspring is a consequence of high exposures attainable only by IV or IP injection;

3. Prenatal effects seen in laboratory animals after high maternal exposures to inorganic arsenicals occur *only* with concomitant maternal toxicity and lethality;

4. Results obtained using arsenicals delivered by environmentally non-relevant routes of exposure are inappropriate for use in assessing the risk to human prenatal development.

Taken together, the preceding conclusions provide the basis for our assessment that, under realistic human exposure scenarios, inorganic arsenic is unlikely to pose a threat to pregnant women or their offspring.

ACKNOWLEDGMENT

The authors thank and recognize Dr. Catherine Jacobson for her critical review and assistance with preparation of this manuscript.

REFERENCES

Ancel, P., Lallemand, S. 1941. Sur l'arrêt de développement du bourgeon caudal obtenu expérimentalement chez l'embryon de poulet. *Arch. Physique Biol.*, **15**, 27–29.

Ancel, P. 1946–1947. Recherche expérimentale sur le spina bifida. *Arch. Anat. Microsc. Morphol. Exp.*, **36**, 45–68.

Aschengrau, A., Zierler, S., Cohen, A. 1989. Quality of community drinking water and the occurrence of spontaneous abortion. *Arch. Environ. Health*, **44**, 283–290.

Baxley, M.N., Hood, R.D., Vedel, G.C., Harrison, W.P., Szczech, G.M. 1981. Prenatal toxicity of orally administered sodium arsenite in mice. *Bull. Environ. Contam. Toxicol.*, **26**, 749–756.

Beaudoin, A.R. 1974. Teratogenicity of sodium arsenate in rats. *Teratology*, **10**, 153–158.

Beckman, L., Nordström, S. 1982. Occupational and environmental risks in and around a smelter in northern Sweden. IX. Fetal mortality among wives of smelter workers. *Hereditas*, **97**, 1–7.

Birge, W.J., Roberts, O.W. 1976. Toxicity of metals to chick embryos. *Bull. Environ. Contam. Toxicol.*, **16**, 319–324.

Börzsönyi, M., Bereczky, A., Rudnai, P., Csanady, M., Horvath, A. 1992. Epidemiological studies on human subjects exposed to arsenic in drinking water in southeast Hungary [letter]. *Arch. Toxicol.*, **66**, 77–78.

Burk, D., Beaudoin, A.R. 1977. Arsenate-induced renal agenesis in rats. *Teratology*, **16**, 247–260.

Carpenter, S.J. 1987. Developmental analysis of cephalic axial dysraphic disorders in arsenic-treated hamster embryos. *Anat. Embryol.*, **176**, 345–366.

Chaineau, E., Binet, S., Pol, D., Chatellier, G., Meininger, V. 1990. Embryotoxic effects of sodium arsenite and sodium arsenate on mouse embryos in culture. *Teratology*, **41**, 105–112.

Cohen, J.T., Beck, B.D., Bowers, T.S., Bornschein, R.L., Calabrese, E.J. 1998. An arsenic exposure model: Probabilistic validation using emperical data. *Human Ecol. Risk Assess.*, **4**, 341–377.

Davis, A., Ruby, M.V., Bergstrom, P.D. 1992. Bioavailability of arsenic and lead in soils from the Butte, Montana district. *Environ. Sci. Technol.*, **26**, 461–468.

DeSesso, J.M., Jacobson, C.F., Scialli, A.R., Farr, C.H., Holson, J.F. 1998a. Inorganic arsenic is not likely to be a developmental toxicant at environmentally relevant exposures. *Teratology*, **57**, 216.

DeSesso, J.M., Jacobson, C.F., Scialli, A.R., Farr, C.H., Holson, J.F. 1998b. An assessment of the developmental toxicity of inorganic arsenic. *Reprod. Toxicol.*, **12**, 385–433.

EPA (U.S. Environmental Protection Agency). 1998. National primary drinking water standards. Web page http://www.epa.gov/OGWDW/wot/appa.html.

Ferm, V.H. and Carpenter, S.J. 1968. Malformations induced by sodium arsenate. *J. Reprod. Fertil.*, **17**, 199–201.

Ferm, V.H., Saxon, A., Smith, B.M. 1971. The teratogenic profile of sodium arsenate in the golden hamster. *Arch. Environ. Health*, **22**, 557–560.

Freeman, G.B., Johnson, J.D., Killinger, J.M., Liao, S.C., Davis, A.O., Ruby, M.V., Chaney, R.L., Lovre, S.C., Bergstrom, P.D. 1993. Bioavailability of arsenic in soil and house dust impacted by smelter activities following oral administration in rabbits. *Fundam. Appl. Toxicol.*, **21**, 83–88.

Freeman, G.B., Schoof, R.A., Ruby, M.V., Davis, A.O., Dill, J.A., Liao, S.C., Lapin, C.A., Bergstrom, P.D. 1995. Bioavailability of arsenic in soil and house dust impacted by smelter activities following oral administration in cynomolgus monkeys. *Fundam. Appl. Toxicol.*, **28**, 215–222.

Gilani, S.H., Alibhai, Y. 1990. Teratogenicity of metals to chick embryos. *J. Toxicol. Environ. Health*, **30**, 23–31.

Gunderson, E.L. 1995. FDA total diet study, July 1986–April 1991, dietary intakes of pesticides, selected elements, and other chemicals. *J. AOAC Int.*, **78**, 1353–1363.

Holmberg, R.E. Jr., Ferm, V.H. 1969. Interrelationships of selenium, cadmium, and arsenic in mammalian teratogenesis. *Arch. Environ. Health,* **18**, 873–877.

Holson, J.F., Stump, D.G., Clevidence, K.J., Knapp, J.F., Farr, C.H. in press a. Evaluation of the reproductive and developmental toxicity of arsenic trioxide in rats. *Food. Chem. Toxicol.*

Holson, J.F., Stump, D.G., Ulrich, C.E., Farr, C.H. in press b. Absence of prenatal developmental toxicity from inhaled arsenic trioxide in rats. *Toxicol. Sci.*

Hood, R.D. 1972. Effects of sodium arsenite on fetal development. *Bull. Environ. Contam. Toxicol.,* **7**, 216–222.

Hood, R.D., Bishop, S.L. 1972. Teratogenic effects of sodium arsenate in mice. *Arch. Environ. Health,* **24**, 62–65.

Hood, R.D., Harrison, W.P. 1982. Effects of prenatal arsenite exposure in the hamster. *Bull. Environ. Contam. Toxicol.,* **29**, 671–678.

Hood, R.D., Pike, C.T. 1972. BAL alleviation of arsenate-induced teratogenesis in mice. *Teratology,* **6**, 235–238.

Hood, R.D., Thacker, G.T., Patterson, B.L. 1977. Effects in the mouse and rat of prenatal exposure to arsenic. *Environ. Health Perspect.,* **19**, 219–222.

Hood, R.D., Thacker, G.T., Patterson, B.L., Szczech, G.M. 1978. Prenatal effects of oral versus intraperitoneal sodium arsenate in mice. *J. Environ. Pathol. Toxicol.,* **1**, 857–864.

Hood, R.D. and Vedel-Macrander, G.C. 1984. Evaluation of the effect of BAL (2,3-dimercaptopropanol) on arsenite-induced teratogenesis in mice. *Toxicol. Appl. Pharmacol.,* **73**, 1–7.

Hunter, F.T., Kip, A.F., Irvine, J.W. 1942. Radioactive tracer studies on arsenic injected as potassium arsenite. *J. Pharmacol. Exp. Therapeutics,* **76**, 207–220.

Ihrig, M.M. 1997. Effect of chronic inhalation of inorganic arsenic on the risk of stillbirth in a community surrounding an agriculture chemical production facility: A hospital-based study. Masters thesis, Texas A & M University.

Ihrig, M.M., Shalat, S.L., Baynes, C. 1998. A hospital-based case-control study of stillbirths and environmental exposure to arsenic using an atmospheric dispersion model linked to a geographical information system. *Epidemiol.,* **9**, 290–294.

James, L.F., Lazar, V.A., Binns, W. 1966. Effects of sublethal doses of certain minerals on pregnant ewes and fetal development. *Am. J. Vet. Res.,* **27**, 132–135.

Kojima, H. 1974. Studies on development pharmacology of arsenic. 2. Effect of arsenite on pregnancy, nutrition and hard tissue. *Folia Pharmacol. Japonica,* **70**, 149–163.

Lewis, T.R., Morrow, P.E., McClellan, R.O., Raabe, O.G., Kennedy, G.L., Schwetz, B.A., Goehl, T.J., Roycroft, J.H., Chhabra, R.S. (1989). Establishing aerosol exposure concentrations for inhalation toxicity studies. *Toxicol. Appl. Pharmacol.,* **99**, 377–383.

Mason, R.W., Edwards, I.R., Fisher, L.C. 1989. Teratogenicity of combinations of sodium dichromate, sodium arsenate and copper sulphate in the rat. *Comp. Biochem. Physiol.,* **93C**, 407–411.

Matsumoto, N., Okino, T., Katsunuma, H., Iijima, S. 1973. Effects of Na-arsenate on the growth and development of the foetal mice. *Teratology,* **8**, 98.

Ménache, M.G., Raabe, O.G., Miller, F.J. 1996. An empirical dosimetry model of aerodynamic particle deposition in the rat respiratory tract. *Inhal. Toxicol.,* **8**, 539–578.

Mirkes, P.E., Cornel, L. 1992. A comparison of sodium arsenite-and hyperthermia-induced stress responses and abnormal development in cultured postimplantation rat embryos. *Teratology,* **46**, 251–259.

Morrissey, R.E., Fowler, B.A., Harris, M.W., Moorman, M.P., Jameson, C.W., Schwetz, B.A. 1990. Arsine: Absence of developmental toxicity in rats and mice. *Fundam. Appl. Toxicol.,* **15**, 350–356.

Morrissey, R.E., Mottet, N.K. 1983. Arsenic-induced exencephaly in the mouse and associated lesions occurring during neurulation. *Teratology,* **28**, 399–411.

Müller, W.U., Streffer, C., Fischer-Lahdo, C. 1986. Toxicity of sodium arsenite in mouse embryos in vitro and its influence on radiation risk. *Arch. Toxicol.,* **59**, 172–175.

Nagymajtényi, L., Selypes, A., Berencsi, G. 1985. Chromosomal aberrations and fetotoxic effects of atmospheric arsenic exposure in mice. *J. Appl. Toxicol.,* **5**, 61–63.

Nemec, M.D., Holson, J.F., Farr, C.H., Hood, R.D. 1998. Developmental toxicity assessment of arsenic acid in mice and rabbits. *Reprod. Toxicol.,* **12**, 647–658.

Nordström, S., Beckman, L., Nordenson, I. 1978a. Occupational and environmental risks in and around a smelter in northern Sweden. I. Variations in birth weight. *Hereditas,* **88**, 43–46.

Nordström, S., Beckman, L., Nordenson, I. 1978b. Occupational and environmental risks in and around a smelter in northern Sweden. III. Frequencies of spontaneous abortion. *Hereditas,* **88**, 51–54.

Nordström, S., Beckman, L., Nordenson, I. 1979a. Occupational and environmental risks in and around a smelter in northern Sweden. V. Spontaneous abortion among female employees and decreased birth weight in their offspring. *Hereditas,* **90**, 291–296.

Nordström, S., Beckman, L., Nordenson, I. 1979b. Occupational and environmental risks in and around a smelter in northern Sweden. VI. Congenital malformations. *Hereditas,* **90**, 297–302.

Peterková, R., Puzanová, L. 1976. Effect of trivalent and pentavalent arsenic on early developmental stages of the chick embryo. *Folia Morphol.,* **24**, 5–13.

Ridgway, L.P., Karnofsky, D.A. 1952. The effects of metals on the chick embryo: Toxicity and production of abnormalities in development. *Ann. NY Acad. Sci.,* **55**, 203–215.

Schlesinger, R.B. 1995. Deposition and clearance of inhaled particles. In *Concepts in Inhalation Toxicology*, 2nd ed. (R.O. McClellan and R.F. Henderson, Ed.), pp. 191–224. Taylor & Francis, Washington.

Schroeder, H.A., Mitchener, M. 1971. Toxic effects of trace elements on the reproduction of mice and rats. *Arch. Environ. Health*, **23**, 102–106.

Seidenberg, J.M., Anderson, D.G., Becker, R.A. 1986. Validation of an in vivo developmental toxicity screen in the mouse. *Teratogen. Carcinogen. Mutagen.*, **6**, 361–374.

Stump, D.G., Fleeman, T.L., Nemec, M.D., Holson, J.F., Farr, C.H. 1998a. Evaluation of the teratogenicity of sodium arsenate and arsenic trioxide following single oral or intraperitoneal administration in rats. *Teratology*, **57**, 217.

Stump, D.G., Clevidence, K.J., Knapp, J.F., Holson, J.F., Farr, C.H. 1998b. An oral developmental toxicity study of arsenic trioxide in rats. *Teratology*, **57**, 216–217.

Stump, D.G., Ulrich, C.E., Holson, J.F., Farr, C.H. 1998c. An inhalation developmental toxicity study of arsenic trioxide in rats. *Teratology*, **57**, 216.

Stump, D.G., Holson, J.F., Fleeman, T.L., Nemec, M.D., Farr, C.H. in press. Comparative effects of single intraperitoneal or oral doses of sodium arsenate or arsenic trioxide during *in utero* development. *Teratology*.

Tabacova, S., Baird, D.D., Balabaeva, L., Lolova, D., Petrov, I. 1994a. Placental arsenic and cadmium in relation to lipid peroxides and glutathione levels in maternal-infant pairs from a copper smelter area. *Placenta*, **15**, 873–881.

Tabacova, S., Little, R.E., Balabaeva, L., Pavlova, S., Petrov, I. 1994b. Complications of pregnancy in relation to maternal lipid peroxides, glutathione, and exposure to metals. *Reprod. Toxicol.*, **8**, 217–224.

Tabacova, S., Hunter, E.S. 3rd, Gladen, B.C. 1996. Developmental toxicity of inorganic arsenic in whole embryo culture: oxidation state, dose, time, and gestational age dependence. *Toxicol. Appl. Pharmacol.*, **138**, 298–307.

Texas Department of Health. 1992. An investigation of a cluster of neural tube defects in Cameron County, Texas. Report.

Umpierre, C.C. 1981. Embryolethal and teratogenic effects of sodium arsenite in rats. *Teratology*, **23**, 66A.

Vahter, M. 1994. Species differences in the metabolism of arsenic compounds. *Appl. Organomet. Chem.*, **8**, 175–182.

Willhite, C.C. 1981. Arsenic-induced axial skeletal (dysraphic) disorders. *Exp. Molec. Pathol.*, **34**, 145–158.

Zierler, S., Theodore, M., Cohen, A., Rothman, K.J. 1988. Chemical quality of maternal drinking water and congenital heart disease. *Int. J. Epidemiol.*, **17**, 589–594.

Arsenic Exposure and Health Effects
W.R. Chappell, C.O. Abernathy and R.L. Calderon (Editors)
© 1999 Elsevier Science B.V. All rights reserved.

Cancer Risks from Arsenic in Drinking Water: Implications for Drinking Water Standards

Allan H. Smith, Mary Lou Biggs, Lee Moore, Reina Haque,
Craig Steinmaus, Joyce Chung, Alex Hernandez, Peggy Lopipero

ABSTRACT

The current drinking water standard for arsenic in the U.S. and much of the world is 50 μg/L. The WHO has recommended lowering permissible concentrations to 10 μg/L, and the U.S. EPA to 2 μg/L, in each case based on extrapolation of skin cancer risks from a population in Taiwan with high levels of arsenic in their drinking water. Evidence from studies in Taiwan, Argentina and Chile is presented in this paper to show that, more important than skin cancer which is usually non-fatal, ingestion of inorganic arsenic in drinking water is also a cause of several internal cancers. For lifetime consumption of inorganic arsenic in drinking water containing around 500 μg/L, it is estimated that on the order of 10% of all deaths in adults would be attributable to ingestion of arsenic, mainly as a consequence of lung and bladder cancer. This extremely high cancer mortality risk estimate is based primarily on investigations in Region II of Chile, but is also supported by studies of other exposed populations, particularly Taiwan. Linear risk extrapolation from 500 μg/L to lifetime consumption of water with an arsenic concentration of 50 μg/L, the current drinking water standard, results in cancer mortality risk estimates reduced by a factor of ten to around 1 in 100 adult deaths being attributable to arsenic. Consideration is given to evidence for possible sub-linearity in the dose–response relationship which would make this estimate excessively high. The evidence is mixed, but neither human epidemiological studies, nor consideration of potential carcinogenic mechanisms, give assurance that the dose–response relationship would be significantly sub-linear in the dose range resulting from consumption of water between 50 and 500 μg/L arsenic in water. Even if marked sub-linearity were present, and risks at 50 μg/L were ten times lower than predicted from linear extrapolation, risk estimates would still be roughly of the order of 1 in 1000 persons dying due to arsenic in drinking water. Since such high cancer risks are unacceptable by any yardstick, it might be thought that the drinking water standard should be drastically reduced, even to lower concentrations than the 2 μg/L suggested by the U.S. E.P.A. However natural food sources become the predominant source of inorganic arsenic ingestion once water arsenic concentrations are reduced to about 10 μg/L and below. It is concluded that although much more research on arsenic is needed, the need for such research should not be used as an excuse to delay implementation of an inorganic arsenic drinking water standard considerably lower than the current 50 μg/L.

Keywords: arsenic, drinking water, epidemiology, cancer, risk assessment

INTRODUCTION

The purpose of this document is to summarize information pertinent to setting arsenic drinking water standards. The current standard in the U.S. has been 50 μg/L since the 1940s (U.S. EPA, 1988). The World Health Organization recently recommended a standard of 10 μg/L (WHO, 1993). Some countries have introduced new arsenic water standards including 25 μg/L in Canada and 7 μg/L in Australia. The U.S. E.P.A. proposed for consideration a standard of 2 μg/L. Both this standard, and that recommended by WHO, were based on skin cancer studies in Taiwan and a risk assessment published by the U.S. E.P.A. ten years ago. In this paper, the evidence that arsenic causes several internal cancers in addition to skin cancer is summarized. Cancer mortality risk estimates for high levels of exposure found in various parts of the world are presented. Consideration is then given to potential cancer risks at lower concentrations of arsenic such as those occurring at the current standard of 50 μg/L. The potential for sub-linearity in the dose–response relationships is also discussed. Finally, recommendations are made for standard setting which include consideration of non-water sources of inorganic arsenic in the diet.

ARSENIC INGESTION AND CANCER: SUMMARY OF RECENT HUMAN EVIDENCE

Until recently, the evidence that ingestion of arsenic is a cause of various cancers other than skin cancer came mainly from studies in Taiwan (Chen et al., 1985, 1988; Chiou et al., 1995; Guo et al., 1997; Wu et al., 1989) and to a lesser extent from two studies in Japan (Tsuda et al., 1990, 1995). A review published in 1992 concluded that these studies strongly suggested that ingested inorganic arsenic causes cancers of the bladder, kidney, lung and liver, and possibly other sites, but that confirmatory studies were needed (Bates et al., 1992). Since then several studies have provided strong additional evidence that arsenic ingestion does indeed cause internal cancers, in particular cancers of the bladder and lung.

A threefold increase in bladder cancer mortality (SMR 3.07; 95% CI 1.01–7.3) was reported after further follow-up of a cohort of 478 patients treated with Fowler's solution (potassium arsenite) in England (Cuzick et al., 1992), strengthening the bladder cancer evidence previously reported for this cohort (Cuzick et al., 1982). With one exception, the bladder cancer cases had received cumulative doses of less than 2000 mg of arsenic. This is a relatively low cumulative dose, equivalent to drinking 2 liters per day of water with an arsenic concentration of 100 μg/L for 30 years. No overall increase in lung cancer was found (SMR 1.00; 0.5–1.7) but a weak dose–response trend for respiratory cancer with cumulative arsenic dose had previously been reported in this cohort (SMRs 0.8, 1.1, 1.4, 1.8, $p = 0.16$) (Cuzick et al., 1982). The most recent publication did not provide comparative respiratory cancer data.

No overall increased risk in bladder cancer was found in a study involving low arsenic exposure levels in the state of Utah (Bates et al., 1995). However among smokers, there were increased trends in time window latency analyses especially in the period 30–39 years prior to cancer diagnosis. Arsenic water levels ranged from 0.5 to 160 μg/L. It was concluded that smoking might potentiate the effect of arsenic on the risk of bladder cancer. Since the risk estimates obtained were higher at these low levels of exposure than predicted from the results of the studies in Taiwan, the investigators concluded that confirmatory studies were needed.

A mortality study in the arsenic exposed region in Cordoba, Argentina, showed increased risks of bladder cancer among both men and women during the study period 1986 to 1991 (Hopenhayn-Rich et al., 1996a). The standardized mortality ratios for the low, medium, and high exposure counties were 0.80, 1.42, 2.14 for males (*p* value test for trend 0.001) and 1.21, 1.58, 1.82 for women (*p* = 0.04), respectively. The high exposure counties also showed increased mortality from lung and kidney cancer, but the findings for liver cancer were equivocal with increased risks in all counties (Hopenhayn-Rich et al., 1998). Evidence was presented showing that smoking did not contribute to the increased risk of deaths from these cancers. The crude estimate of the average concentration of arsenic in drinking water among the water sources containing more than 40 $\mu g/L$ tested 50 years ago was about 180 $\mu g/L$.

Dramatically increased mortality from bladder (SMRs about 7 and 8) and lung cancer (SMRs about 3–4) in Region II of Chile for the period 1989–93 has recently been reported (Smith et al., 1998). Kidney cancer mortality was increased to a lesser extent, but no increases were found for liver cancer. Increased mortality was also reported for skin cancer. Approximately 5–10% of all deaths among adults over the age of 30 were attributable to arsenic; chiefly to lung and bladder cancer. There was no increase in deaths from all other causes combined. Evidence indicated that smoking had not contributed to the increased cancer mortality in the Region. The arsenic levels in drinking water in the peak exposure period from about 1965 to 1980 averaged between 500 and 600 $\mu g/L$.

The studies in Argentina and Chile were conducted with *a priori* hypotheses that internal cancers, in particular bladder, lung, kidney and liver cancer would be increased, based mainly on findings in Taiwan. Given the *a priori* hypotheses, the results concerning lung and bladder cancer, and to a lesser extent kidney cancer, strongly support the evidence in Taiwan that ingestion of arsenic in drinking water is a cause of these cancers. The findings do not support liver cancer as an outcome. In retrospect, it is noteworthy that liver cancer in the Taiwan studies was associated with lower relative risks among the arsenic-exposed than bladder, lung and kidney cancer. It is possible that arsenic does cause liver cancer, but co-factors such as those associated with high liver cancer rates in Asia may be required.

CONCLUSION REGARDING BLADDER CANCER

There is sufficient evidence from several studies in several countries to conclude that ingestion of arsenic is a cause of human bladder cancer. Beyond the findings in Taiwan, the strongest additional evidence comes from large population studies in Chile and Argentina, each conducted with the *a priori* hypothesis that bladder cancer risks would be increased. Both studies found that the highest relative risks for internal cancer mortality associated with arsenic exposure were for bladder cancer. These ecological studies are supplemented by studies with individual data, in particular in Taiwan and in the Fowler's solution study in England. There is therefore ample evidence to conclude that inorganic arsenic ingestion is a cause of human bladder cancer.

CONCLUSION REGARDING LUNG CANCER

Recent studies add to the evidence that ingestion of inorganic arsenic causes increased risks of lung cancer. Clear increased risks were found in ecological studies in both Argentina and in Chile. Confounding due to smoking could be excluded as the explanation in both populations. Increased lung cancer risks had already been reported in a small study in Japan involving drinking water. As yet there are no large studies with individual exposure data. However, the findings in Argentina and especially in Chile where arsenic exposures were higher, provide evidence that the ingestion of arsenic most probably causes increased human lung cancer risks. Biological plausibility that arsenic from ingestion might increase lung cancer risks is strengthened by the fact that it is a confirmed lung carcinogen by inhalation. Taking this into account, there is now sufficient evidence to conclude that ingestion of inorganic arsenic is a cause of human lung cancer.

CONCLUSION REGARDING OTHER INTERNAL CANCERS

While recent studies add to the existing evidence and make it probable that ingestion of arsenic can cause kidney cancer, the findings are not as strong as for bladder and lung cancer. The evidence concerning liver cancer has actually been weakened by recent studies.

POPULATION RISK ESTIMATION FOR HIGH LEVELS OF EXPOSURE

Smith et al. (1998) showed that 5–10% of deaths occurring in adults in Chile were attributable to arsenic exposure chiefly due to bladder and lung cancers. In this population, arsenic in water contributed more to mortality than did cigarette smoking.

Other studies, particularly in Taiwan, are consistent with the very high population risk estimates calculated for Chile. For example, the bladder cancer relative risk estimates in Chile were around 7 and 8 for relatively short exposures (around 15 years) averaging between 500 and 600 µg/l. In Taiwan, the highest exposed populations drank water containing an average of 800 µg/L for longer periods and the relative risk estimates were on the order of 30–60 (Chen et al., 1988; Smith et al., 1992). Lower bladder cancer relative risks on the order of 2 were found in Argentina in association with much lower exposures, probably averaging around 180 µg/L.

Regarding lung cancer, relative risks were again higher in Taiwan where exposures were higher and of longer duration than in Chile. Lower relative risks were found in Argentina where exposures were lower. Population relative risk estimates for bladder and lung cancer thus show a consistent pattern. Since the exposures in Chile were much less than lifetime, with the highest levels occurring over only 15 years, a conservative rough estimate of lifetime mortality from drinking water containing inorganic arsenic at around 500 µg/L, might mean that 10% of adult deaths could result, predominantly due to cancers of the lung and bladder.

The consistency of this estimate derived from Chile with ecological studies in Taiwan can be seen by comparing the estimation of cancer risks for consumption of 1 liter per day of 500 µg/L. A risk assessment in 1992 used linear extrapolation to estimate that consumption of 1 liter per day of water containing 50 µg/L of arsenic might result in 13.4 per 1000 deaths when U.S. background cancer rates were incorporated into the analysis (Smith et al., 1992). Using the same methods, the estimate for 500 µg/L would be 13.4% which is consistent with the 10% estimate derived from Region II of Chile.

DOSE–RESPONSE RELATIONSHIPS: LINEAR OR SUB-LINEAR?

Clear dose–response data are still lacking in epidemiological studies of populations exposed to arsenic in their drinking water. Most studies have employed ecological groupings rather than individual exposure data. The highest priority for arsenic health effects research should be to add to the currently available information concerning dose–response relationships between ingestion of arsenic in drinking water and the risk of various outcomes, including cancer. However, quite extensive dose–response data are available for inhalation of inorganic arsenic and lung cancer risks.

THE DOSE–RESPONSE BETWEEN ARSENIC INHALATION AND HUMAN LUNG CANCER MAY BE LINEAR OR SUPRALINEAR

It is reasonable to propose that the shape of the dose–response curve for lung cancer caused by arsenic inhalation would be similar to that for lung cancer and other cancers caused by ingestion of inorganic arsenic. As far as inhalation is concerned, a reasonable question is whether or not the dose–response relationship might be supralinear or linear (Hertz-Picciotto and Smith, 1993). There is no evidence to suggest sub-linearity in the dose–response relationship. The findings using air measurements of arsenic inhalation were consistent with supralinearity in six studies conducted in three countries. One possible explanation is consistent overestimation of exposure at high air concentrations due to work practices to avoid exposure. This explanation is supported by one study which found supralinearity using air measurements for exposure, but linearity when urine measurements of arsenic were used (Enterline et al., 1987). Urine arsenic concentrations reflecting absorbed dose may

TABLE 1

Arsenic in drinking water cancer risk extrapolation to 50 and 10 μg/L. Estimates of lifetime cancer mortality.

Water arsenic concentration (μg/L)	Actual risk or linear extrapolation	10 times less than linear extrapolation	100 times less than linear extrapolation
500	1 in 10		
50	1 in 100	1 in 1,000	1 in 10,000
10	1 in 500	1 in 5,000	1 in 50,000

give a better estimate of inhaled dose than measurements of air concentrations using fixed samplers. This would occur if workers tended to avoid the most dusty environments as much as possible during their workday. We are of the opinion that this is the most likely explanation, and that the true dose–response relationship between inhaled arsenic dose and lung cancer risks is linear in the observable range, rather than supralinear. Since detailed dose–response data with individual exposure estimation for arsenic ingestion are still lacking, the studies of arsenic inhalation are important in that they do not provide any evidence for sub-linearity in the observed dose range in which lung cancer relative risks increased from less than 2 to more than 5.

EXAMINATION OF EPIDEMIOLOGICAL EVIDENCE FOR A THRESHOLD OR SUBLINEARITY CONCERNING ARSENIC

Two ecological analyses have suggested that the relationship between arsenic water concentrations and cancer occurrence in Taiwan is sub-linear or has a threshold. Brown and Chen (1995) reanalyzed the Taiwanese data and concluded that there could be a threshold or sub-linearity in the arsenic and cancer dose–response relationships. However, the reanalysis appears to have involved re-classifying village exposure and deleting villages according to *post hoc* criteria.

A further ecological analysis has been presented for bladder cancer incidence data in Taiwan (Guo et al., 1997). The investigators used a novel method for ecological data analysis. Superficial examination of the results suggests a threshold for arsenic water levels and bladder cancer. However, the unusual methods used were not accompanied by any results allowing the comparison of findings with other studies in Taiwan. Indeed, they would appear to be in conflict with them. For these reasons, this study provides little, if any, evidence for non-linearity in dose–response relationships for arsenic-induced bladder cancer, let alone evidence of a threshold.

In contrast to these unusual ecological analyses of data from Taiwan, results of other epidemiological studies, including further studies in Taiwan, demonstrate that it is unlikely that there is marked sublinearity and provide no evidence for a threshold. Skin cancer prevalence in Taiwan increased according to duration of residence in the area, duration of consumption of high-arsenic artesian well water, average arsenic water levels, and cumulative dose (Hsueh et al., 1995). Similar findings have been reported for lung and bladder cancer (Chiou et al., 1995). Although variables were for the most part categorized into three levels, the findings generally demonstrated a monotonic dose–response relationship for both cancers by duration of exposure, average arsenic concentration in drinking water, and cumulative exposure.

Apart from two unusual ecological analyses of Taiwanese data, there are no data supporting sub-linearity nor a threshold. This does imply that the results of these analyses should be excluded. However, in the absence of data supporting them, it is important to note that findings in various ecological studies, and limited findings with some individual data studies, support a monotonic dose–response relationship in the ranges of exposure considered thus far.

MECHANISTIC EVIDENCE

The mechanisms for arsenic carcinogenicity are unknown and there appear to be almost as many theories as there are investigators. Because arsenic does not cause point mutations in experimental systems, some investigators have postulated that these results are consistent with theories of sub-linearity for arsenic dose–response relationships. However, inference of sub-linearity from simple toxicological considerations is at best speculative without support from empirical data from human studies. Since there may be several mechanisms involved, multiple interactions with other factors both extrinsic and intrinsic, and variations in genetic susceptibility, inference from *in vitro* experiments and mechanistic theories cannot predict the shape of dose–response relationships for incidence rates of long latency diseases with complex multistage and multifactorial etiologies such as cancer. In addition, no information has been produced to identify the range of arsenic exposures in which meaningful sub-linearity might occur for any postulated theoretical mechanisms.

As with other major causes of human cancer, it is not likely that mechanisms allowing for valid predictions of dose–response relationships for low levels of arsenic will be identified in the foreseeable future. Indeed, mechanistic theories to date do not even predict why such high rates of bladder cancer would occur in humans exposed to arsenic at levels not much higher than the current drinking water standards. Until they do, it is futile to even begin to use such theories to postulate what might be happening below the as yet detectable effect levels in humans. This is not to say that mechanistic research is not important. However, this research involves a long term investment which may take decades and as such will not provide the methods for determining permissible exposure limits for arsenic in drinking water in the near future. It is also noteworthy that for many established causes of human cancer, the dose–response relationships found in epidemiological studies are more or less linear, whether or not point mutations are caused by the particular agents involved.

There is quite extensive human evidence concerning dose–response relationships for arsenic methylation. As discussed in a previous review, there is substantial evidence that inorganic arsenic was present in urine in approximately similar proportions to methylated forms at all levels of exposure from very low to very high (Hopenhayn-Rich et al., 1993). Subsequent studies have confirmed these findings. The largest human study examining methylation patterns in humans as reflected in urine profiles was in a population in the North of Chile. This study showed that the percentage of inorganic arsenic in urine was only slightly greater in the high exposure compared to the low exposure population (Hopenhayn-Rich et al., 1996b). These results were confirmed by an intervention study among highly exposed persons who were provided with arsenic-free water for two months. Total urinary arsenic averages fell from 636 μg/L to 166 μg/L whereas the percentage of the inorganic form changed very little, from 17.8% to 14.6% (Hopenhayn-Rich et al., 1996c). Another study reported very low levels of the metabolite monomethylarsonic acid (MMA) in urine in an isolated population in Argentina, but the levels of inorganic arsenic were similar to those reported in other populations (Vahter et al., 1995). Taking all the evidence into account, it can be concluded that some sub-linearity in cancer dose–response relationships could be supported by the human methylation data if inorganic arsenic is the main carcinogenic agent. However, the sub-linearity would be very slight, and there is no evidence from methylation patterns that would support a threshold below which there would be no cancer risks.

While extensive evidence is now available for urinary arsenic patterns of methylation, a biomarker of exposure, the information to be derived from studies using biomarkers of effect is more limited. A bladder cell micronucleus study in the North of Chile measured micronuclei in exfoliated bladder cells for persons residing in two towns with either high or low exposure (Moore et al., 1997a). Water levels in the high exposure town were on the order of 600 μg/L while actual exposure assessed by measuring urinary arsenic varied over a wide range. Increases in micronucleus prevalence were associated with urinary arsenic levels less than 700 μg/L. Above that level, micronucleus prevalence returned to background levels, perhaps as a result of cytotoxicity. When the population was divided into quintiles according to urinary arsenic concentrations, increased micronucleus prevalence was found at urinary arsenic concentrations on the order of 100 μg/L (range 54–137 μg/L) where there was a doubling of the prevalence of micronuclei (prevalence ratio 2.1, 95% confidence interval 1.4–3.4). An intervention study in a highly exposed sub-set of participants provided further evidence supporting these findings (Moore et al., 1997b). Although confirmatory studies are needed, the aforementioned results suggest that ingested inorganic arsenic might have genotoxic effects in bladder cells at low levels of exposure.

EXTRAPOLATION OF CANCER RISKS TO THE CURRENT DRINKING WATER STANDARD

A major risk assessment undertaking concerning arsenic in drinking water was published in 1992 with linear extrapolation to 50 μg/L (Smith et al., 1992). In the same year, investigators in Taiwan conducted risk extrapolations which produced results of a similar order of magnitude (Chen et al., 1992). It should be noted that the extrapolations are over a short range, much shorter than is usually the case for environmental exposure to carcinogens. Ecological evidence in Taiwan suggests a detectable increased risk in villages with average water levels around 170 μg/L (Chen et al., 1988). In Argentina, the highest exposure counties average estimate was about 180 μg/L where bladder cancer risks were clearly increased. (Hopenhayn-Rich et al., 1996a). Bladder cancer risks were also increased in counties classified as having medium exposure. The relative risks were 1.42 for men and 1.58 for women. It would appear then that detectable increased bladder cancer risks have already been found for levels of arsenic in water only 3 to 4 times that of the current drinking water standard. Of course, in ecological studies it is possible that effects are due to a small proportion of persons having much higher exposure than the average. Even if this were true, it would surely be accepted that there are real effects at 500 μg/L. If this were so, the extrapolation to 50 μg/L only involves a factor of 10.

Another way of considering these risks is in safety factor terms. If approximately 10% of people will die with a given exposure level, and it is not yet clear what the lowest detectable effect in epidemiological studies will be, what safety factor might be appropriate? We might start by allowing a factor of 10 because further epidemiological studies will certainly find effects below exposure levels causing deaths in 10% of people. We might then say we want a safety factor of 10 from that level, plus another factor of 10 to allow for variations in human susceptibility, and sensitive sub-populations. This would bring us down from the 500 μg/L level at which around 10% of people might die from the exposure to 0.5 μg/L. This safety factor approach is presented here because it again demonstrates what a small extrapolation is being made from 500 μg/L to 50 μg/L in the above risk estimation.

IMPLICATIONS OF NON-WATER SOURCES OF ARSENIC

Based on either traditional methods of risk extrapolation, whether using linear or sub-linear models (Table 1), or by considering safety factors, it is apparent that proposed drinking water standards for arsenic would be very low, presumably less than 1 μg/L. Reaching such a standard would involve treating almost all sources of drinking water. Perhaps fortunately, there is good reason to reject such a low drinking water standard without resorting to an assessment of the costs involved.

Arsenic is present everywhere in the earth's crust, in soil, in vegetables, fruits, and meats (IARC, 1987). In fact, it is present in all food sources. Even if all arsenic were removed from water, we would still have food intake which could not be prevented. For this reason, any consideration of human risks at low water concentrations needs to consider risks from all pathways, in particular from food. Increasing data are becoming available on food sources of inorganic arsenic, but the best source of information involves urinary levels of inorganic arsenic and its metabolites in persons whose water has very low arsenic concentrations. Unfortunately, detailed studies of urinary concentrations in persons drinking very low arsenic containing water have not been conducted, and there are considerable uncertainties in trying to base estimates on calculations using food concentrations since only limited data are available. Based on what we know so far however, it is reasonable to conclude that when water arsenic levels are below 10 μg/L, food becomes the main source of intake of inorganic arsenic. If so, there is little to gain from reducing water levels below 10 μg/L when food intake cannot be altered.

Based on the above considerations, revision of the current drinking water standard warrants urgent consideration. A prudent approach might be to make the permissible concentration 10 μg/L, as recommended by WHO, although as noted without technical justification. This would considerably reduce cancer risks which might occur with consumption of water containing 50 μg/L, and perhaps even prevent these risks altogether if there is a threshold or marked sub-linearity in dose–response. Finally, a limit of 10 μg/L would result in a major reduction in human exposure in spite of food sources of inorganic arsenic, something which could not be said about any proposal to reduce the water standard much below 10 μg/L.

SUMMARY RECOMMENDATIONS REGARDING DRINKING WATER STANDARDS

1. Any proposed drinking water standard should take into consideration potential cancer mortality risks due to lung and bladder cancer. At high levels of exposure, arsenic in drinking water results in the highest known population cancer mortality, other than that occurring among cigarette smokers.
2. While there may be sub-linearity in the dose–response relationship, even marked sub-linearity would result in unacceptably high cancer risks at the current drinking water standard of 50 μg/L.
3. The theoretical benefits of an extremely low drinking water standard is offset by the inevitable intake of inorganic arsenic from food. When drinking water levels are below 10 μg/L, food becomes the main source of exposure, so there is little to gain by reducing drinking water levels much below 10 μg/L.
4. While further research is needed concerning variation in individual susceptibility, interactions with other exposures, more precise estimates of the dose–response relationships, and the possibility that arsenic is an essential nutrient, etc., this need should not be used as an excuse to delay prudent public health action, based on what we already know.

REFERENCES

Bates, M.N., Smith, A.H., Hopenhayn-Rich, C. 1982. Arsenic ingestion and internal cancers: A review. *Am. J. Epidemiol.*, **135**(5), 462–474.
Bates, M.N., Smith A.H. Cantor, K.P. 1995. Case-control study of bladder cancer and arsenic in drinking water. *Am. J. Epidemiol.*, **141**, 523–530.

Brown, K. and Chen, C.J. 1995. Significance of exposure assessment to analysis of cancer risks from inorganic arsenic in drinking water in Taiwan. *Risk Analysis,* **15**(4), 475–484.

Chen, C.J., Chuang, Y.-C., Lin, T.-M., et al. 1985. Malignant neoplasms among residents of a blackfoot disease-endemic area in Taiwan: High-arsenic artesian well water and cancers. *Cancer Res.,* **45**, 5895–5899.

Chen, C.J., Kuo, T.L., and Wu, M.M. 1988. Arsenic and Cancer (Letter). *Lancet,* **1**, 414–415

Chen, C.J., Chen, C.W., Wu, M.M., et al. 1992.Cancer potential in liver, lung, bladder and kidney due to ingested inorganic arsenic in drinking water. *Br. J. Cancer,* **66**, 888–92.

Chiou, H.Y., Hsueh, Y.M., Liaw, K.F., et al. 1995. Incidence of internal cancers and ingested inorganic arsenic: a seven- year follow-up study in Taiwan. *Cancer Res.,* **55**, 1296–300.

Cuzick, J., Evans, S., Gillman, M., Price Evans, D.A. 1982. Medicinal arsenic and internal malignancies. *Br. J. Cancer,* **45**, 904–911.

Cuzick, J., Sasieni, P., and Evans, S. 1992. Ingested arsenic, keratoses, and bladder cancer. *Am. J. Epidemiol,* **136**, 417–421.

Guo, H., Chiang, H., Hu, H., et al. 1997. Arsenic in drinking water and incidence of urinary cancers. *Epidemiology,* **8**, 545–550.

Hertz-Picciotto, I. and Smith, A.H. 1993. Observations on the dose–response curve for arsenic exposure and lung cancer. *Scand. J. Work Environ. Health,* **19**(4), 217–226.

Hopenhayn-Rich, C., Smith, A.H., Goeden, H.M. 1993. Human studies do not support the methylation threshold hypothesis for the toxicity of inorganic arsenic. *Environ. Res.,* **60**, 161–177.

Hopenhayn-Rich, C., Biggs, M.L., Fuchs, A. et al. 1996a. Bladder cancer mortality associated with arsenic in drinking water in Argentina. *Epidemiology,* **7**(2), 117–124.

Hopenhayn-Rich, C., Biggs, M.L., Smith, A.H., et al. 1996b. Methylation study of a population environmentally exposed to arsenic in drinking water. *Environ. Health Perspect.,* **104**, 620–628.

Hopenhayn-Rich, C., Biggs, M.L., Kalman, D.A., et al. 1996c. Arsenic methylation patterns before and after changing from high to lower concentrations of arsenic in drinking water. *Environ. Health Perspect.,* **104**, 1200–1207.

Hopenhayn-Rich, C., Biggs, M.L., Smith, A.H. 1998. Lung and kidney cancer mortality associated with arsenic in drinking water in Cordoba, Argentina. *Int. J. Epidemiol.,* **27**(4), 561–569.

Hsueh, Y.M., Cheng, G.S., Wu, M.M., et al. 1995. Multiple risk factors associated with arsenic induced skin cancer: effects of chronic liver disease and malnutrition status. *Br. J. Cancer,* **71**, 109–114.

IARC (International Agency for Research on Cancer) 1987. IARC Monographs on the Evaluation of Carcinogenic Risks to Humans. Overall Evaluations of Carcinogenicity: An Updating of IARC Monographs. Volumes 1 to 42. Supplement 7. International Agency for Research on Cancer, Lyon.

Moore, L.E., Smith, A.H., Hopenhayn-Rich, C., et al. 1997a. Micronuclei in exfoliated bladder cells among individuals chronically exposed to arsenic in drinking water. *Cancer Epidemiol. Biomarkers Prev.,* **6**, 31–6.

Moore, L., Smith, A., Hopenhayn-Rich, C., et al. 1997b. Decrease in bladder cell micronucleus prevalence after intervention to lower the concentration of arsenic in drinking water. *Cancer Epidemiol., Biomakers Prev.,* **6**, 1051–1056.

Smith, A. H., Hopenhayn-Rich, C., Bates, M. N., et al. 1992. Cancer risks from arsenic in drinking water. *Environ. Health Perspect.,* **97**, 259–67.

Smith, A. H., Goycolea, M., Haque, R., et al. 1998. Marked increase in bladder and lung cancer mortality in a region of Northern Chile due to arsenic in drinking water. *Am. J. Epidemiol.,* **147**, 660–9.

Tsuda, T., Nagira, T., Yamamoto, M., and Kume, Y. 1990. An epidemiological study on cancer in certified arsenic poisoning patients in Toroku. *Industr. Health,* **28**, 53–62.

Tsuda, T., Babazono, A., Yamamoto, E., et al. 1995. Ingested arsenic and internal cancer: a historical cohort study followed for 33 years. *Am. J. Epidemiol.,* **141**, 198–209.

U.S. EPA (Environmental Protection Agency) 1988. Special report on inorganic arsenic: Skin cancer; nutritional essentiality, EPA 625/3-87/013. EPA Risk Assessment Forum, Washington, DC.

Vahter, M., Concha, G., Nermell, B., et al. 1995. A unique metabolism of inorganic arsenic in Native Andean women. *Eur. J. Pharmacol.,* **293**, 455–462.

WHO. 1993. Guidelines for Drinking Water Quality: Recommendations. 1. Geneva: World Health Organization, 1993.

Wu, M.M., Kuo, T.L., Hwang, Y.H, Chen, C.J. 1989. Dose–response relation between arsenic concentration in well water and mortality from cancers and vascular diseases. *Am. J. Epidemiol.,* **130**, 1123–1132.

Arsenic Exposure and Health Effects
W.R. Chappell, C.O. Abernathy and R.L. Calderon (Editors)
© 1999 Elsevier Science B.V. All rights reserved.

Preliminary Incidence Analysis in Skin Basalioma Patients Exposed to Arsenic in Environmental and Occupational Settings

Vladimír Bencko, Jiří Rameš, Miloslav Götzl

ABSTRACT

The subject of our analysis was a database of 404 skin basalioma cases collected within 15 years (3 five-year intervals) in a region polluted by emissions from the burning of coal with high arsenic content ranging between 900 and 1,500 g per metric ton of dry coal. The standardized incidence of skin basaliomas (each confirmed histologically) in a district with a population of ~125,000 in non-occupational settings ranged from 39.66 to 39.88 per 100,000 (study base 961,960 man/year) while relevant data for occupational settings (male workers of a power-plant burning arsenic-rich coal) ranged from 175.44 to 493.31 per 100,000 (study base 21,360 man/year). Exposure assessment was based on biological monitoring. Determination of arsenic was done in groups of 10-year-old boys (in non-occupational settings) by analyzing hair and urine samples at different localities situated up to a distance of 30 km from the local power-plant. The results obtained seemed to suggest that arsenic is probably a promoter rather than a true carcinogen—at least in connection with arsenic-exposure-related skin basalioma incidence. The non-threshold concept of arsenic carcinogenicity seems not to be supported by the results of our database analysis.

Keywords: cancer epidemiology, biological monitoring, arsenic toxicity, skin basalioma incidence

INTRODUCTION

The trace element content of coal is known to vary with the specific geological conditions of mines (Bezacinský et al., 1984; Thornton and Farago, 1997; Niu et al., 1997). Ecological aspects of the excessive contamination of the environment by arsenic due to the burning of coal with a high arsenic content, including the extinction of honey bee colonies up to 30 km in the direction of the prevailing winds from the power plant, and a harmful influence of arsenic on the reproductive functions of domestic animals, have been summarized (Bencko, 1977). Neurotoxicity and immunotoxicity phenomena encountered in humans exposed in environmental (Bencko et al., 1977) and occupational settings (Buchancová et al., 1998) have also been described.

Assuming that the arsenic exposure leads to the development of malignant tumors in persons occupationally exposed, we have conducted a retrospective epidemiological study, the objective of which was to verify the anticipated increased rate of tumor mortality among employees of the power-plant under study. The mortality pattern was analyzed among workers of a power-plant, combusting coal with a high level of arsenic and compared with the mortality of deceased employees from three coal-fired power plants, where the arsenic content in coal was "normal". To ensure homogeneity of the investigated groups, this study involved only the male employees of the above power-plants (female employees formed about 20% of all workers).

The ascertained numbers of man/year male employees in power-plants burning arsenic-rich coal during the whole period of study (1960–1978) were 15,768 and 17,363 in controls respectively.

The rate of tumor mortality among the exposed subjects who died before age 60 was 38% (in the control group 23%), among those died after 60 it was 51% (in the controls 43%). This increase in the tumor mortality rates was, in spite of being evident, not statistically significant due to the small numbers involved.

Whereas the youngest case of tumor-caused death in the control group was 45 years old, in the exposed group the youngest case was 32 and in the age category below 45 years there occurred 26.9% of a total of tumor-caused deaths in the exposed group. These differences are statistically significant. The analysis shows that the mean age of cancer mortality in the exposed group is 5 years lower: 55.9 years versus 61.2 years control group ($p > 0.05$) (Bencko et al., 1980).

The basic imperfection of the retrospective study was the difficulty in controlling for other risk factors, such as tobacco smoking. This imperfection stimulated our population based cohort study, beginning in the mid-1970s (Bencko and Götzl, 1993).

MATERIAL AND METHODS

Our population based cohort study, beginning in the mid-1970s (Bencko and Götzl, 1993) covers the entire population of the Prievidza district, Central Slovakia, with the primary goal of following up the incidence of all types of malignancies in this area. Our study attempted to obtain a complete detailed register of malignant tumors within an administrative unit of about 125,000 population. This project was feasible due to our previous national health care system, which operated in this country. Each cancer patient or any person suspected of any malignancy was referred to the district oncologist who was responsible for the final diagnosis and therapy of the patient. Originally, our intention was to perform a 10-year study. However, the data collection efforts and the comprehensive nature of the health care system permitted extending this study to 15 years. The study was initiated in 1976. The results of the first year were eliminated as the system of data collection and trials of how our questionnaire was constructed and implemented were fine-tuned.

The district was divided into two areas marked off by a 7-km circle around the power-plant burning coal with a high arsenic content. This circle was established using biological monitoring of human exposure within the particular locality. The exposure rates were established by analysis of hair and urine samples for arsenic content.

To describe the human exposure in environmental settings arsenic determination was carried out on hair, urine, and blood samples taken from groups of 10-year-old boys, each group numbering 20 to 25 individuals, residing in the region polluted by arsenic (Bencko, 1966; Bencko and Symon, 1977; Obrusník et al., 1979). The samples were taken from the boys living at various residential places up to approximately 30 km away from the source of emissions. In all the materials examined, elevated concentrations of arsenic were found. On the basis of the results obtained, the most advantageous material for estimation of non-occupational exposure, and especially to demonstrate environmental pollution, seems to be hair, in spite of some problems with the decontamination procedure involved. The results corresponded to the theoretical ideas on spreading of emissions from elevated sources in the open air and tend to establish the applicability of arsenic determination in the hair as suitable means for monitoring contamination of the environment by arsenic. Considerable variability among individual arsenic values in the hair makes group examination a necessity (Bencko, 1995). The same applies to the blood and urine sampling, which is complicated by several technical difficulties concerning sampling and storage of the collected samples. Levels in urine reflect the quantities of arsenic inhaled or ingested after their absorption into the blood, and give a more realistic picture of possible total daily intake during recent days. In the region polluted by emissions arising from the arsenic-contaminated coal, elevated values of arsenic were detected in autopsy samples as well (Balázová et al., 1976).

The criterion of higher exposure was arsenic content exceeding, on average, hair concentrations of 3 $\mu g/g$ of arsenic. About two-tenths of the district population under study live within a 7-km radius of the exposed region. Values up to 1 $\mu g/g$ are considered normal (WHO/ IPCS, 1981). For example, the population in Prague showed approximately 0.2 $\mu g/g$, which is less than one-tenth of the mean value, which predominated in this heavily emission-loaded area near Prievidza.

RESULTS AND DISCUSSION

Preliminary analysis of the database assembled (Tables 1 and 2) suggests a significant increase of skin basalioma cancer incidence in the most polluted part of the district compared with the data relevant for the rest of the district during the first five-year period. The incidence of skin basalioma is even markedly influenced by exposure to arsenic in occupational settings (Bencko and Götzl, 1994) as can been seen from Table 3.

Measurements, conducted quite recently (Fabiánová et al., 1993a,b), have revealed that the significantly increased arsenic concentrations exceeding the established hygienic limit (MAC) values for arsenic in occupational settings occur mainly during boiler-cleaning operations. Considering, however, the relatively long period of latency, so frequently described in arsenic-caused cancers, we may assume that the changed tumor mortality pattern was a result of arsenic exposures during the years characterized by the much less favorable hygienic conditions at the workplaces from the end of the 1950s to the mid-1970s.

As the result of radical reduction of emissions the main interests now are the late effects of the previous occupational and environmental exposure to arsenic in the former heavily polluted region (Kapalín, 1966; Medvedová and Cmarko, 1974; Obrusník et al., 1979).

CONCLUSION

Currently, we are performing meta analysis of the database on the malignant tumors obtained by the population-based cohort epidemiological study within EXPASCAN,

TABLE 1

Basalioma incidence in a population living in the vicinity of the power-plant burning coal of high arsenic content and in the rest of the district (females only)

	1977–1981		1982–1986		1987–1991		1977–1991	
	Exposed Cases (m-years)	Rest of Dist. Cases (m-years)	Exposed Cases (m-years)	Rest of Dist. Cases (m-years)	Exposed Cases (m-years)	Rest of Dist. Cases (m-years)	Exposed Cases (m-years)	Rest of Dist. Cases (m-years)
Absolute number	31 (44111)	90 (269061)	26 (41518)	102 (287148)	19 (37073)	138 (305853)	76 (122702)	330 (862062)
Expected number	16.07		17.43		21.24		55.33	
Non-standardized rate	70.28	33.45	62.62	35.52	51.25	45.12	61.94	38.28
Age standardized rate	65.44	33.87	54.53	36.34	42.16	46.44	53.68	39.13

Statistical Parameters (Confidence interval ($p = 0.1$))

		Min	Max		Min	Max		Min	Max		Min	Max
Ratio of standardized rates	1.93	1.37	2.72	1.50	1.04	2.15	0.91	0.61	1.36	1.37	1.07	1.76
Mantel–Haenszel estimate	1.93	1.37	2.72	1.49	1.04	2.14	0.90	0.60	1.34	6.26	1.07	1.76
Chi-square	10.33	S	$p < 0.005$	3.38	S	$p < 0.1$	0.20	NS		6.26	S	$p < 0.025$
Standardized morbidity ratio	1.93			1.49			0.89			1.37		

TABLE 2

Basalioma incidence in a population living in the vicinity of the power-plant burning coal of high arsenic content and in the rest of the district (males only)

	1977–1981		1982–1986		1987–1991		1977–1991	
	Exposed Cases (m-years)	Rest of Dist. Cases (m-years)	Exposed Cases (m-years)	Rest of Dist. Cases (m-years)	Exposed Cases (m-years)	Rest of Dist. Cases (m-years)	Exposed Cases (m-years)	Rest of Dist. Cases (m-years)
Absolute number	29 (44969)	99 (270119)	26 (41980)	109 (286619)	14 (36898)	127 (302734)	69 (123847)	335 (859472)
Expected number	18.99		19.37		19.01		55.63	
Non-standardized rate	64.49	36.65	61.93	38.03	37.94	41.95	55.71	38.98
Age standardized rate	62.15	36.90	55.43	38.73	30.30	43.22	49.28	39.73

Statistical Parameters (Confidence interval ($p = 0.1$))

		Min	Max		Min	Max		Min	Max		Min	Max
Ratio of standardized rates	1.68	1.19	2.38	1.43	1.00	2.05	0.70	0.44	1.11	1.24	0.96	1.61
Mantel–Haenszel estimate	1.68	1.19	2.38	1.42	0.99	2.04	0.71	0.44	1.12	1.24	0.96	1.61
Chi-square	6.19	S	$p < 0.025$	2.65	NS		1.53	NS		2.67	NS	
Standardized morbidity ratio	1.53			1.34			0.74			1.24		

Table 3. Basalioma incidence in male workers of power-plant burning coal of high arsenic content (ENO) and in the rest of the district (males only)

	1977–1981		1982–1986		1987–1991		1977–1991	
	Exposed	Rest of Dist.	Exposed	Rest of Dist.	Exposed	Rest of Dist.	Exposed	Rest of Dist.
	Cases (m-years)	Cases (m-years)	Cases (m-years)	Cases (m-years)	Cases (m-years)	Cases (m-years)	Cases (m-years)	Cases (m-years)
Absolute number	4 (6672)	124 (308416)	6 (7387)	129 (321213)	6 (7301)	133 (332331)	16 (21360)	386 (961960)
Expected number	1.64		1.43		1.69		4.78	
Non-standardized rate	59.95	40.21	81.22	40.16	82.18	40.2	74.91	40.13
Age standardized rate	201.46	39.88	175.44	39.69	493.31	39.66	228.21	39.74

Statistical Parameters (Confidence interval ($p = 0.05$))

		Min	Max		Min	Max		Min	Max		Min	Max
Ratio of standardized rates	5.05	1.36	18.69	4.42	1.15	16.97	12.44	2.09	73.92	5.74	2.54	12.95
Mantel–Haenszel estimate	2.45	0.89	6.7	4.22	1.83	9.76	3.56	1.54	8.26	3.36	2.02	5.59
Chi-square	3.25	NS		13.46	S	$p < 0.0005$	10.03	S	$p < 0.005$	24.5	S	$p < 0.0005$
Standardized morbidity ratio	2.43			4.21			3.56			3.35		

INCO-COPERNICUS project. Future analysis will include all types of malignancies, including lung carcinoma which has already been associated with arsenic exposure (Pershagen et al., 1977; Pershagen, 1985; Léonard and Lauwerys, 1980; WHO/IPCS, 1981; IARC, 1982; EPA, 1984; Isinishi et al., 1986), in spite of expected problems with exposure assessment due to cigarette smoking carefully registered in our study. The main objective of our present activity is making the exposure assessment as precise as possible in skin basalioma cases, especially in occupationally exposed subjects, in collaboration with the district and regional Institutes of Public Health in Prievidza and Banská Bystrica respectively.

REFERENCES

Balázová, G., Rippel, A., Jeník, M., Kemka, R. 1976. Metal levels in necroptic materials related to the environment (in Slovak). *Cs. Hyg.*, **21**, 313–318.

Bencko, V. 1966. Arsenic in hair of non-occupationally exposed population (in Slovak). *Cs. Hyg.*, 11, 539—43. Reprinted in: A collection of studies on health effects of air pollution on children. *US Publ. Hlth. Service*, 3, 948–957.

Bencko, V., Symon, K. 1977. Health aspects of burning coal with a high arsenic content. I. Arsenic in hair, urine, and blood in children residing in a polluted area. *Environ. Res.*, **13**, 378–385.

Bencko, V., Symon, K., Chládek, V., Pihrt, J. 1977. Health aspects of burning coal with a high arsenic content. II. Hearing changes in exposed children. *Environ. Res.*, **13**, 386–395.

Bencko, V., Symon, K., Štálnik, L. et al. 1980. Rate of malignant tumor mortality among coal burning power plant workers occupationally exposed to arsenic. *J. Hyg. Epidemiol. (Praha)*, **24**(3), 278–284.

Bencko, V., Wagner, V., Wagnerová, M., Bátora, J. 1988. Immunological profiles in workers of a power plant burning coal rich in arsenic content. *J. Hyg. Epidemiol (Praha)*, **32**, 137–146.

Bencko, V., Götzl, M. 1993. Incidence of lung and skin cancer in population exposed to emissions from burning coal of high arsenic content. *Proc. 5th International Conference, ISEE, Stockholm, August 1993.*

Bencko, V. , Götzl, M. 1994. Exposure assessment and arsenic related skin basalioma cancer epidemiology. In: *Proc. 14th Asian Conference on Occupational Health, October 1994, Beijing, China*, p. 152.

Bencko, V. 1995. Use of human hair as a biomarker in the assessment of exposure to pollutants in occupational and environmental settings. *Toxicology,* 101, 29–39.

Bencko, V., Wagner, V. 1995. Metals, metalloids and immunity. Methodical approaches and group diagnostics. *Centr. Eur. J. Occup. Environ. Med.,* 1(4), 327–337.

Bencko, V. 1997. Health aspects of burning coal with a high arsenic content: the Central Slovakia experience. In : C.O. Abernathy, R.L. Calderon, and W.R. Chappell (eds.), *Arsenic, Exposure and Health Effects.* Chapman and Hall, New York, pp. 84–92.

Bezacinský, M., Pilátová, B., Jirele, V., Bencko, V. 1984. To the problem of trace elements and hydrocarbons emissions from combustion of coal. *J. Hyg. Epidemiol. (Praha),* 28(2), 129–138.

Buchancová, J., Klimentová, G., Knizková, M., Meško, D., Gáliková, E., Kubík, J., Fabiánová, E., Jakubis, M. 1998. A health status of workers of a thermal power station exposed for prolonged periods to arsenic and other elements from fuel. *Centr. Eur. J. Publ. Hlth.,* 6, 29–36.

Cmarko, V. 1963. Hygienic problems of arsenic emissions of ENO plant (in Slovak). *Cs. Hyg.,* 8, 359–363.

EPA, 1984. Health Assessment Document for Inorganic Arsenic. Final Report. Jacobson-Kram, D. et al. Environmental Criteria and Assessment Office. EPA, Research Triangle Park, N.C.

Fabiánová, E., Hettychová, L., Horvátová, E. et al. 1993a. Health impact from environmental contamination due to industrial technologies in the Central Slovakia. *Proceedings of 2nd International Conference on Environmental Impact Assessment of all Economic Activities,* Vol. 1, Prague 20–23 September, pp. 76–79.

Fabiánová, E., Koppová, K., Skupenová, V., Miškovic, P., Mihalíková, E. 1993b. Health impact of selected industrial technologies in environmental settings (in Slovak). *Idem ibid.:* pp. 87–93.

Fabiánová, E., Hettychová, L'., Hrubá, F. et al. 1994. Occupational exposure assessment and bioavailability of arsenic. Final report. EPRI Research Agreement RP 3370-12, pp.106.

Ishinishi, N., Tsuchiya, K., Vahter, M., Fowler, B.A. 1986. Arsenic. In: L. Friberg, G.F. Nordberg and V.B. Vouk (eds.), *Handbook on the Toxicology of Metals,* 2nd edn, Vol. II. Elsevier, Amsterdam, pp. 43–83.

IARC Monographs 1982. Evaluation of Carcinogenic Risk of Chemicals to Humans. Suppl. 4. Chemicals, Industrial Processes and Industries Associated with Cancer in Humans. International Agency for Research on Cancer, Lyon.

Kapalín, V. 1966. Reflect of the influence of some external conditions of the organism of school children (in Czech). *Cs. Hyg.,* 11(8), 468–472.

Léonard, A., Lauwerys, R.R. 1980. Carcinogenicity, teratogenicity and mutagenicity of arsenic. *Mutat. Res.,* 75, 49–62.

Medvedová, H., Cmarko, V. 1974. Some results of the observation of morbidity of 0–15-year-old children in the area polluted by industrial emissions (in Slovak). *Cs. Hyg.,* 19(3), 142–148.

Niu, S., Cao, S., Shen, E. 1997. The geochemistry of arsenic. In: C.O. Abernathy, R.L. Calderon, and W.R. Chappell (Eds.), Arsenic, Exposure and Health Effects. Chapman and Hall, New York, pp. 78–83.

Obrusník, I., Stárková, B., Blazek, J., Bencko, V. 1979. Instrumental neutron activation analysis of fly ash, aerosols and hair. *J. Radioanal. Chem.,* 54, 311–324.

Paris, J. 1820. *Pharmacologica III.* W. Philips, London 1820, pp. 132–134, quoted in Bencko, V. Carcinogenic, teratogenic and mutagenic effects of arsenic. *Environ. Health Persp.,* 1977, 19, 179–82.

Pershagen, G., Elinder, C.-G., Bolander, A.M. 1977. Mortality in a region surrounding an arsenic emitting plant. *Environ. Hlth. Persp.,* 19, 133–137.

Pershagen, G. 1985. Lung cancer mortality among men living near an arsenic-emitting smelter. *Am. J. Epid.,* 122(4), 684–694.

Thornton, I., Farago, M. 1997. The geochemistry of arsenic. In: C.O. Abernathy, R.L. Calderon, and W.R. Chappell (eds.), *Arsenic, Exposure and Health Effects.* Chapman and Hall, New York, pp. 1–16.

WHO/IPCS 1981. Arsenic. Environmental Health Criteria 18, Geneva, p. 114.

Arsenic Exposure and Health Effects
W.R. Chappell, C.O. Abernathy and R.L. Calderon (Editors)
© 1999 Elsevier Science B.V. All rights reserved.

Model Sensitivity in an Analysis of Arsenic Exposure and Bladder Cancer in Southwestern Taiwan

Knashawn H. Morales, Louise M. Ryan, Kenneth G. Brown,
Tsung-Li Kuo, Chien-Jen Chen, Meei-Maan Wu

ABSTRACT

The Environmental Protection Agency (EPA) is under congressional mandate to revise its current standards for arsenic in drinking water. This chapter addresses issues surrounding model choice in a quantitative risk assessment. The data used were collected from 42 villages in an arsenic-endemic region of Taiwan. Excess lifetime risk estimates based on generalized linear models (GLM) and the multistage Weibull model were calculated. Model sensitivity was examined. Model choice along with potential measurement error may have a large impact on estimates of lifetime risk at low concentrations. These results are not intended to serve as an actual risk assessment.

Keywords: arsenic, bladder cancer, lifetime risk, margin of exposure, multistage Weibull, generalized linear model

INTRODUCTION

In response to the congressional mandate for the Environmental Protection Agency (EPA) to revise its current standards for arsenic in drinking water which stand at 50 $\mu g/L$, several issues concerning the risk assessment process have been debated. Arsenic is unique in being the only element considered a human carcinogen, but not found to be carcinogenic in rodents (Smith et al., 1992). For this reason, risk assessment for arsenic in drinking water must rely almost entirely on epidemiological data.

The EPA's interim arsenic risk assessment (EPA, 1988) was based on data published by Tseng et al. (1968). This cross-sectional study included 37 villages in a region of Taiwan where high concentrations of arsenic in wells had been observed. Subjects were examined for skin lesions and skin cancer. Exposure was assessed based on village concentrations. Individuals were assigned categories based on their resident villages. The observed concentration levels, ranging from 1 $\mu g/L$ to over 1000 $\mu g/L$, were grouped into three exposure categories: low (<300 $\mu g/L$), medium (300–600 $\mu g/L$) and high (>600 $\mu g/L$). Subjects were also classified into four age groups: 0–19, 20–39, 40–59 and 60 and over.

While risk assessments based on epidemiological data have the benefit of not requiring interspecies extrapolation, they have other disadvantages. In particular exposure assessment is often weak in this context. Concerns arise when an ecological study design has been used. This means that subjects are not individually assessed for exposure, but instead are assigned an exposure based on the group to which they belong, resulting in the statistical problem of measurement error (Greenland and Morgenstern, 1989). Depending on its nature, measurement error may lead to biased estimates of dose–response parameters and underestimation of the variance parameters of estimated model parameters (Carroll et al., 1990). Another concern associated with risk assessments based on ecological studies is the potential for bias due to unmeasured confounders. To reduce the bias, the study population should be fairly homogeneous. Epidemiological data is also a very important tool in cancer risk assessments. The criteria for classifying an agent as a human carcinogen need sufficient evidence from epidemiologic studies (NRC, Table 4-1, 1996). Although these are important issues, they will not be formally discussed in this paper. The focus will be directed more to model fit.

Due to increasing evidence that arsenic causes internal cancers as well as skin cancer (Wu et al., 1989), it is of interest to do a new analysis since the interim risk assessment was based on a disease that is generally not fatal (skin cancer). The objective of this paper is to investigate the dose–response relationship between bladder cancer and arsenic in drinking water in the same general region of Taiwan in which the Tseng study was conducted. In particular, we examine the sensitivity of risk estimates to the choice of model for two classes of models (the multistage-Weibull, used by the EPA in its current risk assessment based on skin cancer (1988), and Poisson regression) and then the robustness of risk estimates from those models to selected subsets of data.

Before proceeding, it is important to note that the analyses in this paper are not intended to serve as an actual risk assessment, but are meant to address some issues surrounding the choice of model and its impact on the risk assessment.

BLADDER CANCER DATA

Bladder cancer mortality data were collected from the arsenic endemic region of Taiwan, in particular, from a population of 42 villages on the southwestern coast. The data include the person-years at risk and the number of deaths due to bladder cancer in 5-year age increments for both males and females. Although analyses of these data have been previously reported (Chen et al., 1992), the focus of those papers was more to test for effects, rather than to characterize the dose response. Although it is difficult to say for sure, it is likely that the 42 villages include some, but not all, of the villages studied by Tseng.

In this study, exposure levels were not assessed individually. Instead, an ecological study design was again used wherein subjects were assigned the median concentration level corresponding to the level in their village well water. Arsenic concentration levels for the 42 villages ranged from 10 to 934 μg/L. Separate village levels were kept instead of grouping into three exposure intervals. Twenty of the 42 villages had only one well tested, while the remaining villages had multiple wells tested. There was high variation in some of the villages with multiple wells tested. For example one village ranged in concentration from 10 to 686 μg/L with median being 110 μg/L. This variation raises concern over the potential effect of measurement error, since it is difficult in such circumstances to reliably predict individual exposures.

STATISTICAL METHODS

A central task in cancer risk assessment is to determine the exposure level that yields an "acceptable risk" above background levels. Because the probability of cancer is age-dependent, it is common to base these calculations on the excess lifetime risk of cancer. The additive excess lifetime risk is defined to be the lifetime risk of cancer over background,

$$excess(x) = lr(x) - lr(0)$$

where $lr(x)$ is the lifetime risk at exposure level x. The lifetime risk is calculated as,

$$lr(x) = \sum_t p_\theta(x,t)q_t$$

where $\sum$ denotes sum, $p_\theta(x,t)$ is one minus the exponential of the cumulative cause specific hazard of dying of cancer by age t given exposure to concentration x and q_t, the probability of death for each age group. EPA's new guidelines for cancer risk assessments (1996) introduce a "point of departure" analysis when dealing with linear assumptions. The idea is to estimate a point within the observed range of the data, then extrapolate linearly to lower doses. The lower 95% confidence limit on a dose associated with 10% excess risk (LED_{10}) and the dose associated with 10% excess risk (ED_{10}) are standard points of departure. Often in epidemiological studies, however, an excess risk of 10% is fairly large and occurs only at relatively high doses. Instead, a lower value of 1% or 5% is often chosen. We will use a 1% excess risk for the point of departure. The new guidelines also suggest a "margin of exposure" analysis (MOE). It is defined to be the point of departure divided by the environmental exposure of interest. This approach is the proposed default mode of action when linearity is not the most reasonable assumption (EPA, 1996).

To get to this point, we first must characterize the probability of death from cancer as a function of arsenic concentration and age. As previously stated, for Tseng data, the multistage Weibull (MSW) model was used (model as described by Krewski et al., 1983). The model takes the form,

$$p_\theta(x,t) = 1 - \exp[-(Q_0 + Q_1 x + Q_2 x^2)(t - T_0)_+^C],$$

where $p_\theta(x,t)$ is as defined above and unknown model parameters are $Q_0, Q_1, Q_2, T_0,$ and C. In most cases, the estimated parameters, Q_0, Q_1 and Q_2 are constrained to be positive. The plus sign (+) indicates a truncation on the $(t - T_0)$ term (i.e. if $T_0 > t$ then the term is set to zero). This model assumes that for a fixed age, the effect of dose on the prevalence of cancer can be described by a multistage model. The multistage model has been derived under the assumption that events occur in a single cell before cancer develops. Also, for fixed exposure concentrations, the model assumes a person's age at the time of cancer development follows a Weibull distribution. Appendix A describes how to construct a likelihood function once the form of $p_\theta(x,t)$ has been specified.

To minimize the negative of the log-likelihood (equivalent to maximizing the likelihood), we used the function *nlminb* in Splus (MathSoft, 1993). This function uses a modified algorithm of Newton's method to minimize a specified function and is able to accommodate parameter constraints. However, a limitation is that it does not provide the Hessian matrix (needed for the asymptotic variance of the estimated parameters). In turn we are not able to provide confidence intervals for the risk estimates by conventional methods. As suggested by Geyer (1991), bootstrap techniques were used to calculate confidence intervals. Other techniques have been suggested for constrained optimization settings in general (Self and Liang, 1987), and more specifically in the context of dose–response modeling (Guess and Crump, 1978).

Excess lifetime risk estimates were then calculated using equations (1) and (2). Also, the upper 95% confidence limit for the dose–response curve was calculated using bootstrap techniques. An adjustment was made in the concentration level to control for the differences in weight and drinking rates between Taiwan and the U.S. The EPA assumes a typical Taiwanese male weighs 55 kg and drinks 3.5 L of water per day, while a male living in the U.S. typically weighs 70 kg and drinks 2.0 L per day. Females living in the U.S. are assumed to average the same weight as males and drink the same amount of water, but Taiwanese females weigh on average 50 kg and drink 2.0 L of water per day. To assess whether the parameter constraints were forcing the shape of the dose–response curve at low doses, we also fit the model relaxing the parameter constraints. Sensitivity of the model was assessed by calculating excess lifetime risk estimates at 50 $\mu g/L$ with certain villages eliminated from the data. This tool is used to ensure the results are not driven by outliers. The analysis was done excluding villages with one well tested, villages with multiple wells tested and villages with the highest and lowest concentrations.

For comparison, Poisson regression techniques were used as an alternative to the multistage Weibull model. Because this is a member of the well known class of generalized linear model (GLM), asymptotic properties can be easily derived. The most frequently applied version of the Poisson model characterizes the log of cancer incidence rates as a linear function of covariates. The Poisson model assumes that the number of cancers among subjects exposed to a specific concentration at a particular age follows a Poisson distribution with rate equal to cancer incidence multiplied by the person-years at risk in that age group. Several functions of the covariates were considered, including linear or quadratic in concentration and age, and also interactions. As with the multistage Weibull model, we also assessed the sensitivity of the GLM that provided the best fit.

RESULTS

Table 1 contains excess lifetime risk estimates and upper confidence limits for males and females at 50 $\mu g/L$, for the MSW model with the Qs constrained to be positive. It also contains ED_{01} and LED_{01} estimates. Although not shown here, it was interesting to observe that relaxing the parameter constraints changes the shape of the curve slightly, but not significantly so. The female risk estimates appear to be greater than that of males exposed to the same concentration by about a factor of three. It is possible that a risk assessment based on the male data will not effectively protect the female population. Figure 1 gives a graphical comparison for males and females. Each dot corresponds to the estimated lifetime risk estimates for exposure levels combined into groups of width 100 $\mu g/L$ (0–100, 100–200, etc.). The *x* axis is labeled in three ways in terms of (1) concentration ($\mu g/L$) of arsenic found in Taiwan; (2) micrograms of arsenic consumed per kilogram of body weight (based on assumptions previously mentioned); and (3) equivalent concentration ($\mu g/L$) of arsenic consumed by the U.S. population.

Table 2 contains the results of the sensitivity analysis for the multistage Weibull model for both males and females. The estimates in the sensitivity analysis for males, closely matching

TABLE 1

Excess lifetime risk estimates for males and females (MSW)

	Males	Females
Excess risk ($\times$ 1000) at 50 μg/L	0.0612	0.1883
95% upper confidence limit[*]	1.580	0.564
ED_{01}	641.57	365.58
LED_{01}	304.74	304.21

[*]Based on bootstrap methods.

TABLE 2

Sensitivity analysis (MSW)

Village exclusion criteria	Excess risk ($\times$ 1000) at 50 μg/L			
	Males	N[*]	Females	N[*]
Single measurements only	0.0612	22	0.1883	22
Multiple measurements	0.0914	20	0.1282	20
Highest 5	0.0519	37	0.1216	37
Lowest 5	1.6386	37	0.6024	37

[*]Number of villages included in estimates.

values with no village eliminations, come from excluding the villages with a single measurement and also excluding the villages with the five highest concentrations. Excluding the villages with lower concentrations increased the risk estimates by more than a factor of 25. The estimates for females are fairly consistent except when the villages with lower concentrations were eliminated. The excess lifetime risk estimates increased by nearly a

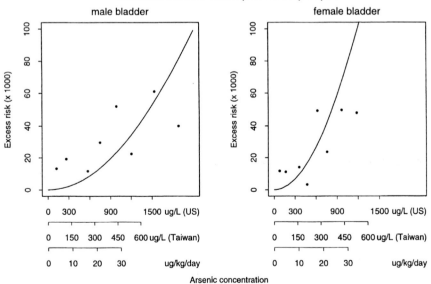

Fig. 1. Estimate excess lifetime risk based on MSW model.

Excess Lifetime risk estimates for different models

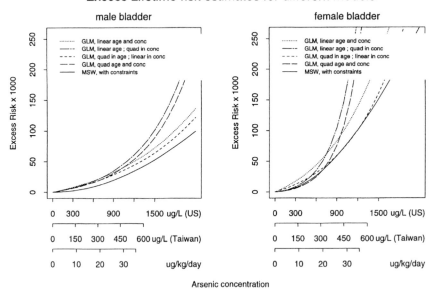

Fig. 2. Estimated excess lifetime risk based on GLM models.

factor of three. The model appears to be highly sensitive to the lower exposure levels, which is another indication of a poor fit.

Figure 2 shows the estimated excess lifetime risk under four different Poisson regression models: (1) linear in concentration and age, (2) linear in age and quadratic in concentration, (3) quadratic in age and linear in concentration, and (4) quadratic in concentration and age. From a likelihood ratio test, we conclude that model 3 seems to yield the best fit. Tables 3 and 4 contain excess lifetime risk estimates at 50 $\mu g/L$ for each model, also including likelihood estimates and excess dose estimates. It is interesting to note that although the shapes are different for models 1 and 2, they are not significantly different. We see the same effect with models 3 and 4.

Table 5 contains the results of the sensitivity analysis for the GLM with quadratic age and linear dose. This model is less sensitive to influential data values than the MSW model. The only inflation of the excess risk occurs when the villages with multiple well readings were excluded. The MOE at 50 $\mu g/L$ is given in Table 6. The lower 95% confidence limit for the excess dose is only above the human exposure of interest by less than a factor of 10. We would expect that if the standard at 50 $\mu g/L$ was set at an excess risk of 10^{-6}, and if the dose response was linear, then the LED_{01} should be a factor of 1000 greater than 50 $\mu g/L$. So in essence, this estimate is very low.

CONCLUSIONS

Using epidemiological data for risk assessments is attractive in that it avoids the issue of interspecies extrapolation, but several other problems can arise. Results from dose–response assessments depend on both the data and the choice of model. A limitation of the data for bladder cancer is that the same "dose", i.e., the arsenic concentration in wells used for drinking water, must be assumed for all persons in the same village, although the arsenic

TABLE 3

GLM Models (males)

Model	LRT*	df	Excess risk at 50 μg/L ($\times$ 1000)	ED_{01}	LED_{01}
Linear in dose and age	3.5731	2	1.1937	362.33	303.21
Linear in age; Quad. in dose	3.4339	1	0.9903	393.11	230.36
Quad. in age; linear in dose	0.1266	1	1.1027	404.14	334.30
Quadratic in dose and age			1.5491	432.47	256.77

*Likelihood ratio test using model with quadratic terms as the reference model.

TABLE 4

GLM models (females)

Model	LRT*	df	Excess risk at 50 μg/L ($\times$ 1000)	ED_{01}	LED_{01}
Linear in dose and age	25.979	2	2.4196	187.10	155.93
Linear in age; Quad. in dose	23.848	1	0.5937	328.97	149.45
Quad. in age; linear in dose	2.1635	1	1.7024	253.06	211.80
Quadratic in dose and age			0.3786	399.13	205.27

*Likelihood ratio test using model with quadratic terms as the reference model.

TABLE 5

Sensitivity analysis (GLM*)

Village exclusion criteria	Excess risk (x 1000) at 50 μg/L			
	Males	N*	Females	N*
Single measurements only	0.9683	22	1.8321	22
Multiple measurements	1.1421	20	1.1369	20
Highest 5	0.9302	37	1.2810	37
Lowest 5	0.9156	37	1.3880	37

*Number of villages included in estimates.

TABLE 6

Margin of exposure for 50 μg/L

	Males		Females	
	MSW	GLM*	MSW	GLM*
ED_{01}	12.831	8.083	7.312	5.061
LED_{01}	6.095	6.686	6.084	4.236

*GLM with quadratic age and linear concentration.

content sometimes varies widely across wells within a village. The sensitivity analysis gave an indication of how the exposure measurements limited the fit of the models. Two classes of models fit to the data produced substantially different lifetime risk estimates at low arsenic concentration in the U.S. For example, at 50 μg/L, the difference is about 20-fold in males. Further, results for multistage-Weibull model are sensitive to exclusion of selected subsets of data while the results for Poisson regression are not. Although it is a difficult concept to assess directly, it seems likely that much of the observed model sensitivity can be explained by measurement error in the median exposure levels assigned to each village. Although formal goodness of fit evaluations suggest relatively little difference between the Multistage

Weibull model and various Poisson models with exposure entered as linear and quadratic terms, the fitted curves tend to look fairly different from each other. For example, in Figure 2 we see dramatic differences in the predicted dose–response curves at high exposure levels, depending on whether or not the model includes a quadratic term in exposure. In contrast, models with linear versus quadratic age effects do not look that different, even though the quadratic model provides a statistically superior fit. Our explanation for this phenomenon is that measurement error in the exposure variable leads to somewhat unstable estimates whereas the estimated age effect is quite stable because this variable is measured without error. It is not known what is the "correct" model, and other choices may produce still different results. We conclude that model choice along with potential measurement error may have a large impact both on estimates of lifetime risk at low arsenic concentrations in the U.S. and on the sensitivity of the estimates to subsets of the data. These may be important considerations in setting a regulatory standard for arsenic in drinking water in the U.S.

ACKNOWLEDGMENTS

This research was supported in part by the David and Lucile Packard Foundation and NIH grants CA48061 and 5F31GM18906.

REFERENCES

Carroll, R.J., Stefanski, L.A. 1990. Approximate quasi-likelihood estimation in models with surrogate predictors. *J. Am. Stat. Assoc.*, **85**, 652–663.

Chen, C.J., Chen, C.W., Wu, M.M., Kuo, T.L. 1992. Cancer potential in liver, lung, bladder and kidney due to ingested inorganic arsenic in drinking water. *Br. J. Cancer*, **66**, 888–892.

EPA (US Environmental Protection Agency). 1988. Special Report of Inorganic Arsenic: Skin Cancer; Nutritional Essentiality. EPA 625/3-87/013. US Environmental Protection Agency, Risk Assessment Forum, Washington, DC.

EPA (US Environmental Protection Agency). 1996. Proposed Guidelines for Carcinogen Risk Assessment. US Environmental Protection Agency, Washington, DC.

Geyer, C.J. 1991. Constrained maximum likelihood exemplified by isotonic convex logistic regression. *J. Am. Stat. Assoc.*, **86**, 717–724.

Greenland, S., Morgenstern, H. 1989. Ecological bias, confounding and effect modification. *Int. J. Epidemiol.*, **18**, 269–274.

Guess, H.A., Crump, K.S. 1978. Maximum likelihood estimation of dose–response functions subject to absolutely monotonic constraints. *Ann. Stat.*, **6**, 101–111.

Krewski, D., Crump, K.S., Farmer, J., Gaylor, D.W., Howe, R., Portier, C., Salsburg, D., Sielken, R.L., Van Ryzin, J. 1983. A comparison of statistical methods for low dose extrapolation utilizing time-to-tumor data. *Fund. Appl. Toxicol.*, **3** 140–160.

Kuo, T.-L. 1968. Arsenic content of artesian well water in endemic area of chronic arsenic poisoning. *Reports, Institute of Pathology, National Taiwan University*, **20**, 7–13.

Laird, N., Olivier, D. 1981. Covariance analysis of censored survival data using log-linear analysis techniques. *J. Am. Stat. Assoc.*, **76**, 231–240.

NCHS (National Center for Health Statistics). 1994. Vital Statistics of the United States, Vol.2, Mortality. National Center for Health Statistics, Hyattsville, MD.

NRC (National Research Council). 1996. *Science and Judgement in Risk Assessment*. Taylor & Francis, Washington, DC.

NRC (National Research Council). 1991. *Environmental Epidemiology*. National Academy Press, Washington, DC.

Self, S.G., Liang, K.Y. 1987. Asymptotic properties of maximum likelihood estimators and likelihood ratio tests under nonstandard conditions. *J. Am. Stat. Assoc.*, **82**, 605–610.

Smith, A.H., Hopenhayn-Rich, C., Bates, M.N., Goeden, H.M., Hertz-Picciotto, I., Duggan, H., Wood, R., Kosnett, M., Smith, M.T. 1992. Cancer risks from arsenic in drinking water. *Environ. Health Perspect.*, **97**, 259–67.

S-PLUS Guide to Statistical and Mathematical Analysis. 1993. MathSoft, Inc. Seattle, Washington.

Tseng W.P., Chu, H.M., How, S.W., Fong, J.M., Lin, C.S., Yeh, S. 1968. Prevalence of skin cancer in an endemic area of chronic arsenicism in Taiwan. *J. Nat. Cancer Inst.*, **40**, 453–63.

Wu, M.M., Kuo, T.L., Hwang, Y.H., Chen, C.J. 1989. Dose–response relation between arsenic concentration in well water and mortality from cancers and vascular diseases. *Am. J. Epidemiol.*, **130**, 1123–1132.

APPENDIX—MAXIMUM LIKELIHOOD ESTIMATION

If the data were in the form of prevalence data (number of subjects alive and the number of those with cancer), a model could be fit by maximizing the following likelihood.

$$L = \prod_{i=1}^{N} p_\theta(x,t)^{y_i} [1 - p_\theta(x,t)]^{(1-y_i)}$$

where N represents the number of subjects and y_i represents the binary indicator of cancer for subject i. This was how the multistage Weibull model was fit to the Tseng data. One approach to handling the incidence data is to model the cause-specific hazard of dying of cancer at age t for someone exposed to arsenic concentration x. According to Laird and Olivier (1981), if we have the number who die of cancer at age t over a specified period of time, $d(x,t)$, and the number at risk at age t during the same time period, $r(x,t)$, then we can assume the number who die conditioned on the number at risk follows a Poisson distribution with rate equal to the number at risk times the cause specific hazard (i.e. $d \mid r \sim \text{Poisson}(rh)$). It follows that the likelihood can be written as

$$L = \prod_x \prod_t h(x,t)^{d(x,t)} \exp[-r(x,t)h(x,t)].$$

There are a variety of choices available for the hazard function, for example a standard proportional hazards model where the hazard is modeled as linear on the log scale. The hazard function for the multistage Weibull used in this analysis is,

$$h(t,x) = \frac{\dfrac{d}{dt} p_\theta(x,t)}{1 - p_\theta(x,t)}$$

$$= C(Q_0 + Q_1 X + Q_2 x^2)(t - T_0)_+^{C-1}$$

Arsenic Exposure and Health Effects
W.R. Chappell, C.O. Abernathy and R.L. Calderon (Editors)

Tumours in Mice Induced by Exposure to Sodium Arsenate in Drinking Water

Jack C. Ng, Alan A. Seawright, Lixia Qi, Corinne M. Garnett,
Barry Chiswell, Michael R. Moore

ABSTRACT

Groups of 90 female C57Bl/6J mice and 140 female metallothionein knock-out transgenic (MT⁻) mice were given drinking water containing sodium arsenate, 500 μg As/L *ad libitum* for up to 26 months. The average intake of arsenic by the test groups was estimated to be approximately 2.0 μg As per day or a daily dose rate of 0.07 mg As/kg body weight for a 30 g mouse. For a mouse surviving two years, it would have consumed about 1.46 mg of arsenic. Preliminary findings indicate that 37/90 (41.1%) C57Bl/6J and 37/140 (26.4%) MT⁻ test mice had one or more tumours. The incidence of which involved the gastrointestinal tract (C57Bl/6J = 14.4%, MT⁻ = 12.9%), lungs (17.8%, 7.1%), liver (7.8%, 5%), spleen (3.3%, 0.7%), bone (2.2%, 0%), skin (3.3%, 1.4%), reproductive system (3.3%, 5%) and eye (1.1%, 0%). No tumours were observed in the control groups. Our results establish for the first time that arsenic induces multiple tumours in mice and that metallothionein does not provide protection from the neoplastic effect of arsenic.

Keywords: arsenic, tumours, cancers, carcinogenic, metallothionein, transgenic

INTRODUCTION

Arsenic is a ubiquitous element in the environment, which is produced commercially by reduction of arsenic trioxide with charcoal. Arsenic trioxide is produced as a by-product of metal smelting operations. It is present in flue dust from the roasting of ores, especially those produced in copper smelting. In the 1960's, the use patterns for arsenic trioxide in the U.S.A. comprised 77% as pesticides, 18% as glass, 4% as industrial chemicals and 1% as medicine. However, the use patterns have changed over the years to include the use of arsenic compounds for timber treatment which has become increasingly popular since the late 1980s. Worldwide usage has been estimated to be 16,000 t (metric tons) As/yr as herbicide, 12,000 t As/yr as cotton desiccant and 16,000 t As/yr in wood preservative (Chilvers and Peterson, 1987). Arsenic pentoxide and arsenic trioxide are used as additives in alloys, particularly with lead and copper. Arsenic and arsenic trioxide are used in the manufacturing of low-melting glasses. High purity arsenic metal is used in semiconductor products. Potassium arsenite as a 1% solution is known as Fowler's solution which was used as a medication for the treatment of chronic myelogenous leukaemia and certain skin lesions (IARC, 1973).

Hutton and Symon (1986) reported that about 5000 t/yr arsenic trioxide is imported into the U.K. for conversion to other arsenic compounds. These processes result in an estimated discharge of 87 t/yr arsenic in manufacturing sludges on landfilled sites. In Australia, arsenic trioxide was widely used for cattle tick (*Boophilus microplus*) control in 1900–1950. Dipping cattle in an arsenical solution was first undertaken at St. Lawrence, in Queensland, in 1895 (Seddon, 1951). There were some 3,700 privately owned dip baths in Queensland. A similar number of dipping baths existed in New South Wales but these were all government owned and operated. The capacity of the New South Wales baths is 2,400 gallons and about 2,800 gallons in Queensland. The dipping solution is 0.2% As_2O_3 solution. As a legacy of cattle tick control alone using arsenic in the past, there are thus thousands of arsenic-contaminated sites in Australia.

Smelting activities generate the largest single anthropogenic input into the atmosphere. They contribute about 40% of the anthropogenic total with coal burning being the next most significant at about 20%. Chilvers and Peterson (1987) have estimated that average atmospheric emission factors of 1.5 kg As/t Cu produced, 0.4 kg As/t Pb produced and 0.65 kg As/t Zn produced. The actual emission factors vary from smelter to smelter and depend on the quality of ores, weather pattern and degree of emission control. These factors affect the degree of soil contamination around any particular smelter.

Elevated concentrations of arsenic in acid sulphate soils in Canada and New Zealand are associated with pyrite (Dudas, 1987). The concentration of arsenic up to 0.5%, through lattice substitution of sulphur in this iron rich pyrite bauxite, have been recorded.

Arsenic is present in the rock phosphate used to manufacture fertilisers and detergents. In 1982, the U.K. imported 1324×10^3 tons of rock phosphate with a total estimated arsenic content of 10.2 tons (Hutton and Symon 1986).

Concentrations of arsenic in seawater are typically less than 2 µg/l. The concentrations of As in unpolluted surface and groundwater are typically in the range of 1–10 µg/L. Elevated concentrations of up to 100–5000 µg/L can be found in areas of sulphide mineralisation and mining (Welch et al., 1988; Fordyce et al., 1995). Elevated arsenic levels (>1 mg As/l) in drinking water of geochemical origins have been found in Taiwan (Chen et al., 1994), West Bengal, India (Chaterjee et al., 1995; Das et al., 1995; Mandal et al., 1996) and more recently in most districts of Bangladesh (Dhar et al., 1997; Biswas et al., 1998). Levels as high as 35 mg As/ L (Kipling, 1977) and 25.7 mg As/L (Tanaka, 1990) associated with hydrothermal activity were reported.

Inorganic arsenic compounds have been classified as carcinogenic to humans based primarily on epidemiological evidence (IARC, 1987). Cancers of the lungs and skin are the predominant target tissues in humans, particularly as a result of occupational exposure and chronic arsenic exposure from contaminated drinking water. Cancers involving the bladder, liver and kidney are also prevalent in endemic areas (Chiou et al., 1995; Brown and Chen, 1995). Non-cancer skin lesions in arsenic patients include melanosis, leucomelanosis and keratosis. It is believed that there are millions of people potentially at risk due to the consumption of arsenic-contaminated drinking water in endemic areas including West Bengal, Bangladesh, Inner Mongolia and Xinjiang province in China. Despite the strong evidence of arsenic carcinogenicity in humans, the evidence of its carcinogenicity in animals is very limited (IARC, 1987). It is the aim of this study to investigate whether or not water containing elevated concentrations of arsenic can induce tumours in the mouse. Such a mouse model could be useful for the study of arsenic carcinogenicity.

EXPERIMENTAL

The experimental protocol was approved by the University of Queensland and Queensland Health Scientific Services Animal Experimental Ethics Committees.

Females C57Bl/6J and metallothionein knock-out transgenic (MT$^-$) mice, aged 4-5 weeks old, were purchased from the Central Animal Breeding House of the University of Queensland. Groups of 90 C57Bl/6J and 140 MT$^-$ mice were given drinking water containing 500 µg As^{5+}/L as sodium arsenate of analytical reagent grade (Ajax Chemicals, Australia) *ad libitum* for up to 26 months. The arsenic solution was prepared every two weeks and stored in a polypropylene container under animal house conditions (see below). Groups of 60 control mice were given normal tap water containing <0.1 µg As/L. The arsenic concentration of the drinking water was measured as total arsenic (presumably all in the arsenate form) by hydride generation atomic absorption spectroscopy (HGAAS) or inductively coupled plasma–mass spectrometry (ICP-MS). Although the concentration of arsenite in the drinking water was not measured, the reduction of arsenate to arsenite was not likely to occur under these experimental conditions. The mice were fed with a commercial rodent diet. All animals were kept in standard polypropylene cages with stainless steel wire-mesh tops equipped with polycarbonate plastic drinking bottles with stainless steel sip-tubes. The proper working order of the sip-tubes was routinely inspected. Evaporation of water via the carbonate drinking bottles was negligible and the loss through the sip-tubes of a mechanical nature was minimal. The animal house was operated at a set temperature range of 21–23°C, 12–13 filtered air changes per hour with a 12/12 light and dark cycle and relative humidity of 60±10% over two years.

The volume of drinking water was measured weekly and each mouse was weighed weekly for the initial 8–9 months and then less frequently (weekly to monthly) for up to 26 months. Animals were sacrificed and necropsied between 24 and 26 months when the experiment was terminated, or earlier if animals were obviously sick. At necropsy, various body tissues were collected and portions of the tissues and tumours were fixed in 10% neutral buffered formalin. Paraffin sections stained with H&E were prepared for microscopic examination by (JCN) and an experienced pathologist (AAS[1]). Remaining tissues were stored at −80°C for future chemical analyses. Unpaired *t*-test (two-samples assuming unequal variances) was used to analyse statistical differences in body weight gains of test and control mice.

---

1   AAS: Consultant toxicologist, Member of Society of Toxicopathologists, Fellow of Royal College of Pathologists and Fellow of American Academy of Veterinary & Comparative Toxicologists.

RESULTS

The average intake of arsenic by the test group was estimated to be about 2.0 µg As per day or a daily dose rate of 0.07 mg As/kg body weight for a 30 g mouse. For a mouse that survived for two years, it would have consumed about 1.46 mg of arsenic. The progressive body weights are shown in Figures 1 and 2.

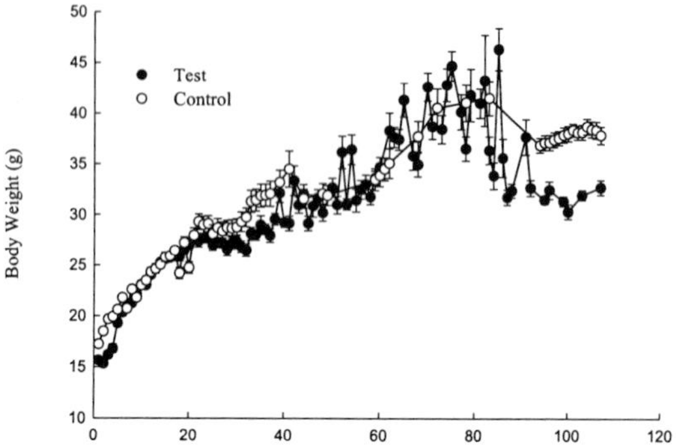

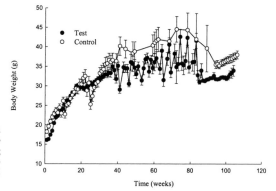

Fig. 1. Progressive body weight of C57/6J mice (mean ± sem) : test group given 500 µg As^{5+}/L as sodium arsenate in drinking water *ad libitum*; control group given normal tap water; significant lower weight gains were observed in test animals from week 90 ($p < 0.001$).

Fig. 2. Progressive body weight of metallothionein knock-out transgenic mice (mean ± sem) : test group given 500 µg As^{5+}/L as sodium arsenate in drinking water *ad libitum*; control group given normal tap water; Significant lower weight gains were observed in test animals from week 40 ($p < 0.001$).

TABLE 1

Incidence of tumours in control, C57Bl/6J and metallothionein knock-out transgenic (MT⁻) mice given sodium arsenate, 500 µg As^{5+}/L in drinking water *ad libitum* for up to 26 months

Organ system	C57Bl/6J control <0.1 µg As/L	C57Bl/6J test 500 µg As/L	MT⁻ control <0.1 µg As/L	MT⁻ test 500 µg As/L
Lung	0%	17.5%	0%	7.1%
Gastrointestinal tract	0%	14.4%	0%	12.9%
Liver	0%	7.8%	0%	5.0%
Spleen	0%	3.3%	0%	0.7%
Reproductive	0%	3.3%	0%	5.0%
Skin	0%	3.3%	0%	1.4%
Bone	0%	2.2%	0%	0%
Eye	0%	1.1%	0%	0%

Seventy-three/90 (81.1%) C57Bl/6J and 103/140 (73.6%) MT⁻ test mice survived for two years compared to their respective controls, 59/60 (98.3%) and 58/60 (96.7%). Preliminary findings indicate that 37/90 (41.1%) C57Bl/6J and 37/140 (26.4%) MT⁻ test mice had one or more tumours. The incidence of tumours involved the gastrointestinal tract, lungs, liver, spleen, bone, skin, reproductive system and eye (Table 1). No tumours were observed in the control groups.

DISCUSSION

The evidence for the carcinogenicity of inorganic arsenic compounds in humans is generally regarded as sufficient (IARC, 1987). Inorganic arsenicals have been strongly implicated cancer-causing agents involving the skin, lung, liver, intestinal tract, kidney, urinary bladder and meninges in humans (IARC, 1987). However, there is very little data available in the literature to show conclusively that arsenic is an animal carcinogen. Rudnay and Borzsonyi (1981) reported lung adenoma in off-spring mice whose mothers were given a single subcutaneous injection of arsenic trioxide at 1.2 mg As/kg body weight during gestation, and then to the 3-day-old pups at 5 μg As/pup by the same route of administration. Pershagen et al. (1984) reported low incidences of carcinomas, adenomas, papillomas and adenomatoid lesions of the respiratory tract in hamsters after they were given 3 mg As/kg of arsenic trioxide using charcoal carbon and 2 mM H_2SO_4 (a carrier to increase retention) by intratracheal instillation once weekly for 15 weeks. Katsnelson et al. (1986) reported adenocarcinomas at the implantation site in the stomach in rats given arsenic trioxide at a dose of 8 mg As in a capsule by surgical implantation. In an arsenic drinking water study (Shirachi et al., 1983), enhanced incidence of renal tumours was observed in rats given an intraperitoneal injection of 30 mg/kg of diethylnitrosamine and sodium arsenite in drinking water at 8–9 mg As/kg for 175 days.

In the studies reported above, dose rates of arsenic used were much higher than realistic environmental levels of exposure. High dosages often lead to high mortality rates and relatively low incidence of tumours in a small number of surviving experimental animals, thus contributing to the weakness of these studies. These results could only provide very limited evidence of arsenic carcinogenicity in animals (IARC, 1987). In our study, we selected C57Bl/6J mice, a strain which has very low incidence (close to 0%) of spontaneous tumours in virgin female mice (Hoag, 1963; Adkinson and Sundberg, 1991). A high incidence of tumours in our test animals is therefore highly significant. The arsenic concentration, 500 μg As^{5+}/kg as sodium arsenate in the drinking water, was similar to the average environmental levels encountered in the contaminated groundwaters of arsenic endemic areas (Chen et al., 1994; Chaterjee et al., 1995; Das et al., 1995; Mandal et al., 1996; Dhar et al., 1997; Biswas et al., 1998). It was estimated that the average consumption of arsenic in a test mouse was approximately 2 μg per day. For a 30 g mouse, this equates to 67 μg /kg body weight. This dosage to our experimental mice is comparable to the actual exposure level currently experienced by adults in West Bengal and Bangladesh. Adults from these endemic areas who work in the rice paddy are estimated to drink an average of 8 litres of water containing 500 μg As/L a day (Chakraborti, personal communication). For a 60 kg adult, he/she would have consumed an equivalent amount of arsenic to that in the present mouse study, namely 67 μg/kg body weight daily.

The selection of a suitable arsenic concentration in the drinking water for this study proved to be a vital factor. In our experience, mice did not drink sufficient water, containing 1 mg As/L of either sodium arsenate or sodium arsenite, to maintain body weight gains and bodily health. Experiments designed to expose mice to this level of arsenic concentration had to be terminated within a month because of apparent unpalatability.

Similar growth rates were observed in test and control groups of both the C57Bl/6J and MT⁻ mice although the test groups had slightly lower body weights for the initial 3 weeks of the dosing period (Figures 1 and 2). After normalisation, there were no significant differences in body weight gains for 80–90 weeks between test and control C57Bl/6J mice (Figure 1). However, lower average weight gains, from week 40 to 90, were observed in the test group of MT⁻ mice (Figure 2). The test groups of both C57Bl/6J and MT⁻ mice had significantly lower body weights ($p < 0.001$) after 90 weeks of arsenic exposure when compared to their corresponding control animals. The weight loss was probably due to the adverse effect of chronic arsenic exposure, as tumours were observed between 18 months to 2 years.

Mortality occurred mostly when mice were approximately 2 years old. The survival rates of control groups were 98.3% and 96.7% for C57Bl/6J and MT⁻ mice respectively. The arsenic dosed C57Bl/6J mice had an 81.1% survival rate whilst the MT⁻ mice had a slightly lower rate of 73.6% at the end of 26 months. Most deaths appeared to be related to the fatal effect of the tumours. The reasonably high survival rate of the MT⁻ mice was somewhat surprising considering that one of the protective mechanisms against heavy metal toxicity, namely the metallothionein gene, has been knocked out in these mice. A previous study (Kreppel et al., 1993) reported that metallothionein in the mouse can be induced by As^{3+}, As^{5+}, MMAA and DMAA. This implies that MT may play a role in the detoxication of arsenicals. However, our results suggest that metallothionein does not provide protection against the chronic effects of arsenic dosing.

Thirty-seven/90 (41.1%) C57Bl/6J and 37/140 (26.4) MT⁻ mice from the test groups had one or more tumours. No tumours were observed in the control groups. The zero incidence of tumours in control animals should be interpreted with caution. For a relatively small number of 60 control mice, spontaneous tumours were not observed in this experiment. The fact that the MT⁻ mice did not have higher incidence of tumours than the C57Bl/6J mice, indicates that metallothionein does not protect the animal from the neoplastic effect of chronic arsenic exposure. Tumours observed in this study involved multiple organ systems including the lung, gastrointestinal tract, liver, spleen, reproductive system, skin, bone and eye. Lungs and the gastrointestinal tract had the highest incidence of tumours. The carcinogenic effect of inorganic arsenic to multiple organ systems is consistent with the epidemiological evidence of the same effect observed in humans (Brown and Chen, 1995; Chiou et al., 1995).

In conclusion, our results confirm the carcinogenicity of inorganic arsenic in animals and that metallothionein does not protect mice from a possible neoplastic effect of chronic arsenic exposure. Tumour studies in animals should at least be based on realistic environmental levels of particular contaminants. This mouse model may provide a useful tool for the study of the mechanism of arsenic carcinogenicity.

ACKNOWLEDGMENT

National Research Centre for Environmental Toxicology (NRCET) is jointly funded by the National Health and Medical Research Council, Queensland Health, Griffith University and the University of Queensland. The supply of the original stock of MT⁻ mice by Murdoch Institute to the Central Animal Breeding House at the University of Queensland via Dr C.T. Dameron (NRCET) is acknowledged.

REFERENCES

Adkinson, D.L., Sundberg, J.P. 1991. "Lipomatous" hamartomas and choristomas in inbred laboratory mice. *Vet. Pathol.*, **28**, 305–312.

Biswas, B.K., Dhar, R.K., Samanta, G., Mandal, B.K., Chakraborti, D., Faruk, I., Islam K.S., Chowdhury, M.M., Islam, A., Roy, S. 1998. Detailed study report of Samta, one of the arsenic-affected villages of Jessore District, Bangladesh. *Current Sci.*, **74**(2), 134–145.

Brown, K.G., Chen, C.J. 1995. Significance of exposure assessment to analysis of cancer risk from inorganic arsenic in drinking water in Taiwan. *Risk Analysis*, **15**(4), 475–484.

Chatterjee, A., Das, D., Mandal, B.K., Chowdhury, T.R., Samanta, G., Chakraborti, D. 1995. Arsenic in ground water in six districts of West Bengal, India: the biggest arsenic calamity in the world. Part 1. Arsenic species in drinking water and urine of the affected people. *Analyst*, **120**, 643–650.

Chen, S.L., Dzeng, S.R., Yang, M.H. 1994. Arsenic species in groundwaters of the blackfoot disease area, Taiwan. *Environ. Sci. Technol.*, **28**, 877–881.

Chilvers, D.C., Peterson, P.J. 1987. In: Hutchinson, T.C. and Meema, K.M. (eds.), *Lead, Mercury, Cadmium and Arsenic in the Environment*. John Wiley, New York. Chapter 17.

Chiou, H.Y., Hsueh, Y.M., Liaw, K.F., Horng, S.F., Chiang, M.H., Pu, Y.S., Lin, J.S.N., Huang, C.H. and Chen, C.J. 1995. Incidence of internal cancers and ingested inorganic arsenic: a seven-year follow-up study in Taiwan.

Das, D., Chatterjee, A., Mandal, B.K., Samanta, G., Chakraborti, D. 1995. Arsenic in ground water in six districts of West Bengal, India: the biggest arsenic calamity in the world. Part 2. Arsenic concentration in drinking water, hair, nails, urine, skin-scale and liver tissue (biopsy) of the affected people. *Analyst*, **120**, 917–924.

Dhar, R.K., Biswas, B.K., Samanta, G., Mandal, B.K., Chakraborti, D., Roy, S., Fafar, A., Islam, A., Ara, G., Kabir, S., Khan, A.W., Ahmed, S.A., Hadi, S.A. 1997. Groundwater arsenic calamity in Bangladesh. *Current Sci.*, **73**(1), 48–59.

Dudas, M.T. 1987. Accumulation of native arsenic in acid sulphate soils in Alberta, Canada. *J. Soil Sci.*, **67**, 317–331.

Fordyce, F.M., Williams, T.M., Palittpapapon, A., Charoenchaisei, P. 1995. Hydrogeochemistry of arsenic in an area of chronic mining-related arsenism, Rono Phibun District. British Geological Survey.

Hoag, W.G. 1963. Spontaneous cancer in mice. *Ann. NY Acad. Sci.*, **108**, 805–831.

Hutton, M., Symon, C. 1986. The quantities of cadmium, lead, mercury and arsenic entering the U.K. environment from human activities. *Sci. Total Eviron.*, **57**, 129–150.

IARC. 1987. *IARC Monographs on Evaluation of Carcinogenic Risks to Humans*. Supplement 7. Lyon France. pp 100–106.

Katsnelson, B.A., Neizvestnova, Y.M., Blokhin, V.A. 1986. Stomach carcinogenesis induction by chronic treatment with arsenic (Russ.). *Vopr. Onkol.*, **32**, 68–73.

Kipling, M.D. 1977. Arsenic, In: J. Lenihan and W.W. Fletcher (eds.), *The Chemical Environment*. Blackie, Glasgow, Chapter 4, pp. 93120.

Kreppel, H., Bauman, J.W., Liu, J., McKim, J.M.Jr., Klaassen C.D. 1993. Induction of metallothionein by arsenicals in mice. *Fundam. Appl. Toxicol.*, **20**, 184–189.

Mandal, B.K., Chowdhury, T.R., Samanta, G., Basu, G.K., Chowdhury, P.P., Chanda, C.R., Lodh, D., Karan, N.K., Dhar, R.K., Tamili, D.K., Das, D., Saha, K.C., Chakraborti, D. 1996. Arsenic in groundwater in seven districts of West Bengal, India—The biggest arsenic calamity in the world. *Current Sci.*, **70**(11), 976–986.

Pershagen, G., Nordberg, G., Bjorklund, N.E. 1984. Carcinomas of the respiratory tract in hamsters given arsenic trioxide and / or benzo[a]pyrene by the pulmonary route. *Environ. Res.*, **34**, 227–241.

Rudnay, P., Borzsonyi, M. 1981. The tumorigenic effect of treatment with arsenic trioxide (Hung.). *Magyar. Onkol.*, **25**, 73–77.

Seddon, H.R. 1951. *Diseases of Domestic Animals in Australia*. Part 3, tick and mite infestations (also animals, miscellaneous insects, etc. harmful to stock). Commonwealth of Australia, Department of Health, Service Publication (Division of Veterinary Hygiene), Commonwealth Government Printer, Canberra. 7, pp. 1–200.

Shirachi, D.Y., Johansen, M.G., McGowan, J.P., Tu, S.H. 1983. Tumorigenic effect of sodium arsenite in rat kidney. *Proc. West. Pharmacol. Soc.*, **26**, 413–415.

Tanaka, T. 1990. Arsenic in the natural environment. Part II: Arsenic concentrations in thermal waters from Japan. *Appl. Organomet. Chem.*, **4**, 197–203.

Vahter, M. 1981. Biotransformation of trivalent and pentavalent inorganic arsenic in mice and rats. *Environ. Res.*, **25**, 286293.

Welch, A.H., Lico, M.S., Hughes, J.L. 1988. Arsenic in groundwater of the western United States. *Ground Water*, **26**(3), 333–347.

Arsenic Exposure and Health Effects
W.R. Chappell, C.O. Abernathy and R.L. Calderon (Editors)
© 1999 Elsevier Science B.V. All rights reserved.

Subchronic Toxicity Study of Sodium Arsenite in Methyl-Deficient Male C57BL/6 Mice

Russell S. Okoji, Joel Leininger, John R. Froines

ABSTRACT

Arsenic is an established human carcinogen. The mechanistic pathway by which arsenic causes cancer is not understood. In preparation for a 24-month chronic study assessing cancer in methyl-deficient male C57BL/6 mice, a 130-day subchronic study of sodium arsenite was undertaken to establish baseline toxicity data. Mice were administered arsenic via drinking water: 0, 2.6, 4.3, 9.5 or 14.6 mg sodium arsenite/kg/day. Dosing continued 7 days a week for the length of the study. Deaths of 3 of the control animals (methyl-sufficient/no arsenic) at Day 111 of the study did not appear to be compound-related. The death of a single animal in the high-dose group did appear to be treatment-dependent. A dose-related reduction was observed for liver weight. Mild to severe fatty infiltration was observed in the livers of methyl-deficient/arsenic treated animals. Severe liver damage was noted in 2 animals from the 2.6 and 4.3 mg/kg/d groups. In addition, hypertrophy/hyperplasia of the bladder was found in 43/60 mice treated with sodium arsenite. No histopathological changes were evident in any other tissue examined. The no observed effect level (NOEL) and no observed adverse effect level (NOAEL) of this study could not be determined as the lowest dose administered produced detrimental effects. The maximum tolerated dose (MTD) for animals maintained on methyl-deficient diets was 2.6 mg/kg/d.

Keywords: arsenic, methyl deficiency, toxicity, cancer, metabolism

INTRODUCTION

The current maximum contaminant level (MCL) for arsenic in drinking water is 50 μg/l. The MCL is based on the U.S. Public Health Service Standard of 1943. Recent epidemiological evidence indicates that the 50 μg/l standard may not be adequately protective of the general population when the risk of cancer is considered. The safe drinking water act of 1996 has therefore mandated the U.S. Environmental Protection Agency (EPA) to reevaluate the uncertainties in arsenic risk assessment.

Arsenic-induced carcinogenesis is unique in that it is the only known agent to cause cancer in humans but not in conventional animal models. This inconsistency emphasizes the need to understand arsenic's biological mode of action. In order to reduce uncertainty, current cancer risk assessment methods require a mechanistic detail of how a compound initiates its toxic effects. There has been considerable emphasis on the need to develop an appropriate animal model to clarify arsenic's mechanism of action, and to determine those factors which predispose certain individuals to arsenic-induced cancer. The current study was undertaken to establish baseline toxicity information for sodium arsenite in order to establish a Maximum Tolerated Dose (MTD) for a 24-month chronic study. The chronic animal bioassay will determine whether environmental factors such as dose of arsenic administered or dietary factors such as methyl-deficiency increase an individual's susceptibility to arsenic-induced cancer.

Metabolism and General Toxicity of Arsenic

Methylation is considered to be the detoxification mechanism for arsenic (Figure 1). Accumulation of inorganic arsenic as a consequence of a saturation or inhibition of the methylation process may result in increased toxicity (Thompson, 1993). Monomethylarsonic acid (MMA) and Dimethylarsinic acid (DMA) are considered less acutely toxic and have lower tissue retention than inorganic forms (Marafante et al. 1985). The chemical structures of arsenite, MMA and DMA are shown in Figure 2.

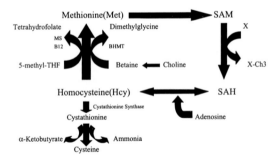

Fig. 1. Methyl Group metabolism: Methionine (Met) can be synthesized from homocysteine (Hcy) by addition of a methyl group from either betaine or 5 methyltetrahydrofolate. S-adenosyl-methionine (SAM) is synthesized from Met via addition of a 1 carbon group. SAM acts as the principal methyl group donor in the cell and in the process is transformed into S-adenosyl-homocysteine (SAH). SAH may also be formed from the condensation of Hcy and adenosine. Transmethylation steps essential to the cell includes the methylation of DNA and the metabolism of xenobiotics. The one carbon metabolic pathway can be disrupted at many points: (A) A deficiency in methionine and subsequently SAM can occur due to dietary shortage of methionine, folate or choline, or due to inhibition of the methionine synthesizing enzymes, methionine synthase (MS) and betaine homocysteine methyl-transferase (BHMT). (B) SAM can be present in suboptimal concentrations due to an increase in transmethylation reaction(s). (C) Excess SAH can inhibit transmethylation reactions by competing with SAM for the active sites on methyltransferases. SAH is formed by the condensation of Hcy and adenosine or due to excess transmethylation reactions. (D) Methylation of DNA can be inhibited by insufficient amounts of SAM or by excess SAH. If a gene is hypomethylated in a critical regulatory region it may allow trans-acting factors and RNApolymerase to bind and the gene to be expressed. Hypomethylation is necessary but not sufficient for the transcription of a DNA.

Fig. 2. Chemical structure of arsenic metabolites.

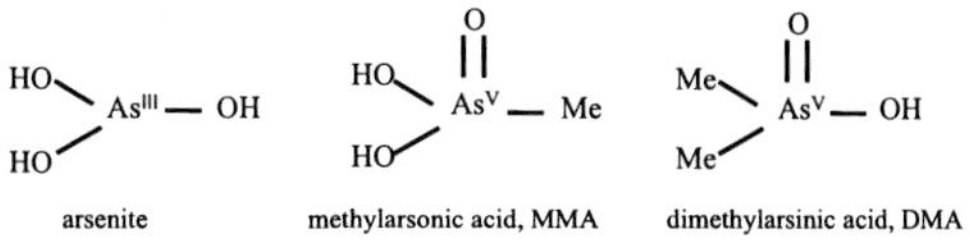

arsenite   methylarsonic acid, MMA   dimethylarsinic acid, DMA

Carlson-Lynch et al. (1994) and Beck et al. (1995) have addressed uncertainties about the relevance of available epidemiological data on the systemic carcinogenesis of arsenic to U.S. populations. In particular, they suggest that the observational studies conducted in Taiwan, from which current risk estimates are based, do not adequately consider nutritional status. The Taiwanese diet is low in protein and methionine and this may decrease the capacity for detoxification of arsenic via methylation. Earlier studies have shown animals deficient in the lipotropic nutrients, folate, choline and methionine have decreased methylation and excretion of arsenic. Vahter and Marafante (1987) have examined the effect of choline, methionine or protein deprivation on the metabolism of arsenite in male rabbits. Their data show a significant decrease in both total excretion of arsenic, as well as percent of DMA excreted in all lipotrope deficient groups as compared to the lipotrope sufficient group. The increase in inorganic arsenic levels results in a concomitant increase in total body retention of arsenic, which may have cytotoxic effects or disrupt normal cellular functions such as the transcription of DNA (Marafante and Vahter, 1986; Vahter and Marafante, 1987).

Studies done by our laboratory and others indicate that exposure to high doses of arsenic results in an alteration in intracellular arsenic methylation. Data from these studies suggest that as the dose of inorganic arsenic, particularly AsIII, approaches levels greater than 500 μg/ d the ratio of excreted monomethylated arsenicals to dimethylated arsenicals significantly increases (U.S. EPA, 1992; Hsu et al., 1997; Buchet et al., 1981a,b; Hopenhayn-Rich et al., 1996a,b). Hughs et al. (1994) have reported saturation of methylating capacity in female B6C3F1 mice. They observed the amount of excreted inorganic arsenic and monomethylated arsenic significantly increased, while the urinary excretion of DMA decreased at high concentrations of arsenic exposure.

Methyl Groups and the Induction of Cancer

Methylation is critical to the normal processes of cellular growth and regulation. Alterations in the methylation patterns of DNA have been shown to increase the expression of oncogenes (Counts and Goodman 1993; Ray et al., 1994) and decrease the expression of tumor suppressor genes (Jones et al., 1992; Herman et al., 1995). DNA methylation is dependent on the availability of methyl-donor stores. The depletion of methyl-donor pools by arsenic or the depletion of SAM through dietary restriction may disturb these patterns. Methylation at specific sites on the 5' upstream promoter region is a putative control point for transcription. Maintenance methylation on cytosine residues helps to regulate the expression status of cell cycle control genes, which are involved in cellular growth and differentiation (Vorce and Goodman,1989). Alterations in a gene's capacity for expression may directly result in a tumor suppressor's inactivation or an oncogene's activation, leading to the transformation of the cell and ultimately to cancer (Herman et al., 1995).

Numerous studies have examined the relationship between methylation and carcino-genesis. Male Fisher rats on methyl-deficient diets display decreased DNA methylation and increased levels of the mRNA of *c-myc, c-fos,* and *c-Ha-ras,* genes known to be involved in the transformation process of many human tumors (Dizik et al., 1991; Christman et al., 1993; Waifan et al., 1989; Waifan and Poirier, 1992). Hsieh and coworkers (1989) have shown a similar association with *c-myc and c-Ha-ras* in B6C3F1 mice. Counts et al. (1996) have shown global DNA hypomethylation in both B6C3F1 and C57Bl/6 mice fed a methyl-deficient diet for one week. B6C3F1 mice displayed a 21% decrease in liver DNA methylation status, while

C57Bl/6 mice exhibited a 9% reduction. James et al. (1997) have suggested that rats fed diets deficient in choline, methionine, or folic acid have altered nucleotide pools, and these imbalances may result in diet-induced carcinogenesis. Herman et al. (1995) have reported that inactivation of the p16 tumor suppressor is frequently associated with aberrant DNA methylation in all common human cancers.

Rats chronically exposed to methyl-deficient diets present with hyperplasia of the lung and liver (Salmon and Copeland, 1954; Salmon et al., 1955; Mikol et al., 1983; Newberne and Rogers 1986; Newberne et al., 1982; Yokoyama et al., 1985; Henning et al., 1997). Thirteen month old Fisher 344 male rats on methyl-deficient diets for 12 months display a 100% incidence of preneoplastic hepatocyte nodules and a 51% incidence of hepatocellular carcinoma (Ghoshal and Farber, 1984). Mice fed choline deficient diets for a period of up to 172 days have increased hepatocelluar proliferation, as well as tumors of the liver (Buckley and Hartroft, 1955; Wilson, 1951). Altered methyl group metabolism may have relevance to carcinogenesis in general and perhaps specifically to arsenic-induced cancer.

The subchronic animal bioassay was intended to establish a maximum tolerated dose for short-term arsenic exposure in C57BL/6 mice on methyl deficient diets. The subchronic study sought to identify potential target organs and adverse effects from which existing human data can then be compared. Given that the mouse metabolizes arsenic by the same mechanistic pathway as man and that both mouse and man accumulate arsenic in similar target organs, the liver, lung, bladder, kidney and skin, we believe an animal model for arsenic-induced toxicity can be constructed.

EXPERIMENTAL DESIGN AND METHODS

Ninety young-adult male C57BL/6 mice approximately 8 weeks old were obtained from Jackson Laboratories and housed 3 to a cage in the UCLA School of Public Health vivarium according to the NIH "Guide for the Care and Use of Laboratory Animals". Cages were polycarbonate, shoebox style with wire lid and filter covers to provide an isolated environment by minimizing variations in temperature, humidity, drafts and CO_2 build up. Mice were maintained on a lab chow diet during a one-week quarantine period in which clinical signs of illness were closely monitored. Clinical examinations included a twice daily check for appearance, morbidity and mortality, deviant activity, posture or behavior. Periodic tests for neurological and ophthalmological response were also conducted. Only clinically healthy mice were used in the study. After the quarantine period, mice were transferred to a negatively adjusted (in relation to the corridor) air pressure room maintained on a 12 hour light/12 hour dark schedule at a constant temperature of $23.3°C \pm 1.1$ and a relative humidity of $40 \pm 5\%$. Mice were then placed randomly into one of six treatment groups and pretreated with either the methyl sufficient or methyl-deficient diet for a period of 21 days. Following the pretreatment period, animals received arsenic as sodium arsenite in drinking water for the remainder of the study.

Treatment Groups
- Group 1 was maintained *ad libitum* on the methyl-sufficient diet for a period of 130 days.
- Group 2 was maintained *ad libitum* on the methyl-deficient diet for a period of 130 days.
- Group 3 concurrently received 2.6 mg/kg/d sodium arsenite (Sigma Chemical Company, St. Louis, MO) via drinking water for a period of 13 weeks.
- Group 4 concurrently received 4.3 mg/kg/d sodium arsenite (Sigma Chemical Company, St. Louis, MO) via drinking water for a period of 13 weeks.
- Group 5 concurrently received 9.5 mg/kg/d sodium arsenite (Sigma Chemical Company, St. Louis, MO) via drinking water for a period of 13 weeks.
- Group 6 concurrently received 14.6 mg/kg/d sodium arsenite (Sigma Chemical Company, St. Louis, MO) via drinking water for a period of 13 weeks

Body weight and water intake was measured weekly. Arsenic water was administered in Nalgene, polycarbonate, 250 ml (8 oz) bottles and was changed at the time of measurement to insure that a significant amount of oxidation did not occur.

Parameters Examined

- Initial body weight (g)
- Mean body weight gain (g)
- Mean water consumption/d (ml)
- Histopathology
- Terminal liver weight (g)

At necropsy, tissues were weighed and sections of livers, lungs, bladders, kidneys and skin were snapped frozen in liquid nitrogen for nucleic acid analysis. Additional tissue samples were fixed in neutral buffered formalin, stored in 70% alcohol and shipped to the National Institute of Environmental Health Sciences (NIEHS) for histopathological examination.

RESULTS

Three mice from the methyl-sufficient control group were found with an apparent infection on Day 111 of the study. Effects are believed not to be treatment-related. Clinical signs include mucous surrounding the eye, slow and shallow breathing, non-response to external stimuli, and absence of movement. Animals were sacrificed and necropsy was conducted. On Day 104 of the study, a single mouse from the methyl-deficient high dose group was sacrificed. Weight loss was evident for approximately 15 days until day of sacrifice when mouse weighed 15 grams. Animal was emaciated with muscular weakness and its equilibrium disturbed as evidenced by unequal gait. With the exception of the four animals noted above, in-life evaluations were unremarkable. The remaining animals exhibited no unusual symptoms or behavior.

Mean body weights over the 130-day period are given in Table 1. Body weights in the 9.4 and 14.6 mg/kg dose groups were significantly lower than that of control groups at the end of the study period. However, because the starting weights of control and arsenic administered animals differed, the rates of weight gain over time, as determined by linear regression, were determined. The 2.6 and 4.3 mg/kg groups had steeper growth rates than controls. The 9.5 and 14.6 mg/kg groups had similar growth curves as compared to controls (data not shown).

TABLE 1

Mean body weight of methyl-sufficient and methyl-deficient mice after exposure to various doses of sodium arsenite

Grp.	Day 0	Day 9	Day 15	Day 30	Day 45	Day 60	Day 75	Day 90	Day 105	Day 130
	24.93±1.39	28.87±2.23	29.53±2.80	29.40±2.29	34.53±2.45	34.90±3.60	37.42±4.26	36.49±5.26	35.45±3.90	34.29±3.61
MSD	100%	100%	100%	100%	100%	100%	100%	100%	100%	100%
	26.93±.960	27.33±.816	27.53±.990	31.00±1.41	33.27±2.55	33.72±3.00	34.90±3.49	34.80±3.40	35.72±3.43	34.43±3.78
MDD	108%	94%	93%	105%	96%	97%	93%	95%	101%	100%
	*20.07±1.10	*20.13±.640	*20.80±1.01	*22.80±2.01	*26.27±3.03	*26.27±3.88	*29.97±4.86	#31.64±5.57	33.96±6.74	34.53±6.80
2.6	81%	70%	70%	78%	76%	75%	80%	87%	96%	101%
	*20.53±.915	*20.60±.828	*23.33±1.40	*23.93±2.22	*26.53±2.53	*27.58±3.90	*30.96±4.50	33.05±5.10	34.73±5.55	35.94±5.82
4.3	82%	71%	79%	81%	77%	79%	83%	91%	98%	105%
	*20.47±.834	*20.53±0.92	*22.13±1.13	*23.13±1.46	*26.20±2.01	*26.91±2.87	*29.76±3.45	*30.40±4.43	*29.30±5.54	*29.16±5.74
9.5	82%	75%	79%	79%	76%	77%	80%	83%	83%	85%
	*21.00±.845	*20.60±.990	*22.47±1.25	*23.60±1.45	*26.33±3.31	*27.29±4.77	*28.97±5.37	*29.53±5.74	*29.83±7.26	*29.88±7.79
14.6	84%	71%	76%	80%	76%	78%	77%	81%	84%	87%

Mean group body weight ± SD in grams. $N=15$. Percent of control body weight .
*Significantly different from MDD control, $P<0.01$. #Significantly different from MDD control, $P<0.05$

TABLE 2

Mean liver weights of C57Bl/6 mice after exposure to varying concentrations of sodium arsenite in drinking water

Group	Liver weight (g)
MSD (N=12)	1.354 ± 0.159
MDD (N=15)	1.239 ± 0.153
2.6 (N=15)	1.358 ± 0.335
4.3 (N=15)	1.277 ± 0.316
9.5 (N=15)	1.175* ± 0.213
14.6 (N=14)	1.146* ± 0.304

*Significantly different from MDD control; $P<0.05$.

Mean liver weights for treated and control mice are given in Table 2. A dose-related decrease in terminal liver weight was observed which was statistically significant in the 9.5 and 14.6 mg/kg/d groups.

Histopathological examination revealed minimal to moderate fatty change in the livers of the majority of animals on methyl-deficient diets. No fatty change was observed in animals on methyl-sufficient diets not exposed to arsenic. Severe damage as indicated by marked loss of hepatocytes (necrosis), collapse of lobules, infiltration of inflammatory cells, proliferation of liver macrophages and lakes of bile-like pigment were observed in two mice, one in the 2.6 and one in the 4.3 mg/kg/d treatment group. Data are presented in Table 3. Histological

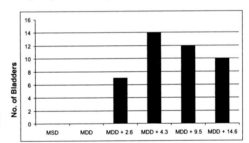

Fig. 3. Hypertrophy/hyperplasia in bladders of methyl-sufficient and deficient mice treated with arsenic.

TABLE 3

Prevalence of histopathological changes in livers of methyl-deficient mice treated with arsenic

	No Change	Hydropic (Grade 1-2)	Hydropic (Grade 3)	Fatty Change (Grade 1-2)	Fatty Change (Grade 2-3)	Fatty Change (Grade 3)	Fatty Change (Grade 3-4)	Severe Damage
MSD	3/15 20%	9/15 60%	0	0	0	0	0	0
MDD	4/15 27%	0	0	9/15 60%	1/15 7%	0	0	0
MDD + 2.6	0	1/15 7%	0	10/15 67%	5/15 33%	0	0	1/15 7%
MDD + 4.3	0	0	1/15 7%	6/15 40%	7/15 47%	1/15 15%	0	1/15 7%
MDD + 9.5	3/15 20%	0	0	10/15 67%	1/15 7%	0	1/15 7%	0
MDD + 14.6	5/15 33%	1/15 7%	0	5/15 33%	2/15 13%	0	1/15 7%	0

Grade 1: Minimal liver damage. Grade 2: Mild liver damage. Grade 3: Moderate liver damage.
*Severe damage: Marked loss of hepatocytes (necrosis) and collapse of lobules. Infiltration of inflammatory cells and proliferation of liver macrophages and lakes of bile-like pigment.

analysis also revealed hypertrophy and hyperplasia of the bladder epithelium in 43 of 60 mice exposed to sodium arsenite (Figure 3). None of the control animals displayed this effect.

DISCUSSION

The principal objective of this investigation was to identify a maximum tolerated dose (MTD) for sodium arsenite in methyl-deficient, male C57Bl/6 mice for purposes of conducting a chronic animal bioassay to develop an animal model for arsenic carcinogenesis. The chronic animal bioassay will also seek to develop mechanistic understanding of the role of methylation in arsenic-related carcinogenesis. The subchronic study was designed to identify an untoward effect of arsenic and methyl deficiency in a 130-day timeframe.

The rate of weight gain for the test animals was greater than that for the methyl sufficient and methyl deficient control animals although the mean body weight was significantly lower in the 9.5 and 14.6 mg/kg/d treatment groups (Table 1). Thus, it is not apparent that the differences in absolute weight are a measure of arsenic toxicity.

A marked effect on the liver of animals treated with methyl-deficient diets alone as well as with arsenic was observed. Fatty infiltration (fatty change) was seen in nearly all the animals treated. Livers appeared light in color and significantly larger by volume. Terminal liver weight was significantly decreased in the high dose arsenic groups. Severe liver damage as evidenced by a significant loss of hepatocytes, collapse of lobules and infiltration of inflammatory cells, was observed in two mice in the 2.6 and 4.3 mg/kg groups.

There were significant pathological changes in animals administered sodium arsenite. The hyperplasia and hypertrophy of the bladder epithelium may be in response to toxic insult or it may be indicative of alterations in the expression or control of cell-cycle genes. There was evidence for hypertrophy/hyperplasia in the test animal groups with no evidence for these changes in controls. The number of animals with hypertrophy/hyperplasia doubled between the 2.6 and 4.3 mg/kg/day groups. Based on these results we conclude that 2.6 mg/kg/day is a reasonable maximum tolerated dose.

Epidemiological studies have concluded that arsenic is a systemic carcinogen, with particular evidence for cancer of the bladder. Our results clearly indicate changes in the bladders of methyl-deficient mice administered arsenic. The chronic study to follow will focus close attention on the bladder of treated methyl-deficient mice, and we shall investigate the methylation status and expression of genes to determine the effects of arsenic and methyl-deficiency on the molecular biology of methylation. Alterations in genomic DNA methylation patterns as well as the gene-specific methylation patterns of cell-cycle control genes will be examined.

REFERENCES

Beck, B., Boardman, P., Hook, G., Rudel, R., Slayton, T., Carlson-Lynch, H. 1995. Response to Smith et al. (letter). *Environ. Health Perspect.*, **103**, 15–17.

Buchet, J.P., Lauwerys, R., Roels, H. 1981a. Comparison of the urinary excretion of arsenic metabolites after a single oral dose of sodium arsenite, monomethylarsonate, or dimethylarsinate in man. *Int. Arch. Occup. Environ. Health*, **48**, 71–79.

Buchet, J.P., Lauwerys, R., Roels, H. 1981b. Urinary excretion of inorganic arsenic and its metabolites after repeated ingestion of sodium metaarsenite by volunteers. *Int. Arch. Occup. Environ. Health*, **48**, 111–118.

Buckley, G.F., Hartroft, W.S. 1955. *Arch. Pathol.*, **59**, 185–197.

Carlson-Lynch, H., Beck, B., Boardman, P. 1994. Arsenic risk assessment. *Environ. Health Perspect.*, **102**, 354.

Christman, J.K., Sheikhnejad, G., Dizik, M., Abileah, S., Waifan, E. 1993. Reversibility of changes in nucleic acid methylation and gene expression induced in rat liver by severe dietary methyl deficiency. *Carcinogenesis*, **14**, 551–557.

Counts, J.L., Goodman, J.I. 1993. Comparative analysis of the methylation status of the 5′ flanking region of Ha-ras in B6C3F1, C3H/He and C57BL/6 mouse liver. *Cancer Lett.*, **75**, 129–136.

Counts, J.L., Sarmiento, J.I., Harbison, M.L., Downing, J.C., McClain, R.M., Goodman, J.I. 1996. Cell proliferation and global methylation status changes in mouse liver after Phenobarbital and /or choline-devoid, methionine-deficient diet administration. *Carcinogenesis*, **17**, 1251–1257.

Dizik, M., Christman, J.K., Waifan, E. 1991. Alterations in expression and methylation of specific genes in livers of rats fed a cancer promoting methyl-deficient diet. *Carcinogenesis*, **12**, 1307–1312.

Ghoshal, A.K., Farber, E. 1984. The induction of liver cancer by dietary deficiency of choline and methionine without added carcinogens. *Carcinogenesis*, **5**, 1367.

Henning, S. Swendseid, M., Coulson, W. 1997. Male rats fed methyl- and folate defined deficient diets with or without niacin develop hepatic carcinomas associated with decreased tissue NAD concentrations and altered poly (ADP-ribose) polymerase activity. *J. Nutrit.*, **127**, 30–36.

Herman, J., Merlo, A., Mao, L., Lapidus, R., Issa, J., Davidson, N., Sidransky, D., Baylin, S. 1995. Inactivation of the CDKN2/p16/MTS1 gene is frequently associated with aberrant DNA methylation in all common human cancers. *Cancer Res.*, **55**, 4525–4530.

Hopenhayn-Rich, C., Biggs, M.L., Smith, A.H., Kalman, D.A., Moore, L.E. 1996a. Methylation study of a population environmentally exposed to arsenic in drinking water. *Environ. Health Perspect.*, **104**, 620.

Hopenhayn-Rich, C., Biggs, M.L., Kalman, D.A., Moore, L.E., Smith, A.H. 1996b. Arsenic methylation patterns before and after changing from high to lower concentrations of arsenic in drinking water. *Environ. Health Perspect.*, **104**, 1200.

Hsieh, L.L., Wainfan, E., Hoshina, S., Dizik, M., Weinstein, I.B. 1989. Altered expression of retrovirus-like sequences and cellular oncogenes in mice fed methyl-deficient diets. *Cancer Res.*, **49**, 3795–3799.

Hsu, K., Froines, J.R., Chen, C.J. 1997. Studies of arsenic ingestion from drinking water in Northeastern Taiwan: Chemical speciation and urinary metabolites.

Hughs, M., Menache, M., Thompson, D. 1994. Dose-dependent disposition of sodium arsenite in mice following acute oral exposure. *Fund. Appl. Toxicol.*, **22**, 80–89.

James, S., Miller, B., Basnakian, A., Pogribny, I., Pogribna, M., Muskhelishvili, L. 1997. Apoptosis and proliferation under conditions of deoxynucleotide pool imbalance in liver of folate/methyl deficient rats. *Carcinogenesis*, **18**, 287–293.

Jones, P.A., Rideout III, W., Shen, C., Spruck, C., Tsai, Y. 1992. Methylation, mutation and cancer. *Bioessays*, **14**, 33.

Marafante, E., Vahter, M., Envall, J. 1985. The role of the methylation in the detoxification of arsenate in the rabbit. *Chem-Biol. Interact.*, **56**, 225–238.

Marafante, E., Vahter, M. 1986. The effect of dietary and chemically induced methylation deficiency on the metabolism of arsenic in the rabbit. *Acta Pharmacol. Toxicol.*, **59** (Suppl. 7), 35–38.

Mikol, Y.B., Hoover, K.L., Creasia, D., Poirier, L.A. 1983. Hepatocarcinogenesis in rats fed methyl-deficient, amino-acid-defined diets. *Carcinogenesis*, **4**, 1619.

Newberne, P.M., Rogers, A.E. 1986. Labile methyl groups and the promotion of cancer. *Ann. Rev. Nutr.*, **6**, 407.

Newberne, P.M., deCamargo, J.L., Clark, A.J. 1982. Choline deficiency, partial hepatectomy and liver tumors in rats and mice. *Toxicol. Pathol.*, **2**, 95–109.

Ray, J.S., Harbison, M.L., McClain, M., Goodman, J.I. 1994. Alterations in the methylation status and expression of the raf oncogene in phenobarbital-induced and spontaneous B6C3F1 mouse liver tumors. *Molec. Carcinogen.*, **9**, 155.

Salmon, W.D., Copeland, D.H. 1954. Liver carcinoma and related lesions in chronic choline deficiency. *Ann. N.Y. Acad. Sci.*, **57**, 664–667.

Salmon, W.D., Copeland, D.H., Burns, M.J. (1955). Hepatomas in choline deficiency. *J. Natl. Cancer Inst.*, **15**, 1549–1567.

Thompson, D.J. 1993. A chemical hypothesis for arsenic methylation in mammals. *Chem.-Biol. Interact.*, **88**, 89–114.

U.S. EPA. 1992. U.S. EPA Criteria document on arsenic. EPA contract number 68-C8-0033.

Vahter, M., Marafante, E. 1987. Effects of low dietary intake of methionine, choline or protein on the biotransformation of arsenite in the rabbit. *Toxicol. Lett.*, **37**, 41–46.

Vorce, R.L., Goodman, J.I. 1989. Altered methylation of ras oncogenes in benzidine-induced B6C3F1 mouse liver tumors. *Toxicol. Appl. Pharmacol.*, **100**, 398.

Waifan, E., Dizek, M., Stender, M., Christman, J. 1989. Rapid appearance of hypomethylated DNA in livers of rats fed cancer-promoting, methyl-deficient diets. *Cancer Res.*, **49**, 4094–4097.

Waifan, E., Poirier, L.A. 1992. Methyl groups in carcinogenesis: Effects on DNA methylation and gene expression. *Cancer Res.* (suppl.) **52**, 2071s.

Wilson, J.W. 1951. *Cancer Res.*, **11**, 290.

Yokoyama, S., Sell, M.A., Reddy, T.V., Lombardy, B. 1985. Hepatogenesis and promoting action of a choline-devoid diet in the rat. *Cancer Res.*, **45**, 2834.

Arsenic Exposure and Health Effects
W.R. Chappell, C.O. Abernathy and R.L. Calderon (Editors)
© 1999 Elsevier Science B.V. All rights reserved.

Arsenite Genotoxicity May Be Mediated by Interference with DNA Damage-inducible Signaling

Toby G. Rossman

ABSTRACT

Although high concentrations of dimethylated arsenic species can be genotoxic in some systems, arsenite is considered the most likely carcinogenic form of arsenic. DNA is not directly damaged by arsenite nor is arsenite significantly mutagenic in mammalian cells at relatively non-toxic concentrations. Arsenite causes chromosome aberrations, aneuploidy, cell transformation, and gene amplification in many cell types. At non-toxic concentrations, arsenite enhances the mutagenicity of UV and small alkylating agents. This suggests that arsenite carcinogenesis may require a partner, such as UV light. Arsenite inhibits the completion of repair of methyl nitrosourea-induced DNA damage. The inhibition by arsenite of the completion of DNA base excision repair appears to occur via effects on DNA ligation, the last step of excision repair. However, neither DNA ligases nor DNA polymerases can be inhibited by arsenite concentrations many-fold higher than those which can inhibit DNA repair in cells. Thus, the effects on DNA repair do not seem to be via inhibition by arsenite of DNA repair enzymes (ligases or polymerases). Rather, arsenite may affect cellular control of DNA repair processes. Our data and that of others is consistent with the hypothesis that arsenite may affect expression of the tumor suppressor gene *p53*. In fact, many of the genotoxic effects of arsenite are consistent with the type of genomic instability that would result from interfering with *p53*-related pathways. Disruption of these pathways has been shown to lead to cellular gene amplification and enhanced mutagenesis.

Keywords: arsenic, mutagenesis, gene amplification, signal transduction

INTRODUCTION

Because arsenite is considered to be the most likely carcinogenic form of arsenic, there is more information on its genotoxicity than on that of other species. In general, arsenate and organic arsenicals are at least an order of magnitude less potent as genotoxicants compared with arsenite (Tinwell et al., 1991; Moore et al., 1997). Cells which do not methylate arsenic compounds, such as fibroblast cell lines, convert arsenate to arsenite and then excrete the arsenite via an efflux pump (Wang et al., 1996). The genetic toxicology of arsenic compounds has been extensively reviewed (Gebhardt and Rossman, 1991; Wang and Rossman, 1996; Rossman, 1998).

There is no evidence that DNA is a target of arsenite or other inorganic arsenic compounds. These compounds neither form DNA adducts, nor do they induce the DNA-protein crosslinks characteristic of another oxyanion, chromate (Zhitkovitch and Costa, 1992). Incubation of supercoiled plasmid DNA with arsenite alone or arsenite with UV light or H_2O_2 failed to induce any DNA strand breaks or alkali-labile sites (Rossman, unpublished data). The inability of arsenite to induce the SOS response (abbreviation for "Save our Ship", a distress signal) in *E. coli* is consistent with its lack of direct genotoxicity (Rossman et al., 1984). Neither arsenite nor arsenate increases the infidelity of DNA polymerization (Tkeshelashvilli et al., 1980).

Arsenite is not generally mutagenic at single gene loci, and the very small numbers of induced mutants usually arise after exposure to highly toxic concentrations. At more relevant concentrations, arsenite induces chromosome aberrations, aneuploidy, and micronuclei (reviewed in Rossman, 1998). Micronuclei (a marker of chromosome damage) are found in the bone marrow of mice treated with arsenite (Tinwell et al., 1991) and in exfoliated bladder cells from exposed humans (Biggs et al., 1997). Arsenite at non-toxic concentrations was found to act as a comutagen with other agents. It enhances the mutagenesis induced by UV in *E. coli* (Rossman, 1981) and by UV, methyl methanesulfonate (MMS), and methyl nitrosourea (MNU) in Chinese hamster cells (Lee et al., 1985b; Li and Rossman, 1989a, 1991; Yang et al., 1992). The comutagenic effects of arsenite are probably the result of inhibition of DNA repair. Arsenite inhibits the repair of DNA damage induced by X-rays and UV (Snyder et al., 1989), postreplication repair of UV-induced damage (Lee-Chen et al., 1992), and completion of repair of MNU-induced DNA damage (Li and Rossman, 1989a). Further support for the idea that arsenite inhibits DNA repair are the findings that it potentiates X-ray and UV-induced chromosomal damage in peripheral human lymphocytes and fibroblasts (Jha et al., 1992), alters the mutational spectrum (but not the strand bias) of UV-irradiated Chinese hamster ovary cells (Yang et al., 1992) and enhances chromosome aberrations induced by diepoxybutane, a DNA crosslinking agent (Wiencke and Yager, 1992).

The inhibition by arsenite of the completion of base excision repair appears to occur via effects on DNA ligation (Li and Rossman 1989b; Lee-Chen et al., 1994). A nick-translation assay for DNA strand breaks or gaps showed that in cells treated with MNU + arsenite, breaks remained open three hours after MNU treatment, whereas in the absence of arsenite, the breaks had closed by that time (Li and Rossman, 1989a). This suggested that either the polymerase or the ligase step of base excision repair had been blocked by arsenite. Since nucleotide excision repair (which repairs UV-induced pyrimidine dimers) also involves polymerase and ligase activity, it seemed reasonable to assay these enzymes for arsenite sensitivity. However, neither DNA ligases nor DNA polymerase α or β can be inhibited by arsenite concentrations many fold higher than those which can inhibit DNA repair in cells (Li, 1989; Li and Rossman, 1989b; Hu et al., 1998). Thus, the effects on DNA repair do not seem to be via inhibition by arsenite of DNA repair enzymes (ligases or polymerases), although effects on accessory proteins (if any) have not been tested. Rather, arsenite may affect cellular control of DNA repair processes.

We report here that arsenite can induce a small number of gene mutations at the *E. coli gpt* locus in an extremely sensitive transgenic cell line, G12 (Klein and Rossman, 1990). In addition, we review some of the comutagenic effects of arsenite in V79 cells, and its effects on gene amplification in a number of cell systems, and suggest a possible mechanism to explain some of the genotoxic effects of arsenite.

METHODOLOGY

Cell Culture

Chinese hamster V79 cells (strain 743-3-6, originally obtained from Dr. Dennis Yep, University of Michigan) were maintained as is Li and Rossman (1989a,b). A clone of cells exhibiting a low spontaneous mutation frequency at the *hprt* locus was isolated, expanded and stored in liquid nitrogen until needed. Cells were thawed and utilized within 4 weeks to ensure a low background mutation frequency in the *hprt* mutation assays.

The G12 cell line was developed in this laboratory (Klein and Rossman, 1990). Briefly, *pSV2gpt*, a plasmid carrying the *E. coli* xanthine-guanine phosphoribosyl transferase gene (*gpt*), was transfected into a non-revertible *hprt⁻* cell line of V79 induced by UV (16 J/m²). Transfectants were selected in HAT medium. Following several generations in the absence of selection transfectants were rechecked for stable *gpt* integration. One transfectant, G12, had a low spontaneous mutation frequency and a single *gpt* insert, as determined by Southern blot. This cell line shows a much higher mutation frequency after X-irradiation, and a slightly higher mutation frequency after UV-irradiation compared with V79 cells at the *hprt* locus (Klein et al., 1994). This is most likely due to the insertion of the *gpt* gene into chromosome 1 (Klein and Snow, 1993), thereby allowing detection of multilocus deletions, a lethal mutation at the X-linked *hprt* locus. G12 cells are stored in liquid nitrogen. The cells were thawed and maintained in HAT medium one week before the mutation assay. The HAT medium was then replaced by F12 medium in the mutation assay.

SV40 transformed human keratinocyte lines were obtained from Dr. Mark Steinberg (City College, CUNY), and handled as previously described (Rossman and Wolosin, 1992).

Test Compounds

Sodium arsenite (>99.9% purity) was purchased from Fisher Scientific Co. (Fairlawn, NJ). A 1 M stock solution was made fresh by weighing and dissolving this compound in water and sterilized using a 0.22 μm syringe filter. The final dilutions were made in serum-free medium immediately prior to use. N-methyl-N-nitrosourea (MNU; Sigma Chemical Co., St. Louis, MO) was dissolved in DMSO and stored at –20°C as a 1 M stock solution.

Ultraviolet Light C (UVC) Treatment

UVC mutation assays were performed by exposure of attached cells in dishes with Earle's balance salt solution (EBSS, Gibco). A 15-W General Electric germicidal lamp (wavelength 254 nm) was used. The cells were irradiated without lids. UV fluences were calculated from the fluence rates calibrated with a radiometer.

Mutation Assays

The *hprt* mutation assay was performed using a modification of the method described by Chang et al. (1978). V79 cells were seeded for treatment in duplicate at 5×10^4 cells per 100 mm dish, and in triplicate for toxicity at 500 cells per 100 mm dish. Following a 4-hour incubation, attached cells were mutagenized as desired. Sodium arsenite was added in EBSS 5 min before UV irradiation or MNU treatment.

After irradiation, the cells were washed twice with EBSS, and sodium arsenite in complete medium, or medium alone, was added for 3 hours. The cells were then washed again and refed with F12 medium for a 5-day expression period, replating once to maintain exponential growth. At that time, the survival plates were fixed and stained and the mutagenesis plates trypsinized and reseeded (10 dishes, 1×10^5 cells/dish) in complete F12 medium containing 10 μg/ml 6TG. For the reseeding survival, five hundred cells were plated concurrently in triplicate into 100 mm dishes containing F12 medium without 6TG and stained after 7 days. Following a 10-day selection period, the mutagenesis plates were fixed and stained. The 6TG mutation frequency per 10^6 surviving cells was calculated, using the reseeding survival values.

The *gpt* mutation assay in G12 cells was performed in a similar manner to the *hprt* mutation assay. The assays for gene amplication have been described previously (Lee et al., 1988; Rossman and Wolosin, 1992). For gene amplification studies, treatments with sodium arsenite were for 3.5 hours in medium.

RESULTS

Mutagenicity of Arsenite in G12 Cells

Treatment of Chinese hamster V79 cells with arsenite failed to cause a significant increase in mutations at the *hprt* locus (Rossman et al., 1980). Although treatment of G12 cells for 24 hours with sodium arsenite appears to result in a dose-dependent increase in mutagenesis at the transgenic *gpt* locus, nevertheless the increase is not significant at $p < 0.05$ (Figure 1). At 15 μM arsenite, which causes more than 60% cell killing, there is almost a two-fold increase ($p < 0.1$).

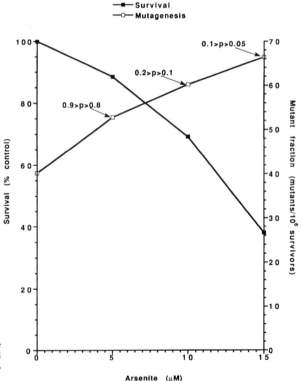

Fig. 1. Mutagenicity of sodium arsenite at the transgenic locus of G12 cells. Cells were treated with arsenite for 24 hours. ■, Survival; □, mutagenesis.

Comutagenesis of Arsenite with UVC and MNU

The enhancement of mutagenesis by two agents at the *hprt* locus in Chinese hamster V79 cells is shown in Figure 2. UVC causes a dose-dependent increase in mutagenesis, as expected (Figure 2A). Treatment with 10 μM sodium arsenite for 3 hours caused no toxicity (data not shown) or mutagenicity. However, at all UVC doses, arsenite caused an enhancement of mutagenesis which was significant at 10 and 15 Joules/m^2 (J/m^2). Similar results are seen when arsenite is combined with MNU (Figure 2B). It is of interest that the enhancement of mutagenesis with UVC increases with the UVC dose, whereas the enhancement of MNU mutagenesis is greater at low MNU doses. DNA damage caused by UVC is repaired predominantly by the nucleotide excision repair pathway, while that of MNU is repaired by the base excision repair pathway.

Effects of Arsenite on Gene Amplification

There have been few studies on the effects of arsenite on gene amplification. These are summarized in Table 1. Lee et al. (1988) were the first to show that arsenite induces gene amplification at the *dhfr* locus of mouse 3T6 cells. We then showed that the effect also occurs in human cells, using SV40-transformed human keratinocytes (Rossman and Wolosin, 1992). The *dhfr* gene is an endogenous gene, and amplification of this gene confers resistance to methotrexate on the cell. Amplification of endogenous genes is a low-frequency event. In contrast, viral genes can also be induced to amplify by many carcinogens. This amplification is detected by increased hybridization using a viral probe, and occurs at high frequency. Surprisingly, arsenite failed to induce amplification of SV40 sequences in either Chinese hamster C11 cells or human keratinocytes (Li, 1989; Rossman and Wolosin, 1992).

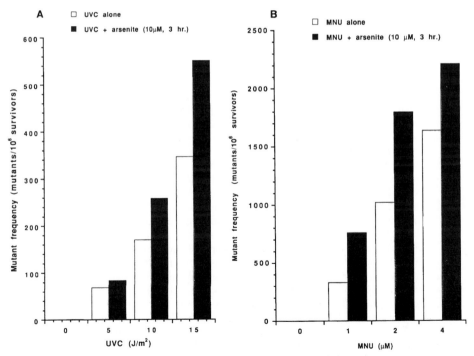

Fig. 2. Comutagenesis by a non-toxic concentration of sodium arsenite. A: Enhancement of UVC-induced mutagenesis by sodium arsenite. B: Enhancement of MNU-induced mutagenesis by sodium arsenite.

TABLE 1

Effects of arsenite on gene amplification

Gene	Cell	Arsenite	Amplification	Reference
dhfr	Mouse 3T6	0.2–6.2 μM	+	Lee et al. (1988)
dhrf	Human keratinocyte AG06	6 μM	+	Rossman and Wolosin (1992)
SV40	Chinese hamster C11	5–20 μM	–	Li and Rossman (unpubl.)
SV40	Human keratinocyte AG06	5–20 μM	–	Rossman and Wolosin (1992)

DISCUSSION AND CONCLUSIONS

Arsenite acts as a weak mutagen at the transgenic *gpt* locus in G12 cells. Mutagenic activity (significant at $p < 0.1$) is not reached until more than 40% of the cells are killed. Similar results are seen in another system capable of detecting large deletions (Moore et al., 1997). When Meng and Hsie (1996) analyzed the mutants resulting from another transgenic cell line treated with high concentrations of arsenite (which also gave mutant fractions only twice background levels), the proportion of deletions was higher than in the spontaneous class. Since the transgenic *gpt* locus of G12 cells is better at detecting deletion compared with the *hprt* locus of V79 cells (Klein et al., 1994), it is possible that many of our mutants were also deletions, but this was not analyzed.

The *gpt* locus of G12 cells can also be silenced by DNA methyation, as occurs in cells treated with carcinogenic nickel compounds (Lee et al., 1993, 1995). However, in the latter case, apparently huge "mutagenic" effects were seen with these compounds in G12 cells (but not in the parental V79 cell or in another transfectant line), which upon analysis turned out to be due to gene silencing. This type of effect is not seen with arsenite treatment of G12 cells.

Arsenite induced gene amplification at the *dhfr* locus in SV40-transformed human keratinocytes, but failed to cause amplification of SV40 sequences (Rossman and Wolosin, 1992). This suggests that arsenite does not induce signaling typical of DNA-damaging agents (which do induce SV40 amplification in this system), but rather affects checkpoint pathways such as those involving p53, whose disruption leads to cellular gene amplification (Livingstone et al., 1992). In fact, it is quite possible that arsenite blocks DNA repair by interfering with cell cycle checkpoints rather than by inhibiting repair enzymes.

The tumor suppressor p53 has a crucial role as "guardian of the genome" in the control of cell cycle progression (Figure 3). If damaged DNA is replicated, it may be mutated or lost due to chromosome breaks. DNA damage results in an accumulation of p53 protein, mainly via post-translational stabilization (Levine and Momand, 1990). p53 protein temporarily halts cell cycle progress, allowing time for DNA repair before replication (Kastan et al., 1991) or else causes apoptosis in heavily damaged cells (Miyashita et al., 1994). Cells with mutant p53 are more likely to continue to divide, and fail to undergo apoptosis, in spite of DNA damage to their chromosomes (Little, 1994). Such cells show greatly elevated rates of chromosome aberrations such as deletions, translocations, amplifications and aneuploidy (Reznikoff et al., 1994; Hainaut, 1995), exactly the classes of genotoxic events induced by arsenite. Li-Fraumeni cells, which are p53-deficient, show reduced excision repair of pyrimidine dimers (Smith et al., 1995). When p53 activity is inactivated by expression of the E6 protein of HPV16 in human cells, UV-induced mutations are elevated about 2-fold and a large increase in deletions is seen (Havre et al., 1995; Yu et al., 1997), suggesting that deletion-prone intermediates, such as strand breaks or gaps, accumulate during faulty repair. Arsenite also increases UV-mutagenesis about 2-fold (Li and Rossman, 1991), increases the proportion of deletions (Meng and Hsie, 1996; Moore et al., 1997) and causes increased accumulation of strand breaks or gaps in cells with DNA damage (Li and Rossman, 1989a).

Fig. 3. Some of the major pathways by which p53 regulates the events following DNA damage in cells.

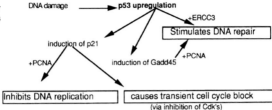

Mass and Wang (1997) have shown that long-term exposure of cells to low concentrations of arsenite resulted in hypermethylation of the *p53* promoter, which is expected to result in blockage of *p53* transcription. Cells with such a blockage would behave as *p53* mutants (i.e. as Li-Faumeni phenocopies). However, short term exposure to arsenite increased p53 protein abundance (Salazar et al., 1997). Evidence suggests that p53 protein is normally degraded through ubiquitin-dependent proteolysis (Maki et al., 1996). Cells which are defective in this pathway show elevated levels of p53 protein (Chowdary et al., 1994). Since arsenite inhibits ubiquitin-dependent proteolysis (Klemperer and Pickart, 1989), the increase in p53 protein after arsenite treatment might be caused by this inhibition. One can speculate that continuous over-expression of p53 protein may result, in the long term, in shutting down its expression by hypermethylation of its promoter.

Spontaneous amplification of endogenous genes is rare in normal cells, but common in tumor cells which have mutated *p53* genes (Livingstone et al., 1992). Double strand breaks have been implicated as a possible cellular signal for gene amplification (Nelson and Kastan, 1994). Arsenite causes *dhfr* gene amplification in human keratinocytes (Table 1). Although these keratinocytes are a highly relevant system in which to study arsenite, the SV40 T-antigen inactivates the p53 protein in these cells, allowing gene amplification. It would be of interest to determine whether long term exposure to arsenite would induce gene amplification in human cells with wild type *p53* genes.

Given its very weak mutagenic activity, the assumption has been made that if arsenite is "non-genotoxic", it must therefore act as a tumor promoter. There is little evidence for this view, as negative results were obtained in a bioassay testing for promotional activity (Milner, 1969). However, arsenite appears to enhance the promoting activity of phorbol ester (Germolec et al., 1997). The arsenite metabolite dimethlyarsinic acid (DMAA) did act as a promoter (Yamamoto et al., 1995), but the significance of this finding for human exposure (which is to inorganic compounds at low concentrations) is questionable.

One of the unexplained facts about arsenic carcinogenesis is the difficulty in finding a good animal model, since most attempts to induce tumors by arsenic compounds in rodents have failed. This might be related to inappropriate dosage or treatment regimens. Arsenite should be tested as a co-carcinogen rather than as a complete carcinogen or promoter. Arsenite-induced genomic instability might develop gradually. Thus, long term arsenite treatment (for example in the drinking water) might be necessary, prior to treatment with a second genotoxic carcinogen.

ACKNOWLEDGEMENTS

We thank Eleanor Cordisco for her expert help in document preparation. This work was supported by United States Public Health Service Grants CA57352 and ES09252, and is part of NYU's Nelson Institute of Environmental Medicine Center programs supported by Grants ES00260 from the National Institute of Environmental Health Sciences.

REFERENCES

Biggs, M.L., Kalman, D.A., Moore, L.E., Hopenhayn-Rich, C., Smith, M.T., Smith, A.H. 1997. Relationship of urinary arsenic to intake estimates and a biomarker of effect, bladder cell micronuclei. *Mutat. Res.*, **386**, 185–195.

Chang, C.C., Castellazzi, M., Glover, T.W., Trosko, J.E. 1978. Effects of harmon and nonharmon on spontaneous and ultraviolet light-induced mutagenesis in cultured Chinese hamster cells. *Cancer Res.*, **38**, 4527–4533.

Chowdary, D., Dermody, J.J., Jha, K.K., Ozer, H.L. 1994. Accumulation of p53 in a mutant cell lines defective in the ubiquitin pathway. *Mol. Cell. Biol.*, **14**, 1997–2003.

Gebhart, E., Rossman, T.G. 1991. Mutagenicity, carcinogenicity, teratogenicity. In: E. Meria (ed.), *Metals and their Compounds in the Environment*, pp. 617–641. VCH Verlagsgesellschaft, Weinheim.

Germolec, D.R., Spalding, J., Boorman, G.A., Wilmer, J.L., Yoshida, T., Simeonova, P.P., Bruccoleri, A., Kayama, F., Gaido, K., Tennant, R., Burleson, F., Dong, W., Lang, R.W., Luster, M.I. 1997. Arsenic can mediate skin neoplasia by chronic stimulation of keratinocyte-derived growth factors. *Mutat. Res.*, **386**, 209–18.

Hainaut, P. 1995. The tumor suppressor protein p53: a receptor to genotoxic stress that controls cell growth and survival. *Curr. Opinion Oncol.*, **7**, 76–82.

Havre, P.A. Yuan, J. Hedrick, L. Cho, K.R., Glazer, P.M. 1995. p53 inactivation by HPV16 E6 results in increased mutagenesis in human cells. *Cancer Res.*, **55**, 4420–4424.

Hu Y., Su L., Snow E.T. 1998. Arsenic toxicity is enzyme specific and arsenic inhibition of DNA repair is not caused by direct inhibition of repair enzymes. *Mutat. Res.*, **408**, 203–218.

Jha, A.N. Noditi, M. Nilsson, R., Natarajan, A.T. 1992. Genotoxic effects of sodium arsenite on human cells. *Mutat. Res.*, **284**, 215–221

Kastan, M.B., Onyekwere, O., Sidransky, D., Vogelstein, B., Craig, R.W. 1991. Participation of p53 protein in the cellular response to DNA damage. *Cancer Res.*, **51**, 6304–6311.

Klein, C.B., Rossman, T.G. 1990. Transgenic Chinese hamster V79 cell lines which exhibit variable levels of gpt mutagenesis. *Environ. Mol. Mut.*, **16**, 1–12.

Klein, C.B., Snow, E.T. 1993. Localization of the gpt sequence in transgenic G12 cells via fluorescent in situ hybridization. *Environ. Mol. Mut.*, **21**(Suppl.22), 35.

Klein, C.B., Su, L., Rossman, T.G. and Snow, E.T. 1994. Transgenic *gpt* V79 cell lines differ in their mutagenic response to clastogens. *Mutat. Res.*, **304**, 217–228.

Klemperer, N.S., Pickart, C.M. (1989) Arsenite inhibits two steps in the ubiquitin-dependent proteolytic pathway. *J. Biol. Chem.*, **264**, 19245–19252.

Lee, T.-C. Huang, R.Y., Jan, K.Y. 1985. Sodium arsenite enhances the cytotoxicity clastogenicity and 6-thioguanine-resistant mutagenicity of ultraviolet light in Chinese hamster ovary cells. *Mutat. Res.*, **148**, 83–89.

Lee, T.C., Tanaka, N., Lamb, P.W., Gilmer, T., Barrett, J.C. 1988. Induction of gene amplification by arsenic. *Science*, **241**, 79–81.

Lee-Chen, S.F. Yu, C.T., Jan, K.Y. 1992. Effect of arsenite on the DNA repair of UV-irradiated Chinese hamster ovary cells. *Mutagenesis*, **7**, 51–55.

Lee, Y.-W., Pons, C., Tummolo, D.M., Klein, C.B., Rossman, T.G., Christie, N.T. 1993. Mutagenicity of soluble and insoluble nickel compounds at the *gpt* locus in G12 Chinese hamster cells. *Environ. Mol. Mut.*, **21**, 365–371.

Lee-Chen, S.F. Yu, C.T. Wu, D.R., Jan, K.Y. 1992. Differential effects of luminol, nickel, and arsenite on the rejoining of ultraviolet light and alkylation-induced DNA breaks. *Environ. Mol. Mut.*, **23**, 116–120.

Lee, Y.-W., Klein, C.B., Kargacin, B., Salnikow, K., Kitahara, J., Dowjat, K., Zhitkovich, A., Christie, N.T. and Costa, M. 1995. Carcinogenic nickel silences gene expression by chromatin condensation and DNA methylation: a new model for epigenetic carcinogens. *Mol. Cell. Biol.*, **15**, 2547–2557.

Levine, A.J., Momand, J. 1990. Tumor suppressor genes: the p53 and retinoblastoma sensitivity genes and gene products, *Biochim. Biophys. Acta*, **1032**, 119–136.

Li, J.-H., Rossman, T.G. 1989a. Mechanism of comutagenesis of sodium arsenite with N-methyl-N-nitrosourea. *Biol. Trace Element Res.*, **21**, 373–381.

Li, J.-H., Rossman, T.G. 1989b. Inhibition of DNA ligase activity by arsenite: A possible mechanism of its comutagenesis. *Mol. Toxicol.*, **2**, 1–9.

Li, J.-H. 1989. Ph.D. Thesis, New York University.

Li, J.H., Rossman, T.G. 1991. Comutagenesis of sodium arsenite with ultraviolet radiation in Chinese hamster V79 cells. *Biol. Metals*, **4**, 197–200.

Little, J.B. 1994. Failla Memorial Lecture: changing views of radiosensitivity. *Radiat. Res.*, **140**, 299–236.

Livingstone, L.R., White, A., Sprouse, J., Livanos, E., Jacks, T., Tlsty, T.D. 1992. Altered cell cycle arrest and gene amplification potential accompany loss of wild-type p53. *Cell*, **70**, 923–935.

Maki, C.G., Huibregtse, J., Howley, P.M. 1996. *In Vivo* ubiquitination and proteosome-mediated degradation of p53. *Cancer. Res.*, **56**, 2649–2654.

Mass, M.J., Wang, L. 1997. Arsenic alters cytosine methylation patterns of the promoter of the tumor suppressor gene p53 in human lung cells: A model for a mechanism of carcinogenesis. *Mutat. Res..*, **386**, 263–277.

Meng, Z., Hsie, A.W. 1996. Polymerase chain reaction-based deletion analysis of spontaneous and arsenite-enhanced gpt mutants in CHO-AS$_{52}$ cells. *Mutat. Res.*, **356**, 255–0259.

Milner, J.E. 1969. The effects of ingested arsenic exposure on methyl cholanthrene induced skin tumours in mice. *Arch. Environ. Health*, **18**, 7–11.

Miyashita, T., Krajewski, S., Krajewska, M., Wang, H.G., Lin, H.K., Liebermann, D.A., Hoffman, B., Reed, J.C. 1994. Tumor suppression p53 is a regulator of bcl-2 and bax gene expression in vitro and in vivo. *Oncogene*, **9**, 1799–1805.

Moore, M.M., Harrington-Brock, K., Doerr, C.L. 1997. Relative genotoxic potency of arsenic and its methylated metabolites. *Mutat. Res.*, **386**, 279–290.

Nelson, W.G., Kastan, M.B. 1994. DNA strand breaks: the DNA template alterations that trigger p53-dependent DNA damage response pathways. *Mol. Cell. Biol.*, **14**, 1815–1823.

Reznikoff, C.A., Belair, C., Savelieva, E., Zhai, Y., Pfeifer, K., Yeager, T., Thompson, K.J., DeVries, S., Bindley, C., Newton, M.A. 1994. Long-term gonome stability and minimal genotypic and phenotypic alterations in HPV16 E7-, but not E6-, immortalized human urepithelial cells. *Genes Dev.*, **8**, 2227–2240.

Rossman, T.G. 1998. Molecular and Genetic Toxicology of Arsenic. In: J. Rose, (ed.), *Environmental Toxicology: Current Developments*, Gordon and Breach Science Publishers, Amsterdam, pp. 171–187.

Rossman, T.G. 1981. Enhancement of UV-mutagenesis by low concentrations of arsenite in *E. coli. Mutat. Res.*, **91**, 207–211.

Rossman, T.G., Wolosin, D. 1992. Differential susceptibility to carcinogen-induced amplification of SV40 and *dhfr* sequences in SV40-transformed human keratinocytes. *Mol. Carcinogen.*, **6**, 203–213.

Rossman, T.G. Molina, M., Meyer, L.W. 1984. The genetic toxicology of metal compounds: I. Induction of λ prophage in *E. coli* WP2$_s$(λ). *Environ. Mutagen.*, **6**, 59–69.

Rossman, T.G., Stone, D., Molina, M., Troll, W. 1980. Absence of arsenite mutagenicity in *E. coli* and Chinese hamster cells. *Environ. Mutagen.*, **2**, **371–379**.

Salazar, A.M., Ostrosky-Wegman, P., Menendex, D., Miranda, E., Garcia-Carranca, A., Rojas, E. (1997) Induction of p53 protein expression by sodium arsenite. *Mutat. Res.*, **381**, 259–265.

Smith, M.L., Chen, I., Zhan, Q., O'Connor, P.M., Fornace, A.J. 1995. Involvement of p53 tumor suppressor in repair of UV-type DNA damage. *Oncogene*, **10**, 1053–1059.

Snyder, R.D., Davis, G.F., Lachmann, P. 1989. Inhibition by metals of x-ray and ultraviolet-induced DNA repair in human cells. *Biol. Trace Element Res.*, **21**, 389–398.

Tinwell, H., Stephens, S.C., Ashby, J. 1991. Arsenite as the probable active species in the human carcinogenicity of arsenic: Mouse micronucleus assays on Na and K arsenite, orpiment, and Fowler's solution. *Environ. Health Perspect.*, **95**, 205–210.

Tkeshelashvili, L.K., Shearman, C.W., Zakour, R.A., Koplitz, R.M., Loeb, L.A. 1980. Effects of arsenic, selenium, and chromium on the fidelity of DNA synthesis. *Cancer Res.*, **40**, 2455–60.

Wang, Z., Rossman, T.G. 1996. The carcinogenicity of arsenic. In: Louis W. Chang (ed.),*Toxicology of Metals*, pp. 219–227 CRC Press, Boca Raton, FL.

Wang, Z., Dey, S., Rosen, B.P., Rossman, T.G. 1996. Efflux-mediated resistance to arsenicals in arsenic-resistant and -hypersensitive Chinese hamster cells. *Toxicol. Appl. Pharmacol.*, **137**, 112–119.

Wiencke, J.K., Yager, J.W. 1992. Specificity of arsenite in potentiating cytogenetic damage induced by the DNA crosslinking agent diepoxybutane. *Environ. Mol. Mutagen.*, **19**, 195–200.

Yamamoto, S., Konishi, Y., Matsuda, T., Murai, T., Shibata, M., Matsui-Yuasa, I., Otani, S., Kuroda, K., Endo, G., Fukushima, S. 1995. Cancer induction by an organic arsenic compound, dimethylarsinic acid (cacodylic acid), in F344/DuCrj rats after pretreatment with five carcinogens. *Cancer Res.*, **55**, 1271–1276.

Yang, J.-L., Chen, M.-F., Wu, C.-W., Lee, T.-C. 1992. Posttreatment with sodium arsenite alters the mutation spectrum induced by ultraviolet light irradiation in Chinese hamster ovary cells. *Environ. Mol. Mutagen.*, **20**, 156–164

Yu, Y., Li, C.-Y., Little, J.B. 1997. Abrogation of p53 function by HPV16 E6 gene delays apoptosis and enhances mutagenesis but does not alter radiosensitivity in K6 human lymphoblast cells. *Oncogene*, 14, 1661–1667.

Zhitkovich, A., Costa, M. 1992. A simple, sensitivity assay to detect DNA-protein-crosslinks in intact cells and *in vivo. Carcinogenesis*, **13**, 1485–1489.

Arsenic Exposure and Health Effects
W.R. Chappell, C.O. Abernathy and R.L. Calderon (Editors)

243

Modulation of DNA Repair and Glutathione Levels in Human Keratinocytes by Micromolar Arsenite

Elizabeth T. Snow, Yu Hu, Chong Chao Yan, Salem Chouchane

ABSTRACT

Arsenic (As) is a human carcinogen, but not a mutagen, although it inhibits DNA repair and is a comutagen. Human AG06 keratinocytes treated with micromolar arsenic exhibit dose and time-dependent loss of DNA ligase function. However, purified human DNA ligase I, ligase III, and other repair enzymes such as DNA polymerase β, are not inhibited by less than millimolar arsenite, As(III), the most toxic form of As found in the environment. DNA ligase activity in extracts from untreated keratinocytes is also insensitive to less than millimolar As. Pyruvate dehydrogenase, on the other hand, *is* inhibited by micromolar As and probably determines As-induced cytotoxicity. Simultaneous treatment of AG06 cells with an alkylating agent, 1-methyl-3-nitro-1-nitrosoguanidine (MNNG), plus As produces a synergistic increase in viability (dye uptake) at low doses and a synergistic increase in toxicity at high doses. Micromolar As also modulates cellular redox levels and induces a variety of cellular stress response genes. Keratinocytes treated with As exhibit both a time- and dose-dependent increase in cellular GSH levels and alterations in the relative activity of several GSH-dependent enzymes. These As-induced changes in cellular redox capacity and DNA repair activity are not directly related to toxicity. Maximal induction of GSH and DNA repair occurs after treatment with sub-toxic concentrations of As. At submicromolar concentrations, arsenic also induces hyperproliferation of keratinocytes, both *in vivo* and *in vitro*. Our results suggest that As modulates DNA repair and redox levels primarily through post-translational or transcriptional mechanisms.

Keywords: arsenate, enzyme inhibition, keratinocytes, cell culture, DNA repair, redox, oxidative stress

INTRODUCTION

Low dose, chronic exposure to inorganic arsenic (As) is a well established human skin, bladder, and lung carcinogen (Snow 1992; Wang and Rossman 1996). Similar low doses of As inhibit generalized DNA repair in cultured mammalian cells (Wang and Rossman, 1996; Hartwig et al., 1997; Lynn et al., 1997). As is the first known non-genotoxic environmental carcinogen to exhibit effects on DNA repair. The goal of our research has been to examine the nature and mechanism of arsenic-induced inhibition of DNA repair *in vitro* using SV40-transformed human keratinocytes (AG06 cells) and normal human epidermal keratinocytes (NHEK cells).

The mechanism of arsenic-induced human cancer is not clear. Nontoxic doses of arsenic do not induce mutations (Rossman et al., 1980) and As is not generally considered a complete carcinogen. Most animal models of arsenic-induced carcinogenesis have focused on arsenic as a tumor promoter or progressor (Cavigelli et al., 1996; Germolec et al., 1997; Ludwig et al., 1998). Yet, arsenic toxicity is multifactorial. Various concentrations of As have been found to: (i) inhibit critical enzyme systems, including mitochondrial respiration and DNA repair, (ii) to induce transcription of a variety of stress response genes, (iii) to promote gene amplification in mammalian cells in culture, and (iv) to induce cytogenetic damage, including clastogenesis.

Acute treatment with As produces a cellular stress response and is cytotoxic; however, the effects of low, physiologically relevant (sub-micromolar or less than 1 ppm), chronic exposures to arsenic are less clear. Low doses of arsenic can induce hyperproliferation of epithelial cells (Germolec et al., 1996) and gene amplification (Barrett and Lee 1991; Rossman and Wolosin 1992). As can also alter levels and patterns of cellular phosphorylation (Mivechi et al., 1994; Barchowsky et al., 1996; Cavigelli et al., 1996; Liu et al., 1996; Kato et al., 1997), possibly by inhibiting specific phosphatases (Cavigelli et al., 1996; Liu et al., 1996). However, our data (Hu et al., 1998) show that most enzymes are not inhibited by less than millimolar concentrations of arsenite [As(III)] and are even less inhibited by arsenate [As(V)]. It is likely that only a few proteins or cellular processes are directly affected by physiological levels of As and that these primary effectors then alter the regulation of secondary responses, such as redox levels and DNA repair.

Cellular redox levels, particularly the levels of GSH and GSSG play an important role in cellular regulation and in the detoxification and excretion of arsenic (Huang et al., 1993; Scott et al., 1993). Conversely, exposure to As can modulate cellular levels of GSH leading to long lasting changes in the redox-dependent control of cellular functions. Glutathione levels may be significantly raised (Li and Chou, 1992) or lowered (Szinicz and Forth, 1988) in arsenic-exposed cells, depending on the time after exposure, the dose, and the cell type. Changes in GSH levels can be seen with As exposures as short as 3 to 6 hours and chronic exposure to low concentrations of As can cause substantial increases in GSH levels.

Chronic and persistent alteration of signal transduction pathways due to alterations in cellular redox levels may play an important role in As-induced oncogenesis. Cytoplasmic signal transduction pathways consist of a cascade of protein phosphorylations mediated by a series of protein kinases and their substrates, and controlled by an additional series of protein phosphatases. The induction of these pathways can produce a very rapid and transient response or a more long-lasting response, or a combination of both. The effects of high (20 to 200 μM) concentrations of As on signal transduction appear to be mediated by As-specific inhibition of one or more protein phosphatases, by alterations in cellular redox levels, or possibly by other as yet unidentified mechanisms. Our data suggest that physiologically relevant, low dose exposure to As can significantly influence cellular redox levels and thereby modulate DNA repair and other cellular control systems.

METHODS

Reagents

Sodium arsenite ($NaAsO_2$), sodium arsenate ($NaHAsO_4$), reduced glutathione (GSH), *N*-tosyl-L-lysine chloromethyl ketone (TLCK), *N*-tosyl-L-phenylalanine chloromethyl ketone (TPCK), aprotinin, leupeptin, pepstatin A, sodium carbenicillin, 1-methyl-3-nitro-1-nitrosoguanidine (MNNG), phenylmethanesulfonic fluoride (PMSF), pyruvic acid, L-cysteine, calf thymus DNA, thiamine pyrophosphate chloride, purified glutathione S-transferase-π (GST) from equine liver, and bovine erythrocyte glutathione peroxidase (GPx) were purchased from Sigma (St. Louis, MO). Glutathione reductase (GR) from yeast was obtained from Boehringer-Mannheim Biochemicals. DNA ligase I antibody was prepared commercially (Alpha Diagnostics, San Antonio, TX) in rabbits using a human ligase I-specific peptide (Prasad et al., 1996). Human AP endonuclease antibody and cDNA probe were a gift from Dr. Bruce Demple. All other reagents were molecular biology grade.

Cellular Toxicity

AG06 cells were obtained from Dr. Mark Steinberg, City College of New York. NHEK cells were obtained from Clonetics and cultured in serum-free keratinocyte media, as recommended by the supplier. For determining relative toxicity, cells were exposed to $NaAsO_2$ or $NaHAsO_4$ in complete medium for various periods of time ranging from 3 to 72 hours and assayed for neutral red dye uptake (Little et al., 1996). The concentration of arsenic at which the dye uptake was reduced by 50% (IC_{50}) was calculated from a least squares fit of the log transformed data obtained for the log-linear portion of the toxicity curve.

Glutathione Measurements

Cellular GSH concentrations were determined enzymatically using the colorimetric method of Tietz (1969), as modified for use in 96-well plates (Kavanagh et al., 1994).

Enzyme Assays

DNA repair enzymes were assayed as described previously (Hu et al., 1998). GST was assayed using the protocol described by Warholm et al. (1985). Glutathione reductase was assayed according to Styblo and Thomas (1995). Glutathione peroxidase activity was determined as described by Wendal et al. (1980).

RESULTS

Arsenic-induced Cytotoxicity in Human Epithelial Keratinocyte Cells

We have used neutral red dye uptake to measure arsenic cytoxicity under a variety of conditions in SV40-transformed AG06 human keratinocytes and normal human epithelial keratinocytes (NHEK). The relative survival of NHEK cells after exposure to As(III) is slightly greater than that of the AG06 cells or other cell types, such as human osteosarcoma (HOS) cells and normal human fibroblasts (WI38 cells) (Table 1). This difference is not related to the initial cellular GSH concentration (not shown), although arsenic toxicity in keratinocytes is strongly affected by GSH levels. Decreasing the cellular GSH concentration, by the inhibition of γ-glutamylcysteine synthetase (γ-GCS) activity with 250 μM L-buthionine-S,R-sulfoximine (BSO) or by blocking GSH function with 1.5 mM 2-chloroethanol (CHE) produced ᵧ·eater than a 10-fold increase in sensitivity to As(III) in AG06 cells. Conversely, pretreatment with 10 mM N-acetylcysteine (NAC) to increase GSH concentration had no effect on As sensitivity (Figure 1). Decreased levels of GSH also sensitize the NHEK cells to As-induced cytotoxicity (not shown). Note that, although pure As(III) or As(V) were used to treat the cells, it is likely that the As(V) was at least partially reduced to As(III) within the cells and thus toxicity

Fig. 1. The effect of GSH on As(III)-induced toxicity in AG06 cells. AG06 cells were plated in 96-well plates then pretreated for 24 hours with 10 mM N-acetyl cysteine (NAC), 1.5 mM 2-chloroethanol (CHE), or 250 μM buthionine sulfoximide (BSO) prior to incubating for 48 hours with As(III). Viability was determined by neutral red dye uptake. Results show the average of 8 wells (standard error <10%).

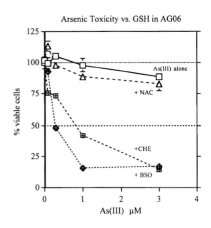

TABLE 1

Relative arsenic-induced cytotoxicity in different human cell cultures. Toxicity was measured by neutral red dye uptake after 72 (NHEK cells) or 96 hours treatment with As. ND = not determined

Cell type	IC$_{50}$ for dye uptake (μM)	
	As(III)	As(V)
HOS	3.5	11
AG06	1.1	16
NHEK	10.8	ND
WI38	8.8	30

attributed to As(V) could, in fact, be mediated by As(III). Likewise, although human keratinocytes methylate arsenic with very low efficiency (Styblo and Thomas, personal communication), it is possible that a small amount of methylated arsenic could contribute to the observed effects.

Enzyme Inhibition by Arsenic

It has been proposed that As(III) exerts cytotoxic as well as genotoxic effects by binding to sulfhydryls causing protein denaturation and inhibiting enzyme activity. A number of enzymes have been suggested as targets for As(III)-mediated enzyme inhibition, including DNA ligases, protein phosphatases, and various mitochondrial enzymes (Mitchell et al., 1971; Rein et al., 1979). In order to test this hypothesis we have examined the effects of As on the activity of several purified enzymes *in vitro*, including enzymes required for DNA repair and some related to GSH metabolism. Except for pyruvate dehydrogenase (PDH), with its lipoic acid cofactor, most purified enzymes we examined, including those listed in Table 2, are not inhibited by physiologically relevant, micromolar concentrations of As. Some were even activated by greater than millimolar As (Hu et al., 1998). The relative sensitivity of a variety of enzymes to As-induced inhibition is summarized in Table 2.

We have also measured the activity of some of these same enzymes in extracts of human cells treated with As in culture. In several cases we have found large differences between the enzyme sensitivity *in vitro* and the activity of the enzyme in the treated cells. Two different protocols were used to examine arsenic sensitivity in cell extracts: (i) the enzyme activity was determined in extracts obtained from untreated cells to which arsenic was added *in vitro*, and (ii) cells were treated in culture with various concentrations of As and then protein extracts

TABLE 2

The IC$_{50}$ for arsenic-induced inhibition of enzyme activity. The IC$_{50}$, that concentration of As required to inhibit enzyme activity by 50%, was determined for a variety of purified enzymes and enzyme activities in extracts from As-treated or untreated AG06 cells. Enzymes and sources: Pol β, purified DNA polymerase β (S. Wilson, NIEHS); DNA ligase I and III, purified cloned human DNA ligase (T. Lindahl, ICRF); PDH, pyruvate dehydrogenase, yeast GSSG reductase, bovine GSH peroxidase (GPx), and GSTπ are from commercial sources.

Enzyme	As(III)	As(V)
Pol β	212 mM	160 mM
Pol I-KF	59 mM	100 mM
DNA Ligase III	18 mM	116 mM
DNA Ligase I	6.3 mM	34 mM
Ligase (*in vitro*)[a]	6.5 mM	26 mM
GSSG Reductase	34 mM	>100 mM
GSH Peroxidase	0.13 mM	>>100 mM
GSTπ	≥1.8 mM	>50 mM
PDH	5.6 μM	206 mM
Ligase (AG06)[b]	14.5 μM	173 μM
GPx (AG06)[b]	2.0 μM	ND

[a]Enzyme activity in extracts from untreated AG06 cells assayed in the presence of As.
[b]Enzyme activity in extracts of AG06 cells pretreated for 24 h with As.
ND = not determined.

were made from the treated cells and assayed for enzyme activity (in the absence of additional As). Using these two protocols we found that: (i) The activity of DNA ligase in extracts from untreated AG06 cells was no more sensitive than the purified enzymes to inhibition by arsenic (Table 2). (ii) There was a significant increase in ligase activity after three hours of treatment with low doses of As, with As(III) giving a more pronounced effect than As(V). And (iii), after 24 hours of treatment with As, AG06 cells show a dose-dependent decrease in ligase activity (Figure 2). However, 50% reduction in ligase activity is seen only after a dose of As sufficient to kill most of the cells (compare Table 1 with Table 2). These results are similar to those previously reported by Li and Rossman (1989). We also found that glutathione peroxidase (GPx) activity is extremely sensitive to treatment of AG06 cells with low doses of As(III), although the purified enzyme is nearly 70 times less affected.

In contrast, we find there is a significant dose-dependent increase in GSH levels in AG06 cells after treatment with As(III) for 24 to 72 hours that is maximal at a dose of 3 μM and after 48 hours of treatment (Figure 3). This increase in GSH concentration is due primarily to an

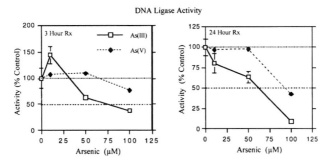

Fig. 2. DNA ligase activity in extracts from AG06 cells after treatment with As for 3 hours (left panel) or 24 hours (right panel). Each experiment was performed in duplicate (or quadruplicate, As(III), 3 hours) and the average (± s.e.m.) is shown.

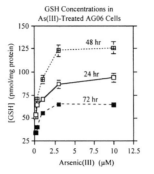
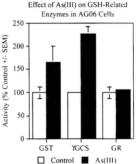

Fig. 3. Up-regulation of GSH and GSH-related enzymes in As(III)-treated AG06 cells. The left panel shows GSH concentrations (± s.e.m.) in AG06 cells plated in 8 replicate wells in a 96-well plate and treated with As(III) for the 24, 48, or 72 hours. The right panel shows the relative specific activity of 3 GSH-related enzymes (glutathione S-transferase-π, γ-glutamylcysteine synthetase, and GSSG reductase) in AG06 cells treated for 48 hours with 3 μM As(III). Triplicate dishes were assayed and the average activity per mg protein for each enzyme (± s.e.m.) was determined and compared to the activity in untreated control cells.

increase in the activity of γ-GCS. The specific activity of glutathione S-transferase (primarily GSTπ) is also increased after 24 hours treatment with As(III), although glutathione reductase (GR) is not affected (Figure 3). These results suggest that cellular treatment with As leads to changes in relative enzyme concentration within the cells or causes inactivation or activation of the enzymes by some means other than by direct As inhibition. We propose that As-induced modulation of these (and possibly other) enzyme activities is mediated by either (reversible) protein phosphorylation or transcriptional control. Preliminary data (not shown) suggest that maximal enzyme induction (in the case of the major human apurinic/apyrimid-inic (AP) endonuclease, hAPE, for example) occurs at subtoxic concentrations of As(III) and is mediated by transcriptional up-regulation.

Synergistic Toxicity Induced by As(III) in Human Keratinocytes

In addition to these changes in ligase function and cellular redox enzymes, we also see evidence of a synergistic interaction between As and DNA damaging agents. This is seen both for agents that cause DNA damage repaired by base excision repair (BER), such as alkylation damage induced by MNNG, and for UVB, which induces DNA damage that is repaired by nucleotide excision repair (NER) (not shown). AG06 cells were treated with As(III) for 24 hours prior to treatment with MNNG. The cells were then allowed to recover for an additional 24 hours and assayed for cytotoxicity by neutral red dye uptake (Figure 4).

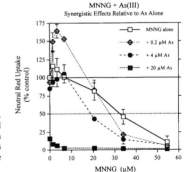

Fig. 4. Synergistic effects of As(III) plus MNNG in AG06 cells. Cells were seeded in 96-well plates for 24 hours, then treated with As(III) for an additional 24 hours prior to treatment with MNNG. After an additional 24 hours recovery, viability was measured by neutral red dye uptake. Each point is the average of 4 measurements (± s.e.m.).

Interestingly, low doses of As(III) (less than 1 μM) and MNNG (less than 4 μM) when given separately, both produce a slight, but reproducible, increase in neutral red dye uptake in the AG06 keratinocytes. However, when the cells are pretreated with a non-toxic dose of 0.2 μM As(III) for 24 hours prior to MNNG treatment the neutral red dye uptake is significantly increased ($p \leq 0.001$). This is an indication of either increased metabolic function (lysosomal activity) or increased cell number. Higher doses of both As(III) and MNNG produce a synergistic increase in toxicity, shown by a significant decrease in dye uptake relative to either treatment alone (Figure 4). These results show that it is not wise to try to predict low-dose effects of As by extrapolation down from a high dose.

DISCUSSION

DNA repair is critically involved in carcinogenesis at many levels: during the initiation stages and during both promotion and progression. DNA nucleotide excision repair takes place in four steps: damage recognition, DNA strand cleavage, resynthesis of the damaged strand, and ligation (Wood, 1996). Damage recognition and strand cleavage follows two separate pathways depending on the type of DNA damage. Base excision repair utilizes DNA glycosylases to recognize and remove base damage induced by most methylating agents, many types of oxidative DNA damage, and other types of endogenous damage. This pathway results in the removal of only one or two nucleotides at the site of damage and uses DNA polymerase β for resynthesis (Singhal et al., 1995). Bulky lesions, such as UV damage, are recognized by the nucleotide excision repair pathway which incises the damaged DNA strand on either side of the lesion leaving a gap of approximately 29 nucleotides. DNA polymerase δ or ϵ then fill the gap and a DNA ligase seals it (Wood, 1996; Nicholl et al., 1997). Ligation and polymerization are common to both NER and BER, although different enzymes may be used for the different pathways. Arsenic has been shown to inhibit DNA repair in bacteria (Rossman et al., 1977) and both BER and NER in mammalian cells in culture (Okui and Fujiwara, 1986; Li and Rossman, 1989; Lee-Chen et al., 1993; Hartmann and Speit, 1996; Hartwig et al., 1997; Lynn et al., 1997). However, it is not clear which steps of these multi-step repair processes are most affected by As.

We show here that treatment of human keratinocyte cells with low concentrations of As(III) can cause a transient increase, followed by a dose-dependent decrease, in DNA ligase activity. We also show a dose-dependent increase in cellular GSH levels and changes in enzyme activity that do not correlate with direct enzyme inhibition by As. Li and Rossman (1989) were the first to show inhibition of ligase activity in Chinese hamster V79 cells. Hartwig et al. (Hartwig et al., 1997) recently showed that DNA repair synthesis is also inhibited by low concentrations of As(III). There is no evidence to indicate that the recognition or incision steps of the repair pathways are likewise affected by As. Yet, as we have shown in Table 2, relevant concentrations of As do not directly inhibit most human DNA repair enzymes, including the major ligases, DNA ligase I or ligase III (Hu et al., 1998). This suggests that As must modulate DNA repair capacity in an indirect manner such as by transcriptional or post-translational regulation of the gene products.

The regulation of DNA repair in mammalian cells is not well understood. Many of the genes responsible for BER have only recently been cloned and characterized (Wood, 1996). In the case of NER dozens of genes have been cloned and their interaction and regulation is the subject of intense research effort. The data that are now available suggest that BER is regulated, at least in part, by stress-induced transcriptional control of the repair genes. NER appears to be regulated primarily via protein–protein interactions with components of the transcription factor TFIIH, p53, and other nuclear proteins. Several proteins essential for various steps in the repair process, e.g., the p34 protein of RPA (Ariza et al., 1996), p53 (Steegenga et al., 1996), and DNA ligase I (Prigent et al., 1992), can also be activated by reversible phosphorylation. With the exception of the recent report by Li et al. (1997) there have been few studies on the regulation of DNA repair in human keratinocytes. The data shown here suggest that arsenic affects the control of repair, and other DNA damage responses. Changes in the repair of DNA damage, such as that induced by UVB, may play an important role in the co-mutagenicity and toxicity of arsenic and may be a critical reason why arsenic functions as a tumor promotor or progressor.

Redox levels may also regulate aspects of nucleotide excision repair. Low doses of As(III) can induce the synthesis of GSH and alter the expression of at least three different GSH metabolizing enzymes. Subtoxic to moderately toxic doses of As(III) are most effective in modulating these responses (not shown). This is quite different from most previous studies showing arsenic-induced alterations in gene expression and protein phosphorylation in cells treated with very toxic doses of arsenic, in the range of 20 to 200 μM (Mivechi et al., 1994; Vietor and Vilcek 1994; Cavigelli et al., 1996; Guyton et al., 1996; Liu et al., 1996). The results presented here have been obtained with doses of As that are relevant to human carcinogenesis. Future investigations will determine the mechanisms by which As regulates these critical cellular processes.

ACKNOWLEDGEMENT

Supported by the Electric Power Research Institute (Agreement No. WO3370-22), the US-EPA (Grant No. 96-NCERQA-14), NYU/NIEHS Center Grant (ES00260), and the NYU Medical Center Kaplan Comprehensive Cancer Center (CA13343).

REFERENCES

Ariza, R.R., Keyse, S.M., Moggs, J.G., Wood, R.D. 1996. Reversible protein phosphorylation modulates nucleotide excision repair of damaged DNA by human cell extracts. *Nucl. Acid Res.*, **24**(3), 433–440.

Barchowsky, A., Dudek, E.J., Treadwell, M.D., Wetterhahn, K.E. 1996. Arsenic induces oxidant stress and NF-kappa B activation in cultured aortic endothelial cells. *Free Radical Biol. Med.*, **21**(6), 783–790.

Barrett, J.C., Lee, T-C. 1991. Mechanisms of arsenic-induced gene amplification. In: R.E. Kellems (ed.), *Gene Amplification in Mammalian Cells: Techniques and Applications*. Marcel Dekker, Inc., New York.

Cavigelli, M., Li, W.W., Lin, A.N., Su, B., Yoshioka, K., Karin, M. 1996. The tumor promoter arsenite stimulates AP-1 activity by inhibiting a JNK phosphatase. *EMBO J.*, **15**(22), 6269–6279.

Germolec, D.R., Spalding, J., Boorman, G.A., Wilmer, J.L., Yoshida, T., Simeonova, P.P., Bruccoleri, A., Kayama, F., Gaido, K., Tennant, R., Burleson, F., Dong, W.M., Lang, R.W., Luster, M.I. 1997. Arsenic can mediate skin neoplasia by chronic stimulation of keratinocyte-derived growth factors. *Mutat. Res. Rev. Mutat. Res.*, **386**(3), 209–218.

Germolec, D.R., Yoshida, T., Gaido, K., Wilmer, J.L., Simeonova, P.P., Kayama, F., Burleson, F., Dong, W.M., Lange, R.W., Luster, M.I. 1996. Arsenic induces overexpression of growth factors in human keratinocytes. *Toxicol. Appl. Pharmacol.*, **141**(1), 308–318.

Guyton, K.Z., Xu, Q.B., Holbrook, N.J. 1996. Induction of the mammalian stress response gene GADD153 by oxidative stress: role of AP-1 element. *Biochem. J.*, **314**(Part 2), 547–554.

Hartmann, A., Speit, G. 1996. Effect of arsenic and cadmium on the persistence of mutagen-induced DNA lesions in human cells. *Environ. Mol. Mutagen.*, **27**(2), 98–104.

Hartwig, A., Groblinghoff, U.D., Beyersmann, D., Natarajan, A.T., Filon, R., Mullenders, L.H.F. 1997. Interaction of arsenic(III) with nucleotide excision repair in UV-irradiated human fibroblasts. *Carcinogenesis*, **18**(2), 399–405.

Hu, Y., Su, L., Snow, E.T. 1998. Arsenic toxicity is enzyme specific and arsenic inhibition of DNA repair is not caused by direct inhibition of repair enzymes. *Mutat. Res.*, **408**, 203–218.

Huang, H., Huang, C., Wu, D., Jinn, C., Jan, K. 1993. Glutathione as a cellular defence against arsenite toxicity in cultured Chinese hamster ovary cells. *Toxicology*, **79**(3), 195–204.

Kato, K., Ito, H., Okamoto, K. 1997. Modulation of the arsenite-induced expression of stress proteins by reducing agents. *Cell Stress Chaperones*, **2**(3), 199–209.

Kavanagh, T.J., Raghu, G., White, C.C., Martin, G.M., Rabinovitch, P.S., Eaton, D.L. 1994. Enhancement of glutathione content in glutathione synthetase-deficient fibroblasts from a patient with 5-oxoprolinuria via metabolic cooperation with normal fibroblasts. *Exp. Cell Res.*, **211**, 69–76.

Lee-Chen, S., Gurr, J., Lin, I., Jan, K. 1993. Arsenite enhances DNA double-strand breaks and cell killing of methyl methanesulfonate-treated cells by inhibiting the excision of alkali-labile sites. *Mutat. Res.*, **294**(1), 21–8.

Li, G., Ho, V.C., Mitchell, D.L., Trotter, M.J., Tron, V.A. 1997. Differentiation-dependent p53 regulation of nucleotide excision repair in keratinocytes. *Am. J. Pathol.*, **150**(4), 1457–1464.

Li, J-H., Rossman, T.G. 1989. Inhibition of DNA ligase activity by arsenite: A possible mechanism of its comutagenesis. *Molec. Toxicol.*, **2**, 1–9.

Li, W., Chou, I-N. 1992. Effects of sodium arsenite on the cytoskeleton and cellular glutathione levels in cultured cells. *Tox. Appl. Pharmacol.*, **114**, 132–139.

Little, M.C., Gawkrodger, D.J., Macneil, S. 1996. Chromium- and nickel-induced cytotoxicity in normal and transformed human keratinocytes: an investigation of pharmacological approaches to the prevention of Cr(VI)-induced cytotoxicity. *Br. J. Dermatol.*, **134**, 199–207.

Liu, Y.S., Guyton, K.Z., Gorospe, M., Xu, Q.B., Lee, J.C., Holbrook, NJ 1996. Differential activation of ERK, JNK/SAPK and p38/CSBP/RK map kinase family members during the cellular response to arsenite. *Free Radical Biol. Med.*, **21**(6), 771–781.

Ludwig, S., Hoffmeyer, A., Goebeler, M., Kilian, K., Hafner, H., Neufeld, B., Han, J.H., Rapp, U.R. 1998. The stress inducer arsenite activates mitogen-activated protein kinases extracellular signal-regulated kinases 1 and 2 via a MAPK kinase 6/p38-dependent pathway. *J. Biol. Chem.*, **273**(4), 1917–1922.

Lynn, S., Lai, H.T., Gurr, J.R., Jan, K.Y. 1997. Arsenite retards DNA break rejoining by inhibiting DNA ligation. *Mutagenesis*, **12**(5), 353–358.

Mitchell, R.A., Change, B.F., Huang, C.H., DeMaster, E.G. 1971. Inhibition of mitochondrial energy-linked functions by arsenate. *Biochemistry*, **10**, 2049–2054.

Mivechi, N.F., Koong, A.C., Giaccia, A.J., Hahn, G.M. 1994. Analysis of HSF-1 phosphorylation in A549 cells treated with a variety of stresses. *Int. J. Hyperthermia*, **10**(3), 371–379.

Nicholl, I.D., Nealon, K., Kenny, M.K. 1997. Reconstitution of human base excision repair with purified proteins. *Biochemistry*, **36**(24), 7557–7566.

Okui, T., Fujiwara, Y. 1986. Inhibition of human excision DNA repair by inorganic arsenic and the comutagenic effect in V79 Chinese hamster cells. *Mutat. Res.*, **172**(1), 69–76.

Prasad, R., Singhal, R., Srivastava, D., Molina, J., Tomkinson, A., Wilson, S. 1996. Specific interaction of DNA polymerase beta and DNA ligase I in a multiprotein base excision repair complex from bovine testis. *J. Biol. Chem.*, **271**(27), 16000–7.

Prigent, C., Lasko, D., Kodama, K., Woodgett, J., Lindahl, T. 1992. Activation of mammalian DNA ligase I through phosphorylation by casein kinase II. *EMBO J.*, **11**(8), 2925–33.

Rein, K.A., Borreback, B., Bremer, J. 1979. Arsenite inhibits oxidation in isolated rat liver mitochondria. *Biochim. Biophys. Acta*, **574**, 487–494.

Rossman, T.G., Meyn, M.S., Troll, W. 1977. Effects of arsenite on DNA repair in *Escherichia coli*. *Environ. Health Perspect.*, **19**, 229–233.

Rossman, T.G., Stone, D., Molina, M., Troll, W. 1980. Absence of arsenite mutagenicity in E. coli and Chinese hamster cells. *Environ. Mutagen.* **2**, 371–379.

Rossman, T.G., Wolosin, D. 1992. Differential susceptibility to carcinogen-induced amplification of SV40 and *dhfr* sequences in SV40-transformed human keratinocytes. *Molec. Carcinogen.*, **6**, 203–213.

Scott, N., Hatlelid, K.M., MacKenzie, N.E., Carter, D.E. 1993. Reactions of arsenic(III) and arsenic(V) species with glutathione. *Chem. Res. Toxicol.*, **6**, 102–106.

Singhal, R., Prasad, R., Wilson, S. 1995. DNA polymerase beta conducts the gap-filling step in uracil-initiated base excision repair in a bovine testis nuclear extract. *J. Biol. Chem.*, **270**(2), 949–57.

Snow, E.T. 1992. Metal carcinogenesis: mechanistic implications. *Pharmacol. Therapeut.*, **53**, 31–65.

Steegenga, W.T., Vandereb, A.J., Jochemsen, A.G. 1996. How phosphorylation regulates the activity of p53. *J. Mol. Biol.*, **263**(2), 103–113.

Styblo, M., Thomas, D. 1995. In vitro inhibition of glutathione reductase by arsenotriglutathione. *Biochem. Pharmacol.*, **49**(7), 971–7.

Szinicz, L., Forth, W. 1988. Affect of As_2O_3 on gluconeogenesis. *Arch. Toxicol.*, **61**, 444–449.

Tietze, E. 1969. Enzymatic method for quantitative determination of nanogram amounts of total and ozidized glutathione: Applications to mammalian blood and other tissues. *Anal. Biochem.*, **27**, 502–522.

Vietor, I., Vilcek, J. 1994. Pathways of heat shock protein 28 phosphorylation by TNF in human fibroblasts. *Lymphokine Cytokine Res.*. **13**(5), 315–323.

Wang, Z., Rossman, T.G. 1996. The Carcinogenicity of Arsenic. In: L. W. Chang (ed.), *Toxicology of Metals*. pp. 219–227. CRC Press, Boca Raton, FL.

Warholm, M., Guthenberg., C., von Bahr., C., Mannervik, B. 1985. Glutathione transferases from human liver. *Meth. Enzymol.* **113**, 499–504.

Wendel, A. 1980. Glutathione peroxidase. (ed.), *Enzymatic Basis of Detoxication*, Vol. 1, pp. 333–353. Academic Press.

Wood, R.D. 1996. DNA repair in eukaryotes. *Annu. Rev. Biochem.*, **65**, 135–167.

Arsenic Exposure and Health Effects
W.R. Chappell, C.O. Abernathy and R.L. Calderon (Editors)

Evaluation of Cell Proliferative Activity in the Rat Urinary Bladder After Feeding High Doses of Cacodylic Acid

Samuel M. Cohen, Lora L. Arnold, Margaret K. St. John, Martin Cano

ABSTRACT

Cacodylic acid fed at relatively high doses in the diet (100 ppm) produces an increased incidence of bladder tumors in rats, with the effect greater in females than in males. No similar urothelial changes are seen in mice. At similar doses, cacodylic acid also enhances bladder tumor formation following prior administration of N-butyl-N-(4-hydroxybutyl)nitrosamine. The weight of the evidence strongly suggests that cacodylic acid does not produce these changes by direct interactions with DNA. Possible increased cell proliferative effects secondary to the dietary administration of cacodylic acid fed for 10 weeks were evaluated in female F344 rats. Proliferative activity was evaluated by light and scanning electron microscopy and by bromodeoxyuridine labeling index. Hyperplasia and significantly increased labeling index occurred at doses of 40 and 100 ppm but not at 2 or 10 ppm. Significant changes of necrosis and proliferation were detectable by scanning electron microscopy at the doses of 40 and 100 ppm. Urinary changes included increased volume with decreased osmolality and creatinine, but urinary calcium was increased. There was no urinary precipitate, microcrystalluria or calculi detected related to the administration of cacodylic acid. Increased calcification occurred in the kidneys. These studies show that orally administered high doses of cacodylic acid produce urothelial toxicity and regeneration in female rats, which likely contribute to the ultimate development of a low incidence of bladder tumors.

Keywords: cacodylic acid, bladder, hyperplasia

INTRODUCTION

Cacodylic acid (dimethylarsinic acid) was found to produce an increased incidence of bladder tumors in rats when fed in the diet in a two year bioassay (van Gemert and Eldan, 1998). Tumors and proliferative lesions were produced at doses of 40 and 100 ppm of the diet, but significant incidences of bladder lesions were not observed at doses of 2 or 10 ppm. The results in rats showed that there was a higher incidence of bladder lesions in females than in males. No bladder tumors or other treatment-related tumors were produced in mice in a two year bioassay. Cacodylic acid has also been shown to increase the incidence of bladder tumors when administered after a brief exposure to a known bladder carcinogen, N-butyl-N-(4-hydroxybutyl)nitrosamine (BBN) (Wanibuchi et al., 1996).

The mechanism by which cacodylic acid produces bladder cancer in rats is not known. However, there is extensive evidence that cacodylic acid and other arsenicals do not appear to react directly with DNA or form DNA adducts (International Agency for Research on Cancer, 1980; Abernathy et al., 1996; Byrd et al., 1996). Extensive research has strongly suggested that a non-linear, possibly threshold-related mode of action is present for the carcinogenicity of these arsenicals, similar to processes known for other non-DNA reactive chemicals (Wang and Rossman, 1996; Abernathy et al., 1996; Eastern Research Group, 1997).

A common mechanism involved with the carcinogenicity of non-DNA reactive chemicals is the production of increased cell proliferation, which can be produced by a variety of processes (Cohen and Ellwein, 1990; 1991). Increased cell births occur secondary to toxicity with consequent cell regenerative hyperplasia or to direct mitogenic stimulation (Cohen, 1997; 1998). Mechanisms of toxicity related to the urothelium include production of urinary solids, such as calculi, microcrystalluria, or precipitate formation, or toxicity due to the administered chemical or a metabolite (Cohen, 1998). Chemicals involved in the production of urinary solids are effective only at high doses, and the effect tends to be greater in rats than in mice (Clayson et al., 1995). Production of a calcium phosphate-containing precipitate following administration of high doses of sodium salts, such as sodium saccharin or sodium ascorbate, appear to be specific to the rat in addition to being a high-dose phenomenon. Other mechanisms for production of toxicity of the bladder include chemical irritation of the urothelium, either producing erosion and ulceration of the urothelium (e.g. tributyl phosphate, acetic acid), or a milder cytotoxicity involving necrosis of the superficial and intermediate cell layers (e.g. ortho-phenylphenol) of the bladder epithelium (Arnold et al., 1997; Smith et al., 1998; Cohen, 1998). Urinary changes secondary to administration of a variety of chemicals can affect the toxicity of those chemicals (Cohen, 1995). This has been examined extensively in the rat, and includes changes such as urinary pH, protein, and calcium.

Toxicity and cell proliferation in the rat bladder can be evaluated by light microscopy and by bromodeoxyuridine labeling index (Cohen et al., 1990). In addition, scanning electron microscopic observation of the surface of the bladder epithelium provides a sensitive technique for assessing more subtle changes of necrosis and hyperplasia (Cano et al., 1993; Cohen et al., 1990). Utilizing these techniques, we have evaluated the various urinary and urothelial changes following administration of different doses of cacodylic acid to female rats. Females were examined since they were more sensitive to the urothelial tumorigenic effects of cacodylic acid than the male rat in the two year bioassay. For our experiment, we fed cacodylic acid to the rats for ten weeks at the same doses fed in the two year rat bioassay. It has been our experience that by ten weeks, toxic and proliferative changes are evident if they are part of the response to an administered chemical (Cohen et al., 1990).

METHODOLOGY

Chemical

Cacodylic acid was received from Luxembourg Industries (Pamol), LTD. (Tel-Aviv, Israel). The purity of the test article was documented by Luxembourg Industries and confirmed by NMR at our facility. It was fed in the diet at levels of 2, 10, 40, and 100 mg/kg.

Diets and Test Animals

The use of the rats was approved by the University of Nebraska Medical Center Institutional Animal Care and Use Committee. Seventy-three female F344 rats, four weeks old at the time of arrival, were purchased from Charles River Breeding Laboratories, Inc. (Raleigh, NC). Three of the rats were ordered for health surveillance and sacrificed during week 8 of the study. One day after arrival, the rats were randomized into five groups by a weight stratification program and placed on pelleted Purina Mills Certified Rodent Lab Chow 5002 (St. Louis, MO). Groups 1 and 5 had 20 rats each and the remaining groups (2–4) had 10 rats each. The rats were housed five per cage in polycarbonate cages with dry corn-cob bedding in a room with a targeted temperature of 71°C and humidity of 50% and a 12 hour light/dark cycle (i.e. light on/off at 0600 and 1800 hours). Administration of the test article at the doses mentioned above in pelleted Purina 5002 (considered as day 0 of the experiment) began after seven days of acclimation. Dietary concentrations and stability were determined by Dr. William Cullen (University of British Columbia) by inductively coupled plasma mass spectroscopy following methanol:water (1:1) extraction and high pressure liquid chromatography. Food and tap water were available *ad libitum* throughout the study. Food and water consumptions were measured during study weeks 2, 6 and 10, and rats were weighed on day 0 of the experiment, at the end of each food and water consumption period, before and after placement in metabolism cages, and on the day of sacrifice. Detailed clinical observations were done on day 0 of the experiment and at the end of each food and water consumption period. Ten rats in group 1 and group 5 were sacrificed during week 9 of the study to determine if treatment effects were present in the urothelium. The remaining ten rats in each group were scheduled for sacrifice after 10 weeks of treatment based on the results of examination of bladders from the week 9 sacrifice.

Urine Collection and Analyses

Fresh voided urine was collected from all rats between 0700 and 0900 hours during weeks 4 and 9 directly into a microcentrifuge tube. Urinary pH was measured using a microelectrode (Fisher et al., 1989). The urines were centrifuged and the supernatants removed. Urinary filters were prepared by reconstituting the precipitate in 10 μl of distilled water and transferring the solution to a 0.22 μM filter (Millipore Corp., Bedford, MA) affixed to an aluminum stub. During week 5 all rats were acclimated to metabolism cages for 48 hours and then 24 hour urine samples were collected for determination of volume and sodium (ion selective electrode) (Tietz et al., 1986), creatinine (coupled enzyme method) (Mauck et al., 1986), calcium (Sundberg and Dappen, 1979), magnesium (Smith-Lewis et al., 1986) and phosphorus (Fiske and Subbarow, 1925) (colorimetric methods) on the Vitros 250 (Johnson & Johnson Clinical Diagnostics, Inc., Rochester, NY).

Animal Sacrifice, Necropsy, and Tissue Processing

Rats were sacrificed with an overdose of Nembutal. One hour prior to anesthesia rats were injected intraperitoneally with bromodeoxyuridine (BrdU) (Sigma Chemical Co., St. Louis, MO), 100 mg/kg body weight. At necropsy the urinary bladder and stomach were inflated *in situ* with Bouin's fixative and then placed in this fixative. Kidneys were removed, weighed and placed in formalin. After fixation the bladders were rinsed with 70% ethanol, cut in half longitudinally, weighed and examined macroscopically for abnormalities. One half of the bladder was processed for examination by scanning electron microscopy (SEM) and classified according to previously established criteria (Cohen, et al., 1990).

Briefly, class one is for bladders that show uniform, large, flat polygonal cells without necrosis or exfoliation; class 2 has occasional, small foci of necrotic and/or exfoliated cells; class 3 has larger and more numerous foci; class 4 has extensive areas of necrosis and exfoliation; and class 5 has piling up of rounded cells (hyperplasia).

The other half of the bladder was cut longitudinally into four strips and processed for microscopic histopathological examination with hematoxylin and eosin staining using criteria described by Cohen (1983). Sections of kidneys were stained with hematoxylin and eosin for histopathologic evaluation. Sections of kidneys and bladders were also stained by the von Kossa stain (Sheehan and Hrapchak, 1980) to evaluate the presence of calcium.

Statistical Analyses

Body and tissue weights, food and water consumptions, urinary pH, osmolality and other chemistries, and the labeling indices were compared by Duncan's multiple range test. Histopathology was compared using the 2 tail, Fisher's exact test, and the scanning electron microscopy results were compared using nonparametric analysis of variance. All analyses were made using a SAS program (SAS Institute, 1996).

RESULTS

The body weights were comparable among the different treatment groups throughout most of the study with a statistically significant increase in the high-dose group at week 10 (Table 1). Water consumption increased in a dose responsive manner as the concentration of cacodylic acid in the diet increased. Food consumption was somewhat increased in Group 5 compared to the control group.

Urine Chemistries

Urinary volume was significantly increased in the 40 and 100 mg/kg groups compared to controls (Group 1) with a corresponding decrease in creatinine concentration (Table 2). Sodium concentration was significantly decreased in the two high-dose groups with a slight, but statistically insignificant decrease in phosphorus and magnesium in these two groups. Calcium concentration was increased in the 40 and 100 ppm groups even though there was an overall urinary dilution. The increase in calcium concentration was statistically significant in the 100 ppm group and was statistically significant at the doses of 40 and 100 ppm when the concentrations are normalized for creatinine concentration (Tables 2 and 3). At week 4 urinary pH was comparable in all groups but by week 9 the pH was significantly increased in the 100 mg/kg group compared to controls (Table 4). There was no evidence of calculi, microcrystalluria or formation of calcium phosphate-containing precipitate in the urine related to the administration of cacodylic acid.

TABLE 1

Body weights, water and food consumption during week 10 of treatment with cacodylic acid

Group	Cacodylic Acid (mg/kg)	Body Weight (g)	Water Consumption (g/rat/day)	Food Consumption (g/rat/day)
1	0	169±2.6	19±0.2	10±0.03
2	2	175±1.3	21±0.7	11±0.33
3	10	172±2.5	22±1.2	11±0.04
4	40	169±3.1	23±1.9	11±0.51
5	100	178±2.2*	26±1.2*	12±0.06*

*Significantly different from control group, $p < 0.05$.

TABLE 2

Twenty-four hour urine chemistries during week 5 of treatment with cacodylic acid

Group	Cacodylic Acid (mg/kg)	Volume (ml)	Creatinine (mg/dl)	Sodium (mEq/L)	Calcium (mg/dl)	Magnesium (mg/dl)	Phosphorus (mg/dl)
1	0	6±0.4	68±3.3	154±7.3	15.6±1.3	40±3.6	149±13.5
2	2	7±0.6	66±5.7	133±8.0*	15.3±0.9	41±4.1	164±17.4
3	10	7±0.4	65±3.8	147±9.3	15.5±0.9	43±3.5	173±9.7
4	40	9±0.5*	52±2.3*	120±5.2*	17.7±1.4	35±3.4	142±8.7
5	100	10±0.5*	48±2.1*	117±5.0*	21.5±2.0*	35±4.1	132±8.7

*Significantly different from control group, $p < 0.05$.

TABLE 3

Sodium and calcium concentrations normalized for creatinine concentration (week 5)

Group	Cacodylic Acid (mg/kg)	Volume (ml)	Sodium/Creatinine (mEq/mg)	Calcium/Creatinine (mg/mg)
1	0	6±0.4	0.23±0.01	0.25±0.03
2	2	7±0.6	0.21±0.01*	0.26±0.02
3	10	7±0.4	0.23±0.01	0.25±0.02
4	40	9±0.5*	0.23±0.01	0.35±0.03*
5	100	10±0.5*	0.25±0.01	0.47±0.05*

*Significantly different from control group, $p < 0.05$.

TABLE 4

pH of fresh voided urine

Group	Cacodylic Acid (mg/kg)	Week 4 (Mean ±S.E.)	Week 9 (Mean ±S.E.)
1	0	6.7 ±0.1	6.8 ±0.1
2	2	6.6 ±0.1	7.2 ±0.1*
3	10	6.6 ±0.1	7.0 ±0.2
4	40	6.5 ±0.1	7.3 ±0.2*
5	100	6.4 ±0.1*	7.2 ±0.1*

*Significantly different from control group, $p < 0.05$.

Bladder Histopathology, Labeling Index and SEM Results

Simple hyperplasia of the bladder epithelium was present in nine of ten rats in the high dose group, ranging in severity from mild to moderate, and it was also present in the bladders of rats fed 40 ppm cacodylic acid (Table 5). Blood was present on the epithelial surface of some bladders in Groups 4 and 5, and there was extensive epithelial vacuolization (Figures 1 and 2). The BrdU labeling index was significantly increased in the bladders from rats at the two highest doses. It was slightly increased at the 10 ppm dose but not statistically significant. SEM examination of the urothelium showed extensive necrosis with exfoliation of the large, flat polygonal cells normally present on the epithelial surface and piling up of the small round cells with some pleomorphic microvilli (Figures 3 and 4). Changes by SEM were clearly present at the two higher doses. Changes seen by SEM in the bladders of rats fed the 2

Fig. 1. Bladder urothelial hyperplasia with focal vacuolization (arrows), some of which contain red cells; from a female rat fed 100 ppm cacodylic acid in Purina diet for 10 weeks, ×400.

Fig. 2. Bladder urothelial simple hyperplasia in a rat fed 100 ppm cacodylic acid in Purina diet for 10 weeks, ×400.

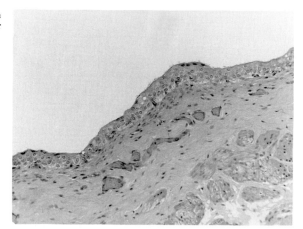

Fig. 3. Bladder surface from a female rat fed 100 ppm cacodylic acid in Purina diet for 10 weeks, showing necrosis and exfoliation, ×106.

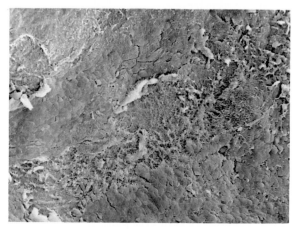

Fig. 4. Bladder surface from a female rat fed 100 ppm cacodylic acid in Purina diet for 10 weeks, showing piling up of round cells indicative of hyperplasia, ×300.

Fig. 5. Kidney showing extensive calcification at the corticomedullary junction; from a female rat fed 100 ppm cacodylic acid in Purina diet for 10 weeks von Kossa stain, ×100.

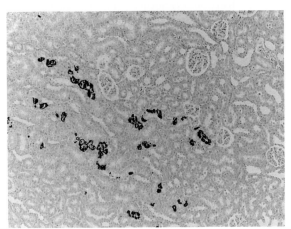

TABLE 5

Bladder histopathology and labeling index after 10 weeks of treatment with cacodylic acid

Group	Cacodylic Acid (mg/kg)	Normal	Simple Hyperplasia	Labeling Index (Mean ±S.E.)
1	0	9	1	0.22±0.04
2	2	10	0	0.20±0.03
3	10	10	0	0.33±0.08
4	40	6	4	0.95±0.15*
5	100	1	9*	0.90±0.11*

*Significantly different from control group, $p < 0.05$.

ppm dose are within the ranges we have normally seen in controls (SEM classes 1 to 3) (Table 6). Kidneys were essentially normal histopathologically except for increased focal calcification of the tubules at the corticomedullary junction, occasionally extending into the cortex, but without inflammation (Figure 5). Stomachs showed no abnormalities.

TABLE 6

Scanning electron microscopic classification of urothelial lesions after 10 weeks of treatment with cacodylic acid

Group	Cacodylic Acid (mg/kg)	SEM Class				
		1	2	3	4	5
1	0	5(–)*	5(7)	–(3)	–(–)	–(–)
2	2	–	4	5	1	–
3	10	–	2	5	3	–
4	40	–	5	3	2	–
5	100	–(–)	–(–)	–(–)	4(2)	6(8)

*Numbers in parentheses are for the rats killed at 9 weeks in groups 1 and 5.

DISCUSSION

Based on our present experiment, it is apparent that cacodylic acid administered in the diet to female rats causes cytotoxicity of the urinary bladder epithelium with a consequent regenerative hyperplasia. Taking into account the observations made by light and scanning electron microscopy and the BrdU labeling index, the toxicity and proliferation occurred significantly at doses of 40 and 100 ppm. The cytotoxic and proliferative changes were similar to those observed with other chemicals producing bladder cancer in rats, such as sodium saccharin, sodium ascorbate, and ortho-phenylphenol (Smith et al., 1998; Cohen, 1998).

Careful examination of the urine, including evaluation by scanning electron microscopy, did not show any evidence of calculous formation, microcrystalluria, or formation of calcium phosphate-containing precipitate. It is unlikely that the formation of solid materials in the urine is the cause of the cytotoxicity following the administration of cacodylic acid in the diet.

There was a dose-responsive increase in urinary volume, possibly secondary to increased water consumption. The increased urinary volume was associated with decreased urinary osmolality and urinary creatinine concentrations. Most of the urinary chemistries that we examined also showed decreased concentrations, as expected for diluted urine secondary to an increased urinary volume. However, there was a striking increase in the calcium concentration in the urine, and when this was normalized to creatinine concentration, as an indicator of the extent of urinary dilution and increased volume, the absolute amount of calcium being excreted was markedly increased in the urine of the rats administered cacodylic acid in the diet. This occurred significantly at doses of 40 ppm and 100 ppm. The increased excretion of calcium in the urine is the likely basis for the marked increased calcium deposition in the kidney of the rats administered cacodylic acid at the two highest doses.

Based on our experiments, the urothelial toxicity does not appear to be secondary to the formation of urinary solids. Other possible mechanisms for the cytotoxicity that occurred following administration of cacodylic acid in the diet to rats include alterations observed in calcium metabolism, including marked changes in the urinary calcium concentrations (Cohen, 1995), or cytotoxic effects of the cacodylic acid itself or one of its metabolites. An effect due to calcium and an effect due to toxicity from cacodylic acid and/or its metabolites are not mutually exclusive. Extensive analysis of a variety of arsenicals, including cacodylic acid, have provided strong evidence that these agents do not react with DNA or form DNA adducts (International Agency for Research on Cancer, 1980; Abernathy et al., 1996; Byrd et al., 1996; Rossman, 1997). Thus, a DNA reactive mechanism for the effects of cacodylic acid in the rat bladder is highly unlikely.

Based on the unlikely possibility of DNA reactivity, most investigators have concluded that the carcinogenic effects of cacodylic acid and other arsenicals will have a significantly non-linear dose response (Abernathy et al., 1996; Eastern Research Group, 1997). The present experiment provides strong support for such a nonlinearity. The carcinogenic effect in the rat bladder is likely a consequence of a long term regenerative hyperplasia secondary to cytotoxicity of the bladder epithelium. The dose response that was present in this experiment was similar to that observed in long term bioassays in rats with neoplasia as the end point. Our findings are similar to those reported by Dr. Shoji Fukushima and his colleagues (Wanibuchi et al., 1996). Their experiments have utilized administration of the cacodylic acid in the drinking water, but the dose response appears similar.

Extensive biochemical research has suggested that methylation is a critical process in the biological effects of arsenicals (Aposhian, 1997). However, although a trimethyl metabolite can be produced from cacodylic acid, it is unlikely that methylation can explain its effects on the urothelium. More importantly, the metabolism of cacodylic acid and other arsenicals appears to be considerably different qualitatively and quantitatively in rats than in humans (Aposhian, 1997), leaving some investigators to suggest that investigations in rats are not appropriate for risk assessment purposes in humans.

ACKNOWLEDGEMENTS

We gratefully acknowledge Dr. William Cullen for the chemical analyses of the diet, Traci Anderson for her dedicated, excellent assistance with these studies, Michelle Moore for her assistance in the preparation of this manuscript, and Dr. Marcia van Gemmert for her expert review and critique of the manuscript.

REFERENCES

Abernathy, C.O., Chappel, W.R., Meek, M.E., Gibb, H., Guo, H.R. 1996. Roundtable Summary — Is Ingested Inorganic Arsenic a "Threshold" Carcinogen? *Fund. Applied Toxicol.*, **29**, 168–175.

Aposhian, H.V. 1997. Enzymatic methylation of arsenic species and other new approaches to arsenic toxicity. *Annu. Rev. Pharmacol. Toxicol.*, **37**, 397–419.

Arnold, L.A., Christenson, R., Cano, M., St. John, M.K., Wahl, B.S., Cohen, S.M. 1997. Tributyl phosphate effects on urine and bladder epithelium in male Sprague-Dawley rats. *Fund. Applied Toxicol.*, **40**, 247–255.

Byrd, D.M., Roegner, M.L., Griffiths, J.C., Lamm, S.H., Grumski, K.S., Wilson, R., Lai, S. 1996. Carcinogenic risks of inorganic arsenic in perspective. *Int. Arch. Occup. Environ. Health*, **68**, 484–494.

Cano, M., Suzuki, T., Cohen, S.M. 1993. Application of scanning electron microscopy and x-ray analysis to urinary tract cancer in animals and humans. *Scanning Microscopy*, **7**, 363–370.

Clayson, D.B., Fishbein, L., Cohen, S.M. 1995. The effect of stones and other physical factors on the induction of rodent bladder cancer. *Fund. Chem. Toxicol.*, **33**, 771–784.

Cohen, S.M. 1983. Pathology of experimental bladder cancer in rodents. In: Cohen, S.M,. and Bryan, G.T. (eds.), *The Pathology of Bladder Cancer*, Vol. II, pp.1–40. CRC Press, Boca Raton, FL.

Cohen, S.M. 1995. The role of urinary physiology and chemistry in bladder carcinogenesis. *Fund. Chem. Toxicol.*, **33**, 715–730.

Cohen, S.M. 1997. The role of cell proliferation in the etiology of neoplasia. In: G.T. Bowden and S.M. Fisher (Eds.), *Chemical Carcinogens and Anticarcinogens, Volume 12, Comprehensive Toxicology*, Sipes, I.G., Gandolfi, A.J., and McQueen, C.A. (Series Eds), Pergamon (Elsevier), Amsterdam, pp. 401–424.

Cohen, S.M. 1998. Urinary bladder carcinogenesis. *Toxicologic Pathol.*, **26**, 121–127.

Cohen S.M., Ellwein, L.B. 1990. Cell proliferation in carcinogenesis. *Science*, **249**, 1007–1011.

Cohen, S.M., Ellwein, L.B. 1991. Genetic errors, cell proliferation, and carcinogenesis. *Cancer Res.*, **51**, 6493–6505.

Cohen, S.M., Fisher, M.J., Sakata, T., Cano, M., Schoenig, G.P., Chappel, C.I., Garland, E.M. 1990. Comparative analysis of the proliferative response of the rat urinary bladder to sodium saccharin by light and scanning electron microscopy and autoradiography. *Scanning Microscopy*, **4**, 135–142.

Cohen, S.M., Cano, M., Johnson, L.S., St. John, M.K., Asamoto, M., Garland, E.M., Thyssen, J.H., Sangha, G.K., Van Goethem, D.L. 1994. Mitogenic effects of propoxur on male rat bladder urothelium. *Carcinogenesis*, **15**, 2593–2597.

Eastern Research Group, 1997. Report on the expert panel on arsenic carcinogenicity: review and workshop. Prepared by Eastern Research Group, Lexington, MA, for the Center for Environmental Assessment, Washington, DC, under Environmental Protection Agency contract no. 68-C6-0041.

Fisher, M.J., Sakata, T., Tibbels, T.S., Smith, R.A., Patil, K., Khachab, M., Johansson, S. and Cohen, S.M. 1989. Effect of saccharin on urinary parameters in rats fed Prolab 3200 or AIN-76 diet. *Fund. Chem. Toxicol.*, **27**, 1–9.

Fiske, C.H., Subbarow, Y. 1925. The colorimetric determination of phosphorus. *J. Biol. Chem.*, **66**: 375–400.

International Agency for Research on Cancer, 1980. IARC Monographs on the Evaluation of the Carcinogenic Risk of Chemicals to Humans. Vol. 23.

Mauck, J.C., Mauck, K.L. Novros, J., Norton, G.E. 1986. Development of a single Kodak Ektachem® thin-film assay for serum and urine creatinine. *Clin. Chem.*, **32**, 1197–1198.

Rossman, T.G. 1997. Molecular and genetic toxicology of arsenic. In Rose, J. (ed.) *Environmental Toxicology*. Gordon and Breach Publishers.

SAS Institute, 1996. SAS User's Guide 1996 Edition, SAS Institute, Raleigh, NC.

Sheehan, D.C. and Hrapchak, B.B. (eds.). 1980. *Pigments and Minerals. Theory and Practice of Histotechnology, Second Edition*. C.V. Mosby Co., St. Louis, MO, p. 227.

Smith, R.A., Christenson, W.R., Bartels, M.J., Arnold, L.L., St. John, M.K., Cano, M., Wahle, B.S., McNett, D.A., Cohen, S.M. 1998. Urinary physiologic and chemical metabolic effects on the urothelial cytotoxicity and potential genotoxicity of o-phenylphenol in male rats. *Toxicol. Appl. Pharmacol.*, **150**, 402–413.

Smith-Lewis, M.J., Babb, B.E., Hilborn, D.A., Mauck, J.C., Norkus, N.S., Toner, J.L., Weaver, M.S. 1986. Thin-film colorimetric assay for magnesium. *Clin. Chem.*, **32**, 1200.

Sundberg, M.W., Dappen, G.M. 1979. An assay for total serum calcium using a coated film. *Clin. Chem.*, **25**, 1140.

Tietz, N.W., Pruden, E.L., Siggaard-Andersen, O. 1986. Electrolytes, blood gases and acid-base balance. In: Tietz, N.W. (ed.), *Textbook of Clinical Chemistry*, pp. 1179–1180. W.W. Saunders Co., Philadelphia, PA.

van Gemert, M., Eldan, M. Chronic carcinogenicity assessment of cacodylic acid. Third International Conference on Arsenic Exposure and Health Effects. San Diego, CA, 1998.

Wang, Z., Rossman, T.G. 1996. The carcinogenicity of arsenic. In Chang, L.W. (ed.) *Toxicology of Metals*. CRC Press, Boca Raton, FL, pp. 219–227.

Wanibuchi, H., Hori, T., Chen, H., Yoshida, K., Yamamoto, S., Endo, G., Fukushima, S. 1996. Promoting effect of dimethylarsinic acid on rat bladder and liver carcinogenesis. *Carcinogenesis*, **37**, 160.

Arsenic Exposure and Health Effects
W.R. Chappell, C.O. Abernathy and R.L. Calderon (Editors)

263

Differences of Promoting Activity and Loss of Heterozygosity Between Dimethylarsinic Acid and Sodium L-ascorbate in F_1 Rat Urinary Bladder Carcinogenesis

Tianxin Chen, Yifei Na, Hideki Wanibuchi, Shinji Yamamoto,
Chyi Chia R. Lee, Shoji Fukushima

ABSTRACT

Dimethylarsinic acid (DMA) is known to have promoting activity on rat urinary bladder carcinogenesis in F344 rats initiated with N-butyl-N-(4-hydroxybutyl)nitrosamine (BBN). Sodium L-ascorbate is also a strong promoter in this animal model. In this study, we used (Lewis×F344)F_1 rats to compare the promoting activity between DMA and sodium L-ascorbate and to find molecular alterations in the urinary bladder tumors. Male, 6-week-old rats were given 0.05% BBN in drinking water for 4 weeks, and then the rats were kept with no treatment for group 1, administered 0.01% DMA in drinking water (group 2) or 5% sodium L-ascorbate in the powdered diet (group 3). Group 4 rats were continuously given BBN alone. At weeks 36 and 44, the rats were sacrificed and the urinary bladders were fixed in 10% phosphate buffered formalin and embedded in paraffin. H&E staining was done for histology, and microdissection was done for loss of heterozygosity (LOH) examination. DMA and sodium L-ascorbate showed promoting activity on urinary bladder carcinogenesis of F_1 rat, however DMA revealed weaker promotion activity than that of sodium L-ascorbate, although doses were different. LOH existed in the urinary bladder tumors treated with DMA, whereas no LOH was detected in the urinary bladder tumors treated with sodium L-ascorbate.

Keywords: F_1 rat, urinary bladder tumor, LOH analysis, BBN, DMA, sodium L-ascorbate

INTRODUCTION

Epidemiological studies indicate that arsenicals are carcinogenic in humans (Chen et al., 1985; Chen and Wang, 1990). A main metabolite of inorganic arsenic is dimethylarsinic acid (DMA) in most mammals (Vahter, 1994). Studies on the fate of DMA administered to humans and rodents have shown that the inorganic arsenicals were excreted as DMA in the urine (Vahter, 1994). DMA is also found in natural waters and bird eggshells (Braman and Foreback, 1973), and in fact was used as a general herbicide or pesticide for many years (Wagner and Weswig, 1974). Recent *in vitro* studies indicate that DMA is a potent clastogenic agent and can induce chromosome aberrations, such as tetraploid formation (Endo et al., 1992; Dong and Luo, 1993). Our laboratory has shown that DMA exerts promoting effects on F344 rat urinary bladder carcinogenesis (Yamamoto et al., 1995; Wanibuchi et al., 1996), and the conclusion whether DMA is a carcinogen or a non-genotoxic promoter, was drawn by these *in vivo* experiments. There are, however, no reports on the *in vivo* carcinogenicity of DMA. On the other hand, sodium L-ascorbate is known to be itself a non-genotoxic strong promoter to rat urinary bladder carcinogenesis (Fukushima et al., 1983; Fukushima et al., 1986). The loss of heterozygosity (LOH) of specific genetic markers suggests inactivation of tumor suppressor genes (Hollingsworth and Lee, 1991).

In the present study, the differences of promoting activity between DMA or sodium L-ascorbate were investigated in N-butyl-N-(4-hydroxybutyl)nitrosamine (BBN) urinary bladder two-stage carcinogenesis using (Lewis×F344)F_1 rat. LOH was also investigated in urinary bladder tumors to find the differences of molecular alteration between DMA and sodium L-ascorbate. Furthermore, the LOH analysis was also applied in urinary bladder tumors induced by the carcinogen BBN continuously for a long period.

METHODOLOGY

A total of 60, 6 week old (Lewis×F344)F_1 male rats were purchased from Charles River, Inc., Shiga, Japan. All animals were randomly divided into 4 groups (15 animals in each group). Animals in groups 1 to 3 were administered 0.05% BBN (obtained from Tokyo Kasei Co., Osaka, Japan) in their drinking water for 4 weeks and then given no chemicals *ad libitum* to group 1, 0.01% dimethylarsinic acid (DMA, Wako Pure Chemical Ind., Osaka, Japan) in drinking water to group 2 or 5% sodium L-ascorbate (Wako Pure Chemical Ind., Osaka, Japan) in powdered diet to group 3, respectively, for 40 weeks. Group 4 rats were continuously given 0.05% BBN for 36 weeks and were killed at 36 weeks because of the toxic effects of BBN. Urinary bladders were fixed in 10% buffered formalin and embedded in paraffin for light microscopic examination (Fukushima et al., 1982) and LOH analysis. For LOH analysis, microdissection technique was used to extract DNA from paraffin sections (Chen et al., 1998). All of the tumors in groups 2 to 4 (46 cases), including 2 carcinomas and 7 papillomas from group 2, 15 carcinomas from group 3 and 22 carcinomas from group 4 were analyzed. One F_1, Lewis and F344 rats (44 weeks old) each were killed as controls without treatment.

The primers (Table 1) used for LOH analysis were randomly chosen from chromosomes X and 1 to 20 (8 and 19 excised). The differences in base pairs between Lewis and Fisher 344 strains were from 8 to 54 and could be separated by 3% TAE agarose gel electrophoresis directly after polymerase chain reaction (PCR). Template DNA (1 μl) was mixed with primers (0.15 μl, 6 mM), dNTP (0.4 μl, 2.5 mM), 10x buffer (0.8 μl, 1.5 mM MgCl$_2$ included), 25 mM MgCl$_2$ (0.1 to 0.5 μl), Ampli Taq Gold (0.1 μl, 5 unit/μl) and 5 μl of distilled water. Initial heating was at 96°C for 10 min, and then 30 or 35 cycles were performed of 95°C for 30 s, 54–60°C for 30 s, and 72°C for 45 s. PCR products were electrophoresed through 3% agarose gels in 1x TAE buffer to detect LOH. With the control DNAs, the Lewis and F344 inbred rats demonstrated only single bands while the (Lewis×F344) F_1 rat had both bands. For informative cases, allelic

TABLE 1

Markers used for LOH analysis

Chromosomes	Loci	Chromosomes	Loci
1	D1Mgh2, D1Mit13	12	D12Mit2, D12Mit4
2	D2Mgh7, D2Mit12, D2Mit14	13	D13Mgh7, D13Mit4
3	D3Mgh2, D3Mgh3, D3Mgh10	14	D14Mit1, D14Mit2
4	ENO2	15	D15Mgh8
5	D5Mgh5, D5Mit4, D5Rjrl	16	D16Mit2
6	D6Mit6, D6Mgh5	17	D17Mgh5, D17Mit4
7	D7Mgh9	18	D18Mit1
9	D9Bro1	20	D20Mgh1
10	D10Mgh7, D10Mit6, D10Mit9	X	DxMgh4, DxMit5
11	D11Mgh1, D11Mgh4, D11Mit1		

loss was scored if the amount of one allele was at least 70% reduced in the tumor DNA as compared with normal F_1 allele. All of the LOH were checked in duplicate or triplicate. For the assessment of differences in lesion incidence, the Fisher exact probability test was used. The Student's *t*-test was applied for analyses of other parameters.

RESULTS AND DISCUSSION

The incidences of the urinary bladder tumors bearing either papillomas and/or carcinomas were 7% in group 1, 47% in group 2, and 100% in groups 3 and 4. Data for tumor number/rat were 0.1±0.3, 0.5±0.7, 3.5±1.2 and 18.3±4.2 in groups 1 to 4, respectively. There were significant differences ($P < 0.05$) between group 1 and groups 2, 3 and 4. The number of tumors in group 3 was about 7-fold that in group 2. Thus, both DMA and Na-AsA exert promoting activity in (Lewis×F344)F_1 rat two-stage urinary bladder carcinogenesis, however DMA revealed weaker promoting activity than that of sodium L-ascorbate.

LOH was detected in 2 papillomas (22%) in group 2; 3 carcinomas (14%) in group 4. One of them is shown in Figure 1. There was no LOH detected in group 3. LOH were randomly distributed on chromosomes 2, 9, 11, 14, 18 and 20 and there was no hot-spot (Table 2). Because there was not more than two deleted regions to be found in the same chromosome, these deletions are not thought to be monosomy and might only have been partial allelic losses. The results point to a difference in the mechanism of promoting activities between DMA and sodium L-ascorbate, although both could affect cell proliferation (Yamamoto et al., 1995; Wanibuchi et al., 1996; Shibata et al., 1989). The question of whether DMA is a carcinogen or a promoter, has not been answered, but the finding in this investigation of LOH in two papillomas, at a similar incidence as that with continuous BBN is relevant. The data suggest that DMA may be carcinogenic to the rat urinary bladder and indeed, we recently found that F344 male rats treated with 0.01% or 0.02% DMA for 2 years without any initiating pretreatment, induced urinary bladder tumors (unpublished data). In summary, we conclude that DMA enhances urinary bladder carcinogenesis in (Lewis×F344) F_1 rats and that this is associated with genetic alteration, in contrast to the case with the promoter sodium L-ascorbate. Therefore we speculate that DMA might be a complete carcinogen.

Fig. 1. LOH analysis in case DT8 (a papilloma in group 2). The Lewis and F344 rats had only single bands and F_1 rat had both bands. LOH was detected at marker D18Mit1 in Lewis rat allele.

TABLE 2

LOH analysis in urinary bladder tumors of (Lewis×F344)F_1 rats

Group	Tumor cases	Loci	PCR products (bp)		Deleted allele
			Lewis	F344	
2	DT1	D9Bro1	160	150	F344
		D14Mit2	193	178	Lewis
	DT8	D11Mgh1	140	156	F344
		D18Mit1	309	255	Lewis
4	BT3	D20Mgh1	236	204	Lewis
	BT6	D11Mgh1	140	156	F344
	BT7	D2Mgh7	220	236	Lewis

ACKNOWLEDGMENTS

This work was supported by Grants-in-Aid for Cancer Research from the Ministry of Health and Welfare and the Ministry of Education, Science, Sports and Culture, Japan, as well as CREST (Core Research for Evolutional Science & Technology) of Japan Science and Technology Corporation (JST).

REFERENCES

Braman, R.S. and Foreback, C.C. 1973. Methylated forms of arsenic in the environment. *Science*, 182, 1247–1249.

Chen, C.-J., Chuang, Y,-C., Lin, T.-M. and Wu, H.-Y. 1985. Malignant neoplasms among residents of a black-foot disease endemic area in Taiwan: high-arsenic artesian well water and cancers. *Cancer Res.*, 45, 5895–5899.

Chen, T.X., Yamamoto, S., Kitano, M., Murai, T., Wanibuchi, H., Matsukuma, S., Nakatsuru, Y., Ishikawa, T. and Fukushima, S. 1998. Possible rare involvement of O^6-methylguanine formation as a significant mutational actor in mouse urinary bladder carcinogenesis models. *Teratogen. Carcin. Mut.*, 18, 101–110.

Chen, C.-J. and Wang, C.-J. 1990. Ecological correlation between arsenic level in well water and age-adjusted mortality from malignant neoplasms. *Cancer Res.*, 53, 5470–5474.

Dong, J.T. and Luo, X.M. 1993. Arsenic-induced DNA-strand breaks associated with DNA-protein crosslinks in human fetal lung fibroblasts. *Mutat. Res.*, 302, 97–102.

Endo, G., Kuroda, K., Okamoto, A. and Horiguchi, S. 1992. Dimethylarsinic acid induces tetraploids in Chinese hamster cells. *Bull. Environ. Contam. Toxicol.*, 48, 131–137.

Fukushima, S., Murasaki, G., Hirose, M., Nakanishi, K., Hasegawa, R. and Ito, N. 1982. Histopathological analysis of preneoplastic changes during N-butyl-N-(4-hydroxybutyl)nitrosamine-induced urinary bladder carcinogenesis in rats. *Acta Pathol. Jpn.*, 32, 243–250.

Fukushima, S., Imaida, K., Sakata, T., Okamura, T., Shibata, M.-A. and Ito, N. 1983. Promoting effects of sodium L-ascorbate on two-stage urinary bladder carcinogenesis in rats. *Cancer Res.*, 43, 4454–4457.

Fukushima, S., Shibata, S., Shirai, T., Tamano, S. and Ito, N. 1986. Roles of urinary sodium ion concentration and pH in promotion by ascorbic acid of urinary bladder carcinogenesis in rats. *Cancer Res.*, 46, 1623–1626.

Hollingsworth, R.E. and Lee, W.-H. 1991. Tumor suppressor genes: new prospect for cancer research. *J. Natl. Cancer Inst.*, 83, 91–96.

Shibata, M.-A., Yamada, M., Asakawa, E., Hagiwara, A. and Fukushima, S. 1989. Responses of rat urine and urothelium to bladder tumor promoters: possible roles of prostaglandin E2 and ascorbic acid synthesis in bladder carcinogenesis. *Carcinogenesis*, 10, 1651–1656.

Vahter, M. 1994. Species differences in the metabolism of arsenic compounds. *Appl. Organomet. Chem.*, 8, 175–182.

Wagner, S.L. and Weswig, P. 1974. Arsenic in blood and urine of forest workers as indices of exposure to cacodylic acid. *Arch. Environ. Health.*, 28, 77–79.

Wanibuchi, H., Yamamoto, S., Chen, H., Yoshida, K., Endo, G., Hori, T. and Fukushima, S. 1996. Promoting effects of dimethylarsinic acid on N-butyl-N-(4-hydroxybutyl)nitrosamine-induced urinary bladder carcinogenesis in rats. *Carcinogenesis*, 17, 2435–2439.

Yamamoto, S., Konishi, Y., Matsuda, T., Murai, T., Shibata, M.A., Matsui-Yuasa, I., Otani, S., Kuroda, K., Endo, G. and Fukushima, S. 1995. Cancer induction by an organic arsenic compound, dimethylarsinic acid (Cacodylic acid), in F344/DuCrj rats after pretreatment with five carcinogens. *Cancer Res.*, 55, 1271–1276.

Arsenic Exposure and Health Effects
W.R. Chappell, C.O. Abernathy and R.L. Calderon (Editors)

Variation in Human Metabolism of Arsenic

Marie Vahter

ABSTRACT

Inorganic arsenic is methylated to methylarsonic acid (MMA) and dimethylarsinic acid (DMA) by alternating reduction of AsV to AsIII and addition of a methyl group from S-adenosylmethionine. The methylated metabolites are less reactive with tissue constituents and more readily excreted in the urine than is inorganic arsenic. Therefore, a low percentage of MMA + DMA in urine correlates with a low rate of excretion of the absorbed arsenic. Although most studies indicate on average 10–30% inorganic arsenic, 10–20% MMA, and 60–70% DMA in urine, there seems to be a polymorphism in arsenic methylation. Our studies on arsenic-exposed native Andean people in north-west Argentina, show only a few percent MMA in the urine, which in most other populations is a rare event, occurring in a few individuals only. However, in several other mammalian species this is a common feature. On the other hand, one study indicates a higher than usual % MMA in the urine of arsenic exposed people in Taiwan. There is also a marked intra-individual variation in the urinary pattern of arsenic metabolites. In experimental studies we have shown that the methylation of arsenic is influenced by the form of arsenic absorbed, the dose level, route of exposure, and nutritional status. Our studies in Argentina indicate that arsenic methylation is influenced by age, dose level and, probably, hormonal status. For example, in children there was a lower % DMA in urine compared to adults, and an increasing % DMA with increasing exposure level.

Keywords: arsenic methylation, polymorphism, dose level, age and gender

INTRODUCTION

Many thousands of people in various parts of the world are exposed to arsenic via drinking water or industrial emissions (see e.g. Das et al., 1995; Thornton and Farago, 1997). Arsenic exposure has been associated with cancer of the skin, lungs, urinary bladder and kidneys, as well as hyperkeratosis, pigmentation changes, and effects on the circulatory and nervous systems (EPA, 1988; Chen et al., 1992; Smith et al., 1998). Arsenic is known to pass the placenta (Concha et al., 1998c), which implies that the exposure may start at the very beginning of life. However, there are very few studies on the reproductive effects in humans.

It has been suggested that there is a variation in the susceptibility to arsenic-induced health effects between human population groups and individuals (Hsueh et al., 1995; 1997; Hsu et al., 1997), but there are very few studies designed to address this issue. Possible reasons for such a variability include age, nutritional status, concurrent exposures to other agents or environmental factors, and metabolic polymorphism. The marked differences in susceptibility to arsenic toxicity between animal species (NRCC, 1978), may partly be explained by the pronounced differences in arsenic metabolism (Vahter and Marafante, 1988; Vahter, 1994 a, b). Thus, variability in the metabolism of arsenic may influence the susceptibility to arsenic also in people.

Most mammals methylate inorganic arsenic to methylarsonic acid (MMA) and dimethylarsinic acid (DMA; see e.g. Vahter and Marafante, 1988; Hopenhayn-Rich et al., 1993; Vahter, 1994b). Compared with inorganic arsenic, the methylated arsenic metabolites are less reactive with tissue constituents, less toxic, and more readily excreted in the urine (Buchet et al., 1981a; Vahter and Marafante, 1983; Vahter et al., 1984; Yamauchi and Yamamura, 1984; Marafante et al., 1987; Moore et al., 1997; Rasmussen and Menzel, 1997; Concha et al., 1998a; Hughes and Kenyon, 1998; Sakurai et al. 1998). Following oral intake of MMA or DMA by human volunteers, about 75% of the ingested dose was excreted in the urine within four days, compared to about 45% of a similar dose of inorganic arsenic (Buchet et al., 1981a). Also, in human volunteers ingesting known amounts of arsenate (AsV) or arsenite (AsIII), the rate of excretion of arsenic (% of ingested dose) increased with increasing methylation efficiency (Figure 1) (Crecelius, 1977; Tam et al., 1979; Buchet et al, 1981a, b; Johnson and Farmer, 1991). Following exposure to inorganic arsenic, AsIII is the main form of arsenic interacting with tissues (Buchet et al., 1981a; Vahter and Marafante, 1983; Bogdan et al., 1994, Styblo et al., 1995). Consequently, a lower capacity to methylate arsenic is associated with higher tissue concentrations of arsenic (Marafante and Vahter, 1984; Marafante et al., 1985; Concha et al., 1998a), which is likely to correspond to an increased risk of toxic effects.

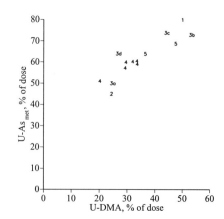

Fig. 1. Relationship between the rate of excretion of ingested arsenite (AsIII) or arsenate (AsV), expressed as the amount (% of dose) of total metabolites of inorganic arsenic (U-As$_{met}$) excreted in the urine within 4–6 days, and the methylation efficiency, expressed as the absolute amount of DMA (% of dose) in the urine. The data represent individual data from (1) Crecelius, 1977: 63 μg AsIII; (2) Buchet et al., 1981a: 500 μg AsIII; (3) Buchet et al., 1981b: 125–1000 μg AsIII a day for 5 days; (4) Tam et al., 1979: 0.01 μg AsV; (5) Johnson and Farmer, 1991: 220 μg AsV.

MECHANISMS OF ARSENIC METHYLATION

The biomethylation of inorganic arsenate to DMA is believed to occur via alternating reduction of pentavalent arsenic to trivalent and addition of a methyl group (Figure 2) (Challenger, 1945; Cullen et al., 1984; Hirata et al., 1990), but little is known about the mechanisms involved. That S-adenosylmethionine is the main methyl donor was indicated by studies showing that inhibition of the SAM-dependent methylation reactions in rabbits resulted in a marked decrease in the urinary excretion of DMA (Marafante and Vahter, 1984). Also, *in vitro* incubation of rat liver preparations with arsenite requires SAM for formation of MMA and DMA (Buchet and Lauwerys, 1988; Styblo et al., 1995). Obviously, a reducing environment is required for the reduction of AsV and MMA(V) prior to the addition of the methyl groups. Glutathione (GSH) and other thiols are believed to serve as reducing agents (Buchet and Lauwerys, 1985; 1988; Hirata et al., 1988; Delnomdeldieu et al., 1994a; Zakharyan et al., 1995; Styblo et al., 1995). Experimental studies have shown that a substantial fraction of absorbed AsV is rapidly reduced to AsIII, most of which is then methylated to MMA and DMA (Vahter and Envall, 1983; Vahter and Marafante, 1985; Marafante et al., 1985). Thus, the equilibrium between AsV and AsIII is driven to the right both by the reducing environment and the cellular uptake of AsIII and subsequent methylation. Whether the reduction of MMA(V) to MMA(III) may result in a release of reactive MMA(III) species (Styblo and Thomas, 1997a; Styblo et al., 1997) in the body is not known. Apparently, MMA(V) is not as easily reduced in the body as is AsV, as indicated by *in vitro* studies on rabbit erythrocytes (Delnomdeldieu et al., 1994b) or rat liver cytosol (Styblo et al, 1995). Also, the biological half-time of MMA(V) is much shorter than that of AsV (Buchet et al., 1981a; Tam et al., 1979), giving evidence against a significant distribution of MMA(III) compounds *in vivo*.

$$H_2As^VO_4^- \Leftrightarrow H_3As^{III}O_3 \Rightarrow CH_3As^VO_3^{2-} \Leftrightarrow CH_3As^{III}O_2^{2-} \Rightarrow (CHL_3)_2As^VO_2^-$$
$$\text{MMA} \qquad\qquad\qquad\qquad \text{DMA}$$

Fig. 2. Proposed mechanism of arsenic methylation.

It has been suggested that arsenite is bound to a carrier protein before the addition of the methyl groups (Hirata et al., 1990; Thompson, 1993), but the detailed reaction sequences have not been elucidated. Little is also known about the methyltransferases involved in arsenic methylation. Purification of arsenic methyltransferase activity from rabbit liver indicates that both the arsenite and the MMA methyltransferase activities are in the same protein (Zahkaryan et al., 1995; Zahkaryan et al., 1996). Both enzyme activities had a molecular mass of 60 kDa, but different pH optima and different saturation concentrations for their substrates.

Experimental studies have indicated that the liver is an important initial site of arsenic methylation, especially following ingestion of arsenic, most of which passes initially through the liver following the absorption (Charbonneau et al., 1979; Vahter, 1981; Marafante et al., 1985). That the liver is a major site of arsenic methylation is supported by studies showing a marked improvement in the methylation in patients with end-stage liver disease following liver transplantation (Geubel et al., 1988). However, most tissues seem to have arsenic methylating capacity. Recent *in vitro* studies using mouse tissue preparations showed that the highest amount of arsenic methylating activity is in the testes, followed by kidney, liver, and lung (Healy et al., 1997; Aposhian, 1997). Thus, arsenite initially bound to tissue constituents can be methylated and released, which would fit with a several compartment model (Pomroy et al., 1980; Marafante et al., 1981; Vahter and Marafante, 1983; Mann et al., 1996 a, b; Menzel, 1997).

There are major species differences in the biotransformation of inorganic arsenic (Vahter, 1994b). For example, most animals excrete more DMA and less MMA in the urine than do humans. While mice and dogs show an efficient methylation of inorganic arsenic and excrete more than 80% of the administered dose in the urine within a few days (Charbonneau et al., 1979; Vahter, 1981), the rat, although being equally efficient in the methylation, retains a major part of the produced DMA in the erythrocytes resulting in a very slow urinary excretion (Odanaka et al., 1980; Lerman et al., 1983; Vahter et al., 1984). Rabbits (Vahter and Marafante, 1983; Maiorino and Aposhian, 1985; DeKimpe et al., 1996) and hamsters (Charbonneau et al., 1980; Yamauchi and Yamamura, 1985; Marafante et al., 1987) show an arsenic methylation more similar to that in humans. The marmoset monkey, tamarin monkey, squirrel monkey, chimpanzee, and guinea pig seem to lack the ability to methylate arsenic (Vahter et al., 1982; Vahter and Marafante, 1985; Vahter et al., 1995b; Zakharyan et al., 1995, 1996; Healy et al., 1997).

BIOMARKERS OF ARSENIC METHYLATION

The major route of excretion of most arsenic compounds is via the kidneys. A few experimental studies on human volunteers, who were given a known dose of arsenite or arsenate, indicate that the efficiency of arsenic methylation may be evaluated by the relative distribution of metabolites (inorganic arsenic, MMA, and DMA) in the urine. In one study, six males ingested ^{74}As-labeled arsenate (about 0.01 μg of AsV per person), and approximately 58% of the dose was excreted in the urine within five days (by Tam et al., 1979). In another study, two subjects ingested mineral water containing 200 μg AsV, and about 66% of the dose was excreted over 7 days (Johnson and Farmer, 1991). Taken together, the results show that the fraction of DMA in the urine (% of total metabolite concentration in the urine, U-As$_{met}$) is highly correlated to the amount of DMA formed in the body and excreted in urine (% of the dose) (Figure 3). A similar relationship is obtained using data from studies on human volunteers ingesting arsenite (Crecelius, 1977; Buchet et al., 1981a, b), but the variation is somewhat more pronounced (Figure 3). Thus, on a group basis, a low percentage of DMA in the urine (% of U-As$_{met}$) indicates that the methylation efficiency is low, that the overall rate of excretion (in percent of the dose) is low, and that more inorganic arsenic is retained in the body. This was also indicated in a study of women and children exposed to arsenic via drinking water in northern Argentina (Concha et al., 1998a). The ratio between arsenic in blood and urine increased significantly with decreasing proportion (%) of DMA in the urine, indicating that more arsenic was bound to blood, probably also to various tissues (Figure 4).

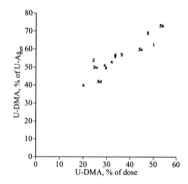

Fig. 3. Association between the relative amount of DMA in the urine (% of total amount of metabolites of inorganic arsenic, U-As$_{met}$) and the methylation efficiency, expressed as the absolute amount of DMA (% of dose) in the urine. The data represent individual data from (1) Crecelius, 1977: 63 μg AsIII; (2) Buchet et al., 1981a: 500 μg AsIII; (3) Buchet et al., 1981b: 125–1000 μg AsIII a day for 5 days; (4) Tam et al., 1979: 0.01 μg AsV; (5) Johnson and Farmer, 1991: 220 μg AsV.

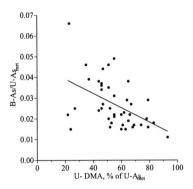

Fig. 4. Association between the ratio of blood arsenic to U-As$_{met}$ and the percentage of DMA in urine of children and women in northern Argentina (from Concha et al., 1998a).

However, it should be noted that the methylated arsenic metabolites are excreted in the urine more easily than inorganic arsenic (Buchet et al., 1981a), which means that a change in the tissue methylation of arsenic, e.g. due to chemical or nutritional inhibition of the methylation (see below), is reflected by a smaller change in the percentage of DMA in the urine.

One component of the apparent variability in the concentrations of arsenic metabolites in urine of human subjects, that is seldom evaluated and reported in the publications, is the analytical variability. A recent interlaboratory comparison exercise, involving seven experienced laboratories that determined arsenic metabolites in a set of human urine samples, showed that the variations between laboratories were 1.3–2.7 μg/L for inorganic arsenic, 1.3–2.7 μg/L for MMA and 5.8–11 μg/L for DMA in a urine sample with low arsenic concentrations; and 3.2–6.0 μg/L inorganic arsenic, 3.5–5.5 μg/L MMA, and 15–22 μg/L DMA for a medium concentration sample (Crecelius and Yager, 1997).

VARIATION IN ARSENIC METHYLATION AMONG POPULATION GROUPS AND INDIVIDUALS

The average relative distribution of arsenic metabolites in the urine of various population groups seems to be fairly constant, irrespective of the type and extent of exposure. A number of studies on human subjects exposed to inorganic arsenic occupationally, experimentally, or environmentally have shown that, in general, U-As$_{met}$ consists of 10–30% inorganic arsenic, 10–20% MMA, and 60–80% DMA (for review, see Hopenhayn-Rich et al., 1993; Vahter, 1994a). However, recently it has become apparent that there are populations with a somewhat different distribution of arsenic metabolites in the urine. In people exposed to arsenic via the drinking water in northern Argentina (Andean natives and natives mixed with Spanish descendants), the fraction of MMA in urine was very low, on average 2–4% (Vahter et al., 1995a; Concha et al., a, c). Further, studies in San Pedro and Toconao in northern Chile showed that about 5% of the study group of 220 individuals had less than 5% MMA in their urine (Hopenhayn-Rich et al., 1996a). The percentage of MMA was significantly lower in Atacameños than in subjects of European descent. By contrast, a study of the exposure to arsenic via drinking water in northeastern Taiwan showed an unusually high percentage of MMA in the urine—on average, 27% (Chiou et al., 1997). Whether the reported variations in urinary arsenic metabolites are genetically determined or due to environmental factors remains to be investigated. It should be noted that polymorphism is reported for a number of other human methyltransferases, e.g., histamine N-methyltransferase, nicotinamide N-methyltransferase, thiopurine S-methyltransferase, catechol O-methyltransferase,

TABLE 1

Inter-individual variation in the relative amounts (median or mean values and total ranges) of arsenic metabolites (inorganic arsenic, MMA, and DMA) in the urine of individuals exposed to inorganic arsenic via drinking water

Drinking-water exposure to arsenic	N	U-As$_{met}$ (μg/L)	Urinary arsenic metabolites (%)			References
			Inorganic arsenic	MMA	DMA	
Argentina (200 μg/L) Women						
S.A. de los Cobres	11	303 127–481	25 6.5–42	2.1 0.6–8.3	74 55–93	Vahter et al. (1995)
Taco Pozo	12	386 90–606	39 18–52	2.2 1.1–3.5	58 46–80	Concha et al. (1998a)
Argentina (200 μg/L) Children						
S.A. de los Cobres	22	323 125–578	49 21–76	3.6 0.9–12	47 22–69	Concha et al. (1998a)
Taco Pozo	12	440 337–621	42 26–54	3.4 1.3–7.9	54 44–68	Concha et al. (1998a)
Chile (600 μg/L) Adults						
San Pedro	122	482 61–1890	18 5.6–39	15 1.7–31	67 42–93	Hopenhayn-Rich et al. (1996a)
Chile (15 μg/L) Adults						
Toconao	98	49 6–267	14 3.6–31	9.7 3.1–24	76 51–92	Hopenhayn-Rich et al. (1996a)
China (<1–>300 μg/L) Adults						
Taiwan	115	173 ±19 (SE)	12 9.8–32	27 19–32	61 56–69	Chiou et al. (1997)
Mexico (415 mg/L) Adults						
Santa Ana	35	544 429–689	31 28–34	11 9.5–13	54 50–58	Del Razo et al. (1997)
USA (300–400 μg/L) Adults						
California	10	130 66–299	24 8–44	18 13–27	55 38–77	Hopenhayn-Rich et al. (1993)

U-As$_{met}$, urinary arsenic metabolites; SE, standard error.

O_6-methylguanine-DNA methyltransferase, and guanidinoacetate methyltransferase (e.g., Scott et al., 1988; Aksoy et al., 1996; Krynetski et al., 1996; Scheller et al., 1996; Stockler et al., 1996; Li et al., 1997; Ganesan et al., 1997; Yates et al., 1997).

Reported data on the relative distribution of the arsenic metabolites in urine of people exposed to arsenic via the drinking water are presented in Table 1. Although there are a limited number of studies published, and several of them include few individuals, it is obvious that there is a marked inter-individual variation in the pattern of urinary metabolites of inorganic arsenic. The variation is an integrated result of the various factors influencing the methylation of arsenic (e.g., genetic, physiological, nutritional, recreational, and analytical), but very little is known about the relative importance of the various factors.

VARIATION WITH AGE AND GENDER

Recently it has been reported that the arsenic methylation efficiency may change with age. Children living in villages in northern Argentina with about 200 μg As/L in the drinking water had a significantly higher percentage of inorganic arsenic and a lower percentage of DMA in urine, compared with adults (see Table 1) (Concha et al., 1998a). As the retention of arsenic seemed to increase with decreasing methylation efficiency, the results indicate that children retain more arsenic than adults, and, thus, may be more susceptible to arsenic toxicity. In the few previous studies on arsenic metabolism in children, age had no influence on the arsenic metabolites found in urine (Buchet et al., 1980; Kalman et al., 1990). However, the arsenic exposure levels were very low in those studies. One recent study from Finland indicates an increase in the proportion of DMA in urine with age also in adults (Kurttio et al., 1998).

In an extensive study on arsenic metabolism in people exposed to arsenic via the drinking water (about 600 μg/L) in northern Chile, gender, ethnicity, length of exposure, and smoking, but not age, seemed to have a small, but significant influence on the relative amounts of urinary arsenic metabolites (Hopenhayn-Rich et al., 1996a). Multiple linear regression indicated that women had about 3% more DMA and less MMA in the urine than men. Smoking 10 cigarettes a day resulted in an increase of a few points in the percentage of MMA and a corresponding decrease in the percentage of DMA. Also, one study of human exposure to arsenic via drinking water (up to 600 μg/L) in northeastern Taiwan indicates that women had a somewhat higher percentage of DMA and lower percentage of MMA in the urine than men (Hsu et al., 1997). However, the concentrations of arsenic in urine were not given. In another study from the northeast of Taiwan (average U-As$_{met}$ 173 μg/L), no gender difference was found in the relative amount of the various urinary metabolites of arsenic (Chiou et al., 1997). Neither was gender a determining factor for the relative distribution of urinary arsenic metabolites in the recent study from Finland (Kurttio et al., 1998).

The observed higher relative amount of DMA in urine in women compared to men in some studies might be related to hormonal effects. Recently, it was reported that pregnant women in the third trimester had more than 90% DMA in plasma and urine, which was significantly higher than in non-pregnant women (Concha et al., 1998b). Therefore, the fraction of pregnant women in a study group will markedly influence the average % DMA in the urine. In the above-mentioned studies, the numbers of pregnant women were not reported. However, most likely, the probability of including pregnant women in a study group is higher in northern Chile than in Finland.

VARIATION WITH EXPOSURE LEVEL

Experimental animal studies have shown that the methylation of arsenic decreases and the tissue concentrations increase at high doses of inorganic arsenic (2 mg As/kg body weight and higher doses; Vahter, 1981; Hughes et al., 1994). Also, in humans acutely intoxicated by very high doses of inorganic arsenic, there is a marked delay in the urinary excretion of DMA (Foà et al., 1984; Mahieu et al., 1981). In one case of attempted suicide a person ingested about 3 g of As$_2$O$_3$ and was admitted to the hospital two hours later (Foà et al., 1984). The relative percentage of urinary DMA increased from about 30% on the eighth day to 78% on the thirteenth day after ingestion. In three subjects who by mistake ingested 500–1000 mg arsenic in the form of As$_2$O$_3$, which they mistook for sugar, DMA increased from about 10% the first days after ingestion to about 75% after a week (Mahieu et al., 1981). This may be a result of an overloading or inhibition of the arsenic methyltransferases. *In vitro* studies on rat liver cytosol have shown that the formation of DMA, but not that of MMA, was inhibited at incubations with high concentrations of arsenite (Buchet and Lauwerys, 1985).

However, the results of a large number of studies on urinary excretion of arsenic metabolites in people exposed to inorganic arsenic occupationally, experimentally, or via drinking water, indicate no difference in the relative distribution of the different urinary metabolites compared to that in people exposed to background concentrations of arsenic in the general environment (for review see Hopenhayn-Rich et al., 1993). A further support for arsenic methylation being rather insensitive to the dose level is a study of arsenic in the urine of people in Nevada with high concentrations of arsenic in the well water (up to $1,300\,\mu g/L$) and on average $750\,\mu g/L$ in the urine. The urinary arsenic consisted of 22% MMA and 58% DMA (Warner et al., 1994). Furthermore, in a man who developed neuropathy after four months of daily consumption of well water containing $25,000\,\mu g$ As/L, urine samples collected before DMSA chelation therapy was initiated showed $5,500\,\mu g$ As/L, out of which 26% was inorganic arsenic and 72% was in the form of MMA and DMA (about 36% of each; Kosnett and Becker, 1988).

The latter study indicates a lower % DMA and higher %MMA in the urine compared to people exposed to lower doses, who, as mentioned above, generally have 10–30% inorganic arsenic, 10–20% MMA, and 60–80% DMA in urine. Also, a few recent studies where people have been exposed to elevated arsenic levels in the drinking water showed a slight decrease in the % DMA and a corresponding increase in the %MMA in the urine, as the exposure to arsenic via drinking water increased (Hopenhayn-Rich et al., 1996a; Del Razo et al., 1997; Hsu et al., 1997). For example, speciation of arsenic in the urine of people exposed to arsenic via drinking water in northern Chile indicated that a $500\,\mu g/L$ increase in total U-As$_{met}$ corresponded to a 2% increase in urinary MMA and a 3% decrease in DMA (Hopenhayn-Rich et al., 1996a). A temporary (2 months) change in the source of water, involving a decrease in arsenic concentration from 600 to $45\,\mu g/L$, resulted in a small decrease (about 3%) in the average % inorganic arsenic, as well as in the ratio MMA/DMA in the urine (Hopenhayn-Rich et al., 1996b). However, changes were not related to the magnitude of the decrease in total U-As$_{met}$.

Thus, the observed effects of the arsenic dose level on the methylation efficiency seem to be small and mainly affecting the ratio of MMA to DMA. The toxicological implication of such a shift in the arsenic metabolites is not known. Available data indicate that the rate of excretion of MMA is similar to that of DMA (about 75% in 4 days; Buchet et al., 1981a) compared to about 50% for inorganic arsenic (Tam et al., 1979; Buchet et al., 1981a). However, the formation of reactive MMA(III) intermediates cannot be excluded (Styblo and Thomas, 1997b; Styblo et al., 1997). Furthermore, the increased excretion of MMA relative to DMA might indicate that the second methylation step is inhibited by inorganic AsIII, which would be an indicator that the concentrations of this form of arsenic are increasing in the tissues (Thompson, 1993).

There are also studies indicating an increase in the percentage of DMA in urine at higher exposure concentrations. In children exposed to arsenic via the drinking water in northern Argentina, the percentage of urinary DMA increased with increasing urinary As$_{met}$ (about 30% increase in % DMA with $400\,\mu g/L$ increase in U-As$_{met}$; Concha et al., 1998a). A similar change in arsenic methylation was not observed among adults, who had significantly higher proportion of DMA in urine than the children. However, in a study from northeastern Taiwan, the percentage of DMA in urine of adult individuals increased and that of MMA decreased with increasing arsenic concentrations in the drinking water (Chiou et al., 1997). Obviously, more research is needed to clarify what small changes in the relative distribution of the various arsenic metabolites in urine mean in terms of retention of arsenic in the tissues.

INFLUENCE OF NUTRITIONAL FACTORS

In areas with severe arsenic-related health effects due to ingestion of drinking water with high arsenic concentrations, i.e. the southwest of Taiwan and the Antofagasta region in northern Chile, the inhabitants were reported to have a low socio-economic level and a poor nutritional status (Borgono et al., 1977; Tseng, 1977; Zaldivar and Guillier, 1977; Hsueh et al., 1995). This may indicate an increased susceptibility towards arsenic toxicity in subjects with poor nutritional status. However, detailed studies on the nutritional intakes are essentially lacking. There may be several reasons for nutritional factors influencing arsenic toxicity. One reason could be an influence on the methylation of arsenic, resulting in increased tissue retention. The experimental studies on mice and rabbits showing that chemical inhibition of SAM-dependent methylation reactions results in decreased methylation of arsenic and increased tissue concentrations (Marafante and Vahter, 1984; Marafante et al., 1985), indicate that the availability of dietary methyl groups may be a rate limiting factor for the methylation of arsenic. In fact, experimental studies on rabbits fed diets with low amounts of methionine, choline, or proteins, have shown a marked decrease in the urinary excretion of DMA (Vahter and Marafante, 1985). For the rabbits, the diet low in methyl groups resulted in 2–3 times increased tissue concentrations of arsenic, especially in the liver. In particular, there was an increase in the arsenic concentrations in the liver microsomes (Vahter and Marafante, 1987), similar to that seen in the marmoset monkey, which is unable to methylate arsenic (Vahter et al., 1982). A decreased urinary excretion of arsenic also was observed in mice fed a choline-deficient diet (Tice et al., 1997). There is a need for further studies on the variation in arsenic methylation by nutritional factors in humans.

The influence of a low dietary intake of vitamin B_{12} or folate on arsenic methylation in people is not known. *In vitro* studies using rat liver preparations have shown that arsenic may be methylated by vitamin B_{12} (cyanocobalamin), its coenzyme, 5'-deoxycobalamin, and methylcobalamin (Buchet and Lauwerys, 1985; Styblo and Thomas, 1997a). Interestingly, the arsenite was methylated by methylcobalamin also in the absence of any enzymes. There is a case report on neurotoxicity of arsenite in a girl with 5,10-methylenetetrahydrofolate reductase (MTHFR) deficiency (Brouwer et al., 1992). MTHFR is necessary for the conversion of 5,10-methylenetetrahydrofolate to 5-methyltetrahydrofolate, the main methyl donor for the remethylation of homocysteine to methionine, and the girl has clearly elevated homocysteine levels in urine. Although the rest of the family has a similar history of exposure to the pesticide copper acetate arsenite, they showed no symptoms.

Dimercapto compounds, especially those with adjacent SH groups, form stable complexes with arsenic and increase the urinary excretion. The administration of 2,3-dimercapto-1-propanesulfonic acid (DMPS) to people in San Pedro in the north of Chile, where the drinking water contained about 600 μg As/L, resulted in a five times increase in the total urinary arsenic excretion within six hours after DMPS administration (Aposhian et al., 1997). During the first 2-hour period following DMPS administration, inorganic arsenic represented 20–22%, MMA 42%, and DMA 37–38% of the total U-As$_{met}$, i.e. much more MMA and less DMA than before the treatment. Interestingly, *in vitro* studies on arsenic methylation in rat liver cytosol showed that both DMSA and DMPS markedly inhibited the methylation of inorganic arsenic to DMA, especially the second step (Buchet and Lauwerys, 1985; 1988). Also, EDTA (1 mM) has been shown to inhibit the DMA formation from AsIII *in vitro* (Styblo and Thomas, 1997a). Thus, chelating agents to humans may result in an inhibition of the second methylation step, releasing more MMA in the body for excretion in urine. These results indicate that dithiols occurring naturally in various types of food, e.g. garlic, may influence arsenic methylation. Further research is needed to elucidate the effects of dietary thiols on arsenic methylation.

REFERENCES

Aksoy, S., Raftogianis, R. and Weinshilboum, R. 1996. Human histamine N-methyl-transferase gene: structural characterization and chromosomal location. *Biochem. Biophys. Res. Comm.*, **219**(2), 548–554.

Aposhian, H.V. 1997. Enzymatic methylation of arsenic species and other new approaches to arsenic toxicity. *Ann. Rev. Pharmacol. Toxicol.*, **37**, 397–419.

Aposhian, H.V., Zakharyan, R.A., Wu, Y., Healy, S. and Aposhian, M.M. 1997. Enzymatic methylation of arsenic compounds: II. An overview. In: C.O. Abernathy, R.L. Calderon and W.R. Chappell (eds.), *Arsenic Exposure and Health Effects II*, pp. 296–321. Chapman & Hall, New York.

Bogdan, G.M., Sampayo-Reyes, A. and Aposhian, H.V. 1994. Arsenic binding proteins of mammalian systems: I. Isolation of three arsenite-binding proteins of rabbit liver. *Toxicology*, **93**, 175–193.

Borgono, J.M., Vincent, P., Venturino, H. and Infante, A. 1977. Arsenic in the drinking water of the city of Antofagasta: epidemiological and clinical study before and after the installation of the treatment plant. *Environ. Health Perspect.*, **19**, 103–105.

Brouwer, O.F., Onkenhout, W., Edelbroek, P.M., de Kom, J.F.M., de Wolff, F.A. and Peters, A.C.B.. 1992. Increased neurotoxicity of arsenic in methylenetetrahydrofolate reductase deficiency. *Clin. Neurol. Neurosurgery*, **94**, 307–310.

Buchet, J.P. and Lauwerys, R. 1985. Study of inorganic arsenic methylation by rat liver *in vitro*: Relevance for the interpretation of observations in man. *Arch. Toxicol.*, **57**, 125–129.

Buchet, J.P., and Lauwerys, R. 1988. Role of thiols in the *in vitro* methylation of inorganic arsenic by rat liver cytosol. *Biochem. Pharmacol.*, **37**, 3149–3153.

Buchet, J.P., Lauwerys, R. and Roels, H. 1980. Comparison of several methods for the determination of arsenic compounds in water and in urine. *Int. Arch. Occup. Environ. Health*, **46**, 11–29.

Buchet, J.P., Lauwerys, R. and Roels, H. 1981a. Comparison of the urinary excretion of arsenic metabolites after a single dose of sodium arsenite, monomethylarsonate or dimethylarsinate in man. *Int. Arch. Occup. Environ. Health*, **48**, 71–79.

Buchet, J.P., Lauwerys, R. and Roels, H. 1981b. Urinary excretion of inorganic arsenic and its metabolites after repeated ingestion of sodium metaarsenite by volunteers. *Int. Arch. Occup. Environ. Health*, **48**, 111–118.

Buchet, J.P., Geubel, A., Pauwels, S., Mahieu, P. and Lauwerys, R. 1984. The influence of liver disease on the methylation of arsenite in humans. *Arch. Toxicol.*, **55**, 151–154.

Challenger, F. 1945. Biological methylation. *Chem. Rev.*, **36**, 315–361.

Charbonneau, S.M., Tam, G.K.H., Bryce, F., Zawidzka, Z. and Sandi, E. 1979. Metabolism of orally administered inorganic arsenic in the dog. *Toxicol. Lett.*, **3**, 107–113.

Charbonneau, S.M., Hollins, J.G., Tam, G.K.H., Bryce, F., Ridgeway, J.M. and Willes, R.F. 1980. Whole-body retention, excretion and metabolism of [^{74}As]arsenic in the hamster. *Toxicol. Lett.*, **5**, 175–182.

Chen, C.J., Chen, C.W., Wu, M.M. and Kuo, T.L. 1992. Cancer potential in liver, lung, bladder and kidney due to ingested inorganic arsenic in drinking water. *Br. J. Cancer*, **66**, 888–892.

Chiou H.-Y., Hsueh, Y.-M., Hsieh, L.-L., Hsu, L.-I., Hsu, Y.-H., Hsieh, F.-I., Wei, M.-L., Chen, H.-C., Yang, H.-T., Leu, L.-C., Chu, T.-H., Chen-Wu, C., Yang, M.-H. and Chen, C.-J. 1997. Arsenic methylation capacity, body retention, and null genotypes of glutathione S-transferase M1 and T1 among current arsenic-exposed residents in Taiwan. *Mutat. Res.*, **386**, 197–207.

Concha, G., Vogler, G., Nermell, B. and Vahter, M. 1998a. Metabolism of inorganic arsenic in children with chronic high arsenic exposure in northern Argentina. *Environ. Health Perspect.*, **106**(6), 355–359.

Concha, G., Vogler, G., Lezcano, D., Nermell, B. and Vahter, M. 1998b. Exposure to inorganic arsenic metabolites during early human development. *Toxicol. Sci.*, **44**(2), 185–190.

Concha, G., Vogler, G., Nermell, B. and Vahter, M. 1998c. Low arsenic excretion in breast milk of native Andean women exposed to high levels of arsenic in the drinking water. *Int. Arch. Occup. Environ. Health* **71**, 42–46.

Crecelius, E.A. 1977. Changes in the chemical speciation of arsenic following ingestion by man. *Environ. Health Perspect.*, **19**, 147–150.

Crecelius E. and Yager J. (1997). Intercomparison of analytical methods for arsenic speciation in human urine. *Environ. Health Perspect.*, **105**(6), 650–653.

Cullen, W.R., McBride, B.C. and Reglinski, J. 1984. The reaction of methyl arsenicals with thiols: some biological implications. *J. Inorg. Biochem.*, **21**, 179–193.

Das, D., Chatterjee, A., Mandal, B.K., Samanta, G., Chakraborti, D. and Chanda, B. 1995. Arsenic in ground water in six districts of West Bengal, India: the biggest arsenic calamity in the world. part 2. Arsenic concentration in drinking water, hair, nails, urine, skin-scale and liver tissue (biopsy) of the affected people. *Analyst*, **120**, 917–924.

De Kimpe J, Cornelis, R., Mees, L. and Vanholder, R. 1996. Basal metabolism of intraperitoneally injected carrier-free ^{74}As-labeled arsenate in rabbits. *Fund. Appl. Toxicol.*, **34**, 240–248.

Delnomdedieu, M., Basti, M.M., Otvos, J.D. and Thomas, D.J. 1994a. Reduction and binding of arsenate and dimethylarsinate by glutathione: a magnetic resonance study. *Chem.-Biol. Interact.*, **90**, 139–155.

Delnomdedieu, M., Basti, M.M., Styblo, M., Otvos, J.D. and Thomas, D.J. 1994b. Complexation of arsenic species in rabbit erythrocytes. *Chem. Res. Toxicol.*, **7**, 681–687.

Del Razo, L.M., Garcia-Vargas, G.G., Vargas, H., Albores, A., Gonsebatt, M.E., Montero, R., Ostrosky-Wegman, P., Kelsh, M. and Cebrián, M.E. 1997. Altered profile of urinary arsenic metabolites in adults with chronic arsenicism: A pilot study. *Arch. Toxicol.*, **71**(4), 211–217.

EPA (U.S. Environmental Protection Agency). 1988. Special Report on Inorganic Arsenic: Skin Cancer; Nutritional Essentiality. EPA 625/3-87/013. U.S. Environmental Protection Agency, Risk Assessment Forum, Washington, D.C.

Foà, V., Colombi, A., Maroni, M., Buratti, M. and Calzaferri, G. 1984. The speciation of the chemical forms of arsenic in the biological monitoring of exposure to inorganic arsenic. *Sci. Total Environ.*, **34**, 241–259.

Ganesan, V., Johnson, A., Connelly, A., Eckhardt, S. and Surtees, R.A. 1997. Guanidinoacetate methyltransferase deficiency: new clinical features. *Pediatr. Neurol.*, **17**(2), 155–157.

Geubel, A.P., Mairlot, M.C., Buchet, J.P. and Lauwerys, R. 1988. Abnormal methylation capacity in human liver cirrhosis. *Int. J. Clin. Pharmacol. Res.*, **8**(2), 117–122.

Healy, S.M., Zakharyan, R.A. and Aposhian, H.V. 1997. Enzymatic methylation of arsenic compounds: IV. *In vitro* and *in vivo* deficiency of the methylation of arsenite and monomethylarsonic acid in the guinea pig. *Mutat. Res.*, 386, 229–239.

Hirata M, Hisanaga A, Tanaka A, Ishinishi N. 1988. Glutathione and methylation of inorganic arsenic in hamsters. *Appl. Organomet. Chem.*, **2**, 315–321.

Hirata, M., Tanaka, A., Hisanaga, A. and Ishinishi, N. 1990. Effects of glutathione depletion on the acute nephrotoxic potential of arsenite and on the arsenic metabolism in the hamster. *Toxicol. Appl. Pharmacol.*, **106**, 469–481.

Hopenhayn-Rich, C., Smith, A.H. and Goeden, H.M. 1993. Human studies do not support the methylation threshold hypothesis for the toxicity of inorganic arsenic. *Environ. Res.*, **60**, 161–177.

Hopenhayn-Rich, C., Biggs, M.L., Smith, A.H., Kalman, D.A. and Moore, L.E. 1996a. Methylation study of a population environmentally exposed to arsenic in drinking water. *Environ. Health Perspect.*, **104**, 620–628.

Hopenhayn-Rich, C., Biggs, M.L., Kalman, D.A., Moore, L.E. and Smith, A.H. 1996b. Arsenic methylation patterns before and after changing from high to lower concentrations of arsenic in drinking water. *Environ. Health Perspect.*, **104**, 1200–1207.

Hsu, K.-H., Froines, J.R. and Chen, C.-J. 1997. Studies of arsenic ingestion from drinking-water in northeastern Taiwan: Chemical speciation and urinary metabolites. In: C.O. Abernathy, R.L. Calderon, and W.R. Chappell (eds.), *Arsenic Exposure and Health Effects II*, pp. 190–209. Chapman & Hall, New York.

Hsueh, Y.M., Cheng, G.S., Wu, M.M., Kuo, T.L. and Chen, C.J. 1995. Multiple risk factors associated with arsenic-induced skin cancer: effects of chronic liver disease and malnutrional status. *Br. J. Cancer*, **71**, 109–114.

Hsueh, Y.M., Chiou, H.Y., Huang, Y.L., Wu, W.L., Huang, C.C., Yang, M.H., Lue, L.C., Chen, G.C. and Chen, C.J. 1997. Serum beta-carotene level, arsenic methylation capability, and incidence of skin cancer. *Cancer Epidemiol. Biomarkers Prev.*, **6** (8), 589–596.

Hughes, M.F. and Kenyon, E.M. 1998. Dose-dependent effects on the disposition of monomethylarsonic acid and dimethylarsinic acid in the mouse after intravenous administration. *J. Toxicol. Environ. Health*, **53**(2), 95–112.

Hughes, M.F., Menache, M. and Thompson, D.J. 1994. Dose-dependent disposition of sodium arsenite in mice following acute oral exposure. *Fundam. Appl. Toxicol.*, **22**, 80–89.

Johnson, L.R., and Farmer, J.G. 1991. Use of human metabolic studies and urinary arsenic speciation in assessing arsenic exposure. *Bull. Environ. Contam. Toxicol.*, **46**, 53–61.

Kalman, D.A., Hughes, J., van Belle, G., Burbacher, T., Bolgiano, D., Coble, K., Mottet, N.K. and Polissar, L. 1990. The effect of variable environmental arsenic contamination on urinary concentrations of arsenic species. *Environ. Health Perspect.*, **89**, 145–151.

Kosnett, M.J. and Becker, C.E. 1988. Dimercaptosuccinic acid: Utility in acute and chronic arsenic poisoning. *Vet. Hum. Toxicol.*, **30**(4), 369 (Abstract).

Krynetski, E.Y., Tai, H.L., Yates, C.R., Fessing, M.Y., Loennechen, T., Schuetz, J.D., Relling, M.V. and Evans, W.E. 1996. Genetic polymorphism of thiopurine S-methyltransferase: clinical importance and molecular mechanisms. *Pharmacogenetics*, **6**(4), 279–290.

Kurttio, P., Komulainen, H., Hakala, E., Kahelin, H. and Pekkanen, J. 1998. Urinary excretion of arsenic species after exposure to arsenic present in drinking water. *Arch. Environ. Contam. Toxicol.*, **34**(3), 297–305.

Lerman, S.A., Clarkson, T.W. and Gerson, R.J. 1983. Arsenic uptake and metabolism by liver cells is dependent on arsenic oxidation state. *Chem.-Biol. Interact.*, **45**, 401–406.

Li, T., Vallada, H., Curtis, D., Arranz, M., Xu, K., Cai, G., Deng, H., Liu, J., Murray, R., Liu, X., and Collier, D.A. 1997. Catecol-O-methyltransferase *Val 158Met* polymorphism: frequency analysis in Han Chinese subjects and allelic association of the low activity allele with bipolar affective disorder. *Pharmacogenetics*, **7**, 349–353.

Mahieu, P., Buchet, J.P., Roels, H.A. and Lauwerys, R. 1981. The metabolism of arsenic in humans acutely intoxicated by As$_2$O$_3$. Its significance for the duration of BAL therapy. *Clin. Toxicol.*, **18**, 1067–1075.

Maiorino, R.M. and Aposhian, H.V. 1985. Dimercaptan metal-binding agents influence the biotransformation of arsenite in the rabbit. *Toxicol. Appl. Pharmacol.*, **77**, 240–250.

Maiorino, R.M., Dart, R.C., Carter, D.E. and Aposhian, H.V. 1991. Determination and metabolism of dithiol chelating agents. XII. Metabolism and pharmacokinetics of sodium 2,3-dimercaptopropane-1-sulfonate in humans. *J. Pharmacol. Exp. Ther.*, **259**(2), 808–814.

Mann, S., Droz, P.O. and Vahter, M. 1996a. A physiologically based pharmacokinetic model for arsenic exposure. I. Development in hamsters and rabbits. *Toxicol. Appl. Pharmacol.*, **137**, 8–22.

Mann, S., Droz, P.O. and Vahter, M. 1996b. A physiologically based pharmacokinetic model for arsenic exposure. II. Validation and applications in humans. *Toxicol. Appl. Pharmacol.*, **140**, 471–486.

Marafante, E. and Vahter, M. 1984. The effect of methyltransferase inhibition on the metabolism of (^{74}As) arsenite in mice and rabbits. *Chem.-Biol. Interact.*, **50**, 49–57.

Marafante, E., Rade, J. and Sabbioni, E. 1981. Intracellular interaction and metabolic fate of arsenite in the rabbit. *Clin. Toxicol.*, **18**, 1335–1341.

Marafante, E., Vahter, M. and Envall, J. 1985. The role of the methylation in the detoxication of arsenate in the rabbit. *Chem.-Biol. Interact.*, **56**, 225–238.

Marafante, E., Vahter, M., Norin, H., Envall, J., Sandström, M., Christakopoulos, A. and Ryhage, R. 1987. Biotransformation of dimethylarsinic acid in mouse, hamster and man. *J. Appl. Toxicol.*, **7**(2), 111–117.

Menzel, D.B. 1997. Some results of a physiological based pharmacokinetic modeling approach to estimating arsenic body burdens. In: C.O. Abernathy, R.L. Calderon and W.R. Chappell (eds.), *Arsenic Exposure and Health Effects*, pp. 349–368. Chapman & Hall, London.

Moore, M.M., Harrington-Brock, K. and Doerr, C.L. 1997. Relative genotoxic potency of arsenic and its methylated metabolites. *Mutat. Res.*, **386**, 279–290.

NRCC. National Research Council of Canada, Associate Committee on Scientific Criteria for Environmental Quality. 1978. *Effects of Arsenic in the Canadian Environment*. National Research Council of Canada, pp. 1–349.

Odanaka, Y., Matano, O. and Goto, S. 1980. Biomethylation of inorganic arsenic by the rat and some laboratory animals. *Bull. Environ. Contam. Toxicol.*, **24**, 452–459.

Pomroy, C., Charbonneau, S.M., McCullough, R.S. and Tam, G.K.H. 1980. Human retention studies with ^{74}As. *Toxicol. Appl. Pharmacol.*, **53**, 550–556.

Rasmussen, R.E. and Menzel, D.B. 1997. Variation in arsenic-induced sister chromatid exchange in human lymphocytes and lymphoblastoid cell lines. *Mutat. Res.*, **386**, 299–306.

Sakurai, T., Kaise, T. and Matsubara, C. 1998. Inorganic and methylated arsenic compounds induce cell death in murine macrophages via different mechanisms. *Chem. Res. Toxicol.*, **11**(4), 273–283.

Scheller, T., Orgacka, H., Szumlaski, C.L. and Weinshilboum, R.M. 1996. Mouse liver nicotinamide N-methyltransferase pharmacogenetics: biochemical properties and variation in activity inbred strains. *Pharmacogenetics*, **6**(1), 43–53.

Scott, M. C., Van Loon, J.A. and Weinshilboum, R.M. 1988. Pharmacogenetics of N-methylation: Heritability of human erythrocyte histamine N-methyltransferase activity. *Clin. Pharmacol. Ther.*, **43**, 256.

Smith, A.H., Goycolea, M., Haque, R. and Biggs, M.L. 1998. Marked increase in bladder and lung cancer mortality in a region of northern Chile due to arsenic in drinking water. *Am. J. Epidemiol.*, **147** (7), 660–669.

Stockler, S., Isbrandt, D., Hanefeld, F., Schmidt, B. and Figura, K. 1996. Guanidinoacetate methyl-transferase deficiency: the first inborn error of creatine metabolism in man. *Am. J. Hum. Genet.*, **58**(5), 914–922.

Styblo, M. and Thomas, D.J. 1997a. Factors influencing *in vitro* methylation of arsenicals in rat liver cytosol. In: C.O. Abernathy, R.L. Calderon, and W.R. Chappell (eds.), *Arsenic Exposure and Health Effects II*, pp. 283–295. Chapman & Hall, New York.

Styblo, M. and Thomas, D.J. 1997b. Binding of arsenicals to proteins in an *in vitro* methylation system. *Toxicol. Appl. Pharmacol.*, **147**(1), 1–8.

Styblo, M., Yamauchi, H. and Thomas, D. 1995. Comparative *in vitro* methylation of trivalent and pentavalent arsenicals. *Toxicol. Appl. Pharmacol.*, **135**, 172–178.

Styblo, M., Serves, S.V., Cullen, W.R. and Thomas, D.J. 1997. Comparative inhibition of yeast glutathione reductase by arsenicals and arsenothiols. *Chem. Res. Toxicol.* **10**(1), 27–33.

Tam, G.K.H., Charbonneau, S.M., Bryce, F., Pomroy, C. and Sandi, E. 1979. Metabolism of inorganic arsenic (^{74}As) in humans following oral ingestion. *Toxicol. Appl. Pharmacol.*, **50**, 319–322.

Thompson, D.J. 1993. A chemical hypothesis for arsenic methylation in mammals. *Chem.-Biol. Interact.*, **88**, 89–114.

Thornton, I. and Farago, M. 1997. The geochemistry of arsenic. In: C.O. Abernathy, R.L. Calderon and W.R. Chappell (eds.), *Arsenic Exposure and Health Effects II*, pp. 1–16. Chapman & Hall, New York. Tice, R.R., Yager, J.W., Andrews, P. and Crecelius, E. 1997. Effect of hepatic methyl donor status on urinary excretion and DNA damage in B6C3F1 mice treated with sodium arsenite. *Mutat. Res.*, **386**, 315–334.

Tseng, W.-P. 1977. Effects and dose–response relationships of skin cancer and blackfoot disease with arsenic. *Environ. Health Perspect.*, **19**, 109–119.

Vahter M. 1981. Biotransformation of trivalent and pentavalent inorganic arsenic in mice and rats. *Environ. Res.*, **25**, 286–293.

Vahter, M. 1994a. What are the chemical forms of arsenic in urine, and what can they tell us about exposure? *Clin. Chem.*, **40**(5), 679–680.

Vahter M. 1994b. Species differences in the metabolism of arsenic. In: W.R. Chappell, C.O. Abernathy and C.R. Cothern (eds.), *Arsenic Exposure and Health.* pp. 171–179. Science and Technology Letters, Northwood.

Vahter, M. and Envall, J. 1983. *In vivo* reduction of arsenate in mice and rabbits. *Environ. Res.*, **32**, 14–24.

Vahter, M. and Marafante, E. 1983. Intracellular interaction and metabolic fate of arsenite and arsenate in mice and rabbits. *Chem.-Biol. Interact.*, **47**, 29–44.

Vahter, M. and Marafante, E. 1985. Reduction and binding of arsenate in marmoset monkeys. *Arch. Toxicol.*, **57**, 119–124.

Vahter, M. and Marafante, E. 1987. Effects of low dietary intake of methionine, choline or proteins on the biotransformation of arsenite in the rabbit. *Toxicol. Lett.*, **37**, 41–46.

Vahter, M. and Marafante, E. 1988. *In vivo* methylation and detoxication of arsenic. In: P.J. Craig and F. Glockling (eds.), *The Biological Alkylation of Heavy Elements*, pp. 105–119. Royal Society of Chemistry, London.

Vahter, M., Marafante, E., Lindgren, A. and Dencker, L. 1982. Tissue distribution and subcellular binding of arsenic in marmoset monkeys after injection of ^{74}As-arsenite. *Arch. Toxicol.*, **51**, 65–77.

Vahter, M., Marafante, E. and Dencker, L. 1984. Tissue distribution and retention of ^{74}As-dimethylarsinic acid in mice and rats. Arch. *Environ. Contam. Toxicol.*, **13**, 259–264.

Vahter, M., Concha, G., Nermell, B., Nilsson, R., Dulout, F. and Natarajan, A.T. 1995a. A unique metabolism of inorganic arsenic in native Andean women. *Eur. J. Pharmacol.*, **293**, 455–462.

Vahter, M., Couch, R., Nermell, B. and Nilsson, R. 1995b. Lack of methylation of inorganic arsenic in the chimpanzee. *Toxicol. Appl. Pharmacol.*, **133**, 262–268.

Warner, M.L., Moore, L.E., Smith, M.T., Kalman, D.A., Fanning, E. and Smith, A.H. 1994. Increased micronuclei in exfoliated bladder cells of persons who chronically ingested arsenic-contaminated water in Nevada. *Cancer Epidemiol. Biomarkers Prev.*, **3**, 583–590.

Yamauchi, H. and Yamamura, Y. 1984. Metabolism and excretion of orally administered dimethylarsinic acid in the hamster. *Toxicol. Appl. Pharmacol.*, **74**, 134–140.

Yamauchi and Yamamura 1985. Yamauchi, H., and Y. Yamamura. 1985. Metabolism and excretion of orally administrated arsenic trioxide in the hamster. *Toxicology*, **34**, 113–121.

Yates, C.R., Krynetski, E.Y., Loennechen, T., Fessing, M.Y., Tai, H.L., Pui, C.H., Relling, M.V. and Evans, W.E. (1997) Molecular diagnosis of thiopurine S-methyltransferase deficiency: genetic basis for azathioprine and mercaptopurine intolerance. *Ann. Intern. Med.*, **126**(8), 608–614.

Zakharyan, R.A., Y. Wu, G.M. Bogdan, and H.V. Aposhian. 1995. Enzymatic methylation of arsenic compounds. I: Assay, partial purification, and properties of arsenite methyltransferase and monomethylarsonic acid methyltransferase of rabbit liver. *Chem. Res. Toxicol.*, **8**, 1029–1038.

Zakharyan, R.A., Wildfang, E. and Aposhian, H.V. 1996. Enzymatic methylation of arsenic compounds: III. The marmoset and tamarin, but not the rhesus, monkey are deficient in methyltransferases that methylate inorganic arsenic. *Toxicol. Appl. Pharmacol.*, **140**, 77–84.

Zaldivar, R. and Guillier, A. 1977. Environmental and clinical investigations on the endemic chronic arsenic poisoning in infants and children. *Zbl. Bakt. Hyg., I. Abt. Orig. B.*, **165**, 226–234.

Arsenic Exposure and Health Effects
W.R. Chappell, C.O. Abernathy and R.L. Calderon (Editors)
© 1999 Elsevier Science B.V. All rights reserved.

Profile of Urinary Arsenic Metabolites in Children Chronically Exposed to Inorganic Arsenic in Mexico

Luz María Del Razo, Gonzalo G. García-Vargas, María C. Hernández,
Arístides Gómez-Muñoz, Mariano E. Cebrián

ABSTRACT

The urinary arsenic (As) excretion pattern was evaluated in 71 children aged 3–10 in Region Lagunera, Mexico. The high exposure group (43) was chosen from Ampueros, a town that had 0.415 mgAs/L in its drinking water. The low exposure group (28) was from Nazareno, whose As concentration was 0.025 mg/L. Most As was in its inorganic (As_i) form and more than 92% was in its pentavalent state. The average values of total arsenic (TAs) in children (TAs = 1,641.7 μgAs/gcreat; CI = 965.9–2317.0) were higher than those reported for adults (TAs geometric mean = 561 μgAs/gcreat; 95% IC = 450–689) exposed to similar As concentrations in drinking water. Highly exposed children had significant increases in the average proportion of As_i in urine. There was an increasing percentage of As_i and methylarsonic acid (MMA) and a corresponding decrease of dimethylarsinic acid (DMA) excreted in urine, associated to increasing concentrations (100 to 1,250 μgAs/L) of TAs. This suggests that high chronic arsenic exposure also decreases children's ability to methylate As_i to MMA and DMA. Our results were not generally consistent with those reported for northern Argentinean children who had a decreasing percentage of urinary As_i and increased DMA associated to increasing TAs (range 125–578 μg/L. However, we found some children with low proportions (<5%) of urinary As_i or MMA, in spite of having TAs levels as high as 2,500 μg/L, illustrating the considerable interindividual variation in children's As methylation efficiency.

Keywords: arsenic, methylation, children, chronic exposure, urine

INTRODUCTION

Arsenic (As) and some of its compounds are naturally present in the environment and are insidious environmental pollutants in many industrialized cities (Fergusson, 1990). High environmental exposure of human beings to As is mainly resultant from natural contamination of well water supplies by As-rich geologic strata (Chen et al., 1985; De Sastre et al., 1992; Sancha et al., 1992; Cebrián et al., 1994; Mazumder et al., 1998) or from industrial activities, such as nonferrous smelters where workers and nearby residents are exposed both through air and atmospheric deposition in soil, dust and water (WHO, 1981; Lagerkvist and Zetterlund, 1994). In many species, including humans, inorganic arsenic (As_i) is enzymatically methylated to methylarsonic acid (MMA) and dimethylarsinic acid (DMA) (Styblo et al., 1995). The methylation of As_i has been traditionally considered a detoxification process, because when methylated compounds are compared with As_i, they are: (i) less acutely toxic (Yamauchi and Fowler, 1994); (ii) less mutagenic (Moore et al., 1997); (iii) less reactive with tissue components (Vahter and Marafante, 1983); and (iv) excreted faster in urine (Buchet et al., 1981a; Hughes and Kenyon, 1998). Nevertheless, there is growing evidence indicating that As methylated metabolites, specifically DMA, are able to damage DNA and promote tumors in experimental animals (Yamanaka et al., 1997).

Humans excrete arsenic mainly via kidney. Therefore, urinary As concentrations have been considered a useful biomarker of exposure (Crecelius, 1977; Buchet et al., 1980). Knowledge about the nature and concentration of As species in urine provides information on As exposure, the type of As compounds ingested and the methylating capacity of individuals. This information is relevant because the action of arsenic as a toxin and carcinogen is fundamentally associated with its metabolism and disposition. Previous studies in experimental models have shown that As methylation capacity is influenced by the dose level (Vahter, 1981; Buchet et al., 1981b). In addition, chronically exposed humans have shown an altered urinary arsenic excretion pattern (Del Razo et al., 1994). However, little information is available about the methylation capability of children chronically exposed to As via drinking water. This would provide useful background information for assessing the risk of arsenic exposure to human health. Thus, the aim of this paper is to report on the urinary excretion of As metabolites in children living in Region Lagunera, Mexico, where chronic arsenic poisoning is endemic.

METHODOLOGY

Group Selection

The study was conducted in Region Lagunera located in the central part of north Mexico. The high exposure group (43 children) was chosen from Ampueros, Coahuila, a town which had an average of 0.415 mgAs/L (range 0.369–0.470 mg/l; $n = 10$) in its drinking water. The low exposure group (28 children) was chosen from Nazareno, Durango, whose average concentration was 0.025 mgAs/L (range 0.020–0.040 mg/L; $n = 9$). In both rural towns, most As was in its inorganic form and more than 92% was in its pentavalent state. Participants were between 3 and 10 years old and had lived in their respective towns all their lives. Data on socioeconomic variables of families, clinical state of children and length of exposure were obtained by questionnaire and physical examination. First morning voided urine samples were collected from each child and stored in stoppered polyethylene bottles at –15°C until analyses were performed.

Reagents

Sodium arsenate, sodium arsenite and DMA were bought from Sigma Chemical Co. (St. Louis, MO), whereas MMA was obtained from Vineland Chemical Co. (Vineland, NJ). Stock solutions containing 1 mgAs/mL were used to prepare working standards for analysis each day. Sodium borohydride was obtained from Merck (Darmstadt, Germany). Water (SRM 1643c) and freeze-dried urine (SRM 2670) standard reference material for toxic metals were obtained from the US National Institute of Standards and Technology (Gaithersburg, MD). All other chemicals were of analytical grade. Doubly distilled and deionized water was used for all analytical work, glassware was soaked in 10% nitric acid rinsed with water and dried before use.

Urine Analysis

Arsenic was determined by hydride generation atomic absorption, using a Perkin Elmer 3100, equipped with a FIAS-200 flow injection atomic spectroscopy system. All measurements were made using an arsenic electrodeless discharge lamp at 197.3 nm in a heated quartz cell. Total As (TAs) in urine was determined in samples previously wet digested with nitric, sulfuric and perchloric acids (Cox, 1980). Separation of As species (As$_i$, MMA and DMA) was performed by ion exchange chromatography as previously reported (Del Razo et al., 1994). Creatinine in urine was measured according to Lim (1982) and As concentrations in urine were adjusted for creatinine.

Statistical Procedures

Data analyses were carried out using STATA 5.0® statistical program package. The concentrations of As species, their proportions and ratios were transformed to a log scale in order to calculate means and confidence intervals. The values were transformed back to arithmetic scale for reporting purposes. The significance of differences between means was analyzed using Student's test.

RESULTS

As expected, average urinary concentrations of arsenic species and TAs, were significantly higher (8 to 16 times) in children exposed to high As concentrations in drinking water (0.415 mg/L), as compared to those exposed to lower concentrations (0.025 mg/L). The highly exposed children had significant increases in the average proportion of As$_i$ in urine, but no significant differences were found in the proportions of MMA or DMA (Table 1). There was a good agreement between the values obtained for TAs and the sum of species, suggesting that TAs was not influenced by seafood arsenic in this study.

TABLE 1

Total arsenic in urine and As species in children. Region Lagunera, Mexico. Mean (CI)[1]

	Nazareno Low Exposure		Ampueros High Exposure	
	Arithmetic mean	%	Arithmetic mean	%
As$_i$ (μg/g creatinine)	19.6	10.3	229.7*	15.7*
	(5.9–33.3)	(7.6–13.2)	(119.6–339.8)	(12.2–19.1)
MMA (μg/g creatinine)	18.1	13.7	257.4*	12.5
	(9.8–26.4)	(10.8–16.5)	(114.7– 400.0)	(9.8–15.1)
DMA (μg/g creatinine)	153.3	75.8	1154.6*	71.8
	(59.2–247.4)	(71.5–80.2)	(681.0–1628.2)	(66.9–76.6)
Total As (μg/g creatinine)	191.9		1641.7*	
	(78.6–303.5)		(965.9–2317.0)	
Creatinine μg/l	976.6		1143.8	
	(770.7–1182.6)		(956.4–1331.2)	
N	28		43	

*$p < 0.05$ Student's *t*-test. High exposure vs low exposure.
CI[1] = 95% Confidence interval.

As shown in Figure 1, the magnitude of exposure altered the proportions of urinary As species and their relationships with the concentration of TAs excreted in urine. In control children, defined as those with TAs in urine lower than 100 μg/L, there were no significant linear relationships between As species and TAs. In contrast, there was a significant positive

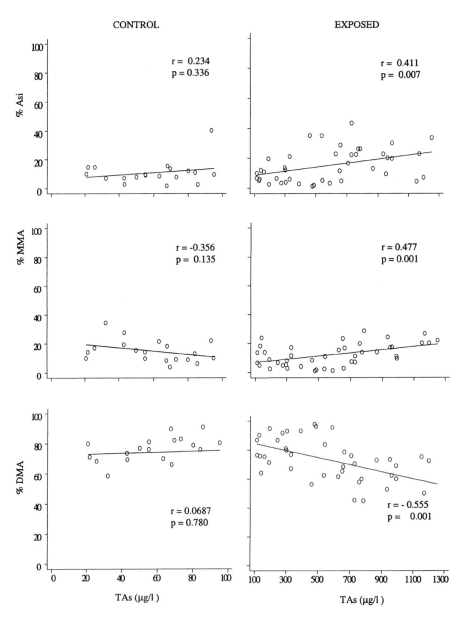

Fig. 1. Distribution of urinary arsenic metabolites (%) in relation to urinary total arsenic (TAs) in children from Region Lagunera, Mexico.

TABLE 2

Relationships between the ratios of urinary arsenic species and TAs concentration in children. Region Lagunera, Mexico

	≤ 100 AsT Arithmetic mean (95% confidence interval)	>100 AsT Arithmetic mean (95% confidence interval)
MMA/As$_i$	2.19 (1.08–3.29)	1.18* (0.87–1.48)
DMA/MMA	6.83 (4.54–9.11)	13.97 (7.87–20.06)
Organic As/As$_i$	14.22 (7.48–20.95)	13.15 (8.41–17.88)
N	19	42

*$p < 0.05$ Student's t-test.

association between the percentages of As$_i$ and MMA with TAs in children whose urinary concentrations were between 100 μg/L and 1300 μg/L. A corresponding negative association between DMA percentage and TAs was also observed. However, we also found a few children with low proportions (<5%) of urinary As$_i$ or MMA, in spite of having TAs levels as high as 500 to 700. Three of those children had TAs levels as high as 2,500 μg/L.

The efficiency of the methylation process was also assessed by the ratios between urinary concentrations of putative products and putative substrates of the As metabolism pathway. Table 2 shows a significant decrease in the values of the MMA/As$_i$ ratio in the highly exposed children.

The association between arsenic exposure and alterations in the profile of As species in urine remained significant when potential confounding factors were controlled. No cutaneous signs of chronic arsenic poisoning were found in the studied children. No significant differences in diet quantity or quality were observed between the populations studied.

DISCUSSION

The concentration of As in the drinking water of the exposed town (Ampueros) was about 40 times higher than the provisional WHO drinking water guideline of 10 μgAs/L (WHO, 1993), whereas that of the low exposure town (Nazareno) was about 2.5 times higher. Although little information is available on the concentration of As in urine from exposed children, the average values of As species and TAs in children from Region Lagunera here reported are among the highest reported in the literature. In fact, average TAs concentration in children's urine was higher than that reported for adults (TAs geometric mean = 561 μgAs/gcreat; 95% IC = 450–689 μgAs/gcreat) exposed to similar As concentrations in drinking water in the same contaminated area (Del Razo et al., 1997). A possible explanation for this difference would be associated to age-related progressive decreases in the dose of As ingested (mg/kg), since children require more water than adults (mL/kg). This would be reflected as an increased urinary As output in children.

The present work has shown that chronically exposed children also have an altered urinary excretion pattern of arsenic species. The main changes were significant increases in the average proportion of As$_i$ excreted, without significant alterations in the proportions of MMA or DMA. The magnitude of As$_i$ increase was consistent with that observed in our previous studies on As-exposed adults from the same area (Del Razo et al., 1997).

There are only three previous reports on the speciation of As in children. Buchet et al. (1980) reported average values of 12% As$_i$, 28% MMA and 60% DMA in Belgium (n = 14). Kalman et al. (1990) reported average values of 13% As$_i$, 16% MMA and 71% DMA in United States children. However in these studies TAs was low (<20 μg/L). In the most recent published study Concha et al. (1998), found that children exposed to ~200 μgAs/L via drinking water with an average TAs of 323 μgAs/L in their urine, had a considerably higher average proportion of As_i (~47%), exceedingly low MMA (3.6 %) and a low proportion of DMA (47%). The low percentage of MMA in children and women was similar to that reported by the same research group in native Andean women (Vahter et al., 1995).

An interesting finding in the present study was the increasing percentage of As_i and MMA excreted in urine and the corresponding decrease of DMA associated to increasing concentrations (100 to 1,250 μg/L) of TAs in urine. The efficiency of the methylation process was also assessed by the ratios between urinary concentrations of putative products and putative substrates of the As metabolism pathway. Highly exposed children showed significant decreases in the MMA/As_i ratio without significant changes in the DMA/MMA ratio. These alterations suggest that high chronic arsenic exposure decreases the body's ability to methylate As_i to MMA and DMA, suggesting that arsenic metabolism in children is similar to that in adults in the mestizo Mexican population of Region Lagunera. In general terms, our results were not consistent with those reported for northern Argentinean children by Concha et al. (1998), who found a decreasing percentage of urinary As_i and a corresponding increase in the percentage of DMA in exposed children, but not in exposed women, associated to increasing TAs (sum of As metabolites) in the range 125–578 μg/L. However, we found some children with low proportions (<5%) of urinary As_i or MMA, in spite of having TAs levels as high as 2,500 μg/L, illustrating the considerable interindividual variation in As methylation. Further studies are needed on the importance of genetically determined variation, since changes in the activities of other methyltransferases have been explained by the presence of genetic polymorphism (Weinshilboum, 1992). The role of high As exposure in children in the appearance of cancer at later ages should also be investigated.

ACKNOWLEDGEMENTS

This work was partially supported from Consejo Nacional de Ciencia y Tecnología, México (CONACYT 211085-5-4082) and by USEPA and ECO/PAHO under the program "Environmental epidemiology: a project for Latin America and the Caribbean". The secretarial assistance of Ms. Rosalinda Flores M. is also acknowledged.

REFERENCES

Buchet, J.P., Roels, H., Lauwerys, R., Bruaux, P., Claeys-Thoreau, F., Lafontaine, A., Verduyn, G. 1980. Repeated surveillance of exposure to cadmium, manganese, and arsenic in school-age children living in rural urban and nonferrous smelter areas in Belgium. *Environ. Res.*, 22, 95–108.

Buchet, J.P., Lauwerys, R., Roels, H. 1981a. Comparison of the urinary excretion of organic metabolites after a single oral dose of sodium arsenite, monomethylarsonate, or dimethylarsinate in man. *Int. Arch. Occup. Environ. Health*, 48, 71–79.

Buchet, J.P., Lauwerys, R., Roels, H. 1981b. Urinary excretion of inorganic arsenic and its metabolites after repeated ingestion of sodium metaarsenite by volunteers. *Int. Arch. Occup. Environ. Health*, 48, 111–118.

Cebrián, M.E., Albores, A., García-Vargas,G.G., Del Razo, L.M. and Ostrosky-Wegman, P. 1994. Chronic Arsenic Poisoning in Humans: The case of Mexico. In: J.O. Nriagu (ed.), *Arsenic in the Environment. Part II: Human Health and Ecosystem Effects*, pp. 93–107. John Wiley & Sons, New York.

Chen, C.J., Chuang, Y.C., Lin, T.M., Wu, T.M. 1985. Malignant neoplasm among residents of a Blackfoot disease-endemic area in Taiwan: High arsenic artesian well water and cancers. *Cancer Res.*, 45, 5895–5899.

Crecelius, E.A. 1977. Changes in the chemical speciation of arsenic following ingestion by man. *Environ. Health Perspect.*, 19, 147–150.

Concha, G., Barbro, N., Vahter, M. 1998. Metabolism of inorganic arsenic in children with chronic high arsenic exposure in Northern Argentina. *Environ. Health Perspect.*, 106, 355–359.

Cox, D.H. 1980. Arsine evolution-electrothermal atomic absorption method for the determination of nanogram levels of total arsenic in urine and water. *J. Anal. Toxicol.*, 4, 207–211.

Del Razo, L.M., Hernández, J.L., García-Vargas, G.G., Ostrosky-Wegman, P., Cortinas, C., Cebrián, M.E. 1994. Urinary excretion of arsenic species in a human population chronically exposed to arsenic via drinking water. A pilot study. In: Chappell W.R., Abernathy C.O., Cothern, C.R. (eds.) *Exposure and Health*, pp. 91–100. Norwood, UK.

Del Razo, L.M., García-Vargas, G.G., Vargas, H., Albores, A., Gonsebatt, M.E., Montero, R., Ostrosky-Wegman, P., Kelsh, M., Cebrián, M.E. 1997. Altered profile of urinary arsenic metabolites in adults with chronic arsenicism. A pilot study. *Arch. Toxicol.*, **71**, 211–217.

De Sastre, M.S., Varillas A.E., Kirschbaum, P. 1992. Arsenic content in water in the northwest area of Argentina. In: *International Seminar Proceedings. Arsenic in the Environment and its Incidence on Health*. Universidad de Chile (ed.) pp. 123–130. Santiago, Chile.

Fergusson, J.E. 1990. The heavy elements: Chemistry, Environmental Impact and Health Effects. *Environ. Chem.*, 3A, 59–107.

Hughes M.F., Kenyon E.M.1998. Dose-dependent effects on the disposition of monomethylarsonic acid and dimethylarsinic acid in the mouse after intravenous administration. *J. Toxicol. Environ. Health, A*, **53**, 95–112.

Kalman, D.A., Hughes, J., van Belle, G., Burbacher, T., Bolgiano, O., Coble, K., Mottet, N.K., Polissar, L. 1990. The effect of variable environmental arsenic contamination on urinary concentrations of arsenic species. *Environ. Health Perspect.*, **89**, 145–151.

Lagerkvist, B.J., Zetterlund, B. (1994). Assessment of exposure to arsenic among smelter workers: A five-year follow-up. *Am. J. Ind. Med.*, **25**, 477–488.

Lim, C.K.1982. Some routine applications of high performance liquid chromatography in clinical chemistry In: R. Kaiser, F. Gabl, M.M. Muler, and M. Bayer (ed) pp 957. *XI International Congress of Clinical Chemistry*. Walter de Gruyter and Co., Berlin.

Mazumder, D.N.G., Haque, R., Ghosh, N., De, B.K., Santra, A., Charkraborty, D., Smith A.H. 1998. Arsenic levels in drinking water and the prevalence of skin lesions in West Bengal, India. *Int. J. Epidemiol.*, **27**, 871–877.

Moore ,M.M, Harrington-Brock, K., Doerr, C.L. 1997. Relative genotoxic potency of arsenic and its methylated metabolites. *Mutat. Res.*, **386**, 279–290.

Sancha, A.M. , Vega, F., Venturino, H., Fuentes, S., Salazar, AM., Moreno, V., Baron, A.M., Rodriguez D. 1992. The arsenic health problem in northern Chile evaluation and control. A case study preliminary report. In: *International Seminar Proceedings. Arsenic in the Environment and its Incidence on Health*. Universidad de Chile (ed.) pp. 187–202, Santiago, Chile.

Styblo, M., Delnomdedieu, M., Thomas D.J. 1995. Biological mechanisms and toxicological consequences of the methylation of arsenic. In: R.A. Goyer and M.G. Cherian (ed.) *Toxicology of Metals-Biochemical Aspects*, Handbook of Experimental Pharmacology Vol. 115, pp. 407–433. Springer-Verlag, Berlin.

Vahter, M. 1981. Biotransformation of trivalent and pentavalent inorganic arsenic in mice and rats. *Environ. Res.*, **25**, 286–293.

Vahter, M., Marafante, E. 1983. Intracellular interaction and metabolic fate of arsenite and arsenate in mice and rabbits. *Chem. Biol. Interact.*, **47**, 29–44.

Vahter, M., Concha, G., Nermell, B., Nilsson, R., Dulout, F., Natarajan, A.T. 1995. A unique metabolism of inorganic arsenic in native Andean women. *Eur. J. Pharmacol.*, **293**, 455–462.

Weinshilboum, R.M. 1992. Methylation pharmacogenetics; thiopurine methyltransferase as a model system. *Xenobiotica*, **22**, 1055–1071.

WHO. World Health Organization 1981. Environmental Health Criteria 18. Arsenic. Geneva.

WHO. World Health Organization. 1993. Guidelines for drinking-water quality, 2nd ed. Geneva.

Yamanaka, K., Hayashi H., Tachikawa, M., Kato, K., Hasegawa, A., Oku, N., Okada, S. 1997. Metabolic methylation is a possible genotoxicity-enhancing process of inorganic arsenic. *Mut. Res.*, **394**, 95–101.

Yamauchi, H., Fowler, B.A. 1994. Toxicity and metabolism of inorganic and methylated arsenicals. In: J.O. Nriagu (ed.), *Arsenic in the Environment. Part II: Human Health and Ecosystem Effects*, pp. 35–43. John Wiley, New York.

Arsenic Exposure and Health Effects
W.R. Chappell, C.O. Abernathy and R.L. Calderon (Editors)
© 1999 Elsevier Science B.V. All rights reserved.

How Is Inorganic Arsenic Detoxified?

H. Vasken Aposhian, Robert A. Zakharyan, Eric K. Wildfang,
Sheila M. Healy, Jürgen Gailer, Timothy R. Radabaugh,
Gregory M. Bogdan, LaTanya A. Powell, Mary M. Aposhian

ABSTRACT

A flow chart to help understand the detoxification and biotransformation of inorganic arsenic is presented. Arsenate is reduced to arsenite enzymatically by arsenate reductase and non-enzymatically by GSH. An early step in the detoxification appears to be the formation of the Gailer compound, seleno-bis(S-glutathionyl) arsinium ion, which is rapidly formed and excreted in the bile. Arsenite-binding proteins initially may prevent or enhance the accumulation of toxic levels of arsenite. As these binding sites become saturated, the arsenite may be released for methylation, a biotransformation process which results in the increase of urinary arsenic. Methylation of arsenic species can occur via SAM and methyltransferases and/or nonenzymatically with methylvitamin B_{12}, GSH and selenite. Methylation by the methylvitamin B_{12} system has been shown *in vitro* only. The substrate for DMA production appears to be MMA^{III}. The lack of methyltransferases in many primates strongly indicates that methylation may not be the primary detoxification pathway for inorganic arsenic. In fact, the EPA classifies dimethylarsinic acid, the final urinary metabolite for arsenic in humans, as a probable human carcinogen. The determination of the amino acid sequences of the arsenic methyltransferases needs to be accomplished so that gene probes can be constructed to better study arsenic methyltransferase polymorphism as it relates to the various responses of people to inorganic arsenic.

Keywords: methylvitamin B_{12}, MMA^{III}, arsenic detoxification

INTRODUCTION

A number of recent research findings in arsenic toxicology have stimulated a reexamination of how inorganic arsenic is detoxified. Many papers published in the past need to be reconsidered. Methylation of inorganic arsenic is clearly a biotransformation process but may not be a detoxification mechanism as has been believed in the past. The putative pathways (Figure 1) for the biotransformation of inorganic arsenic to its methylated derivatives, monomethylarsonic acid (MMA^V) and dimethylarsinic acid (DMA), were based originally on the results gathered from *in vitro* experiments using crude extracts of fungi or bacteria, not from mammalian systems (Challenger, 1951; Cullen et al., 1989; for a review, see Cullen and Reimer, 1989). These early studies were followed by experiments in which inorganic arsenic was given to laboratory animals or humans and the urines collected and analyzed for MMA and DMA (for a review see Vahter, 1994). As the result of the development of a rapid, relatively simple, assay for mammalian arsenic methyltransferases (Zakharyan et al., 1995), the purification and study of the properties of these unusual enzyme activities became possible (for a review, see Aposhian, 1997). These studies have allowed an expansion of earlier studies (Buchet and Lauwerys, 1985, 1988; Styblo et al., 1995, 1996).

Arsenate and Arsenite

A flow chart for the detoxification of inorganic arsenic in mammals (Figure 2) usually begins with the reduction of arsenate to arsenite even though the latter is more toxic than the former. In the blood this reaction can proceed non-enzymatically, using GSH as the reductant, as well as enzymatically (Winski and Carter, 1995). Arsenate reductase is being purified in our laboratory from human liver in order to understand its properties. A thiol, such as GSH, plus a heat stable, small molecular weight, unknown cofactor are required for activity. The unknown cofactor does not appear to be a metal ion. The arsenate and arsenite are interconvertible. The administration of one will result in the appearance of the other in the urine (Vahter, 1981; Vahter and Marafante, 1988).

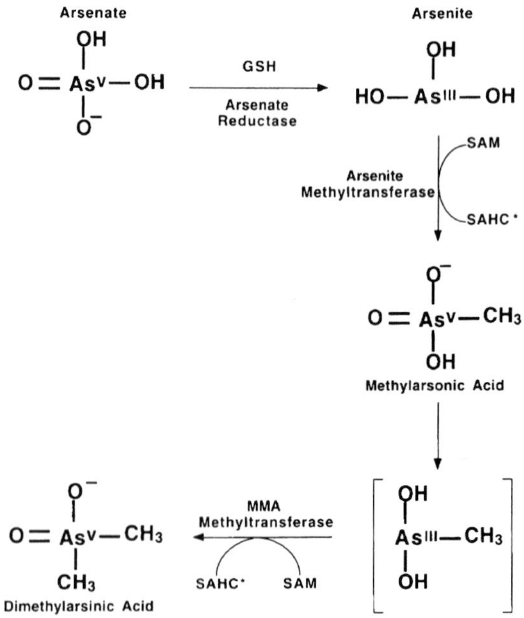

Fig. 1. Putative pathway for the biotransformation of inorganic arsenic. Abbreviations: GSH (glutathione), SAM (S-adenosyl-methionine), SAHC (S-adenosyl-homocysteine).

Fig. 2. Flow chart for detoxification of inorganic arsenic. Abbreviations: GSH (glutathione), CySH (L-cysteine), DMPS (2,3-dimercaptopropane-1-sulfonate), SAM (S-adenosyl-methionine), CH$_3$-vitamin B$_{12}$ (methylcobalamin), MMA^V (methylarsonic acid), MMA^{III} (methylarsonous acid), DMA (dimethylarsinic acid).

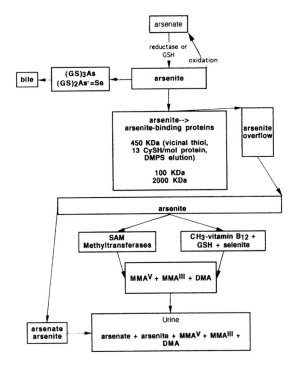

Biliary Arsenic Excretion

It has been known for many years that as the arsenic concentration in the bile increases so do GSH (Gyurasics, et al., 1991) and selenium (Levander and Baumann, 1966). The formation of (GS)$_3$As has been reported (Scott et al., 1993; Delnomdedieu et al., 1993) and it has been detected in biological systems (Carter, personal communications).

A new glutathione–arsenic–selenium compound has been detected recently in rabbit bile by Dr. Jürgen Gailer of our group. He injected rabbits intravenously with sodium selenite followed immediately with intravenous sodium arsenite. Within 25 min a compound containing arsenic and selenium (Figure 2) was excreted in the bile. This newly discovered adduct, the seleno-bis(S-glutathionyl) arsinium ion, that we call the Gailer compound explains for the first time a likely mechanism by which selenium prevents arsenic toxicity. Remember that selenium is a nutritionally required element, is ubiquitous in all cells and has been known for many years to be an antagonist of arsenic toxicity. The formation of the Gailer compound occurs rapidly. It may be one of the first mechanisms by which arsenite is removed from the mammalian body. It has very broad ramifications. For example, what is the selenium status of humans with chronic arsenism? Many areas of the world in which chronic arsenic exposure occurs are low in selenium. Many questions remain about the Gailer compound including its toxicity, metabolism and fate in the mammalian body. The discovery of the Gailer compound and the proof of its structure (Gailer et al., to be submitted) by X-ray absorption spectroscopy at the Stanford Synchroton Radiation Laboratory is probably one of the most important discoveries in metal/metalloid toxicology this year.

Protein Binding

Another means of decreasing metal or metalloid toxicity is by protein binding. For example, cadmium, zinc and mercury bind strongly to metallothionein, the small protein of which

33% of its amino acids are cysteine. In addition, a lead-binding protein has been studied by Bruce Fowler's group (Fowler and DuVal, 1991). Dr. Greg Bogdan, when he was a graduate student in our laboratory, found three arsenite-binding proteins in rabbit liver (Bogdan et al., 1994). They were 450 kDa, 100 kDa and <2000 kDa in size. Inorganic arsenite was firmly bound to them. The affinity of arsenite for these proteins was 20 times greater than for arsenate. Bogdan devoted most of his efforts to studying the 450 kDa protein (manuscript has been submitted). He found that it was firmly bound to an arsenic-affinity column from which it was eluted by 2,3-dimercaptopropane-1-sulfonate (DMPS) a vicinal dithiol. It was not eluted by cysteine. Amino acid analysis of the 450 kDa protein indicated it contained only 0.4% CySH or 13 CySH residues/mol protein. Compared to the 33% cysteine content of metallothionein, the cysteine content of the 450 kDa protein is very small.

Arsenite Overflow of Protein Binding Sites

Our thinking about these arsenic binding proteins is that they are perhaps the initial reservoirs in which arsenite first accumulates so that the crucial enzymes having thiols in their active centers, are not overwhelmed by toxic concentrations of arsenite. As these protein "reservoirs" fill up, eventually an overflow of arsenite may occur. This may explain why methylated arsenic species do not appear in the urine immediately after arsenic exposure. We do not know how the release of arsenite from these proteins is controlled or regulated. This area of research is very complicated, since protein binding for detoxification has to be separated from protein binding for intoxication, both operationally and intellectually. This problem is not a small one and deserves more intensive study. However, protein binding appears to be the primary mechanism for arsenite detoxification as will be discussed more fully below. This is further supported by the studies of Vahter et al. (1982), who showed that when marmosets were given inorganic arsenic, no methylated arsenic compounds appeared in the urine and the labeled arsenic was found bound to the rough microsomal fraction.

Methylation with S-adenosylmethionine as the Methyl Donor and Methyltransferase as the Enzyme

Experiments with highly purified arsenic methyltransferase from rabbit liver clearly demonstrated that methylation of arsenic species, when it occurs, usually involves S-adenosylmethionine (SAM) as the methyl donor and arsenic methyltransferase as the enzyme (Zakharyan et al., 1995; Aposhian, 1997). Our laboratory has purified the arsenic methyltransferases from livers of rabbits (Zakharyan et al., 1995), hamsters (Wildfang et al., 1998), rhesus monkeys (Zakharyan et al., 1996) and from Chang human hepatocytes, a continuous cell line (Zakharyan et al., in prep.). We can summarize some of these important properties as follows: Arsenite and MMA methyltransferase activities of rabbit liver have been purified ~4000 fold. Both activities appear to be on the same protein. The molecular weight of this protein is ~60,000 and the protein appears to be made up of 2 monomers. The protein having arsenite and MMA methyltransferase activities is being sequenced. From rabbit liver, two isoenzymes of arsenite methyltransferase have been purified ~4000 fold. The methylation of arsenite and MMA appears to be catalyzed by a single protein with different active sites (Zakharyan et al., 1995). Other methyltransferases that catalyze successive methylation have been reported (Ridgway and Vance, 1988). We have emphasized studying the rabbit liver enzyme because the rabbit appears to be a good model for the human as far as arsenic is concerned (Maiorino and Aposhian, 1989; Vahter and Marafante, 1983) and rabbit livers are easier to obtain than human livers.

The first very careful study of MMA^{III} in a purified mammalian enzyme system using Michaelis–Menten kinetics has just been completed in our laboratory (Zakharyan et al., in prep.). The smaller the K_m of an enzyme, the greater the affinity of the enzyme for the substrate. The K_m for rabbit liver MMA methyltransferase with MMA^{III} is 9.2×10^{-6} and for

TABLE 1

Nonenzymatic methylation of arsenite by CH_3B_{12} and GSH

	ng
Complete	136
Minus Mg^{++}	122
Minus SAM	116
Minus B_{12}	115
Minus CH_3B_{12}	not detectable
Minus GSH	not detectable

Complete system consisted of: 0.05 M Tris-HCl buffer (pH 7.8); 10 mM GSH; 1 mM $MgCl_2$; 1 mM S-adenosyl-L-methionine; 24 μg B_{12}; 24 μg CH_3B_{12}; $^{73}As^{III}$, 6.6 μg/186,153 cpm; in total volume of 325 μl, pH 7.8. Incubation was for 90 min at 37°C. HPLC assay.

Reaction studied: $^{73}As^{III} \xrightarrow[\text{GSH}]{CH_3B_{12}} MMA$

MMA^V it is 1×10^{-4}. Based on their K_m values, MMA^{III} and not MMA^V is the true substrate for the MMA methyltransferase, confirming the early proposed pathway in which MMA^{III} was considered to be the substrate for the next methylation to DMA (Cullen and Reimer, 1989).

Methylvitamin B_{12} as a Methyl Donor to Inorganic Arsenite

Another *in vitro* mechanism for methylation of inorganic arsenic appears to be via methyl-cobalamin (methylvitamin B_{12}, CH_3B_{12}), and this reaction does not appear to need an enzyme (Buchet and Lauwerys, 1985). A reducing environment, however, is required (Table 1) (Zakharyan and Aposhian, 1999). While selenite is not required in this cell-free system for methylation, it will increase the methylation many fold (Figure 3). DMPS will also increase the reaction rate. The rate of this methylation reaction involving CH_3B_{12} is such that it needs to be considered when methylation of inorganic arsenic is being discussed or investigated.

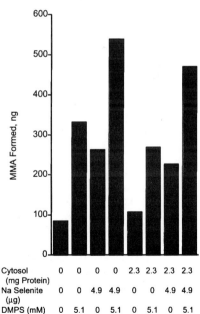

Cytosol (mg Protein)	0	0	0	0	2.3	2.3	2.3	2.3
Na Selenite (μg)	0	0	4.9	4.9	0	0	4.9	4.9
DMPS (mM)	0	5.1	0	5.1	0	5.1	0	5.1

Fig. 3. DMPS and/or sodium selenite stimulated non-enzymic methylation of arsenite by CH_3 cobalamin. DMPS = (2,3-dimercaptopropane-1-sulfonate).

The relationships of SAM, GSH, methylvitamin B_{12} and selenium point out the importance of the nutritional condition of humans chronically exposed to arsenic and the need for careful nutritional evaluations of them, especially by the use of the duplicate plate technique rather than by the questionnaires often used by epidemiologists. It should be kept in mind that vitamin B_{12} does not occur in plants and that for the economically poor who are chronically exposed to arsenic, meat is seldom available.

Diversity of Methylation

Animals such as the guinea pig (Healy et al., 1997), marmoset monkey (Zakharyan et al., 1996), chimpanzee (Wildfang, in prep.), tamarin monkey (Zakharyan et al., 1996), gorilla (Wildfang, in prep.), orangutan (Wildfang, in prep.) and many prosimians (Wildfang, in prep.) lack the arsenic methyltransferase enzymes. In fact, we now know of more animals lacking the enzyme than those having it (Table 2). The observations that the urine of some animals does not contain MMA or DMA (Table 2) and the livers of these animals lack arsenic methyltransferase activity are mutually confirmatory. We have been unable to detect this arsenic methyltransferase activity in human livers even when they were frozen within 10 min following removal. In collaboration with Felix Ayala-Fierro, a predoctoral student in Professor Dean Carter's laboratory, we have shown that Chang human hepatocytes in culture have substantial amounts of arsenic methyltransferase activity and we have partially purified the human enzyme from such cultured cells. The enzyme, when compared to the rabbit liver enzyme, does not seem unusual. The K_m of the Chang human hepatocytes arsenite methyltransferase is 3.2×10^{-6} and for MMA^{III} methyltransferase 3×10^{-6} as compared to 5.5×10^{-6} and 9×10^{-6} for the rabbit enzyme, respectively, indicating no great difference. We do not know as yet why we and others (Buchet and Lauwerys, 1985) cannot detect these methyltransferases in extracts of human liver. There is also tissue diversity of the arsenic methyltransferases (Healy et al., 1998). The liver is not the only site of methylation since the enzyme activity can be detected in mouse testes, kidney, liver and lung.

TABLE 2

Diversity of arsenic methyltransferases. Is methylation a primary detoxification pathway for inorganic arsenic?

Species	MMA and DMA in urine	Liver MeTase
Mouse	Yes	Yes
Rabbit	Yes	Yes
Rat	Yes	Yes
Rhesus	Yes	Yes
Marmoset	No	No
Chimpanzee	No*	No
Guinea Pig	No**	No
Gorilla	No	No
Orangutan	No	No
Human	Yes	***

*MMA and DMA were reported to be absent in the urines of two chimpanzees challenged with inorganic arsenic (Vahter et al., 1995), but MMA and DMA were found in a chimpanzee urine who was not given an arsenic challenge (Aposhian and Cebrian, unpublished).

**Five of six guinea pigs did not have MMA or DMA in their urine. One guinea pig had a very small amount of DMA (Healy et al., 1997).

***Although arsenic methyltransferase has not been detected in human liver, it has been detected in Chang human hepatocyte cells grown in culture (Zakharyan et al., submitted).

Is Methylation a Detoxification Mechanism?

The first questions about this were raised by the lack of methylated arsenicals in the urine of the marmoset monkey and chimpanzee (Vahter et al., 1982, 1995). Whether methylation using SAM and the arsenic methyltransferase is a detoxification mechanism for inorganic arsenic is becoming increasingly controversial. Although methylation does increase the water solubility of arsenite and therefore the rate of the urinary excretion of arsenic, the supposed decrease in the toxicity of the methylated arsenicals MMA and DMA, as compared to inorganic arsenic, is based on LD_{50} determinations. This is a relatively narrow, unsophisticated measurement and criterion of toxicity, usually based on lethality. However, when carcinogenicity and other measures of toxicity are the criteria we do not know the relative potencies, if any, of inorganic As, MMA and DMA in inducing various forms of cancer observed after inorganic As exposure. The manufacturers of MMA and DMA, however, as long ago as 1981 reported to the US EPA that DMA, when given to rats, caused urinary bladder carcinomas. At the present time, the US EPA classifies DMA as a probable human carcinogen. Yamanaka et al. (1993) have shown crosslinking between DMA and nuclear protein in a human lung cell line. These effects may be related to active oxygen species formation. Methylation does not appear to be a detoxification mechanism for inorganic arsenic unless you want to believe that cancer is a benign disease. This may, however, be a problem in semantics and what is meant by detoxification. It is true that methylation of inorganic arsenite increases its excretion rate, but it also converts inorganic arsenite to compounds with carcinogenic potential. Is this what detoxification means? Perhaps it would be more acceptable to say that the methylation of inorganic arsenite is a biotransformation process by which the urinary excretion of arsenic is increased.

Methylation of arsenite may not be a detoxification mechanism because not only have an increasing number of mammals been found to be deficient in the arsenic methyltransferases (Table 2) but in addition MMA^{III} (Fig. 1) is more toxic than arsenite and MMA^{V} for fungal and bacterial systems, as shown by Professor Cullen of the University of British Columbia (Cullen et al., 1989). The toxicity of MMA^{III} in mammalian systems is now under investigation.

Urinary Excretion

Finally, arsenate, arsenite, MMA^{V}, DMA and probably MMA^{III} (Figure 2) are excreted in the urine. This is true for the human, mouse, rat, hamster, and a few other animals. It certainly is not true for the guinea pig, chimpanzee, marmoset monkey, orangutan or gorilla (Table 2 and its legend). There has not been a study to determine whether MMA^{III} is or is not found in human urine. Very few investigators have made an effort to determine whether the MMA species being excreted is MMA^{V}, MMA^{III} or both. The identity of urinary MMA requires a great deal more attention than it has received.

In studies performed in Northeastern Chile (Aposhian et al., 1997), the administration of DMPS, an orally effective chelating agent, resulted in a remarkable change in the urinary profile of arsenic species. The MMA percent increased from 14 to 42% indicating a block in the methylation of MMA to DMA (Figure 4). The results were essentially the same in San Pedro de Atacama, the village where drinking water contains 600 μg As/L and Toconao (control village).

We hope this flow chart (Figure 2) will stimulate a collective attempt by all of us to add to it and correct it. In addition, we are beginning a flow chart to try to delineate and understand arsenic intoxication. Important contributions have been made already by the groups of Dr. Marc Mass (Mass and Wang, 1997) and Dr. Michael Waalkes (Zhao et al., 1997) using molecular biology approaches to better understand arsenic carcinogenesis and toxicity. A greater use of molecular biology technology by those studying arsenic methylation might be expected to clarify many questions in arsenic toxicology.

Fig. 4. Arsenic species in human urine expressed as a percentage of urinary total arsenic at time periods before and after DMPS administration. DMPS, 300 mg, was given by mouth at 0 time. The first numbers given in the bar graphs are the percentages for the arsenic exposed San Pedro de Atacama, Chile group. The numbers in the parentheses are for the Toconao, Chile group (controls).

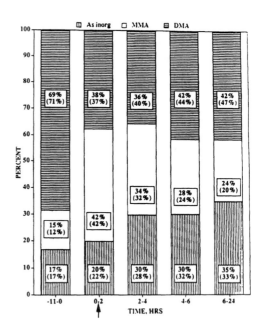

ACKNOWLEDGMENTS

These studies were supported in part by the Superfund Basic Research Program NIEHS Grant ES-04940.

REFERENCES

Aposhian, H.V. 1997. Enzymatic methylation of arsenic species and other new approaches to arsenic toxicity. *Annu. Rev. Pharmacol. Toxicol.*, **37**, 397–419.

Aposhian, H.V., Arroyo, A., Cebrian, M.E., Del Razo, L.M., Hurlbut, K.M., Dart, R.C., Gonzalez-Ramirez, D., Kreppel, H., Speisky, H., Smith, A., Gonsebatt, M.E., Ostrosky-Wegman, P., Aposhian, M.M. 1997. DMPS-arsenic challenge test. I: Increased urinary excretion of monomethylarsonic acid in humans given dimercaptopropane sulfonate. *J. Pharmacol. Exp. Ther.*, **282**, 192–200.

Bogdan, G.M., Sampayo-Reyes, A., Aposhian, H.V. 1994. Arsenic binding proteins of mammalian systems. I. Isolation of three arsenite-binding proteins of rabbit liver. *Toxicology*, **93**, 175–193.

Buchet, J.P., Lauwerys, R. 1985. Study of inorganic arsenic methylation by rat liver in vitro: Relevance for the interpretation of observations in man. *Arch. Toxicol.*, **57**, 125–129.

Buchet, J.P., Lauwerys, R. 1988. Role of thiols in the in vitro methylation of inorganic arsenic by rat liver cytosol. *Biochem. Pharmacol.*, **37**, 3149–3153.

Challenger, F. 1951. Biological methylation. *Adv. Enzymol.*, **12**, 429–491.

Cullen, W.R., McBride, B.C., Manji, H., Pickett, A.W., Reglinski, J. 1989. The metabolism of methylarsine oxide and sulfide. *Appl. Organometall. Chem.*, **3**, 71–78.

Cullen, W.R., Reimer, K.J. 1989. Arsenic speciation in the environment. *Chem. Rev.*, **89**, 713–764.

Delnomdedieu, M., Basti, M.M., Otvos, J.D., Thomas, D.J. 1993. Transfer of arsenite from glutathione to dithiols: A model of interaction. *Chem. Res. Toxicol.*, **6**, 598–602.

Fowler, B.A., DuVal, G. 1991. Effects of lead on the kidney: Roles of high–affinity lead-binding proteins. *Environ. Health Perspect.*, **91**, 77–80.

Gailer, J., George, G.N., Pickering, I.J., Prince, R.C., Ringwald, S.C., Pemberton, J.E., Glass, R.S., Younis, H.S., DeYoung, D.W., Aposhian, H.V. 1999. The seleno-bis(S-glutathionyl) arsinium ion: A metabolic link between As(III) and Se(IV). In preparation.

Gyurasics, A., Varga, F., Gregus, Z. 1991. Effect of arsenicals on biliary excretion of endogenous glutathione and xenobiotics with glutathione-dependent hepatobiliary transport. *Biochem. Pharmacol.*, **41**, 937–944.

Healy, S.M., Zakharyan, R.A., Aposhian, H.V. 1997. Enzymatic methylation of arsenic compounds: IV. In vitro and in vivo deficiency of the methylation of arsenite and monomethylarsonic acid in the guinea pig. *Mutation Res.*, **386**, 229–239.

Healy, S.M., Casarez, E.A., Ayala-Fierro, F., Aposhian, H.V. 1998. Enzymatic methylation of arsenic compounds. V. Arsenite methyltransferase activity in tissues of mice. *Toxicol. Appl. Pharmacol.*, **148**, 65–70.

Levander, O.A., Baumann, C.A. 1966. Selenium metabolism VI: Effect of arsenic on the excretion of selenium in the bile. *Toxicol. Appl. Pharmacol.*, **9**, 106–115.

Maiorino, R.M., Aposhian, H.V. 1989. Determination and metabolism of dithiol chelating agents: IV. Urinary excretion of meso-2,3-dimercaptosuccinic acid and mercaptosuccinic acid in rabbits given meso-2,3-dimercaptosuccinic acid. *Biochem. Pharmacol.*, **38**, 1147–1154.

Mass, M.J., Wang, L. 1997. Arsenic alters cytosine methylation patterns of the promoter of the tumor suppressor gene p53 in human lung cells: A model for a mechanism of carcinogenesis. *Mutation Res.*, **386**, 263–277.

Ridgway, N.D., Vance, D.E. 1988. Kinetic mechanism of phosphatidylethanolamine *N*-methyltransferase. *J. Biol. Chem.*, **263**, 16864–16871.

Scott, N., Hatlelid, K.M., MacKenzie, N.E., Carter, D.E. 1993. Reactions of arsenic(III) and arsenic(V) species with glutathione. *Chem. Res. Toxicol.*, **6**, 102–106.

Styblo, M., Yamauchi, H., and Thomas, D. J. (1995). Comparative in vitro methylation of trivalent and pentavalent arsenicals. *Toxicol. Appl. Pharmacol.*, **135**, 172–178.

Styblo, M., Delnomdedieu, M., and Thomas, D. J. (1996). Mono- and dimethylation of arsenic in rat liver cytosol in vitro. *Chem.-Biol. Interact.*, **99**, 147–164.

Vahter, M. 1981. Biotransformation of trivalent and pentavalent inorganic arsenic in mice and rats. *Environ. Res.*, **25**, 286–293.

Vahter, M. 1994. Species differences in the metabolism of arsenic compounds. *Appl. Organomet. Chem.*, **8**, 175–182.

Vahter, M., Marafante, E., Lindgren, A., Dencker, L. 1982. Tissue distribution and subcellular binding of arsenic in marmoset monkeys after injection of ^{74}As-arsenite. *Arch. Toxicol.*, **51**, 65–77.

Vahter, M., Marafante, E. 1985. Reduction and binding of arsenate in marmoset monkeys. *Arch. Toxicol.*, **57**, 119–124.

Vahter, M., Marafante, E. 1983. Intracellular interaction and metabolic fate of arsenite and arsenate in mice and rabbits. *Chem.-Biol. Interact.*, **47**, 29–44.

Vahter, M., Marafante, E. 1988. In vivo methylation and detoxification of arsenic. In *The Biological Alkylation of Heavy Elements*, Special Publication No. 66 (P.J. Craig and F. Glockling, eds.) pp. 105–119, Royal Society of Chemistry, London.

Vahter, M., Couch, R., Nermell, B., Nilsson, R. 1995. Lack of methylation of inorganic arsenic in the chimpanzee. *Toxicol. Appl. Pharmacol.*, **133**, 262–268.

Wildfang, E., Zakharyan, R.A., Aposhian, H.V. 1998. Enzymatic methylation of arsenic compounds: VI. Arsenite and methylarsonic acid methyltransferase kinetics. *Toxicol. Appl. Pharmacol.*, **152**, 366–375.

Winski, S.L., Carter, D.E. 1995. Interactions of rat red blood cell sulfhydryls with arsenate and arsenite. *J. Toxicol. Environ. Health*, **46**, 379–397.

Yamanaka, K., Tezuka, M., Kato, K., Hasegawa, A., Okada, S. 1993. Crosslink formation between DNA and nuclear proteins by in vivo and in vitro exposure of cells to dimethylarsinic acid. *Biochem. Biophys. Res. Comm.*, **191**, 1184–1191.

Zakharyan, R.A., Aposhian, H.V. 1999. Arsenite methylation by methylvitamin B_{12} and glutathione does not require an enzyme. *Toxicol. Appl. Pharmacol.*, **154**, 287–291.

Zakharyan, R.A., Wildfang, E., Aposhian, H.V. 1996. Enzymatic methylation of arsenic compounds: III. The marmoset and tamarin, but not the rhesus, monkey are deficient in methyltransferases that methylate inorganic arsenic. *Toxicol. Appl. Pharmacol.*, **140**, 77–84.

Zakharyan, R.A., Wu, Y., Bogdan, G.M., Aposhian, H.V. 1995. Enzymatic methylation of arsenic compounds. I: Assay, partial purification, and properties of arsenite methyltransferase and monomethylarsonic acid methyltransferase of rabbit liver. *Chem. Res. Toxicol.*, **8**, 1029–1038.

Zhao, C.Q., Young, M.R., Diwan, B.A., Coogan, T.P., Waalkes, M.P. 1997. Association of arsenic-induced malignant transformation with DNA hypomethylation and aberrant gene expression. *Proc. Natl. Acad. Sci. USA*, **94**, 10907–10912.

Arsenic Exposure and Health Effects
W.R. Chappell, C.O. Abernathy and R.L. Calderon (Editors)

Arsenic Metabolism After Pulmonary Exposure

Dean E. Carter, Marjorie A. Peraza, Felix Ayala-Fierro,
Elizabeth Casarez, David S. Barber, Shannon L. Winski

ABSTRACT

Inorganic arsenic compounds are oxidized, reduced, methylated and complexed with glutathione *in vivo*. Aposhian and coworkers showed that arsenate reduction and arsenite and MMA methylation activities were different in each organ and animal species. Arsenate reduction was found in all organs and animal species, but methylation activity appeared to be absent in some species. Where present, methylation activity was found in all organs studied except the red blood cell. Methylation of inorganic arsenicals has been associated with decreased acute toxicity, while reduction has been associated with increased toxicity. Oxidation and glutathione complexation of arsenite have not been characterized. Since each organ has some capacity to "metabolize" arsenic, absorption from pulmonary exposure would be accompanied by a "first pass effect" from the lung. In our studies, lung tissue metabolized arsenite or arsenate to the major metabolites of inorganic arsenic: arsenite, arsenate, monomethylarsonate (MMA) and dimethylarsinate (DMA). Using rat and guinea pig lung homogenates, the rates of arsenite $\rightarrow$ arsenate, arsenate $\rightarrow$ arsenite, arsenite $\rightarrow$ MMA, and MMA $\rightarrow$ DMA were measured and modeled using SIMUSOLV. The model was tested by comparing the predicted concentration-time curve with measured concentrations, with induction of heat shock protein 32 (hsp 32) at 4 h, and with a LC_{50} (lethal concentration) in BEAS-2B cells at 24 h. In each case, there was a reasonable fit of predicted arsenite concentration with the effect or measured concentration. In addition, an arsenite-glutathione complex was detected in tissue homogenates.

Keywords: arsenic, metabolism, pulmonary exposure

INTRODUCTION

Arsenic compounds are important environmental and industrial toxicants that cause acute and chronic effects in animals and man. The effects depend on the chemical form of the arsenic and are specific for certain organs. There are four stable arsenic compounds that are important in human toxicity: arsenate (AsV), arsenite (AsIII), monomethylarsonate (MMA) and dimethylarsinate (DMA). Each of these arsenic species has been used commercially and each has its own toxicity. Assigning toxic effects to a particular species has been complicated by the finding that the body can, in some cases, metabolize one arsenic species to other species. This is a particular problem for AsV and AsIII because they can be converted to each other in biological systems and both are excreted in the urine. Some of the rates of these metabolic reactions have been determined recently (Zakharyan et al., 1996) and these rates can be used to predict the concentrations of the four arsenic species as a function of time (Mann et al., 1996a and b). These findings should be important in the risk assessment of arsenic compounds because they may provide information about the identity of toxic species of arsenic and the organ concentrations of those species.

Less stable arsenic species that play a role in arsenic metabolism have been identified recently. These include $As(SG)_3^{-3}$, monomethylarsonite $((CH_3)As(OH)_2)$ and dimethylarsinite $((CH_3)_2As(OH))$ (Cullen et al., 1984; Delnomdedieu et al., 1994a and b; Scott et al., 1993; Styblo et al., 1995). These species appear to be formed transiently and may be substrates for the arsenic methyltransferase enzymes but their role in toxicity is unclear. NMR has detected their presence but they have never been isolated from biological samples.

Inhalation is a significant route of exposure to arsenic compounds and the lung is a target organ for toxicity. Lung cancer incidence was increased in arsenic smelter workers, particularly among smokers (Enterline et al., 1987). Some animal models have also shown lung tumors after intratracheal administration of arsenic compounds (Ishinishi et al., 1977; Inamasu et al., 1982; Pershagen and Bjorklund, 1985). It is probable that the concentration of arsenic would be higher in the lung and blood than other target organs like the skin, liver and kidney because the lung is in direct contact with arsenic compounds in the air. The capability of lung to metabolize arsenic compounds may be related to the biological effects observed and these may be different from other target organs. The objective of this study is to examine the disposition and metabolism of AsIII and AsV in the lung and to construct a mathematical model to predict the concentrations of the stable arsenic species as a function of time.

METHODOLOGY

Chemicals

Sodium arsenite, ACS certified, was purchased from Fisher Scientific (Fair Lawn, NJ). Sodium arsenate, ACS certified, was purchased from J.T. Baker Chemical Co. (Phillisburg, NJ). Sodium monomethylarsonate (MMA) was purchased from Pfaltz and Bauer (Stamford, CT). S-Adenosyl-L-[methyl-^{3}H] Methionine (^{3}H-SAMe) was purchased from Amersham Life Science (#TRK581,Arlington Heights, IL). Ketamine-Xylazine-Acepromazine (KRA) is a mixture of 40 mg/ml Ketamine (Ketaset®, Fort Dodge Labs, Inc., Ft. Dodge, IA.), 5 mg/ml Xylazine (Rompun®, Miles, Inc., Shawnee Mission, KS.), and 2.5 mg/ml Acepromazine (Acepromazine Maleate, Fermenta Animal Health Co., Kansas City, MO.) Strong cation exchange resin (AG 50W-X8, 100-200 mesh, hydrogen form), strong anion exchange resin (Dowex 1-1X2, 50-100 mesh, chloride form), and all chemicals for polyacrylamide gel electrophoresis (PAGE) were purchased from Bio-Rad Laboratories (Richmond, CA). Sodium dimethylarsinate (DMA), glutathione (GSH), periodate oxidized adenosine (PAD), S-Adenosyl-L-homocysteine (SAH), Sigma diagnostic kit 525-A for total hemoglobin, and all other chemicals were purchased from Sigma Chemical Co. (St. Louis, MO).

Animals

Male Syrian golden hamsters (130 ± 15 g body weight) were purchased from Charles River Breeding Laboratories (Boston, MA) and allowed to acclimate for at least one week prior to dosing. Water and food (Wayne lab blox) were provided ad libitum throughout the studies. A normal diurnal cycle was maintained with artificial lighting. All animals were dosed at approximately 8:00 a.m. for each study. Male Sprague-Dawley rats (200–250 g, obtained from Sasco, Inc., Omaha, NE) and male Hartley outbred guinea pigs (500–600 g, obtained from Harlan Sprague-Dawley, Indianapolis, IN) were maintained as described for the hamsters.

Intratracheal Administration to Hamsters

The preparation of the dosing solutions and the intratracheal instillation were performed as described by Brain et al., (1976) and revised by Rosner and Carter (1987).

In Vitro Whole Rat Blood Incubations with AsV and AsIII

Blood collection and treatment and the determination of non-protein sulfhydryls (NPSH) were performed as described by Winski and Carter (1995).

Lung Metabolism Studies

(A) *Preparation of lung homogenate in rat and guinea pigs.* Male Sprague-Dawley rats (250–350 g) or male Hartley outbred guinea pigs (500–600 g) were anesthetized by KRA injection and killed by exsanguination (cutting the inferior vena cava). Lungs were perfused through the left ventricle of the heart with cold saline solution (40 mL) to remove blood from the lungs. Lungs were removed intact, weighed, and diced. Lungs were then homogenized with 6 passes in 4× weight volumes of PBS (20% w/v homogenate) using a teflon glass homogenizer.

(B) *Preparation of cytosol for methylation experiments.* Male Sprague-Dawley rats (200–300 g) were anesthetized with KRA and killed by exsanguination. Lungs were perfused through the right ventricle of the heart with 30 mL of ice-cold sterile PBS. Lungs were then removed, weighed, minced, and homogenized in 4 volumes of ice-cold sterile PBS by 7 passes with a teflon-glass homogenizer. Cytosol was prepared from homogenates by ultracentrifugation at 105,000 × g for 60 min. Supernatant from this centrifugation step was considered cytosol.

(C) *Assay for arsenic methylation by cytosol.* Methylation experiments were carried out as described by Zakharyan et al. (1995) at pH 8.0.

(D) *Arsenic speciation.* Arsenic species from incubations containing methylated arsenicals were separated using the mixed bed ion exchange method described by Maiorino and Aposhian (1985). Arsenic species from reduction/oxidation studies were separated by the method of Winski and Carter (1995).

(E) *Homogenate incubations for reduction/oxidation studies.* Reduction and oxidation studies were carried out in rat and guinea pig lung homogenates with methylation inhibited by a mixture of periodate oxidized adenosine (PAD) and S-adenosyl-homocysteine (SAH). PAD is a general methyltransferase inhibitor that works by inhibiting the SAH hydrolase and causing SAH to build up in the incubation. Increased concentrations of SAH inhibit many methyltransferases. A combination of 100 nmol PAD plus 1 mmol SAH was found to be an effective inhibitor of arsenic methyltransferase activity in lung cytosols. This treatment was used for all reduction/oxidation assays to prevent methylation from occurring. 0.25 mL homogenate was mixed with 0.25 mL of twice the final concentration of arsenic and incubated at 37°C. A metabolic model was generated using the modeling and simulation software SIMUSOLV (version 3.0, Dow Chemical Co., Midland, MI).

Formation and Identification of As(SG)$_3$$^{-3}$ in Tissues

(A) *Formation of As(SG)$_3$$^{-3}$ in Rat Lung Homogenates.* Male Sprague-Dawley rats (200–300 g) were used. Animals were anesthetized and lungs perfused with 40 mL of cold saline through the left ventricle of the heart. After perfusion, lungs were removed and trachea and connective tissue were removed. The lungs were homogenized in 4 volumes of cold PBS to make a 20% (w/v) homogenate; 950 μl of homogenate was mixed with 50 μl of 20 mM arsenite (dissolved in PBS and pH adjusted to 7.0 prior to use). Samples were incubated in a 37°C water bath. Incubations were terminated by addition of 10 μl of trifluoroacetic acid (TFA) (final TFA concentration = 1%) which also precipitated protein. Samples were cleared by centrifugation at 16000 × g for 10 min. Supernatant was collected and filtered (0.2 μm) prior to analysis.

(B) *Synthesis of As(SG)$_3^{-3}$ Standard.* As(SG)$_3^{-3}$ standard was synthesized by mixing sodium arsenite and GSH in a 1:3.1 molar ratio in a minimal amount of Milli-Q water. This solution was stirred for 60 min at room temperature. The final product was precipitated with 10 volumes of cold methanol and then was collected by centrifugation. The product was dried by lyophilization and identity was confirmed by ^{13}C NMR using published peak shifts (Scott et al., 1993).

(C) *Separation and Detection of As(SG)$_3^{-3}$.* The standard solutions of As(SG)$_3^{-3}$ in water (injection volume was 50 μl) were analyzed by flow injection electrospray mass spectrometry (ESI-MS). The instrumentation consisted of an HP1050 HPLC (Hewlett Packard, Palo Alto, CA) interfaced to a Finnigan TSQ 7000 mass spectrometer (San Jose, CA). The mobile phase consisted of water containing 0.1% TFA at a flow rate of 0.5 ml/min. The mass spectrometer conditions were as follows: ESI voltage was 4.5 kV, heated capillary was at 200°C, and scans were acquired from 200 to 2000 AMU per second. The As(SG)$_3^{-3}$ complex was separated and detected in rat lung homogenates by HPLC-ESI-MS as described above. The column was a 4.6 mm × 250 mm, 218 TP54 C18 column (Vydac, Hesperia, CA). The mobile phase consisted of water, methanol and acetic acid. The initial conditions were 97% water, 2% methanol and 1% acetic acid held for 9 min, then linearly ramped to 100% methanol in the next 14 min and held for 5 min after which data acquisition was stopped. Finally the column was linearly programmed to initial conditions over 5 min, and allowed to re-equilibrate for an additional 5 min. The flow rate was held constant at 1.0 ml/min. The mass spectrometer conditions were as described above.

Cell Culture

(A) *BEAS-2B Cell Line.* This human bronchial epithelial cell line was obtained from ATCC (Rockville, MD #9609-CRL). Cells were received at passage 37 and were used between passages 40 and 60. Cells were grown in serum-free modified LHC-9 media (Lechner and LaVeck, 1985) at 37°C in a humidified 5% CO_2 atmosphere.

B) *LC$_{50}$ Determined by the XTT Assay.* Cell viability was determined by the reduction of 2,3-bis[2-methoxy-4-nitro-5-sulfophenyl]-2H-tetrazolium-5-carboxanilide inner salt (XTT, X-4251, Sigma Chemical Co., St. Louis, MO) as described by Roehm, et al., (1991) in 96-well plates containing 8×10^3 cells/well. Twenty-four hours after plating, cells were dosed for 20 h (50 μL, 3X concentrated), then XTT [3 mg/6 ml media + 3.5 μl of 30.6 mg/ml PBS of phenazine methosulfate (PMS, P-5812, Sigma Chem. Co., St. Louis, MO)] was added to each well (50 μl/well). Cells were incubated for 4 more hours and viability was determined by measuring OD480 on a Biolinks 2.20 plate reader (Dynatech laboratories, Inc.).

(C) *Heat Shock Protein 32 (hsp32) Assay.* BEAS-2B cells were grown to 90% confluency in 25cm^2 flasks (Comstar Corp., Cambridge, MA). Cells were treated with arsenicals in modified LHC-9 media for 4 h. Then the media was removed and cells were rinsed with 2 ml of sterile PBS (Ca^{2+}, Mg^{2+} free). Cells were scraped into 0.2 ml of sterile PBS and sonicated for 10 seconds to lyse. Proteins were determined by BCA protein assay kit (Pierce Chemical Co., Rockford, IL). Proteins were separated by SDS-PAGE using the method of Laemmli (1970). 10μg of lysate protein was loaded for each sample. Fifty ng of rat recombinant hsp32 (Stressgen, Victoria, BC, Canada) was used as the standard. Gels were 10% acrylamide with 37.5:1 ratio of acrylamide:bis-acrylamide. Gels were cast using mini-protean II apparatus (Bio-Rad Lab., Hercules,CA) and run at 50 milliamps. Western blotting was performed as described by Burnette (1981).

Statistics.

All experiments were performed at least three times in independent experiments, and sample size (n) refer to number of animals. Data are presented as the arithmetic mean ± standard deviation. Analysis of variance (ANOVA) and Student's t-test were performed where appropriate, and differences in data were considered significant only if calculated p values were less than 0.05 (Microsoft Excel statistical package, Redmond, WA).

RESULTS

Intratracheal (i.t.) Administration to Hamsters

The daily excretion rate of arsenic was different after i.t. administration of AsV as compared to AsIII administration during the four day sample collection. After AsV administration, more of the arsenic was excreted in the first day followed by less on the succeeding days. After AsIII was given, slightly more of the arsenic was excreted in the second day, but approximately equal amounts were excreted on days 1, 2 and 4. These excretion rate differences were observed for total arsenic in urine and for the metabolites (AsV, AsIII and total methylated arsenicals) in urine (Figure 1). The total As excreted in urine after four days was equal (AsV administration and AsIII administration 48.5%) but the ratio of the

A.

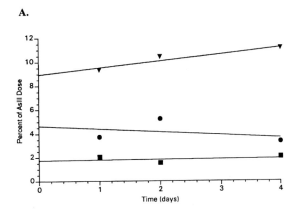

B.

Fig. 1. Percent Dose Excreted in Urine After i.t. Administration of 5.0 mg As/Kg as As (III) (A) or As (V) (B). Urine was analyzed by arsine generation for AsV (■), AsIII (●), and methylated arsenicals (▼) after separation by ion exchange and wet digestion. Metabolite results were expressed as percent of total arsenic in urine found as the various arsenicals.

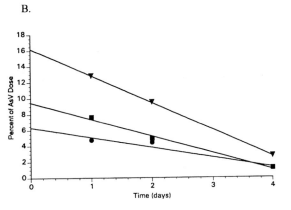

TABLE 1

Total arsenic in the different tissues after intratracheal administration of AsV (A) or AsIII (B). Tissue levels expressed as As equivalents per gram of tissue as determined by direct hydride FAAS. Animals were dosed intratracheally with 5 mg/kg As equivalents of the appropriate compound. Each value represents the mean ± S.D. of $n = 4$.

Total As (μg/g)	Day 1	Day 2	Day 3
(A) Tissue Total As After Intratracheal AsV			
Blood	0.31	0.19	0.07
Lung	1.7	0.7	0.4
Liver	0.75	0.7	0.39
Kidney	3.4	3.7	2.2
(B) Tissue Total As After Intratracheal AsIII			
Blood	0.60	0.47	0.16
Lung	2.9	0.90	0.48
Liver	3.3	0.95	0.58
Kidney	3.8	3.8	1.4

metabolites was different. The methylated arsenicals and the administered compound were in highest concentration in the urine. For example, more AsV and methylated arsenicals and less AsIII were excreted in the urine after AsV administration (data not shown).

The clearance of arsenic from the blood, liver, and lung after AsIII administration did not follow the excretion rate pattern. The total arsenic concentration decreased steadily with time in those organs. Kidney As levels were different in comparison to the other organs after both AsV and AsIII administration. The kidney had approximately equal total As levels for the first two days but the level decreased on the fourth day. Tissue As levels were higher after AsIII administration as compared to AsV administration except for the kidney where the levels were comparable (Table 1).

In Vitro Whole Rat Blood Incubations with AsV and AsIII

Whole blood oxidized AsIII and reduced AsV but did not methylate As (data not shown). Thiols reduce AsV to AsIII and form complexes with AsIII. Non-protein thiol (NPSH) levels were affected differently by AsV as compared to AsIII and indicated that these reactions occurred at different rates. Intracellular NPSH levels decreased slowly over the 5-h incubations with 10mM AsV (Figure 2), and AsV was detected during the entire incubation time (data not shown). After AsIII incubation, NPSH decreased rapidly during the first 20

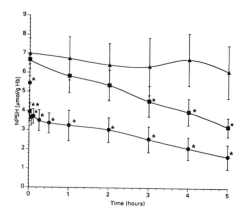

Fig. 2. Depletion of nonprotein reduced sulfhydryls. Time-dependent depletion of nonprotein reduced sulfhydryls (NPSH) in rat blood incubated with 10 mM AsIII (●) or 10 mM AsV (■). Nonprotein reduced sulfhydryl levels in protein-free supernatants were determined by spectrophotometry and normalized to hemoglobin content (μmol NPSH/g Hb). Levels that were significantly lower than time-matched controls (▲) are denoted with an * ($p < 0.05$, $n = 5$).

Fig. 3. Time course of As(SG)$_3^{-3}$ formation from AsIII and AsV. As(SG)$_3^{-3}$ formation in rat lung homogenates (incubated at 37°C) was investigated using 1 mM AsV (■) or 1 mM AsIII (●) as a substrate. The standard used was 10 μM As(SG)$_3^{-3}$ in pH 7.4 buffer (▲). Values are mean ± range (*n* = 2).

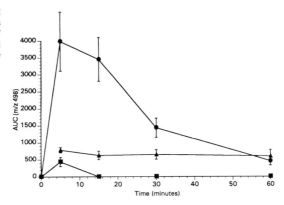

min followed by a slow decline over the remaining period. The rate of decline of NPSH was slightly faster after AsV treatment.

Formation of As(SG)$_3^{-3}$ in Tissues

As(SG)$_3^{-3}$ was isolated from lung homogenates using HPLC and identified by mass spectrometry. Tissue levels were measured using selective-ion mass spectrometry, and they were compared with synthesized standard. As(SG)$_3^{-3}$ was formed within 5 min after the addition of 1mM AsIII but its levels rapidly decreased between 15 and 60 min when the experiment ended. A small amount of the complex appeared at 5 min after the addition of 1mM AsV but it was undetectable by 15 min. Standard As(SG)$_3^{-3}$ levels in buffer solution were unchanged over the 60-min experiment (Figure 3).

Metabolism Model Applied to Biological Effects

A simplified model of the rates of metabolism between AsV, AsIII, MMA and DMA was devised to predict the arsenic species concentrations in a single organ. The model used was: AsV ↔ AsIII → MMA → DMA. In rat lung preparations, AsV was reduced to AsIII [first order rate constant of 0.0104/min]; AsIII was oxidized to AsV [first order rate constant of 0.005/min], methylated to MMA [K$_m$ = 5.383μM, V$_{max}$ = 0.00031 μmol/liter/min/mg], and complexed with GSH; and MMA was converted to DMA [K$_m$ = 63.4 μM, V$_{max}$ = 0.0000384 μmol/liter/min/mg]. Data from which these rates were calculated are published in Barber (1997). Specific inhibitors of methylation were used in the measurement of the oxidation of AsIII to AsV. This oxidation rate was determined again in the guinea pig lung homogenates for comparison, because the guinea pig lacks the enzyme(s) to methylate AsIII (Healy et al., 1997). The results confirmed the value determined in the rat lung (data not shown). The pulmonary arsenic metabolism model used these rates and the program SIMUSOLV to generate equations to predict the concentrations of AsV, AsIII, MMA and DMA as a function of time.

The toxicity of AsIII, AsV, MMA, and DMA was assessed by measuring cell viability and hsp32 induction in the BEAS-2B pulmonary cell line. If all species of arsenic did not produce individual effects, it would be possible to deduce an "active" form of arsenic from these studies. For hsp32 protein induction at 4 h (Table 2A), 1 μM AsIII initial concentration (predicted concentration 0.91 μM after 4 h) was the minimum for induction; 10μM AsIII (9.2μM) produced maximal induction; 10μM AsV (predicted concentration 0.6μM AsIII) did not cause induction; 20 μM AsV (1.2 μM AsIII) was the minimum for induction; and 100 μM AsV (6 μM AsIII) produced a strong induction (Table 2A). MMA and DMA did not induce hsp32 (data not shown). This data suggested that hsp32 induction occurred at a minimum of 0.9 μM AsIII and that the source could be AsIII or AsV.

TABLE 2

Correlation of arsenic metabolism and toxicity. (A) Correlation of predicted AsIII concentration at 4 h with hsp32 induction (4 h is the time of maximal induction according to preliminary experiments). (B) Correlation of predicted AsIII concentration at 24 h with effects on cell viability. Cell viability was performed by the XTT assay. Each value represents the mean of at least three independent experiments.

Initial conditions	AsIII concentration at 4 h
(A) Induction of hsp32	
1 μM AsIII	0.91 μM[a]
10 μM AsIII	9.2 μM[b]
10 μM AsV	0.6 μM[c]
20 μM AsV	1.2 μM[a]
100 μM AsV	

[a]Minimum concentration at which induction was observed.
[b]Conditions at which maximal induction was observed.
[c]Conditions under which induction was not observed.

Initial conditions	AsIII concentration at 24 h
(B) LC$_{50}$ in BEAS-2B cells treated for 24 h	
40 μM AsIII	30 μM
120 μM AsV	31 μM

Cell viability after a 24-h treatment, as determined by the XTT assay, was tested and the LC$_{50}$ for AsIII was 40 μM and for AsV was 120 μM. The modeled AsIII concentration was 30 μM for the AsIII treatment and 31 μM for the AsV treatment (Table 2B). MMA and DMA had LC$_{50}$s of 2 mM and 8 mM, respectively, and did not predict the formation of any AsIII. This data suggested that AsV cell toxicity could result from its conversion to AsIII and that the methylated arsenical compounds had a different potency and perhaps a different mechanism.

DISCUSSION AND CONCLUSIONS

The EPA risk assessments are for inorganic arsenic compounds and do not distinguish between the arsenite and arsenate forms of inorganic arsenic in drinking water. In addition, there are separate cancer risk assessments for water and air exposures. However, distinguishing between different routes of exposure is not simple because biological organisms can convert one arsenic compound to other arsenic compounds and also because there are some common target organs following different routes of exposure. The toxic arsenic species following arsenite or arsenate exposure has not yet been identified. In addition, urinary arsenic species have been used as a biomarker of exposure and the relative amounts of these species as an indicator of potential hazard. The justification for these assumptions has been incompletely tested.

Dose–response determinations have shown arsenite (AsIII) to be the most acutely toxic arsenic species followed by arsenate and by MMA and DMA. This is complicated by the findings that arsenate and arsenite can be converted to each other and that arsenite must be formed before methylation to MMA and DMA can occur. When comparing toxicities, it is unclear if there is a common toxic species, if there is a common toxic mechanism but different potencies, or if there are different toxic mechanisms for the four compounds. Since the same chemical species of arsenic are formed, the rates of formation and disposition of those species may be important in the explanation of those findings.

Our *in vivo* experiments with hamsters given arsenite or arsenate by intratracheal administration showed that the conversion of AsV to AsIII was slow in the lung and blood because excretion rates are different over the time period in days. Arsenate was excreted more rapidly than arsenite, and the excretion of methylated compounds was related to the total arsenic in the body. The slow arsenite excretion indicated that there was a storage depot in the body for arsenic, but the depot must be in organs that were not examined, because the blood, kidney, liver, and lung concentrations were not related to the urinary concentration. These organs were chosen for analysis because they were the major organs for excretion and previous work showed that they contained the most arsenic (Marafante and Vahter, 1987). The kidney contained the most arsenic of the organs tested on a $\mu g/g$ basis. Although the rate of decrease of kidney concentration was somewhat slower than blood, lung, or liver, the differences were insufficient for the kidney to be considered as a depot.

These findings are somewhat different from those reported by Marafante and Vahter (1987) who administered 2 mg/kg AsIII and AsV to hamsters by intratracheal instillation. Although our urinary recovery and AsIII levels were similar, the percent of AsV and DMA were quite different. In our study, there was substantially less AsV and more methylated metabolites excreted in the urine. It is unclear whether our higher dose (5 mg/kg) and/or longer experiment time is responsible for these differences.

AsIII in urine did not follow blood arsenic. This suggests that arsenic metabolism in the kidney played a role in determining the composition of the arsenic species in the urine. Thus, it is possible that the ratios of the arsenic metabolites may be determined by metabolism in the kidney and not by total body metabolism. These findings may also be related to the high dose (5 mg/kg) used in these experiments and that some clearance mechanism may have been saturated.

The changes observed in NPSH levels relate to the metabolism of arsenic. NPSH levels are very low in plasma so changes in those levels reflect reactions in red blood cells where concentrations are approximately 7 $\mu mol/g$ hemoglobin (Winski and Carter, 1995). Therefore, NPSH changes may reflect the rate of arsenical compound transport across the cell membrane and/or the rate of any reactions between arsenic and NPSH. In our experiments, AsIII appeared to cross the RBC membrane within a few minutes and reacted quickly with NPSH. This was followed by a slow decrease in NPSH levels. The rapid reduction in NPSH levels was probably from the formation of $As(SG)_3^{-3}$ and the slower decrease from the formation of As–protein complexes (Figure 2). The reaction rate between AsV and NPSH was much different and showed a slow decrease in NPSH levels over the entire time period. The presence of AsV in the RBC indicated that transport across the membrane was not rate-limiting but that the reduction of AsV to AsIII was slow.

AsV + 2GSH $\rightarrow$ AsIII + GSSG

The results with AsIII showed that subsequent reactions were rapid. The absence of methylated metabolites in the red cells showed that only redox reactions and complexation occurred.

AsIII probably passed the RBC cell membrane by diffusion in the form of uncharged $HAsO_2$. Once inside the cell, AsIII complexes with GSH to form $As(SG)_3^{-3}$. Our experiments identified $As(SG)_3^{-3}$ in lung homogenates using HPLC-MS and found that the levels decreased with time. The decrease may be due to a decline in GSH levels that has been observed in homogenates (Barber, 1997) and the lower GSH levels result in dissociation of the complex. The fate of $As(SG)_3^{-3}$ is probably to exchange GSH with protein thiols and form mixed complexes. It can also be methylated to MMA and oxidized to AsV.

The reaction rates for the steps in the lung metabolism of arsenic were measured and a model developed. We only used the arsenic species that could be rapidly measured (AsV, AsIII, MMA and DMA) so early time points could be determined. The model used was: AsV $\leftrightarrow$ AsIII $\rightarrow$ MMA $\rightarrow$ DMA. The rates that were determined were: reduction of AsV to AsIII; oxidation of AsIII to AsV; methylation of AsIII to MMA; and methylation of MMA to DMA.

The rate equations were generated using the modeling and simulation software, SIMUSOLV. The output of these equations is the concentration of arsenate, arsenite, MMA and DMA at a particular time for a set of initial concentrations. Since any reactions involving AsIII would involve competing reactions, the oxidation of AsIII and the reduction of AsV were measured in two ways: (1) periodate oxidized methionine and S-adenosyl homocysteine were added to rat lung homogenates to block methylation; or (2) guinea pig lung homogenates were used because they have no methylation activity (Healy et al., 1997). The measured rates were comparable in these two systems. The rates of the reduction of MMA to monomethylarsonite and its methylation to DMA were combined into a single rate for the overall reaction of MMA to DMA. Since the formation of MMA and DMA cannot be reversed, the system never reaches a steady state. Of course, the model makes no allowance for excretion, and the endpoint of the model is the formation of DMA from all the arsenic in the system.

A whole-body physiological pharmacokinetic model based on urinary metabolites has been developed for rabbits, hamsters and humans (Mann et al., 1996a and b). Our results showed rates that corresponded with those found for humans. The human rate of metabolism was lower than those found for rabbits and hamsters. Our findings for a slower rate in hamsters may relate to their being limited to the lung and not for the entire body.

The findings for hsp32 induction and cell cytotoxicity in the BEAS-2B cells supported the hypothesis that arsenite was responsible for these effects. The concentrations of arsenic species present in toxicity studies were predicted with this model, and AsIII levels correlated to observed effects. There was good correlation between reduction of AsV to AsIII with toxicity and hsp32 induction. The amounts of MMA and DMA that were formed during the time of these experiments were too small to cause the hsp 32 effects (Barber, 1997). The model did not correlate the MMA and DMA levels with cell death, but they were substantially less potent than either AsV or AsIII.

It is concluded that pulmonary exposure to arsenical compounds would result in arsenic metabolism in the lung before reaching the systemic circulation. Once in the circulation, the arsenical compounds are oxidized and reduced by the red blood cell before they reach the other organs. Uptake and excretion rates appear to be more rapid than methylation and complexation with glutathione was found to be important inside the cell. For hsp32 induction in BEAS-2B cells cytotoxicity, the toxic species appears to be arsenite. Arsenic disposition kinetic models will be important to identify the toxic species of As, but they will require whole body models with metabolism from each organ.

REFERENCES

Barber, D.S. 1997. Correlation of Pulmonary Arsenic Metabolism and Toxicity. Ph.D. Thesis, University of Arizona.

Brain, J.D., Knudson, D.E., Sorokin, S.P., Davis, M.A. 1976. Pulmonary distribution of particles given by intratracheal instillation or by aerosol inhalation. *Environ. Res.*, **11**, 13–33.

Burnette, W.N. 1981. "Western Blotting": electrophoretic transfer of proteins from sodium dodecyl sulfate-polyacrylamide gels to unmodified nitrocellulose and radiographic detection with antibody and radioiodinated protein A. *Anal. Biological Methylation.*, **122**, 195.

Cullen, W., McBride B., Reglinski, J. 1984. The reaction of methylarsenicals with thiols: some biological implications. *J. Inorg. Chem.*, **21**, 179–194.

Delnomdedieu, M., Basti, M., Otvos, J., Thomas, D. 1994a. Reduction and binding of arsenate and dimethylarsinate by glutathione: a magnetic resonance study. *Chem. Biol. Interact.*, **90**, 139–155.

Delnomdedieu, M., Basti, M., Styblo, M., Otvos, J., Thomas, D. 1994b. Complexation of arsenic species in rabbit erythrocytes. *Chem. Res. Toxicol.*, **7**, 621–627.

Enterline, P.E., Marsh, G.M., Esmen, N.A., Henderson, V.L., Callahan, C.M. Paik, M. 1987. Some effects of cigarette smoking, arsenic, and SO_2 on mortality among US copper smelter workers. *J. Occup. Med.*, **29**, 831–838.

Healy, S.M., Zakharyan, R.A., Aposhian, H.V. 1997. Enzymatic methylation of arsenic compounds: IV. *In vitro* and *in vivo* deficiency of the methylation of arsenite and monomethylarsonic acid in the guinea pig. *Mutation Research.*, **386**, 229–239.

Inamasu, T., Hisanaga, A., Ishinishi, N. 1982. Comparison of arsenic trioxide and calcium arsenate retention in the rat lung after intratracheal instillation. *Toxicol. Lett.*, **12**, 1–5.

Ishinishi, N., Kodama, Y., Nobutomo, K., Hisanaga, A. 1977. Preliminary experimental study on carcinogenicity of arsenic trioxide in rat lung. *Environ. Health Perspect.*, **19**, 191–196.

Laemmli, U.K. 1970. Cleavage of structural proteins during the assembly of the head of a bacteriophage T4. *Nature*, **227**, 680.

Lechner, J.F., LaVeck, M.A. 1985. A serum-free method for culturing normal human bronchial epithelial cells at clonal density. *J. Tissue Culture Methods*, **9** (2), 43–48.

Maiorino, R.M., Aposhian, H.V. 1985. Dimercaptan metal-binding agents influence the biotransformation of arsenite in the rabbit. *Toxicol. Appl. Pharmacol.*, **77**, 240–250.

Mann, S., Droz, P.O., Vahter, M. 1996(a). A physiological based pharmacokinetic model for arsenic exposure. *Toxicol. Appl. Pharmacol.*, **140**, 471–486.

Mann, S., Droz, O., Vahter, M. 1996(b). A physiologically based pharmacokinetic model for arsenic exposure. I. Development in hamsters and rabbits. *Toxicol. Appl. Pharmacol.*, **137**, 8–22.

Marafante, E., Vahter, M. 1987. Solubility, retention, and metabolism of intratracheally and orally administered inorganic arsenic compounds in the hamster. *Environ. Res.*, **42**, 72–82.

Pershagen, G., Bjorklund, N.E. 1985. On the pulmonary tumorigenicity of arsenic trisulfide and calcium arsenate in hamsters. *Cancer Letters*, **27**, 99–104.

Roehm, N., Rodgers, G., Hatfield, S., Glasebrook, A. 1991. An improved colorimetric assay for cell proliferation and viability utilizing the tetrazolium salt XTT. *J. Immunological Methods.*, **142**, 257–265.

Rosner, M.H., Carter, D.E. 1987. Metabolism and excretion of gallium arsenide and arsenic oxides by hamsters following intratracheal instillation. *Fund. Appl. Toxicol.*, **9**, 730–737.

Scott, N., Hatlelid, K.M., Mackenzie, N.E. Carter, D.E. 1993. Reactions of Arsenic(III) and Arsenic(V) Species with Glutathione. *Chem. Res. Toxicol.*, **6**(1), 102–106.

Styblo, M., Yamauchi, H., Thomas, D.J. 1995. Comparative in vitro methylation of trivalent and pentavalent arsenicals. *Toxicol. Appl. Pharmacol.*, **135**, 172–178.

Winski, S.L., Carter, D.E. 1995. Interactions of rat red blood cell sulfhydryls with arsenate and arsenite. *J. Toxicol. Environ. Health*, **46**, 379–397.

Zakharyan, R., Wu, Y., Bogdan, G., Aposhian, H.V. 1995. Enzymatic methylation of arsenic compounds: assay, partial purification, and properties of arsenite methytransferase and monomethylarsonic acid methyltransferase of rabbit liver. *Chem. Res. Toxicol.*, **8**, 1029–1038.

Zakharyan, R., Wildfang, E., Aposhian, H.V. 1996. Enzymatic Methylation of Arsenic Compounds. III. The Marmoset and Tamarin, but not the Rhesus, Monkeys are Deficient in Methyltransferases That Methylate Inorganic Arsenic. *Toxicol. Appl. Pharm.*, **140**, 77–84.

Arsenic Exposure and Health Effects
W.R. Chappell, C.O. Abernathy and R.L. Calderon (Editors)
1999 Elsevier Science B.V.

Metabolism and Toxicity of Arsenicals in Cultured Cells

Miroslav Styblo, Libia Vega, Dori R. Germolec, Michael I. Luster,
Luz Maria Del Razo, Changqing Wang, William R. Cullen,
David J. Thomas

ABSTRACT

The metabolism and toxicities of arsenite, arsenate and trivalent and pentavalent methylated arsenicals have been examined in primary rat hepatocytes and in cells derived from human liver, skin, urinary bladder, and cervix. Among the cell lines examined, primary rat hepatocytes exhibited the greatest capacity for methylation of arsenicals. Trivalent arsenicals, arsenite, diiodomethylarsine or methylarsine oxide were better substrates for the methylation reactions than were pentavalent arsenate and methylarsenate. Compared to primary rat hepatocytes, the capacity for methylation of arsenicals was significantly lower in primary human hepatocytes. Even lower capacity for arsenic methylation was found in HeLa (human cervical adenocarcinoma cells) and normal human epidermal keratinocytes. The Urotsa cell line, an SV-40 transformed human urinary bladder cell line, did not methylate any arsenical tested. In primary rat hepatocytes incubated with 0.1 to 1 μM arsenite, dimethylarsenic (DMAs) was the major methylated metabolite and was found mainly in culture media. Small amounts of monomethylarsenic (MAs) were detected in cells. Incubation of primary rat hepatocytes with 4 to 20 μM arsenite resulted in partial inhibition of the methylation reactions, a decreased DMAs/MAs ratio, and the release of significant amounts of MAs from cells. In cell lines with low capacities for arsenic methylation, inorganic arsenic and/or MAs accumulated in the cells, suggesting that complete methylation (dimethylation) is a prerequisite for clearance of arsenic from cells. Addition of glutathione, glutathione ethyl ester or N-acetylcysteine to culture media stimulated the efflux of MAs from cells decreasing the DMAs/MAs ratio. For all cell lines examined, trivalent mono- and dimethylated arsenicals were more toxic than was arsenite. There was no correlation between methylation capacity of cell lines and resistance to the cytotoxicity of trivalent arsenicals. These results suggest that (i) for human tissues, capacity of cells for methylation of arsenicals varies significantly; (ii) methylation is inhibited by high concentrations of inorganic arsenic; (iii) trivalent methylated metabolites are more cytotoxic than inorganic arsenicals, and (iv) high methylation capacity does not protect cells from the acute toxicity of trivalent arsenicals.

Keywords: arsenic, metabolism, methylation, toxicity, cell, culture

INTRODUCTION

The classification of arsenic as a carcinogen (IARC, 1987) has been based exclusively on epidemiological studies carried out among residents of arsenic endemic areas in Taiwan and elsewhere who were exposed to inorganic forms of arsenic (iAs), arsenate (iAs^V) and/or arsenite (iAs^{III}), from consumption of contaminated drinking water. This exposure has been associated with cancer of skin, lung, and urinary bladder (Smith et al., 1992). The mechanism by which iAs induces cancer is unknown. Efforts to investigate the mechanistic basis of arsenic carcinogenesis in laboratory animals have been greatly impeded by the lack of a reliable and reproducible animal model. An alternative approach to studies of the mechanism of arsenic carcinogenicity and toxicity has used human cell lines. This approach has provided important information on the genotoxic effects of arsenic in human cells (Moore et al., 1997; Rassmussen and Menzel, 1997), on arsenic-induced mutations (Wiencke et al., 1997), on the suppression of cell programming by arsenic (Kachinskas et al., 1997), on arsenic-induced changes in methylation status of DNA (Mass and Wang, 1997; Zhao et al., 1997), on the inhibition by arsenic of enzymes involved in DNA methylation (Zhao et al., 1997) and repair (Yager and Wiencke, 1997), and on arsenic-induced expression of genes involved in the regulation of cellular growth and proliferation (Salazar et al., 1997; Burleson et al., 1996; Germolec et al., 1997). Data obtained in cell culture work have linked exposure to arsenic directly to processes fundamental to carcinogenesis. However, most of these studies have focused exclusively on the effects induced by iAs^V and/or iAs^{III}. Little information has been provided about the metabolic fate of iAs in cultured human cells. Scant attention has been paid to the adverse effects associated with exposure to methylated arsenicals, the products of the methylation of iAs.

In humans as in most mammals, iAs is enzymatically methylated yielding mono-, di- and possibly trimethylated arsenicals (for reviews see Styblo et al., 1995a; Aposhian, 1997). According to the oxidative methylation scheme proposed by Cullen et al. (1984a,b), both trivalent and pentavalent methylated arsenicals are intermediates or final metabolites of iAs. It has been shown that the acute toxicities of methylated pentavalent arsenicals, i.e., methylarsonic acid (MAs^V) dimethylarsinic acid (DMAs^V), and trimethylarsinoxide (TMAs^{V}O) were significantly lower than those of iAs^V, and particularly iAs^{III} (Yamauchi and Fowler, 1994). Because the methylated pentavalent arsenicals are less acutely toxic than either iAs^V or iAs^{III}, most investigators have chosen to focus on the effects of iAs. The toxicities of putative trivalent methylated metabolites have never been directly examined. Considering the high reactivity of trivalent arsenic, particularly its high affinity for thiols, trivalent methylated arsenicals are likely to be at least as biologically active as iAs^{III}. We have recently shown that trivalent methylarsonous acid (MAs^{III}) inhibits glutathione reductase (GR) (Styblo et al., 1997), a key enzyme of redox metabolism of glutathione (GSH). The arsinothiol MAs^{III}(GS)$_2$ was by two orders of magnitude a more potent inhibitor of GR than was iAs^{III}. We have also shown that MAs^{III} and dimethylarsinous acid (DMAs^{III}), unlike their pentavalent analogs, display high affinity for specific cellular proteins (Styblo et al., 1996a; Styblo and Thomas, 1997). These data suggest that trivalent methylated metabolites may be partly responsible for adverse effects associated with exposure to iAs.

The experimental work reported here examines the metabolism and cytotoxicity of iAs and both trivalent and pentavalent methylated arsenicals in cultured cells. We have focused on cell lines derived from human tissues that are a major site for the metabolism of iAs (liver) or targets for its carcinogenic effects (skin, bladder). This study demonstrates that the capacity for arsenic methylation varies among cell lines and that trivalent methylated arsenicals, putative metabolites of iAs, are more cytotoxic than iAs^{III}. The pharmacokinetic behavior of iAs and the kinetics of methylation reactions in selected cell lines are also described.

METHODOLOGY

Arsenicals

iAs^V and iAs^{III} (sodium salts) were purchased from Sigma (St. Louis, MO). MAs^V (sodium salt) was obtained from Chem Service (West Chester, PA) and DMAs^V from Strem (Newburyport, MA). Trivalent methylated arsenicals, diiodomethylarsine (MAs^{III}I$_2$), monomethylarsine oxide (MAs^{III}O), iododimethylarsine (DMAs^{III}I), and complex of DMAs^{III} with GSH (DMAs^{III}GS), were synthesized in the Department of Chemistry, University of British Columbia, using previously-described methods (Cullen et al., 1984a,b; Styblo et al., 1997a). Radiolabeled [^{73}As]iAs^V was purchased from Los Alamos Meson Production Facility (Los Alamos, NM). [^{73}As]iAs^{III} was prepared from [^{73}As]iAs^V by reduction with metabisulfite/thiosulfate reagent (MTR) (Reay and Asher, 1977; Styblo et al., 1995b).

Animal and Human Cell Lines

Primary rat hepatocytes were prepared at the Advanced Cell Technologies and Tissue Engineering Facility, School of Medicine, University of North Carolina at Chapel Hill. Cells were isolated from adult male Fischer 344 rats using a previously-described two-step perfusion technique (Seglen, 1973). Cells were plated for 2 hours in collagen-coated culture dishes in William's medium E that contained 10% fetal bovine serum (FBS), glutamine (2 mM), penicillin (100 U/ml), streptomycin (100 μg/ml), insulin (5 μg/ml), transferrin (5 μg/ml), sodium selenite (5 ng/ml), and dexamethasone (0.5 μM). The William's medium E with the same additives but without FBS was used to culture cells for up to 5 days. Normal human epidermal keratinocytes from adult female breast tissue were obtained from Clonetics Corp. (San Diego, CA). Keratinocytes were cultured in Keratinocyte Growth Medium using previously-described procedures (Burleson et al., 1996; Germolec et al., 1997). Urotsa cells, a SV-40 transformed epithelial cell line derived from normal human urinary bladder, were kindly provided by Dr. Nyseo Unimye, Department of Urology, School of Medicine, West Virginia University. These cells were cultured in RPMI 1640 medium in presence of 10% FBS, glutamine (2 mM), penicillin (50 U/ml), and streptomycin (50 μg/ml). The HeLa (human cervical adenocarcinoma) cells were obtained from the American Type Culture Collection and cultured in MEM medium with 10% FBS, in presence of glutamine (2 mM), penicillin (50 U/ml), and streptomycin (50 μg/ml). Primary human hepatocytes were kindly provided by Dr. Edward L. LeCluyse, School of Pharmacy, University of North Carolina at Chapel Hill. Cells were isolated from normal hepatic tissue obtained through the Human Liver Transplant Program using previously-described procedures (Strom et al., 1996) and cultured in the FBS-free William's medium E as described for primary rat hepatocytes. All cell lines were grown at 37°C in a humidified incubator in an atmosphere of 95% air and 5% CO$_2$.

Treatment of Cultured Cells with Arsenicals

Stock solutions of MAs^{III}I$_2$ and DMAs^{III}I (200 mM) were prepared in 70% ethanol. Stock solutions of the other arsenicals (200 mM) were prepared in sterile PBS. To prevent oxidation of trivalent arsenicals, all stock solutions were stored at –80° for no longer than 2 weeks. Dilutions of the stock solutions were prepared in sterile PBS shortly before each experiment and kept at 0°C before addition to cell cultures. Cultured cells were treated with arsenicals at final concentrations of 0.1 to 20 μM for up to 24 hours. For metabolic studies, cells were incubated with [^{73}As]iAs^{III} or [^{73}As]iAs^V and radiolabeled metabolites were analyzed using TLC as described below. To examine metabolism of monomethylated arsenicals, MAs^V or MAs^{III}I$_2$ (1 μM) were added to the culture medium and metabolites were analyzed by hydride generation atomic absorption spectrometry (HG-AAS). To examine effects of thiols on metabolism of arsenicals, GSH (Sigma), GSH-ethylester (GSH-Et) (Sigma) or *N*-acetyl-cysteine (NAC) (Sigma) were added into the culture media in some experiments.

Analysis of Radiolabeled Metabolites by TLC

For analysis of radiolabeled arsenic metabolites, culture medium was removed and cells were harvested by trypsination and scraping. To release protein-bound arsenicals, culture media and cells were treated with 0.2 M CuCl (pH 1) and heated in a water bath at 100°C for 5 min (Styblo et al., 1996). The denatured proteins were removed by centrifugation and the supernates were oxidized by H_2O_2 to facilitate further analysis. Aliquots of the oxidized supernates were analyzed by TLC on PEI-F cellulose following the previously-described procedure (Styblo et al., 1995b, 1996, 1997b). The distribution of the radioactivity on the developed TLC plates was analyzed with an AMBIS 4000 imaging detector.

Analysis of Arsenic Metabolites by HG-AAS

For analysis of arsenic metabolites by HG-AAS, the whole cell culture (cells plus medium) was wet digested in 2 M HCl at 80°C for 3 hours. Arsenic metabolites were then analyzed using a Perkin Elmer 5100 atomic absorption spectrometer equipped with a reaction vessel for the reduction of arsenic species to volatile arsines and with a liquid nitrogen-cooled gas chromatographic trap (Crecelius et al., 1986). Arsenic species in the digested samples were converted to the corresponding arsines in the reaction vessel upon addition of sodium boro-hydride (EM Science, Gibbstown, NJ) at pH 1–2. Arsines generated by this procedure were carried by a continuous flow of helium gas (150 ml/min) into the liquid nitrogen-cooled trap. The warming of the trap allowed separation of the arsines by boiling points. The air/hydrogen flame was used for atomization of arsines in the atomic absorption detector. The reliability of the analysis was assessed by spiking samples with known amounts of various arsenicals. Recoveries ranged from 92% to 107% with coefficients of variation between 3 and 11%.

Examination of Cytotoxicity of Arsenicals

Effects of arsenicals on viability of cultured cells were examined using MTT (thiazolyl blue) assay (Carmichael et al., 1987). After 24-hour exposure to arsenicals, arsenic-containing culture medium was removed, cells were washed with PBS and corresponding phenol red-free medium was added into the culture. MTT (Sigma) was added at final concentration of 0.5 mg/ml and cells were placed into the incubator for 3–4 hours. After incubation, medium was removed and cells washed with PBS. Insoluble purple formazan, a product of MTT cleavage by dehydrogenases of viable cells, was then dissolved in acidic isopropanol. Absorbance of the dye was measured at 570 nm with background subtraction at 630 nm. Cytotoxic effects of arsenicals in some cell lines were also evaluated by neutral red (toluylene red) assay (Bonrenfreund and Puerner, 1985) using Sigma Neutral Red Based Assay Kit.

Statistical Evaluation

Student's two-sided t-test (SlideWrite Plus program package) was used to evaluate differences between experimental groups.

RESULTS

Metabolism of Arsenicals in Primary Rat hepatocytes

The uptake of iAs was examined in primary cultures of rat hepatocytes incubated with 0.1 μM of iAs^{III} or 0.1 μM of iAs^V (Figure 1) for up to 24 hours. The uptake of iAs^{III} by hepatocytes was six-fold greater than the uptake of iAs^V. About 24% of arsenic from iAs^{III} was associated with cells during the first 1 to 3 hours of incubation compared with 4% of iAs^V. Beyond 3 hours of incubation, the amount of arsenic in cells exposed to iAs^{III} continuously decreased. The amount of arsenic retained in cells exposed to iAs^V did not change significantly over the 24-h incubation period. Arsenic metabolites in cells and medium were analyzed in rat hepatocyte culture exposed to 0.1 μM iAs^{III} for up to 24 hours (Figure 2). About 95% of iAs^{III} was methylated during a 12- to 24-h incubation interval. DMA was a major metabolite found mainly in culture media. A small amount of MAs was detected in cells and in medium at short

Fig. 1. Uptake of iAs^{III} (O) and iAs^{V} (●) by primary rat hepatocytes (mean ± SD, $n = 4$). Cells plated in 24-well plate (2.10^5 cells/well) were incubated with 0.1 µM iAs^{III} or 0.1 µM iAs^{V} for up to 24 hours.

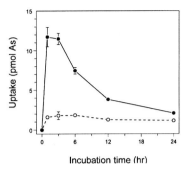

time intervals. After a 24-h incubation, about 4 to 5% of the original amount of iAs^{III} was not methylated; about one third of this was associated with cells. In contrast, less than 3% of 0.1 µM iAs^{V} was methylated in rat hepatocytes yielding mainly DMAs (data not shown). To examine the concentration dependence of methylation capacity in rat hepatocytes, cells were exposed to 0.1 to 20 µM iAs^{III}. Table 1 shows the amounts of MAs and DMAs detected in the culture (cells + medium) after a 24-h incubation. In hepatocytes incubated with 0.1 to 1 µM iAs^{III}, about 90 to 95% of iAs was methylated yielding almost exclusively DMAs. The DMAs/MAs ratio increased from 116 at 0.1 µM iAs^{III} to 275 at 1 µM iAs^{III}. In cells exposed to 4 µM iAs^{III}, a 24-h methylation yield represented only 33% of total arsenic in culture. DMAs remained the main methylated metabolite. However, significant amounts of MAs were detected in both cells and culture medium. DMAs/MAs ratio decreased dramatically to 4.1. Hepatocytes incubated with 10 and 20 µM iAs^{III} produced less methylated metabolites than cells cultured in the presence of 4 µM iAs^{III}. DMAs/MAs ratio further decreased reaching 0.7 at 20 µM iAs^{III}. The portion of MAs released from cells into the culture media increased under these conditions (data not shown). The capacity of rat hepatocytes to methylate MAs was also examined. Cells were incubated with 1 µM MAs^{V} or 1 µM MAs^{III}I$_2$ (Figure 3). During a 24-h incubation period, more than 90% of MAs^{III}I$_2$ was converted to DMAs as compared with 5% of MAs^{V}.

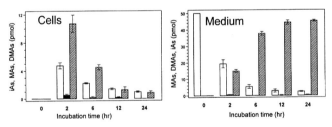

Fig. 2. Methylation of iAs^{III} in primary rat hepatocytes. iAs (white bars), MAs (black bars), and DMAs (diagonal lined bars) in cells and in medium (mean ± SD, $n = 4$). Cells plated in 24-well plate (6.10^4 cells/well) were incubated with 0.1 µM iAs^{III} for up to 24 hours.

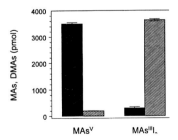

Fig. 3. Methylation of MAs^{V} and MAs^{III}I$_2$ in rat primary hepatocytes. MAs (black bars) and DMAs (diagonal lined bars) in cell culture (cells + medium). Mean and range of duplicates are shown. Cells plated in 6-well plate (9.10^5 cells/well) were incubated with 1 µM MAs^{V} or 1 µM MAs^{III}I$_2$ for 24 hours.

Fig. 4. Methylation of iAs^{III} in primary human hepatocytes. MAs (black bars) and DMAs (diagonal lined bars) in
cell culture (cells + medium); mean ± SD, n = 4. Cells
plated in 24-well plate (1.10^5 cells/well) were incubated
with 0.1 μM iAs^{III} for 24 hours.

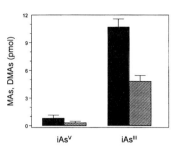

TABLE 1

Effect of iAs^{III} concentration on methylation yield in primary rat hepatocytes[1]

[iAs^{III}] (μM)	Total As (pmol)	MAs (pmol)[2]	DMAs (pmol)[2]	MAs + DMAs (pmol)[2]	MAs + DMAs (% of total As)	DMAs/MAs ratio
0.1	50	0.4 ± 0.19	46.2 ± 0.35	46.6 ± 0.30	93.2	116
0.4	200	1.2 ± 0.95	184.6 ± 0.60	185.8 ± 1.45	92.9	154
1	500	1.7 ± 2.90	464.3 ± 7.27	465.9 ± 7.46	93.2	275
4	2000	130.5 ± 13.70	537.0 ± 90.31	667.5 ± 102.75	33.4	4.1
10	5000	203.8 ± 34.72	251.3 ± 55.41	455.0 ± 80.15	9.1	1.2
20	10000	245.0 ± 63.64	163.5 ± 37.48	408.5 ± 6.65	4.1	0.7

[1]Cells plated in 24-well plates (6.10^4 cells/well) were incubated with iAs^{III} for 24 hours.
[2]Mean ± SD, n = 4.

Metabolism of Arsenicals in Human Cell Lines

Hepatocytes

Like rat hepatocytes, the primary human hepatocytes accumulated several fold more iAs^{III}
than iAs^V (data not shown). Methylation yield in cells incubated with 0.1 μM iAs^{III} for 24
hours did not exceed 30%. Unlike in rat hepatocytes, MAs was the major methylated
metabolite (Figure 4) detected almost exclusively in cultured cells. The DMAs/MAs ratio did
not exceed 0.4. The 24-h methylation yield from 0.1 μM iAs^V was less than 3%.

Keratinocytes

Metabolism of 0.05 μM iAs^{III} and 0.05 μM iAs^V was examined in normal human keratinocytes
obtained from two donors: #2199 and #4021 (Figure 5). In general, cells incubated with iAs^{III}
in Keratinocyte Growth Medium (KGM) for 48 hours retained more arsenic than did cells
incubated with iAs^V. Following exposure to iAs^{III}, cells from donor #4021 accumulated more
arsenic than did cells from donor #2199. The latter cell line methylated iAs^{III} better than the
former one. However, the methylation yield did not exceed 2.5%. Methylation yields in cells
incubated with iAs^V were significantly smaller. With either substrate, MAs was the only
methylated metabolite found in keratinocytes cultured in KGM. Keratinocytes cultured in
this medium methylated neither MAs^V nor MAs^{III}I$_2$ (data not shown). Incubation of
keratinocytes with iAs^{III} in William's medium E did not increase total methylation yield.
However, a small amount of DMAs was detected in the culture under these conditions (data
not shown).

HeLa Cells

Like normal human epidermal keratinocytes, HeLa cells produced relatively small amounts
of methylated metabolites when incubated with 0.05 μM iAs^{III} for 24 hours (Figure 6).
Notably, both MAs and DMAs were found in iAs^{III}-treated HeLa cells. The DMAs/MAs ratio
exceeded 0.6 at the end of 24-h incubation period. Almost 80% of MAs was retained in cells

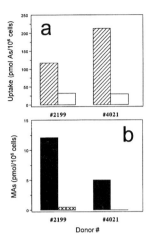
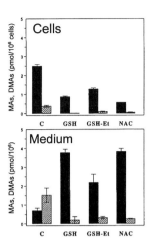

Left: Fig. 5. Uptake and metabolism of iAs^{III} and iAs^V by normal human epidermal keratinocytes obtained from two donors: a, uptake of iAs^{III} (diagonal lined bars) and iAs^V (white bars); b, production of MAs from iAs^{III} (black bars) and iAs^V (cross-hatched bars). Cells plated in 25 cm^2 culture flasks (1–2.10^6 cells/flask) were incubated with 0.05 μM iAs^{III} or 0.05 μM iAs^V for 48 hours.

Right: Fig. 6. Methylation of iAs^{III} in HeLa. MAs (black bars) and DMAs (diagonal lined bars) in cells and in medium. Mean and range of duplicates are shown. Cells plated in 25 cm^2 culture flasks (4.10^6 cells/flask) were incubated with 0.05 μM iAs^{III} for 22 hour without (C) or with addition of 2.5 mM GSH, 2.5 mM GSH-Et or 2.5 mM NAC.

but most DMAs was released into the culture medium. Incubation of cells with iAs^{III} in presence of GSH, GSH-Et or NAC increased efflux of MAs from cells to culture medium without changing the total methylation yield. The DMAs/MAs ratio dramatically decreased under these conditions ranging from 0.06 to 0.1.

Urotsa Cells

This cell line did not methylate either iAs^{III}, iAs^V, MAs^{III}I$_2$ or MAs^V when cultured in RPMI 1640 medium or William's medium E. Like other cell lines, Urotsa cells retained several-fold greater amounts of arsenic from iAs^{III} than iAs^V (data not shown).

Cytotoxicity of Arsenicals

The effects of trivalent arsenicals on the viability of primary rat hepatocytes after 24-h incubation are shown in Figure 7. Based on results of the MTT assay, MAs^{III} was the most toxic species among arsenicals examined followed by DMAs^{III}GS and iAs^{III}. Incubation of rat hepatocytes with as low as 0.4 μM MAs^{III}O resulted in significant decrease in cell viability. At 4 μM MAs^{III}, more than 95% of cells were not viable. Cytotoxicity of MAs^{III}I$_2$ was comparable with that of MAs^{III}O (data not shown). In contrast, 10 μM iAs^{III} or DMAs^{III}GS was needed to decrease significantly the viability of rat hepatocytes. To determine whether GSH in DMAs^{III}GS or I$^-$ anion from MAs^{III}I$_2$ contributed to the cytotoxicities of these arsenicals, hepatocytes were incubated with up to 20 μM GSH or KI. No cytotoxic effects were observed. Pentavalent arsenicals (iAs^V, MAs^V, and DMAs^V) were not toxic for hepatocytes at concentrations up to 20 μM (data not shown).

Toxic effects of arsenicals in normal human epidermal keratinocytes from donor #2199 were examined using both MTT and neutral red assays (Figure 8). Regardless of the method used, MAs^{III}O was found to be the most toxic arsenical followed by DMAs^{III}I, DMAs^{III}GS, and iAs^{III}. For this cell type, the neutral red assay appeared to be more sensitive than the MTT assay in monitoring cytotoxicity of trivalent arsenicals. Using the former assay, MAs^{III}O at

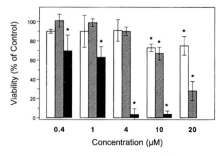

Fig. 7. Cytotoxicity of trivalent arsenicals in primary rat hepatocytes. Cells plated in 96-well plate (15.10³ cells/well) were incubated with iAs^III (white bars), MAs^IIIO (black bars), and DMAsGS (diagonal lined bars) for 24 hours. Viability of cells was then determined by MTT assay (mean ± SD, n = 4). *Viability is statistically different (p < 0.05) from viability of control (untreated) cells.

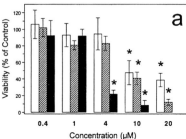

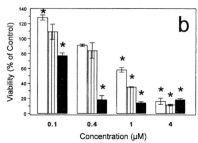

Fig. 8. Cytotoxicity of trivalent arsenicals in primary normal human epidermal keratinocytes (donor #2199). Cells plated in 96-well plate (15.10³ cells/well) were incubated with iAs^III (white bars), MAs^IIIO (black bars), DMAsGS (diagonal lined bars), or DMAsI (vertical lined bars) for 24 hours. Viability of cells was then determined by (a) MTT assay (mean ± SD, n = 4) and (b) neutral red assay (mean ± SD, n = 3). *Viability is statistically different (p < 0.05) from viability of control (untreated) cells.

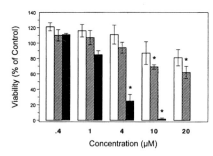

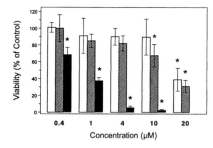

Left: Fig. 9. Cytotoxicity of trivalent arsenicals in HeLa. Cells plated in 96-well plate (25.10³ cells/well) were incubated with iAs^III (white bars), MAs^IIIO (black bars), or DMAsGS (diagonal lined bars) for 24 hours. Viability of cells was then determined by MTT assay (mean ± SD, n = 4). *Viability is statistically different (p < 0.05) from viability of control (untreated) cells.

Right: Fig. 10. Cytotoxicity of trivalent arsenicals in Urotsa. Cells plated in 96-well plate (2.10⁴ cells/well) were incubated with iAs^III (white bars), MAs^IIIO (black bars), or DMAsGS (diagonal lined bars) for 24 hours. Viability of cells was then determined by MTT assay (mean ± SD, n = 4). *Viability is statistically different (p < 0.05) from viability of control (untreated) cells.

concentration as low as 0.1 μM was found toxic for human keratinocytes. In contrast, 4 μM MAs^IIIO was needed to observe a significant effect on viability of cells by MTT assay. MAs^V, DMAs^V, GSH and KI were not toxic for human keratinocytes at concentrations up to 20 μM (data not shown). Incubation with 10 or 20 μM iAs^V resulted in about 45% decrease in viability of cells as indicated by MTT but not by neutral red assay (data not shown).

Based on MTT data, MAs^IIIO was also the most toxic arsenical for HeLa (Figure 9) and Urotsa (Figure 10) cells with significant effects on cell viability at concentrations of 4 and 0.4

μM, respectively. DMAs^{III}GS (or DMAs^{III}I; not shown) was less toxic than MAs^{III}O but more toxic than iAs^{III} for these cell lines. Pentavalent arsenicals, GSH, and KI were not toxic at concentrations up to 20 μM.

DISCUSSION

The goals of this study were to examine both the metabolic conversion and the acute toxicity of trivalent and pentavalent arsenicals in selected human cell lines. This work focussed on cells derived from human tissues that are known or expected to metabolize iAs (liver) or which are targets for its carcinogenic effects (skin and urinary bladder). Because *in vitro* metabolism (methylation) of arsenicals has been described in detail in rat liver (Buchet and Lauwerys, 1985, 1988; Styblo et al., 1995b, 1996b), primary rat hepatocytes which are competent to methylate arsenicals were used as a positive control. Metabolism of iAs in cultured primary rat hepatocytes was briefly examined by Lerman and coworkers (1983). They found that the cellular uptake of iAs^V was much smaller than that of iAs^{III} and concluded that iAs^V, which unlike iAs^{III} is ionized at physiological pH cannot enter hepatic cells by diffusion. Hence, liver was thought to be an unlikely site for the biomethylation of iAs^V. In the present work, a difference in the uptake of these two arsenicals was found not only in rat and human primary hepatocytes but also in cells derived from human skin, urinary bladder and cervix. These data contradict results of a number of *in vivo* experiments that demonstrated relatively rapid metabolism of iAs^V in various laboratory animals (for review see Styblo et al., 1995a). Because phosphate oxyanion interferes with the uptake of iAs^V by rabbit erythrocytes (Thomas, D.J., unpublished data) it is likely that the high concentration of phosphate in cell culture media antagonizes the uptake of iAs^V by cells. Similarly, the uptake of MAs^V may be inhibited by phosphate in the culture media, preventing methylation of MAs^V to DMAs. As shown in this study, the DMAs yield from MAs^{III} in primary rat liver hepatocytes was significantly higher than the DMAs yield from MAs^V. Previous work in a cell-free *in vitro* assay system that contained rat liver cytosol and which methylated arsenic demonstrated that pentavalent arsenicals (iAs^V, MAs^V) were not as good substrates for methylation reactions as were their trivalent analogs (iAs^{III}, MAs^{III}) (Styblo et al., 1995b). This indicates that reduction of pentavalent arsenicals to trivalency plays a critical role in arsenic metabolism *in vivo*. The presence of phosphates in blood and in the extracellular matrix may be an important factor involved in regulation of the cellular uptake of iAs^V in tissues.

It has been previously reported that high concentration of iAs^{III} (above 10 μM) inhibits the production of MAs and especially DMAs in an *in vitro* system that contained rat liver cytosol (Styblo et al., 1996b). In the present study, production of methylated metabolites by rat hepatocytes increased in the range of iAs^{III} concentration between 0.1 and 4 μM with sharp decrease in the DMAs/MAs ratio between 1 and 4 μM iAs^{III}. This decrease is likely a consequence of the preferential inhibition of DMAs synthesis by iAs^{III} and accumulation of MAs in cells. At higher concentrations, the total methylation yield decreased, suggesting that saturation or, more likely, inhibition of methylation reactions occurred. The decrease in production of methylated metabolites in hepatocytes incubated with high concentration of iAs^{III} was accompanied by the release of significant amounts of MAs into the culture media. The increased release of MAs from cells may be an early cytotoxic effect of iAs^{III} and/or MAs^{III} that results in increased permeability of the cellular membrane.

In this study, primary human hepatocytes did not appear to be as good methylators of iAs as were primary rat hepatocytes. In contrast to rat hepatocytes, human hepatocytes produced more MAs than DMAs even at low concentrations of iAs. Because DMAs has been shown to be a major urinary metabolite of iAs in humans and because liver is considered the major metabolic site for iAs, it is likely that the methylation patterns found in human hepatocytes in the present experiments are not typical for intact human hepatic tissue. It is possible that low methylation yield and decreased DMAs/MAs ratio is a consequence of the handling of the tissue (e.g., up to 24-h storage in preservation medium) and delay between surgery and isolation of hepatocytes. The culture conditions may also be responsible for altered ability of human hepatocytes to methylate iAs.

Compared to hepatocytes, normal human keratinocytes and HeLa cells were poor methylators of either iAs[III] or iAs[V]. Differences in the uptake of iAs, in the yield of MAs (the only methylated product), and in the distribution of MAs between cells and medium for the two keratinocyte lines examined (# 2199, #4021) suggest that interindividual differences in the kinetic and dynamic behavior of iAs could occur in human skin *in vivo*. The relationship between such interindividual differences and differences in the susceptibility to arsenic-induced skin cancer is problematic.

HeLa cells were used in this study as an example of a rapidly proliferating transformed human cell line. The human cervical tissue from which these cells are derived is not known to play a significant role in arsenic metabolism or in manifestation of arsenic toxicity or carcino-genicity. HeLa cells served as a model for examination of effects of GSH, GSH-Et and NAC on cellular metabolism of iAs. GSH participates in metabolism of iAs as a donor of electrons for reduction of arsenicals from pentavalency to trivalency (Cullen et al. 1984a,b; Buchet et al., 1988, Delnomdedieu et al., 1994; Styblo et al., 1996b). GSH-Et and NAC have been previously used in cell cultures to increase intracellular concentration of GSH (Harjit and Thornalley, 1995; Flanagan and Meredith, 1991). Unlike GSH that cannot cross the cellular membrane, GSH-Et and NAC are readily accumulated by cells and converted into GSH by two different mechanisms. GSH-Et undergoes deesterification yielding GSH (Anderson et al., 1985); NAC is deacetylated providing cysteine for *de novo* synthesis of GSH (Sjodin et al., 1989; Cotgreave et al., 1991). The results reported here suggest that GSH plays an important role in regulation of arsenic distribution in cellular environment. In particular, increased concentrations of GSH out and/or inside of cells induce the release of MAs. The efflux of MAs results in decreasing production of DMAs. This action of GSH, however, does not increase the cytotoxicity of iAs. In fact, GSH fully protects cultured cells from toxicity of iAs[III], MAs[III] or DMAs[III] (Styblo, unpublished results).

The results of metabolic experiments carried out in Urotsa cells indicate that these cells accumulate iAs from culture media but do not methylate iAs to either MAs or DMAs. The conversion of iAs to DMAs appears to be a prerequisite for release of arsenic from cells. The relatively high retention of iAs and lack of methylation capacity may make urinary bladder cells more vulnerable to its toxic or carcinogenic effect. Notably, the bladder is a target tissue for cancer in an arsenic-exposed population (Hopenhayn-Rich et al., 1996; Smith et al., 1998).

Our examination of cytotoxicity of arsenicals provides novel data on acute toxic effects of trivalent methylated arsenicals that are the likely intermediary metabolites of iAs (Cullen et al., 1984a,b). The results clearly show that MAs[III] is more toxic for all cell lines examined than is iAs[III]. DMAs[III] was found to be as toxic or more toxic than iAs[III] for the cultured cells. This information strongly contradicts the current understanding of the biomethylation as a mechanism for the detoxification of iAs. Rather, it is possible that intermediary trivalent methylated metabolites contribute significantly to the toxic and possibly carcinogenic effects associated with exposure to iAs. As we have shown, the high methylation capacity of rat hepatocytes does not protect against the toxic effects of either iAs[III] or MAs[III]. The susceptibilities of cells that can effectively methylate iAs[III] and MAs[III] to the acutely toxic effects of trivalent arsenicals were comparable to those of cells with limited or no capacity for iAs methylation (Table 2). To evaluate the risk associated with possible toxic effects of trivalent methylated metabolites of iAs *in vivo*, detailed information is needed about concentration and valency of MAs and DMAs in tissues of laboratory animals or humans exposed to iAs. Because most *in vivo* studies have focused exclusively on analysis of arsenic

TABLE 2

Methylation capacities and susceptibility of cells to toxic effects of trivalent arsenicals

Cell line	Methylation capacity (pmol $iAs^{III}/10^6$ cells/h)	Estimated IC_{50} Values[1] (μM) for Arsenicals		
		iAs^{III}	MAs^{III}[2]	$DMAs^{III}$[3]
MTT Assay:				
Rat Hepatocytes	460	5.1–>20	1.6–2.8	2.6–14.5
HeLa	0.25	>20	2.8	>20
Human Keratinocytes[4]	0.2	9.6–>10	2.3–3.3	9.2–10.2
Urotsa	~0	5.5–17.7	0.8–1.9	3.5–14.8
Neutral Red Assay:				
Human Keratinocytes[4]	0.2	1.4–1.6	0.2–0.8	0.4–0.8

[1]IC_{50} is defined as a concentration of an arsenical that results in 50% decrease in viability of cells over a 24-hour incubation period.
[2]IC_{50} values determined for $MAs^{III}O$ and $MAs^{III}I_2$ are shown.
[3]IC_{50} values determined for $DMAs^{III}GS$ and $DMAs^{III}I$ are shown.
[4]Donor #2199.

metabolites in urine, data on arsenic speciation in tissues are sparse. In fact, analytical methods capable of determination of the valency of MAs and DMAs in biological samples have not been developed. However, it has been previously shown that trivalent methylated arsenicals unlike pentavalent methylated arsenicals have high affinity for binding sites in tissue proteins (Styblo et al., 1996a, Styblo and Thomas, 1997). Thus, the amounts of MAs^{III} and $DMAs^{III}$ in tissues could be estimated as amounts of protein-bound MAs and DMAs.

The MTT assay has been successfully used to monitor toxicity of arsenicals in all cell lines used in this study. However, neutral red assay proved to be more sensitive to the arsenic-induced toxic effects in human keratinocytes. Technically, MTT assay follows conversion of MTT to formazan by mitochondrial dehydrogenases of viable cells. In contrast, neutral red assay is based on the active transport of the dye across the cellular membrane and its incorporation into lysosomes. It is possible that trivalent arsenicals directly interfere with the mechanisms that are responsible for the transport and/or lysosomal retention of neutral red in keratinocytes. Interestingly, neutral red assay failed to provide reliable information about toxic effects of arsenicals in the other cell lines used in this study.

CONCLUSIONS

The results presented here suggest that (i) capacity of cells for methylation of arsenicals varies greatly in different human tissues; (ii) metabolic patterns (uptake and methylation) depend on valency and concentration of arsenicals; (iii) trivalent methylated arsenicals are more cytotoxic than is iAs^{III}, suggesting that methylation is not simply a detoxification process, and (iv) high methylation capacity does not protect cells from acute toxicity of trivalent arsenicals.

ACKNOWLEDGEMENT

The authors would like to thank Dr. Nyseo Unimye and Dr. Edward LeCluyse for providing Urotsa cells and primary human hepatocytes for this study and Ms. Felecia Walton for her excellent technical assistance. Dr. Del Razo was a visiting scientist at the Environmental Research Center (US EPA) and was supported by a fellowship from the Pan American Health Organization. This work was in part funded by a Drinking Water STAR Grant R826136-01-0 from the US Environmental Protection Agency. This article has been reviewed in accordance with the policy of the National Health and Environmental Effects Research Laboratory, U.S. Environmental Protection Agency, and approved for publication. Approval does not signify that the contents necessarily reflect the views and policies of the Agency, nor does mention of trade names or commercial products constitute endorsement or recommendation for use.

REFERENCES

Anderson, M.E., Powrie, F., Puri, R.N., Meister, A. 1985. Glutathione monoethyl ester: preparation, uptake by tissues and conversion to glutathione. *Arch. Biochem. Biophys.*, **239**, 538–548.

Aposhian, H.V. 1997. Enzymatic methylation of arsenic species and other new approaches to arsenic toxicity. *Annu. Rev. Pharmacol. Toxicol.*, **37**, 397–419.

Borenfreund, E., Puerner, J. 1985. Toxicity determined in vitro by morphological alterations and neutral red absorption. *Toxicol. Lett.*, **24**, 119–124.

Buchet, J.P., Lauwerys, R. 1985. Study of inorganic arsenic methylation by rat in vitro: relevance for the interpretation of observations in man. *Arch. Toxicol.*, **57**, 125–129.

Buchet, J.P., Lauwerys, R. 1988. Role of thiols in the in vitro methylation of inorganic arsenic by rat liver cytosol. *Biochem. Pharmacol.*, **37**, 3149–3153.

Burleson, F.G., Simeonova, P.P., Germolec, D.R., Luster, M.I. 1996. Dermatotoxic chemical stimulate of c-jun and c-fos transcription and AP-1 DNA binding in human keratinocytes. *Res. Commun. Mol. Pathol. Pharmacol.*, **93**, 131–148.

Carmichael, J., DeGraff, W.G., Gazdar, A.F., Minna, J.D., Mitchell, J.B. 1987. Evaluation of a tetrazolium-based semiautomated colorimetric assay: assessment of chemosensitivity testing. *Cancer Res.*, **47**, 936–942.

Cotgreave, I., Moldeus, P., Schuppe, I. 1991. The metabolism of N-acetylcysteine in human endothelial cells. *Biochem. Pharmacol.*, **42**, 13–21.

Crecelius, E.A., Bloom, N.S., Cowan, C.E., Jenne, E.A. 1986. Determination of arsenic species in limnological samples by hydride generation atomic absorption spectroscopy. In: Speciation of Selenium and Arsenic in natural Waters and sediments. Vol. 2: Arsenic Speciation, pp. 1–28. Electric Power Research Institute Ed., Palo Alto, California, EA-4641, Project 2020-2.

Cullen, W.R., McBride, B.C., Reglinski, J. 1984a. The reaction of methylarsenicals with thiols: Some biological implications. *J. Inorg. Biochem.*, **21**, 179–194.

Cullen, W.R., McBride, B.C., Reglinski, J. 1984b. The reduction of trimethylarsine oxide to trimethylarsine by thiols: a mechanistic model for the biological reduction of arsenicals. *J. Inorg. Biochem.*, **21**, 45–60.

Delnomdedieu, M., Basti, M.M., Otvos, J.D., Thomas, D.J. 1994. Reduction and binding of arsenate and dimethylarsinate by glutathione: a magnetic resonance study. *Chem.-Biol. Interact.*, **90**, 139–155.

Flanagan, R.J., Meredith, T.J. 1991. Use of N-acetylcysteine in clinical toxicology. *Am. J. Med.*, **91** (Suppl. 3C), 131S–139S.

Germolec, D.R., Spaldings, J., Boorman, G.A., Wilmer, J.L., Yoshida, T., Simeonova, P.P., Bruccoleri, A., Kayama, F., Gaido, K., Tennant, R., Burleson, F., Dong, W., Lang, R.W., Luster, M.I. 1997. Arsenic can mediate skin neoplasia by chronic stimulation of keratinocyte-derived growth factors. *Mutation Res.*, **386**, 209–218.

Harjit, S.M., Thornalley P.J. 1995. Comparison of the delivery of reduced glutathione into P388D₁ cells by reduced glutathione and its mono- and diethyl ester derivatives. *Biochem. Pharmacol.*, **49**, 1475–1482.

Hopenhayn-Rich, C., Biggs, M.L., Fuchs, A., Bergolio, R., Tello, E.E., Nicolli, H., Smith, A.H. 1996. Bladder cancer mortality associated with arsenic in the drinking water in Argentina. *Epidemiology*, **7**, 117–124.

IARC (International Agency for Research on Cancer) 1987. In: IARC Monograph on the Evaluation of Carcinogenic Risk to Humans — Overall Evaluation of Carcinogenicity: an update of *IARC Monographs*, 1 to 42, (Suppl. 7), p. 100, Lyon.

Kachinskas, D.J., Qin, Q., Phillips, M.A., Rice, R.H. 1997. Arsenate suppression of human keratinocyte programming. *Mutation Res.*, **386**, 253–261.

Lerman, S.A., Clarkson, T.W., Gerson, R.J. 1983. Arsenic uptake and metabolism by liver cells is dependent on arsenic oxidation state. *Chem.-Biol. Interact.*, **45**, 401–406.

Mass, M.J., Wang, L. 1997. Arsenic alters cytosine methylation patterns of the promoter of the tumor suppressor gene p53 in human lung cells: a model for a mechanism of carcinogenesis. *Mutation Res.*, **386**, 263–277.

Moore, M.M., Harrington-Brock, K., Doerr, C.L. 1997. Relative genotoxic potency of arsenic and its methylated metabolites. *Mutation Res.*, **386**, 279–290.

Rasmussen, R.E., Menzel, D.B. 1997. Variation in arsenic-induced sister chromatid exchange in human lymphocytes and lymphoblastoid cell lines. *Mutation Res.*, **386**, 299–306.

Reay, P.F., Asher, C.J. 1977. Preparation and purification of ⁷⁴As-labeled arsenate and arsenite for use in biological experiments. *Anal. Biochem.*, **78**, 557–560.

Salazar, A.M., Ostrowsky-Wegman, P., Menedez, D., Miranda, E., Garcia-Carranca, A., Rojas, E. 1997. Induction of p53 protein expression by sodium arsenite. *Mutation Res.*, **381**, 259–265.

Seglen, P.O. 1973. Preparation of rat liver cells. *Meth. Cell. Biol.*, **13**, 29–83.

Sjodin,, K., Nilsson, E., Hallberg, A., Tunek, A. 1989. Metabolism of N-acetyl-L-cysteine: some structural requirements for the deacetylation and consequences for the oral bioavailability. *Biochem. Pharmacol.*, **38**, 3981–3996.

Smith, A.H., Hopenhayn-Rich, C. Bates, M.N., Goeden, H.M., Hertz-Picciotto, I., Duggan, H.M., Wood, R., Kosnett, M.J., Smyth, M.T. 1992. Cancer risks from arsenic in drinking water. *Environ. Health Perspect.*, **97**, 259–267.

Smith, A.H., Goycolea, M., Haque, R., Biggs, M.L. 1998. Marked increase in bladder and lung cancer in a region of northern Chile due to arsenic in drinking water. *Am. J. Epidemiol.*, **147**, 660–669.

Strom, S.C., Pisarov, L.A., Dorko, K., Thompson, M.T., Schuetz, J.D., Schuetz, E.G. 1996. Use of human hepatocytes to study P450 gene induction. *Meth. Enzymol.*, **272**, 388–401.

Styblo, M., Delnomdedieu, M., Thomas D.J. 1995a. Biological mechanisms and toxicological consequences of the methylation of arsenic. In: R.A. Goyer and M.G. Cherian (eds.), *Toxicology of Metals – Biochemical Aspects, Handbook of Experimental Pharmacology, Vol. 115*, pp. 407–433. Springer-Verlag, Berlin.

Styblo, M., Yamauchi, H., Thomas, D.J. 1995b. Comparative methylation of trivalent and pentavalent arsenicals. *Toxicol. Appl. Pharmacol.*, **135**, 172–178.

Styblo, M., Hughes, M.F., Thomas, D.J, 1996a. Liberation and analysis of protein-bound arsenicals. *J. Chromatogr. B*, **677**, 161–166.

Styblo, M., Delnomdedieu, M., Thomas, D.J. 1996b. Mono- and dimethylation of arsenic in rat liver cytosol in vitro. *Chem.-Biol. Interact.*, **99**, 147–164.

Styblo, M., Serves, S.V., Cullen, W.R., Thomas, D.J. 1997. Comparative inhibition of yeast glutathione reductase by arsenicals and arsenothiols. *Chem. Res. Toxicol.*, **10**, 27–33.

Styblo, M., Thomas, D.J. 1997. Binding of arsenicals to proteins in an *in vitro* methylation system. *Toxicol. Appl. Pharmacol.*, **147**, 1–8.

Wiencke, J.K., Yager, J.W., Varkonyi, A., Hultner, M., Lutze, L.H. 1997. Study of arsenic mutagenesis using the plasmid shuttle vector pZ189 propagated in DNA repair proficient human cells. *Mutation Res.*, **386**, 335–344.

Yager, J.W., Wiencke, J.K. 1997. Inhibition of poly(ADP-ribose) polymerase by arsenite. *Mutation Res.*, **386**, 345–351.

Yamauchi, H., Fowler, B.A. 1994. Toxicity and metabolism of inorganic and methylated arsenicals. In: J.O. Nriagu (ed.), *Arsenic in the Environment, Part II: Human Health and Ecosystem Effects*, pp. 35–43. Wiley, New York.

Zhao, C.O., Young, M.R., Diwan, B.A., Coogan, T.P., Waalkes, M.P. 1997. Association of arsenic-induced malignant transformation with DNA hypomethylation and aberrant gene expression. *Proc. Natl. Acad. Sci. USA*, **94**, 10907–10912.

Arsenic Exposure and Health Effects
W.R. Chappell, C.O. Abernathy and R.L. Calderon (Editors)
© 1999 Elsevier Science B.V. All rights reserved.

Proportions of Arsenic Species in Human Urine

Margaret E. Farago, Peter Kavanagh

ABSTRACT

The threshold hypothesis for arsenic toxicity based on methylation capacity has been discussed in the literature, however, a later analysis of data on the urinary metabolites (As_i, MMAA and DMAA) of different populations reported in published papers, suggested that on average 20–25% inorganic arsenic remains unmethylated regardless of the exposure level. It was concluded that the data did not support the methylation threshold hypothesis. Our results on urinary arsenic in exposed and unexposed populations in SW England, and others in the literature do not support these latter findings at low exposures. We conclude that these proportions are limiting values, and at low exposure, and thus low urinary arsenic, the proportion on DMAA is much higher.

Keywords: arsenic metabolites, DMAA, MMAA, inorganic arsenic, arsenic exposure

INTRODUCTION

Arsenic is methylated in the human liver as a detoxifying mechanism (Vahter, 1994). In this process inorganic arsenic is reduced to As(III), the substrate for methylating enzymes which produce the less toxic metabolites (Goyer, 1991; Buchet and Lauwerys, 1994). The methylated metabolites (MMAA, monomethylarsonic acid and DMAA, dimethylarsinic acid) are considered to be less reactive than inorganic arsenic with tissue components (Tatken and Lewis, 1983; Yamauchi et al., 1983; 1990) and are excreted in the urine more readily than inorganic arsenic (Buchet et al., 1981a, b; Vahter et al., 1984). It is the As(III) species that is most reactive with tissue components (Vahter and Marafante, 1983) and as a consequence, factors that effect the methylation process may thus affect arsenic toxicity.

The proportions of metabolites of inorganic arsenic are different in various mammalian species, and differ between human groups and individuals. The majority of experimental animals have been found to excrete arsenic efficiently, usually as DMAA. There are some mammalian exceptions, such as the guinea pig, marmoset and chimpanzee, which do not appear to methylate inorganic arsenic (Vahter, 1994). Of the mammals studied including humans, it has been shown that only humans excrete significant concentrations of MMAA (Vahter, 1997).

The hypothesis that there is a threshold for arsenic toxicity based on methylation capacity has been discussed in the literature. Petito and Beck (1990) suggested that when this detoxification mechanism is "overwhelmed" the levels of circulating unmethylated inorganic arsenic will rise, leading to increased potential for interactions with target organs. Since arsenic does not interact directly with DNA, this suggests a threshold for genotoxicity, which would act subsequently to any metabolic thresholds influencing the relationship between the dose to the target tissue and the carcinogenic response. In their review, Petito and Beck (1990) cite three lines of evidence for the concept of non-linear dose–response curves for arsenic: data from arsenic metabolism and detoxification; evidence from epidemiological studies of exposures to arsenic from drinking water, occupational exposure and other sources; and data from genotoxicity studies which indicate that arsenic does not react directly with DNA. The threshold would be related to two aspects of the methylation process: the inhibition of the methylation process by an excess of As(III) and the saturation of the enzymic conversion of MMAA to DMAA. The reduction of inorganic As(V) to As(III) is fast in all species examined (Vahter and Marafante, 1985). This reduced arsenic is bound to tissues, if it is not rapidly methylated. The methylated arsenic is rapidly excreted, whereas As(III) accumulates in certain body tissues (Vahter, 1985). Thus, inorganic arsenic, As_i, levels below the threshold level, will produce 80–90% methylation (Petito and Beck, 1990). The suggested threshold level varies between 259 μg/day to 500 μg/day for ingestion of inorganic arsenic (Petito and Beck, 1990; Storer, 1991).

Buchet and Lauwerys (1994) demonstrated, under experimental conditions in which volunteers were acutely exposed to known concentrations of As_i, as As_2O_3, that while the excretion of As_i and MMAA are linearly related to the dose administered, the excretion of DMAA levels off at the highest dose, indicating a possible saturation of the methylation capacity. The proportion of the three species of urinary arsenic changed markedly over time. In the first 48 to 96 hours after ingestion, arsenic was excreted mainly as the unmetabolised As_i but this was quickly followed by a progressive increase of the proportion excreted as MMAA and DMAA. The period at which the organic metabolites of arsenic are excreted is dependent on the severity of the dose but in all cases, more than 95 % of the excreted arsenic in the organic form was found to be DMAA after 216 hours.

In order to investigate the threshold hypothesis, Hopenhayn-Rich et al. (1993) analysed data on the urinary metabolites (As_i, MMAA and DMAA) of different populations from published papers, ranging from unexposed to highly occupationally and environmentally exposed. These authors focused on the concentrations of inorganic arsenic, As_i, found for different exposure levels and found on average, that 20–25% inorganic arsenic remains unmethylated regardless of the exposure level. They concluded that the data did not support the methylation threshold hypothesis. The results would appear to indicate that ingested inorganic arsenic is excreted in the urine in population groups with a relative distribution of As_i, MMAA and DMAA of about 20:20:60 (Buchet et al., 1981 b; Crecelius, 1977; Tam et al., 1979; Vahter, 1997) with extensive inter-individual variation (Hopenhayn-Rich et al., 1993; Petito and Beck, 1990). Mushak and Crosetti (1995) have suggested that because the percentage of As_i excreted in the urine often does not vary with increasing exposure (as reviewed by Hopenhayn-Rich et al. (1993) the hypothesis that the methylation of As_i becomes saturated at high As doses is implausible.

However, several studies have reported higher MMAA/DMAA ratios in exposed populations compared to control groups, indicating that humans may not be able to convert MMAA to DMAA efficiently at high As_i doses and suggesting a saturation of the methylation at higher exposures (Froines, 1994; Del Razo, 1994; Hseuh et al., 1995; Yamauchi et al., 1995).

Further work by Hopenhayn-Rich et al. (1996) reported that there was no evidence of a threshold for methylation capacity even at very high exposures, and that inter-individual differences were large. These authors suggested that the significance of the MMAA/DMAA ratio needs further investigation. The same authors, in an investigation of the change in the pattern of metabolites when high arsenic drinking water was changed to that with a lower As concentration, concluded that, although the percentage of inorganic arsenic fell from 17.8% to 14.6% and the MMAA/DMAA ratios dropped from 0.23 to 0.18 these results did not support an exposure-based threshold for arsenic methylation in humans.

It is against this background that we discuss recent data for arsenic metabolites in urine obtained from populations living in South West England.

URINARY ARSENIC STUDIES IN SOUTH WEST ENGLAND

The south-western peninsula of England (the South West) consists of the counties of Cornwall to the west, and Devon. The River Tamar forms the boundary between the two counties. This area is extensively contaminated with heavy metals arising from centuries of mining activity in the region. From about 1860 to 1900, this region was the world's major producer of arsenic. The principal minerals of economic importance were arsenopyrite (FeAsS), chalcopyrite ($CuFeS_2$) and galena (PbS). Other local ores were casserite (SnO_2) and stannite ($CuSnS_4$). Mining and smelting activities have left a legacy of contaminated land, with As- and Cu-rich mine tailings and other wastes. Further extensive areas of land were contaminated with fallout from the smelting process and over the area some 700 km^2 of land are affected (Abrahams and Thornton, 1987). Most of the contaminated area is agricultural with villages and small towns; urban development has sometimes taken place on contaminated land. Sources of arsenic in the region and some aspects of the exposure of local populations have been discussed (Thornton,1994; Mitchell and Barr, 1995; Farago et al., 1997; Kavanagh et al., 1998; Kavanagh, 1998).

The area under investigation is in the Tamar Valley. On the east (Devon) side lies the abandoned Devon Great Consols Mine, where mining and smelting of the ores was carried out. On the other (Cornwall) side of the river lies the village of Gunnislake, which is also in close proximity to abandoned waste sites. There is a small number of houses on the Devon Great Consols Mine, close to the waste tips. These were investigated, together with houses from Gunnislake and Cargreen villages. The latter was taken as a control area, being further down the river and away from past mining and smelting activities. The total exposure of residents of the Tamar Valley was assessed and it was concluded that exposure resulted from high arsenic-containing dusts and soils in both Gunnislake and the Devon Great Consols area (Table 1).

TABLE 1

As (μg/g) in garden soils and housedusts in the Tamar Valley (Kavanagh, 1998)

Site	Soils			Dusts		
	n	Mean[a]	Range	n	Mean	Range
Gunnislake	71	365	120–1695	9	217	33–1160[b]
Devon Great Consols	15[c]	4499	345–52600	13	1167	24–3740
Cargreen	18	37	16–198	4	49	20–114

[a]Geometric mean; [b]outlying value of 16700 μg/g ignored; [c]some samples contain mine wastes.

TABLE 2

Concentration ranges of arsenic species detected in urine samples. From Kavanagh et al. (1998)

	Cargreen (n = 7); Ages 4–7 yr (4 boys); 45–56 yr (3 adults)	Gunnislake (n = 17); Ages 3–8 yr (8 boys); 30–43 (adults 9)	Devon G C (n = 7); Ages 4 yr (1 boy); 18–65 (6 adults)
$As_T(As_i$ + DMAA + MMAA) μg/g creatinine			
Range	2.5–32.7 (2.5–5.3)*	2.7–58.9	5.1–17.6
Geometric mean	5.4 (4.0)*	10.5	10.8
Median	4.7 (4.5)*	9.2	10.0
Arsenite (As III) μg/g creatinine			
Range	BDL**– 0.6 (BDL–0.6)*	BDL–8.5	0.6–1.8
Median	BDL (BDL)*	1.7	0.9
Number detected	1	14	7
Arsenate (As V) μg/g creatinine			
Range	BDL ** (BDL)*	BDL–2.95	BDL–2.06
Median	BDL (BDL)	0.9	1.34
Number detected		13	6
DMAA μg/g creatinine			
Range	2.5–32.7 (2.5–5.4)*	1.9 –54.3	3.3 – 15.5
Median	4.7 (4.2)*	5.6	8.5
Number detected	7	17	7
MMAA μg/g creatinine			
range	BDL **	BDL–3.8	BDL–0.9
median	BDL (BDL)*	0.3	0.7
Number detected		2	2
MMAA/DMAA ratios		0.05	0.08

*Indicates that statistics based on data with outlier omitted.
**Where As in urine was below detection limit (BDL) of 0.5 μg L^{-1}, the value was taken as zero i.e. not detected.

The concentrations of urinary arsenic and its organo-metabolites in the populations were assessed in a pilot study (Kavanagh et al., 1998) and the results are shown in Table 2.

From these results we conclude that populations in both Gunnislake and Devon Great Consols are chronically exposed to inorganic arsenic, since inorganic arsenic appears in the urine, and that in these populations, like those from Glasgow, measured by Farmer and Johnson, few excrete MMAA. The data also indicate that chronic exposure results from soil and dust ingestion of arsenic in a partially available form, since dust and soil appear to be the only significant exposure route (Farago et al., 1997).

TABLE 3

Numbers of urine samples showing detected arsenic species in the urine of UK populations (adults + children)

Location	n	As(III)	As(V)	DMAA	MMAA	Reference
Cargreen	7	1	0	7	0	Kavanagh et al. (1998)
Glasgow	50	7	0	50	6	Johnson and Farmer (1989)
Tamar V*	24	21	19	24	4	Kavanagh et al. (1998)
Cornwall	37	17	0	37	33	Johnson and Farmer (1989)

*Gunnislake and Devon Great Consols.

Of the seven individuals in our control population, only one had detectable inorganic arsenic in the urine and none had MMAA in the urine. These results are very similar to those reported by Johnson and Farmer (1989) for an unexposed population of 40 adults from Glasgow, where the geometric mean of As_T (the sum of As_i + DMAA + MMAA) was 4.4 $\mu g/g$ creatinine. The detected species for UK populations are shown in Table 3.

COMPARISON OF DATA FROM DIFFERENT STUDIES

Difficulties arise when comparing literature data from a number of studies (as inferred by Hopenhayn-Rich et al., 1993). Some of the varying analytical, computational and pre-sentational difficulties and questions to be asked are:
- Are the results be presented as: the arithmetic mean? geometric mean? corrected for creatinine?
- At concentrations below the detection limit, have these been presented as: zero? half the detection limit? the detection limit?
- Are the samples 24 h or first void?
- Are the samples from acute dosage or chronic exposure?
- Has seafood consumption been taken into account?
- Are the comparisons between genetically different populations?
- Are there intra-individual variations (e.g. time, diet)?
- Are there inter-individual variations within a population?

Some of these points make a considerable difference to the perceived proportions of metabolites (Table 4).

Similarly, our results for total urinary arsenic concentrations in samples from residents of the Tamar Valley can be presented in a number of ways (Table 5).

Table 6 presents a collection of data from the literature showing the proportions of metabolites in human urine, including those from Hopenhayn-Rich et al. (1993), in which the value of zero is taken for concentrations below the detection limit. From this data it appears that at low levels of exposure, with concomitant low concentrations of "urinary arsenic" (i.e.

TABLE 4

Percentages of DMAA in urinary arsenic (sum of As_i + MMAA + DMAA), from industrially exposed workers, when species which are not detected (ND) are zero or at the detection limit of 0.5 $\mu g/L$, i.e. the extreme cases (Farmer, 1998)

	ND = 0	ND = 0.5
Controls	97.6	74.0
Semiconductors	96.7	70.0
Electronics	75.4	70.3
Glass	69.7	64.5
Timber	67.4	65.4

TABLE 5

Urinary arsenic concentrations (sum of As$_i$ + MMAA + DMAA) concentrations in samples from residents of the Tamar Valley (ND = 0)

	Cargreen		Devon Great Consols		Gunnislake	
	µg/L	µg/g c	µg/L	µg/g c	µg/L	µg/g c
Mean	4.3	4.2	17.2	11.0	20.2	14.4
Median	4.65	4.5	23.5	10.0	12.4	9.2
Geometric mean	4.2	4.0	13.2	10.8	12.4	10.5

µg/g c = g/g creatinine.

TABLE 6

Mean urinary arsenic concentrations (µg/L) (sum of As$_i$ + MMAA + DMAA) and percentage of each metabolite

n	Mean Urinary As (µg/L)	% of metabolite			References
		As$_i$	MMAA	DMAA	
7	4.2	2	0	98	Kavanagh et al. (1998)
40	4.4	2	1	97	Farmer and Johnson (1989,1990)
16	4.7[a]	18	4	78	Buchet et al. (1981a)
4	8.0[b]	18	4	78	Buchet et al. (1981a)
7	9.2	22	3	75	Kavanagh et al. (1998)
53	9.6	15	12	73	Kalman et al. (1990)
17	10.0	28	1	71	Kavanagh et al. (1998)
30	10.2	15	15	70	Farmer and Johnson (1991)
557	19.6	13	17	66	Kalman et al. (1990)
38	30.7	23	15	62	Foa et al. (1984)
15	38.3	24	9	67	Yamauchi et al. (1989)
18	45.4	24	7	69	Yamauchi et al. (1989)
39	45.4	23	11	66	Yamauchi et al. (1989)
5	47.9	19	4	67	Farmer and Johnson (1990)
23	49.6	14	20	66	Smith et al. (1977)
102	50.1	23	7	70	Yamauchi et al. (1989)
20	57.2	12	20	68	Yamamura and Yamauchi (1980)
28	79.4	14	18	68	Farmer and Johnson (1990)
36	96.6	12	22	66	Smith et al. (1977)
6	120	25	9	66	Yamamura and Yamauchi (1980)
11	238	14	15	61	Yamauchi et al. (1989)
24	245	18	18	63	Farmer and Johnson (1990)

[a]Reported as 7.1 µg/24 h. Divided by 1.5 L by Hopenhayn-Rich (1993) to give µg/L.
[b] Reported as 12.0µg/24 h. Divided by 1.5 L by Hopenhayn-Rich (1993) to give µg/L.

the sum of As$_i$ + MMAA + DMAA) the concentration of inorganic arsenic is near zero and that of MMAA is very low, the arsenic in the urine being near 100% DMAA. As the exposure, together with the concentrations of "urinary arsenic" rise, the percentage of DMAA falls to reach a limiting value of around 60%. At the same time the percentage of the sum of As$_i$ + MMAA rises to around 40%. However it can be seen that where the exposure is low and species are below the detection limit use of zero concentrations for these may overestimate the percentages of DMAA. However even if these concentrations are taken at the detection limit this trend may still be evident.

If the concentrations of DMAA, is plotted against the total mean urinary arsenic concentrations (sum of As$_i$ + MMAA + DMAA), then the 100 times the slope of this plot gives the percentage of DMAA. Figure 1 shows this plot for the literature mean values shown in

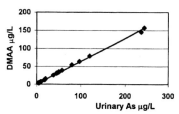

Fig. 1. Plot of mean concentrations of DMAA, versus total mean urinary arsenic concentrations (sum of As$_i$ + MMAA + DMAA) for populations shown in Table 6.

Table 6, and with the mean percentage of DMAA of 64% (R^2 = 0.997) in accordance with previous suggestions in the literature. When the values for individual members of the populations are plotted the percentages of DMAA for the high exposure situations do fall in the region of 60% (Figure 2). Figure 2a shows the plot for results from exposure from a copper smelter (Yamauchi,1989) with 60% DMAA (R^2 = 0.746) from this population; in Figure 2b the percentage of DMAA is 61 (R^2 = 0.627) in a population exposed in an arsenic acid plant (Yamamura and Yamauchi, 1980); and Figure 2c gives 51% (R^2 = 0.862) for a population exposed to arsenic in drinking water (Hoppenhayn-Rich et al., 1993).

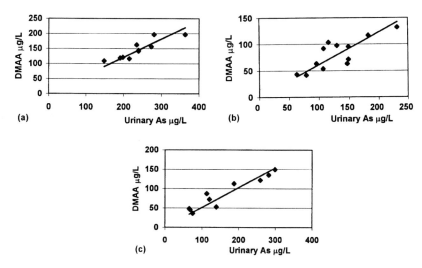

Fig. 2. Plots of concentrations of DMAA, versus total urinary arsenic concentrations (sum of As$_i$ + MMAA + DMAA) for individuals in populations; (a) copper smelter workers (Yamauchi et al., 1989; Hoppenhayn-Rich et al., 1993); (b) arsenic acid plant workers (Yamamura and Yamauchi,1980; Hoppenhayn-Rich et al., 1993); (c) exposed to arsenic in drinking water (Hoppenhayn-Rich et al., 1993).

TABLE 7

Percentages of DMAA in urine from plots of DMAA concentrations versus "urinary arsenic" (As$_i$ + MMAA + DMAA) (concentrations in μg/L) where non detected species, ND, are taken as zero or 0.5 μg/L. R^2 values in parentheses.

	ND = 0	ND = 0.5 μg/L
Gunnislake (n = 17)	85 (0.989)	84 (0.988)
Devon Great Consols (n = 7)	79 (0.930)	78 (0.935)
Cargreen (n = 7)	96 (0.913)	81 (0.963)

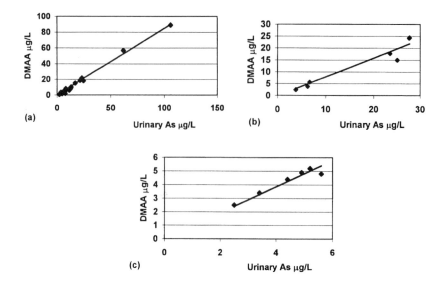

Fig. 3. Plots of concentrations of DMAA, versus total urinary arsenic concentrations (sum of As$_i$ + MMAA + DMAA) for individuals in populations from SW England; (a) Gunnislake; (b) Devon Great Consols; (c) Cargreen.

When our values for low exposure are plotted in the same way (Figure 3), much higher percentages of DMAA are shown, 96% for Cargreen, the unexposed area, 85% for Gunnislake and 79% for the Devon great Consols area. These percentages were obtained using values of zero for those species below the detection limit. When values of 0.5 μg/L are used to calculate the total urinary arsenic, then the correlation is still good, but the percentages drop slightly (Table 7).

These latter results must be the lowest limit for the percentages of DMAA. These results appear to indicate:

1. at low exposure, with concomitant low concentrations of "urinary arsenic" (As$_i$ + MMAA + DMAA) the concentrations of As$_i$ and MMAA are low, a large percentage of the arsenic in the urine being in the form of DMAA;

2. as the exposure, and the concentrations of "urinary arsenic" (As$_i$ + MMAA + DMAA) rise, the percentage of DMAA falls to reach a limiting value near 60% and the percentage of the sum of As$_i$ + MMAA rises to near 40%. There are population differences, both inter- and intra-, in the proportions of these two latter arsenic species;

3. many reports in the literature of the percentages of the urinary arsenic species refer to exposure large enough to reach the limiting values;

4. because of the intra-population differences in the proportions of As$_i$ and MMAA the ratio of MMAA to DMAA is not useful.

If this analysis is correct, then does that "urinary arsenic" concentration (As$_i$ + MMAA + DMAA) at which the limiting values are reached represent the exposure to inorganic arsenic at which the detoxification mechanism is saturated? If so, this value appears to be around 20 μg/L. Assuming a daily urine output of urine of 1.5 L, the urine excreted would be would be 30 μg/day. If 40–60% of daily arsenic intake is excreted each day, then the intake which would produce saturation would be around 60 μg/day. We suggest that more careful work on human populations exposed to low levels of arsenic is needed in order to clarify these issues.

REFERENCES

Abrahams, P. and Thornton, I., 1987. Distribution and extent of land contaminated by arsenic and associated metals in mining regions of south west England. *Trans. Inst. Mining Metall. (Sheet B: Appl. Earth Sci.)*, 6, B1–B8.

Buchet, J.P. and Lauwerys, R. 1994. Inorganic arsenic metabolism in humans. In: W.R. Chappell, C.O., Abernathy and C. Cothern (eds.), *Arsenic; Exposure and Health*, Science Technology Letters, 1994, Northwood, Middlesex, pp. 181–190.

Buchet, J.P., Lauwerys, R., and Roels, H., 1981a. Urinary excretion of inorganic arsenic and its metabolites after repeated ingestion of sodium meta-arsenite by volunteers. *Int. Arch. Occupat. Environ. Health*, 48, 111–118.

Buchet, J.P., Lauwerys, R., and Roels, H., 1981b. Urinary excretion of inorganic arsenic and its metabolites after repeated ingestion of sodium arsenite by volunteers. *Int. Arch. Occupat. Environ. Health*, 48, 71–79.

Crecelius, E.A., 1977. Changes in the chemical speciation of arsenic following ingestion by man. *Environ. Health Perspect.*, 19, 147–150.

Del Razo, L.M., Hernandez, J.L., Garcia-Vargas, G.G., Ostrosky-Wegman, P., de Nava C.C., and Cebrian, M.E. 1994. Urinary excretion of arsenic species in a human population chronically exposed to arsenic via drinking water, a pilot study. In: W.R. Chappell, C.O. Abernathy, C.R. Cothern (eds.), *Arsenic; Exposure and Health*, Science and Technology Letters, Northwood, England, pp. 91–100.

Farago, M.E., Thornton, I., Kavanagh, P., Elliott, P., and Leonardi, G., 1997. Health aspects of human exposure to high arsenic concentrations in soil in south-west England. In: C.O. Abernathy, R.L. Calderon, W.R Chappell, (eds), *Arsenic; Exposure and Health Effects*, Chapman and Hall, London, pp. 191–209.

Farmer, J.G., and Johnson, L.R., 1990. Assessment of occupational exposure to inorganic arsenic based on urinary concentrations and speciation of arsenic. *Br. J. Indust. Med.*, 47, 342–348.

Froines, J., 1994. Studies of arsenic ingestion from drinking water in northeastern Taiwan: chemical speciation and urinary metabolites. Presented at the workshop on *Arsenic Epidemiology and PBPK Modeling*, Annapolis, MD, 27–28 June 1994.

Goyer, R.A., 1991. Toxic effects of metals. In: M.O. Amdur, J. Doull and C.D. Klaassen (eds.), *Toxicology*, 4th edn., Pergammon, New York, pp. 629–633.

Hopenhayn-Rich, C., Biggs, M.L., Smith, A.H., Kalman, D.A., and Moore, L.E. 1996. Methylation study of a population environmentally exposed to arsenic in drinking water. *Environ. Health Perspect.*, 104, 620–628.

Hopenhayn-Rich, C., Biggs, M.L., Smith, A.H., Moore, L.E., and Kalman, D.A. 1996. Arsenic methylation patterns before and after changing from high to lower concentrations of arsenic in drinking water. *Environ. Health Perspect.*, 104, 1200–1207.

Hopenhayn-Rich, C., Smith, A.H., and Goeden, H.M. 1993. Human studies do not support the methylation threshold hypothesis for the toxicity of inorganic arsenic. *Environ. Res.*, 60, 161–177.

Hseuh, Y.M., Huang, Y.L., Wu, W.L., Huang, C.C., Yang, M.H. and Chen, G.S.. 1995. Serum β-carotene level, arsenic methylation capability and risk of skin cancer. Presented at the *Second International Conference on Arsenic Exposure and Health Effects*, San Diego, CA, 12–14 June 1995.

Johnson, L.R. and Farmer, J.G. 1989. Urinary arsenic concentrations and speciation in Cornwall residents. *Environ. Geochem. Health*, 11, 39–44.

Kalman, D.A., Hughes, J., van Belle, G., Burbacher, T, Bolgiano, D., Koble, K., Mottet, N.K., and Pollisar, L., 1990. The effect of variable environmental arsenic contamination on urinary concentrations of arsenic species. *Environ. Health Perspect.*, 89, 145–151.

Kavanagh, P., 1998. Impacts of high arsenic concentrations in south west England. Ph.D. thesis, Imperial College, University of London.

Kavanagh, P., Farago, M.E., Thornton, I., Goessler, W., Kuehnelt, D., Schlagenhaufen, C. and Irgolic, K.J. 1998. Urinary Arsenic Species in Devon and Cornwall Residents, UK. *The Analyst*, 123 (1), 27–30.

Mitchell, P. and Barr, D. 1995. The nature and significance of public exposure to arsenic: a review of its relevance to South West England. *Environ. Geochem. Health*, 17 (2): 57–82.

Mushak, P. and Crocetti, A.F. 1995. Risk and revisionism in arsenic cancer risk assessment. *Environ. Health Perspect.*, 103, 684–689.

Petito, C.T. and Beck, B.D. 1990. Evaluation of evidence for nonlinearities in the dose–response curve for arsenic carcinogenisis. In: D.D. Hemphull and C.R. Cothern (eds.), *Trace Substances in Environmental Health XXIV*. Science Reviews Limited, Northwood, pp. 143–176.

Smith, T.J., Crecelius, E.A., and Reading, J.C. 1977. Airborne arsenic exposure and excretion of methylated arsenic compounds. *Environ. Health Perspect.*, 19, 89–93.

Storer, G. 1991. Arsenic: Opportunity for risk assessment. *Arch. Toxicol.*, 65, 525–531.

Tam, G.K.H., Charbonneau, S.M., Bryce, F., Pomoroy, C., and Sandi, E. 1979. Metabolism of inorganic arsenic (74As) in humans following oral ingestion. *Toxicol. Appl. Pharmacol.*, 50, 319–322.

Tatken, R.L. and Lewis, R.J. (eds.), 1983. *Registry of Toxic Effects of Chemical Substances*, 1981–1982. US Department of Health and Human Services, Cincinnati, OH.

Thornton, I., 1994. Sources and pathways of arsenic in south-west England; health implications. In: W.R. Chappell, C.O. Abernathy and C.R. Cothern (eds.), *Arsenic; Exposure and Health*. Science and Technology Letters, Northwood, pp. 61–70.

Vahter, M.E., 1994. Species differences in the metabolism of arsenic. In: W.R. Chappell, C.O. Abernathy and C.R. Cothern (eds.), *Arsenic; Exposure and Health*. Science and Technology Letters, Northwood, pp. 171–180.

Vahter, M.E., Marafante, E., and Dencker, L. 1984. Tissue distribution and retention of ^{74}As-dimethylarsinic acid in mice and rats. *Arch. Environ. Contamin. Toxicol.*, **13**, 259–264.

Vahter, 1997. Arsenic metabolism in humans. Lecture given at Imperial College, 3rd December, 1997.

Vahter, M.E. 1988. Arsenic. In: T.W. Clarkson, L. Frieberg, G.F. Nordberg and P.R. Sager (eds.), *Biological Monitoring of Toxic Metals*. Plenum, New York, pp. 303–321.

Vahter, M.E. and Marafante, E. 1983. Intracellular interaction and metabolic fate of arsenite and arsenate in mice and rabbits. *Chem. Biolog. Interact.*, **47**, 29–44.

Yamamura, Y. and Yamauchi, H. 1980 Arsenic metabolites in hair, blood and urine in workers exposed to arsenic trioxide. *Ind. Health*, **18**, 303–210.

Yamauchi, H., Takahashi, K., Mashiko, M., and Yamamura, Y. 1989. Biological monitoring of arsenic exposure of gallium arsenide- and inorganic arsenic-exposed workers by determination of inorganic arsenic and its metabolites in urine and hair. *Am. Indust. Hyg. Assoc. J.*, **50**, 606–612.

Yamauchi, H. and Yamamura, Y. 1983. Concentration and chemical species of arsenic in human tissue. *Bull. Environ. Contamin. Toxicol.*, **31**, 267–277.

Yamauchi, H., Kaise, T., Takahashi, K. and Yamaura, Y. 1990. Toxicity and metabolism of trimethylarsine in mice and hamsters. *Fund. Appl. Toxicol.*, **14**, 399–407

Yamauchi, H., Takahashi, K., Mashiko, M., and Yamamura, Y. 1995. Biological monitoring of arsenic exposure of gallium arsenide- and inorganic arsenic-exposed workers by determination of inorganic arsenic and its metabolites in urine and hair. *Am. Indust. Hyg. Assoc. J.*, **50**, 606–612.

Arsenic Exposure and Health Effects
W.R. Chappell, C.O. Abernathy and R.L. Calderon (Editors)
© 1999 Elsevier Science B.V. All rights reserved.

Chronic Arsenic Toxicity: Epidemiology, Natural History and Treatment

D.N. Guha Mazumder, B.K. De, A. Santra, J. Dasgupta, N. Ghosh,
B.K. Roy, U.C. Ghoshal, J. Saha, A. Chatterjee, S. Dutta, R. Haque,
A.H. Smith, D. Chakraborty, C.R. Angle, J.A. Centeno

ABSTRACT

Chronic arsenic toxicity due to drinking arsenic-contaminated water has been one of the worst environmental health hazards affecting eight districts of West Bengal since the early eighties. Detailed clinical examination and investigation of 248 such patients revealed protean clinical manifestations of such toxicity. Over and above hyperpigmentation and keratosis, weakness, anaemia, burning sensation of eyes, solid swelling of legs, liver fibrosis, chronic lung disease, gangrene of toes, neuropathy, and skin cancer are some of the other manifestations. A cross-sectional survey conducted in one of the severely affected districts of West Bengal revealed that the age adjusted prevalence of keratosis and hyperpigmentation were strongly related to water arsenic levels. Follow up study of a cohort of 24 patients with chronic arsenic toxicity revealed drinking of arsenic-free water resulted in partial improvement of hyperpigmentation and keratosis in 45 and 46% cases respectively. But liver enlargement persisted in 80% of cases and signs of lung disease appeared in 41.6% of cases. Therapy with chelating agent DMSA was not found to be superior to placebo effect. However, drinking of arsenic-free water, rest, nutritious diet and symptomatic treatment could reduce nearly forty percent of the patients' symptoms significantly.

Keywords: arsenic, keratosis, hyperpigmentation, India, cross-sectional study, drinking water, DMSA

TABLE 1

Clinical characteristics of 248 patients in West Bengal

Presenting features	No. of patients	Percentage
1. Rain-drop pigmentation	234	94.35
2. Weakness	163	65.73
3. Keratosis (sole and palm)	162	65.32
4. Dyspepsia	165	66.53
5. Cough (± expectoration)	154	62.10
6. Burning sensation of eyes	74	29.84
7. Anemia	109	43.95
8. Hepatomegaly (firm, Non-tender 2–6 cm below costal margin)	190	76.61
9. Splenomegaly (1.5–8 cm below costal margin)	73	29.43
10. Crepitation ± Ronchi	75	30.24
11. Polyneuropathy	74	29.83
12. Pedal oedema (non-pitting)	23	9.27
13. Blackfoot disease (Gangrene)	3	1.20
14. Skin cancer	5	2.02
15. Kidney cancer	1	0.4

INTRODUCTION

Chronic arsenic toxicity due to drinking of arsenic-contaminated water has been reported from many countries. But reports of a large number of affected people in West Bengal, India and Bangladesh are unprecedented. In West Bengal, arsenic contamination of ground water has been reported in 777 villages of eight districts having a population of 38 million. It is suspected that about 6 million people are exposed to arsenic-contaminated drinking water (As level >0.05 mg/L) in 68 blocks of those 8 districts (UNICEF, 1998). Arsenic contamination above 0.05 mg/L was found in the water in 45% of about 20,000 tube wells analysed (Mondal et al., 1996). The source of contamination was suspected to be geological. It is estimated that about a million people have been consuming arsenic contaminated water (Bagla and Kaiser, 1996). Most of the affected villages are located along the eastern side of the river Ganges. Cases of chronic arsenic toxicity were characterized by typical spotty skin pigmentation and keratosis and were reported from West Bengal by many investigators since 1984 (Garai et al., 1984; Chakraborty and Saha, 1987; Guha Mazumder et al., 1988, 1992, 1997, 1998). However, Datta et al. (1976) reported a few cases of liver disease characterized by non cirrhotic portal fibrosis at Chandigarh as early as 1976 due to chronic arsenic toxicity. They found arsenic contamination in 29.49% of water samples in wells, 50% in tube wells and 35.7% in spring water in and around Chandigarh (Datta and Kaul, 1976).

Chronic exposure to high concentrations of arsenic ($>50 \mu g/L$) is of prime importance to people in India. Skin manifestations like spotty pigmentation or depigmentation and keratosis are the two most common clinical manifestations of chronic arsenic toxicity (arsenicosis) described by most people from different parts of the world. However the features as observed in West Bengal, India are protean (De et al., 1999) (Table 1).

A. EPIDEMIOLOGICAL STUDY

To determine the prevalence of the various health effects associated with arsenic, a cross-sectional study was conducted in one of the most affected districts of West Bengal, the South 24 Parganas (Guha Mazumder, 1998b) (Figure 1). This district was a suitable location for this survey because of the heterogeneity in exposure which enabled the investigators to collect exposure response data. The drinking water arsenic levels in this district ranged from nondetectable to 3400 $\mu g/L$.

Fig. 1. Map of West Bengal (India) showing different arsenic-affected districts. Epidemiological study was done in 24 Parganas (south) (shaded area).

METHODS

Study Area and Population

Two particular areas within this district were targeted for the survey. The first area was selected because high levels of arsenic were detected in some of the shallow tube wells as determined in a prior study (Mondal et al., 1996; Guha Mazumder et al., 1988). The second area included the remaining part of the district where people also used shallow tube well water for drinking purposes. No survey was carried out by anybody earlier in this area and no reports of elevated arsenic levels were available before the survey. The two areas combined contain a total population of 150,457. A total of 7818 individuals participated in the drinking water study. Water arsenic levels were obtained from 7683 of them (4093 females and 3590 males) who constitute the study subjects. Ethical clearance was obtained for carrying on human studies from the institutional ethical committee.

The high exposure region included 25 villages within 5 administrative blocks. Convenience sampling was used for this study which involved remote rural areas. An interview was administered and a brief medical examination was conducted. Sampling continued house to house in a village until 50 to 150 participants were recruited.

The second region included 32 villages within 16 administrative blocks. Sampling in this region was restricted to villages with more than 100 houses. One or more villages were selected at random from each of the 16 blocks depending on the population size. Only one village was selected for sampling from a small block, but two or three villages were selected if the block was larger.

Water Sampling and Arsenic Measurement

Water samples were collected from private and public tube wells used for drinking and cooking purposes by each recruited household. Arsenic levels were measured by flow-injection hydride generation atomic absorption spectrophotometry.

Statistical Analysis

The estimation of the prevalence of keratosis and hyper pigmentation was stratified by age, and calculated for each sex separately. Tests for trends in proportions were based on the chi-square distribution using the mid-points of each grouping of arsenic water levels (Breslow and Day, 1987).

The prevalence of skin lesions was also examined by tertiles of daily arsenic dose per body weight (μg/kg/day). Using all subjects, the cut points for the tertiles were: 3.2 and 14.9. The highest dose per body weight found was 73.9 μg/kg/day. Tests for trend in proportions were conducted using the mid-points of each dose tertile (1.6, 9.1, and 44.4 μg/kg/day).

As a rough indicator of nutritional status, we examined body weight in relation to standard weight. Using those who had adequate nutrition status as a referent group, standardized morbidity ratios (observed cases with skin lesions divided by expected cases) were calculated for each tertile of dose per body weight. Statistical tests of significance were based on the Poisson distribution and 95% confidence intervals were calculated using exact methods (Breslow and Day, 1987).

RESULTS

Tube Well Water Concentration

The age, sex and water-arsenic level distribution of the study population is presented in Table 2. The tube well water concentration in the villages ranged up to 3400 μg/L, but 88% (6781/7683) of the participants with recorded arsenic water concentrations were exposed to levels less than 500 μg/L. It was observed that pigmentation (8.82%) and keratosis (3.64%) were the most specific diagnostic parameters of chronic arsenicosis, as none of the people who were drinking water having an arsenic level less than 10 μg/L had these features. Keratosis prevalence was examined by arsenic water levels. Of the 4093 female participants, 48 had keratotic skin lesions. A clear relationship was apparent between water levels of arsenic and the prevalence of keratosis (Figure 2). The test for trend yielded a *p* value less than 0.001 (Table 3). Similar findings were found for males and for hyper pigmentation in both males and females (Figure 2).

Although liver enlargement was found in a higher number of cases (10.21%) in the arsenic-exposed population, its incidence in the unexposed population (As level <50 μg/l) was 3.5%. Thus, though liver enlargement has a higher degree of sensitivity, it is less specific for diagnosis of chronic arsenicosis. Similar was the observation in regard to lung involvement. Symptoms of peripheral nerve disease were also found in a higher number of the arsenic exposed people compared to the control population. There was also a strong exposure–response relationship related to water arsenic level in regard to prevalence of liver enlargement, lung disease and symptoms of generalized weakness and peripheral neuropathy.

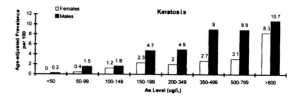

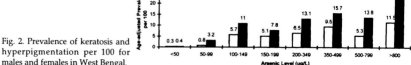

Fig. 2. Prevalence of keratosis and hyperpigmentation per 100 for males and females in West Bengal.

TABLE 2

Age, sex, and arsenic water level (μg/L) distribution of the study population

Age group	As Level (μg/L)								
	<50	50–99	100–149	150–199	200–349	350–499	500–799	>800	Total
Females									
≤9	194	3	53	23	84	50	75	26	536
10–19	400	74	58	54	117	57	65	26	851
20–29	577	120	99	74	135	63	83	24	1157
30–39	308	79	48	46	79	40	44	15	659
40–49	175	331	23	27	36	21	28	10	353
50–59	157	38	23	18	27	18	30	11	322
≥60	97	29	9	17	27	20	10	6	215
All ages	1908	386	313	259	505	269	335	118	4093
Males									
≤9	220	64	65	27	77	51	81	28	613
10–19	330	73	49	56	96	51	64	29	748
20–29	356	79	56	52	79	43	59	25	749
30–39	246	631	38	40	75	44	53	18	577
40–49	160	43	29	24	53	22	25	12	368
50–59	121	34	20	21	27	16	15	6	265
≥60	126	34	20	21	27	16	15	6	265
All ages	1559	385	274	235	442	246	320	129	3590

From Guha Mazumder et al. (1998b).

TABLE 3

Age-adjusted keratosis prevalence per 100 by dose body weight (μg/kg/day) for females and males with number of cases in parentheses (the cut points of tertiles were 3.2 and 14.9)

Age group	Tertile 1	Tertile 2	Tertile 3
Females			
≤9	0.0 (0)	0.0 (0)	3.1 (1)
10–19	0.0 (0)	0.0 (0)	3.5 (2)
20–29	0.0 (0)	1.5 (1)	2.9 (2)
30–39	2.1 (1)	2.3 (1)	2.5 (1)
40–49	0.0 (0)	2.5 (1)	0.0 (0)
50–59	0.0 (0)	6.3 (1)	10.5 (2)
≥60	8.3 (1)	13.3 (2)	6.3 (1)
All ages	0.7 (2)	2.3 (6)	3.5 (9)*
Age-adjusted	0.8	2.2	3.5
Males			
≤9	0.0 (0)	0.0 (0)	2.9 (1)
10–19	0.0 (0)	1.6 (1)	6.6 (4)
20–29	0.0 (0)	5.9 (3)	11.1 (6)
30–39	2.5 (1)	10.3 (3)	21.2 (4)
40–49	0.0 (0)	9.7 (3)	20.0 (6)
50–59	0.0 (0)	5.0 (1)	20.0 (5)
≥60	5.9 (1)	0.0 (0)	0.0 (0)
All ages	0.8 (2)	4.2 (11)	12.7 (36)**
Age-adjusted	0.8	4.6	11.0

*P value 0.028; **P value < 0.01.
From Guha Mazumder et al. (1998b).

Dose per Body Weight

 The prevalence of the skin lesions were also examined by dose per body weight (μg/kg/day). The age-adjusted prevalence of keratosis in females rose from 0.8 per 100 in the lowest tertile (less than 3.2 μg/kg/day) to 3.5 per 100 in the highest tertile (more than 14.9 μg/kg/day). Using the unadjusted data, the one tailed p-value test for trend for females was 0.028. A steeper pattern was apparent with men ($p < 0.001$). In the lowest tertile, the prevalence of keratosis was 0.8 per 100 among males, in the highest tertile the prevalence reached 11.0 per 100 (Table 3). The findings for hyper pigmentation paralleled those for keratosis with males again showing higher prevalence based on estimation of dose per body weight.

Findings Among Those with Low Body Weight

 Of the 2320 females with known body weights, 690 (30%) were below the standard weight by 20% or more. Of the 2123 males with known values, 808 (38%) were below the standard weight by 20% or more.

 Compared to those with adequate nutrition, subjects 20% or more below the standard weight had a higher age-adjusted prevalence of keratosis. The overall Standardised Morbidity Ratio (SMR) for keratosis was 2.1 for females (95% confidence interval: 0.8–4.6, $p = 0.07$) indicating that the age-adjusted keratosis prevalence among females with potentially poor nutrition was approximately twice that of females considered to have adequate nutrition. The overall SMR for males was 1.5 (95% CI: 0.9–2.4, $p = 0.08$). The combined SMR for both sexes was 1.6 (95% CI: 1.0–2.4, $p = 0.02$) (Table 4).

 Weaker findings were found for hyper pigmentation. The overall SMR for females was 1.8 (95% CI: 0.8–3.5, $p = 0.09$). Thus, women with poor nutrition status had an age-adjusted hyper pigmentation prevalence nearly twice that of females considered to have adequate nutrition. This increase was less apparent in males, where the increase in the age-adjusted prevalence was only 10% greater among those with poor nutrition (SMR = 1.1, 0.7–1.7, $p = 0.39$). Combining men and women, the SMR was 1.2 (0.8–1.8, $p = 0.17$) (Table 5).

TABLE 4

Standardized morbidity ratios (SMRs) and 95% confidence intervals (CI) for keratosis by tertiles (the cut points of tertiles were 3.2 and 14.9) of dose per body weight (μg/kg/day) comparing subjects below 80% the standard weight to those above 80% for each sex separately

	Tertile 1	Tertile 2	Tertile 3	Overall
Females				
Observed/Expected	1/0.3	1/0.7	4/1.9	6/2.8
SMR	3.1	1.5	2.2	2.1
95% CI	(0.1–17.4)	(0.04–8.3)	(0.65.5)	(0.8–4.6)
p value	0.27	0.49	0.12	0.07
Males				
Observed/Expected	0/0.6	2/1.2	14/9.0	16/10.8
SMR	0	1.7	1.6	1.5
95% CI	(0–4.7)	(0.2–6.0)	(0.9–2.6)	(0.9–2.4)
p value	0.34	0.07	0.08	
Both sexes				
Observed/Expected	1/0.96	3/1.87	18/10.8	22/13.7
SMR	1.0	1.6	1.7	1.6
95% CI	(0.03–5.80)	(0.3–4.7)	(1.0–2.6)	(1.0–2.4)
p value	0.62	0.29	0.03	0.02

From Guha Mazumder et al. (1998b).

TABLE 5

Standardized morbidity ratios (SMRs) and 95% confidence intervals (CI) for hyperpigmentation by tertiles of dose per body weight comparing subjects below 80% the standard weight to those above 80% for each sex separately

	Tertile 1	Tertile 2	Tertile 3	Overall
Females				
Observed/Expected	0/0	2/0.7	6/3.8	8/4.5
SMR	0	2.8	1.6	1.8
95% CI		(0.4–10.3)	(0.6–3.4)	(0.8–3.5)
p value		0.16	0.19	0.09
Males				
Observed/Expected	0/0.4	4/2.6	15/14.5	19/17.1
SMR	0	1.5	1.0	1.1
95% CI	(0–7.1)	(0.4–3.9)	(0.6–1.7)	(0.7–1.7)
p value		0.26	0.49	0.39
Both sexes				
Observed/Expected	0/0.42	6/3.3	21/18.3	27/22.0
SMR	0	1.8	1.1	1.2
95% CI	(0.0–7.1)	(0.7–4.0)	(0.7–1.8)	(0.8–1.8)
p value		0.12	0.29	0.17

From Guha Mazumder et al. (1998b).

DISCUSSION

This is the first population study assessing water levels of arsenic and skin lesions in India in a structured population survey. Clear exposure–response relationships were found with the prevalence of skin effects. The steepest exposure–response relationships were found for males (Figure 2); this finding is not explained by males having a greater water consumption because this pattern was also apparent using the identical categorization of dose per body weight for both sexes. The most striking finding was that 144 cases with these skin lesions were found among those whose drinking water contained less than 200 $\mu g/L$ of arsenic. Studies conducted in other countries have also investigated the prevalence of hyper pigmentation and keratosis in regions with elevated arsenic levels in drinking water; however, they either lacked individual exposure data or had small numbers. For instance, arsenic levels in Taiwan were reported in villages (Tseng et al., 1968; Tseng, 1977; Chen et al., 1992). Mean arsenic levels were reported for an entire affected village in Mexico (Cebrian et al., 1983) and China (Huang et al., 1985), and by towns in Chile (Borgono et al., 1977). Thus, a major strength of this study is that it is the first large population-based study with individual exposure data, which can provide critical information to characterize the exposure–response relationship.

The overall SMR for keratosis suggested that those with poor nutritional status had an age-adjusted prevalence that was 1.6 times greater than those considered to be adequately nourished (SMR = 1.6, 95% CI: 1.0–2.4, p = 0.02). The overall SMR for hyper pigmentation for both sexes combined was 1.2 (95% CI: 0.8–1.8, p = 0.17). These small differences do not suggest that malnutrition is the reason for the high prevalence of skin lesions in West Bengal. Nevertheless, it is still possible that some dietary factors affect the susceptibility of the whole population, malnourished or not.

In conclusion, this study demonstrates a clear exposure–response relationship between the prevalence of skin lesions and both arsenic water levels and dose per body weight, with males showing greater prevalences of both keratosis and hyper pigmentation. Based on limited exposure assessment, some cases appear to be occurring at surprisingly low levels of exposure. There was evidence that the risks were somewhat greater for those who might be malnourished. Further studies are needed to confirm the apparent low exposure effects, to determine why males have more skin effects than females at the same doses per body weight and to identify susceptibility factors which may be present in this population.

B. NATURAL HISTORY

Not much information is available in the literature regarding the long-term effect of chronic arsenic toxicity after stoppage of drinking arsenic-free water. Arguello et al. (1938) reported that keratodermia appeared insidiously between the second and third years of intoxication and did not disappear after cessation of exposure. Some individuals were followed up for more than 30 years after termination of exposure.

To ascertain the effect of providing safe water to the affected people, a cohort of 24 patients with chronic arsenicosis were reexamined after drinking arsenic-free water (As < 10 μg/L) for a period varying from 2–10 years (13 patients 10 years, 11 patients 2–5 years). These people had been drinking arsenic contaminated water (130 to 2000 μg/L) for 4 to 15 years. Weakness and anaemia were present in 91.6% and 58.3% of cases initially and was persistent in 60.8% and 33% of cases respectively on repeat examination. Partial improvement of pigmentation and keratosis were observed in 45% and 46% of patients respectively. But liver enlargement was persistent in 86% of cases. However, the most distressing observation was the new appearance of signs of chronic lung disease (shortness of breath and chest signs) in 41.6% of cases. There was a slight reduction in clinical symptoms of neuropathy. It was present in 11 cases (45.8%) at the time of initial examination and in 8 cases (33.8%) during the subsequent period ($p < 0.5$). No new case of neuropathy was detected in any of the follow-up patients. However, diminished hearing was observed in 5 cases during follow-up examination though it was present in 2 cases initially. Similarly 3 patients complained of dimness of vision during follow-up examination though none had such a symptom earlier. None of these three patients had cataract or any other abnormality on fundoscopy. From the above it becomes apparent that not only many of the clinical manifestations of chronic arsenicosis persist for long duration in spite of stoppage of taking arsenic free water, but new symptoms may appear in some of them (Guha Mazumder et al., 1998c).

TREATMENT

Chronic arsenicosis leads to irreversible damage in several vital organs and is an established carcinogen (Bencko et al., 1997). Despite the magnitude of this potentially fatal toxicity, there is no effective therapy for this disease; patients once affected may not recover even after remediation of the arsenic contaminated water (Tseng et al., 1968). The need for an effective therapy for chronic arsenicosis is obvious.

Chelation therapy for chronic arsenic toxicity is thought to be the specific therapy for relief of systemic clinical manifestation and reduction of arsenic stores in the body, reducing subsequent cancer risk. Chelation therapy is presumed to be more effective with early features of the toxicity, as severe manifestation of poly neuropathy, chronic lung and liver disease, swelling of hand and legs, defect of hearing and vision are less likely to respond to this therapy. Chelating agents like D-penicillamine, DMSA (Dimercaptosuccinic Acid), DMPS (Dimercaptopropane succinate) are being currently recommended for use in chronic arsenic toxicity. D-Penicillamine is a costly drug with associated toxic side effects in 20% to 30% of patients. Therapy with D-penicillamin in a dose of 250 mg thrice daily for 15 days in a group of 5 patients when followed up by us for 2–5 years, did not show any difference from control patients (Guha Mazumder et al., 1998c). However, the effect of long-term treatment with this agent needs to be studied to ascertain whether such therapy could alter the natural course of chronic arsenic toxicity.

2,3-Dimercaptosuccinic acid (DMSA), a chelating agent, has been used in the therapy for lead and mercury poisoning in humans (Angle, 1995). There are reports of its efficacy in acute arsenic poisoning in mice (Kreppel et al., 1993) and the chronic arsenic poisoning in rats (Flora et al., 1995). There are no studies on its use in chronic arsenic poisoning in humans although mobilization of arsenic from the human body has been shown in acute arsenic poisoning (Shum et al., 1995). Recently we have carried out a prospective randomised controlled trial to evaluate the efficacy and safety of dimercaptosuccinic acid (DMSA) to chronic arsenicosis patients (Guha Mazumder et al., 1998d).

Patients and Methods

Twenty-one consecutive patients with chronic arsenicosis were randomized into 2 groups. Eleven patients (10 males, ages 25.5 ± 8.0 years) received DMSA 1400 mg/d (1000 mg/m^2) in 4 divided doses in the first week and then 1050 mg/d (750 mg/m^2) in 3 divided doses during the next 2 weeks. The same was repeated after 3 weeks during which no drug was administered. The other 10 patients (all males, ages 32.2 ± 9.7 years) were given placebo capsules (resembling DMSA) in the same schedule. The patients were blinded about the nature of treatment being given. The patients included in the study were selected from the arsenic clinic on the basis of history of drinking arsenic-contaminated water (≥ 0.05 mg/L) for 2 years or more and clinical symptoms and signs of chronic arsenicosis. The symptoms and signs of patients were evaluated by a scoring system before and after treatment. The scoring system followed is summarized in Table 6. Any possible therapy-related side effect was monitored in every patient. All the patients were kept hospitalized during the study period. Skin was

TABLE 6

The system of clinical scoring of the symptoms and signs before and after therapy with DMSA and placebo

Symptoms and signs	Nil	Present		
		Mild	Moderate	Severe
Weakness	0	1		
Cough	0	1		
Dyspnoea	0	1	2	3
Rales, ronchii	0	1		
Hepatomegaly	0	1 (14 cm span)	2 (16 cm)	3 (>16 cm)
Splenomegaly	0	1 (2 cm)	2 (4 cm)	3 (>4 cm)
Pigmentation	0	1 (Diffuse)	2 (Spotty)	3 (Blotchy)
Keratosis	0	1 (Thickening)	2 (Few nodules)	3 (Multiple nodules)
Flushing of face	0	1		
Conjunctivitis, non pitting	0	1		
Edema leg/hand	0	1		
Abdominal pain	0	1		
Anorexia	0	1		
Nausea	0	1		
Diarrhoea	0	1		
Hearing defect	0	1		
Claudication	0	1		
Hand/leg ulcers	0	1		
Paresthesia	0	1 (Only legs)	2 (Legs + hands)	
Pallor	0	1		
Ascites	0	1		
Loss of ankle jerk	0	1		

Maximum score 33. From Guha Mazumder et al. (1998d), p. 685 by courtesy of Marcel Dekker. Inc.

biopsied from unexposed areas by punch biopsy technique for histologic evaluation before and after treatment. Urine samples were collected for 2 consecutive days before, and then at 48 and 72 hours after starting the drug or placebo. Urine arsenic was determined by graphite furnace atomic absorption with Zeeman-background correction.

The significance of the differences between the parametric data obtained in the 2 groups was calculated by student's *t*-test. The clinical scores of the patients before and after therapy were compared by one-way ANOVA. The concentration of arsenic before and after therapy was compared by Wilcoxon's rank sum test, as the data were not expected to have normal distribution. For nonparametric data Chi-square with Yates' correction, as applicable, was used.

RESULTS

Demographic and clinical data of patients in each group are summarized in Table 7. There were no differences in age, sex, duration of exposure to the arsenic contaminated water, arsenic concentration in the drinking water, duration of drinking arsenic free water before inclusion in the study, and clinical score of symptoms and signs between patients in the drug and in controls (Table 7).

Therapy with DMSA did not cause any significant clinical improvement as compared to patients treated with placebo. The clinical score improved after therapy with DMSA, but similar improvement was observed in patients treated with placebo (Table 8) (Figure 3).

TABLE 7

Demographic, clinical and laboratory parameters in patients treated with DMSA and placebo

	DMSA group ($n = 11$)	Placebo group ($n = 10$)	p value
Age (years)	25.5 ± 8.0	32.2 ± 9.7	$p = ns^*$
Sex (M:F)	10:1	10:0	
Clinical features:			
Pigmentation	11	10	
Keratosis	11	10	
Hepatomegaly	2	4	
Vasculopathy	4	1	
Clinical score	9.3 ± 3.3	10.6 ± 3.2	$p = ns^{**}$
As concentration in drinking water	0.66 ± 0.39	0.65 ± 0.34	$p = ns^*$
Duration of exposure (years)	15.25 ± 10.7	21.6 ± 11.95	$p = ns^*$
Duration of drinking As-free water before entry (months)	0.54 ± 1.37	1.67 ± 1.22	$p = ns^*$
Portal hypertension	0	0	
Liver histology (fibrosis)	6/7	5/5	

*Student's *t*-test; **one-way ANOVA.
From Guha Mazumder et al. (1998d), p. 685 by courtesy of Marcel Dekker. Inc.

TABLE 8

Clinical scores of patients before and after therapy

	Before	After	p value
DMSA (n = 11)	9.33 ± 3.33	6.2 ± 2.11	0.017
Control ($n = 10$)	10.6 ± 3.20	6.7 ± 1.70	0.003

*One-way ANOVA.
From Guha Mazumder et al. (1998d), p. 685 by courtesy of Marcel Dekker. Inc.

Fig. 3. Clinical score of patients pre and
post therapy with DMSA and placebo.

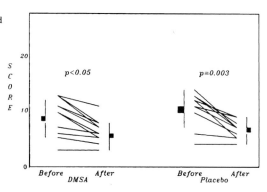

Excretion of arsenic in urine before treatment, at 48 h and 72 h after treatment in DMSA
and placebo group was comparable. There was no difference in the results of the liver
function tests and arsenic concentration in hair and nails before and after treatment (Table 8).
No patient developed any therapy-related side effects. The histologic abnormalities in skin
biopsy did not show any difference in patients treated with DMSA and placebo before and
after therapy.

TABLE 9

Laboratory parameters before and after therapy in patients treated with DMSA and placebo

	DMSA		Placebo		p value
	Before	After	Before	After	
Liver functions (n = D: 11; P = 10)					ns
Bilirubin (mg/dL)	0.7 ± 0.12	0.6 ± 0.08	0.7 ± 0.09	0.7 ± 0.11	
ALT (IU/L)	36 ± 18.8	46 ± 13	38.5 ± 16.8	50.1 ± 28.1	
SAP (IU/L)	241.3 ± 122.8	254.6 ± 66.9	258.2 ± 93.7	342.0 ± 201.3	
Albumin (g/L)	3.8 ± 0.5	4.0 ± 0.6	4.0 ± 0.8	3.9 ± 0.6	
Urine As (μg/L)* (n = D: 9; P = 4)					
Before drug	5.18 (<DL-16.3)	11.3 (<DL-29.8)			
48 h post-drug	—	3.4 (<DL-35.6)	—	1.56 (<DL-6)	ns
72 h post-drug	—	1.83 (<DL-10.8)	—	4.3 (<DL-19.1)	
Hair As concentration (mg/kg) (n = D: 11; P: 10)	2.9 ± 1.8	2.8 ± 3.4	3.5 ± 3.2	2.6 ± 1.6	ns
As concentration in nail (mg/kg) (n = D:9; P:9)	7.5 ± 4.9	7.3 ± 4.9	6.9 ± 4.8	6.5 ± 5.5	ns

As: arsenic; D: drug (DMSA) group; P: placebo group; ALT: alanine aminotransferase; SAP: serum alkaline
phosphatase; DL: detection limit.
*In one patient, of the two days' measurement before starting DMSA, one day's sample resulted in an arsenic level of
202 μg/L; it was analyzed three times with the same result. It was thought that the endogenous arsenic level was
unlikely to be so high. So, the other day's value (2.26 μg/L) was taken for analysis of the results.
From Guha Mazumder et al. (1998d), p. 685 by courtesy of Marcel Dekker. Inc.

DISCUSSION

In this study, we did not find DMSA for 2 courses at 3-week intervals to have any clinical or biochemical benefit in patients with chronic arsenicosis. To the best of our knowledge, this is the first randomized placebo-controlled trial on the use of this chelating agent in the therapy of chronic arsenicosis.

Shum and Whitehead (1995) reported that treatment of an adult who had ingested 80 g methane arsenate with DMSA 30 mg/kg/d for 5 d over 1 month reduced serum arsenic from 2871 μg/L to 6 μg/L. Lenz et al. (1981) also found DMSA to be effective in man. However, Kew et al. (1993) found no improvement in peripheral neuropathy of 4 months duration after DMPS 300 g/d for 3 weeks and DMSA 1.2 g/d for 2 weeks. There was no improvement of neuropathy following treatment with DMSA in our patients. Further, our study has conclusively shown that DMSA is globally ineffective in the therapy of chronic arsenic toxicity in man.

That the patients symptom score did improve significantly following rest and symptomatic treatment in hospital with a good nutritious hospital diet in placebo control cases suggest that such supportive treatment should be given to all patients suffering from chronic arsenic toxicity. Though no curative treatment could be ensured to these ill-fated people, such supporting treatment could help a lot to ease their suffering. Our follow-up study further highlighted that drinking arsenic free water could improve skin manifestations, weakness, anemia and neuropathy in a significant number of cases. Whether these could reduce the incidence of cancer would need a follow-up study of a cohort of such patients for a long period.

ACKNOWLEDGEMENTS

The epidemiological study was funded by the Rajib Gandhi National Drinking Water Mission, Ministry of Rural Development, Government of India, research grants W-11046/2/4/96-TM II (R & D). Support for analysis and preparation for publication was also received from the US Environmental Protection Agency National Center for Environmental Assessment and from research grants P30-ES01896 and P42-ES04705 from the National Institute of Environmental Health Sciences, NIH. Its contents are solely the responsibility of the authors and do not necessarily represent the official views of the Rajiv Gandhi National Drinking Water Mission, the NIEHS, the NIH, or the EPA. Additional support came from the American Water Works Association Research Foundation, the University of California Toxic Substances Research and Teaching Program and the Center for Occupational and Environmental Health..

The authors acknowledge Cilag Ltd., Switzerland, for providing a DMSA capsule as a gift for this study. The authors also acknowledge Dr. Michael J. Kosnett, Division of Clinical Pharmacology and Toxicology, University of Colorado Health Sciences Center, Denver, Colorado, for his help.

The authors are grateful to the Director and Surgeon Superintendent of the Institute of Post Graduate Medical Education and Research and SSKM Hospital, Calcutta for their help in carrying out this work.

REFERENCES

Angle, C.R. 1995. Organ-specific therapeutic intervention. In: M. Waalkes, R. Goyer and C. Klasen (eds.), *Metal Toxicology*. California Academic Press, San Diego, CA, pp. 71–110.

Arguello, R.A., Cenget, D.D., Tello, E.E. 1938. Cancer and endemic arsenism in the Cordoba Region. *Rev. Argent. Dermatosifilgr.*, **22**, 461–487.

Bagla, P., Kaiser, J. 1996. India's spreading health crises draws global arsenic experts. *Science*, **274**, 174–175.

Bencko, V., Gr ötzl, M., and Rames, J. 1997. Human Arsenic exposure related skin Basalioma Cancer Epidemiology. Proc. of Workshop Arsenic: Health Effects, Mechanisms of actions and Research issues organised by NCI, NIEHS and EPA, Hunt Valley, MD, USA. Sept. 22–24, p. S-20.

Borgono, J.M., Vicent, P., Venturino, H., Infante, A. 1977. Arsenic in the drinking water of the city of Antofagasta: Epidemiological and clinical study before and after the installation of a treatment plant. *Environ. Health Perspect.*, **19**, 103–105.

Breslow, N.E., Day, N.E. 1987. *Statistical Methods in Cancer Research. Vol. 1. The Analysis of Case-Control Studies.* IARC Sci. Publ.

Cebrian, M.E., Albores, A., Aguilar, M., Blakely, E. 1983. Chronic arsenic poisoning in the north of Mexico. *Hum. Toxicol.*, **2**, 121–133.

Chakraborty, A.K., Saha, K.C. 1987. Arsenical dermatosis from tubewell water in West Bengal. *Indian J. Med. Res.*, **85**, 326–334.

Chen, C.J., Chen, C.W., We, M.M., Kuo, T.L. 1992. Cancer potential in liver, lung, bladder and kidney due to ingested inorganic arsenic on drinking water. *Br. J. Cancer*, **66**, 888–892.

Datta, D.V., Kaul, M.K. 1976. Arsenic content of drinking water in a village in northern India. A concept of arsenicosis. *J. Assoc. Phys. India*, **24**, 599–604.

De, B.K., Santra, A., Dasgupta, J., Ghosh, N., Guha Mazumder, D.N. 1999. Chronic Environmental Arsenic Toxicity—Global scenario with particular reference to West Bengal. In: M.M. Singh (ed.), *Medicine Update.* Association of Physicians of India, New Delhi, pp. 451–462.

Flora, S.J.S., Dube, S.N., Arora, U., Kannan, G.M., Shukla, M.K., Malhotra, P.R. 1995. Therapeutic potential of meso 2,3-dimercapto succinic acid or 2,3-dimercaptopropane sulfonate in chronic arsenic intoxication in rats. *Bio Metals*, **8**, 111–116.

Garai, R., Chakraborty, A.K., Dey, S.B. et al. 1984. Chronic arsenic poisoning from tubewell water. *J. Indian Med. Assoc.*, **82**, 34–35.

Guha Mazumder, D.N., Chakraborty, A.K., Ghose, A. et al. 1988. Chronic arsenic toxicity from drinking tubewell water in rural West Bengal. *Bull. World Health Org.*, **66**, 499–506.

Guha Mazumder, D.N., Das Gupta, J., Chakraborty, A.K. et al. 1992. Environmental pollution and chronic arsenicosis in south Calcutta. *Bull. World Health Org.*, **70**, 481–485.

Guha Mazumder, D.N., Das Gupta, J., Santra, A. et al. 1997. Non cancer effects of chronic arsenicosis with special reference to liver damage. In: C.O. Abernathy, R.L. Calderon and W.R. Chappel (eds.), *Arsenic: Exposure and Health Effects.* Chapman Hall, London, pp. 112–123.

Guha Mazumder, D.N., Dasgupta, J., Santra, A. et al. 1998. Chronic Arsenic Toxicity in West Bengal—The worst calamity in the world. *J. Indian Med. Assoc.*, **96**, 4–7.

Guha Mazumder, D.N., Haque, R., Ghosh, N., De, B.K., Santra, A., Chakraborty, D., Smith, A. 1998b. Arsenic levels in drinking water and the prevalence of skin lesions in West Bengal, India. *Int. J. Epidemiol.*, **27**, 871–877.

Guha Mazumder, D.N., De, B.K., Santra, A., Dasgupta, J., Ghose, A., Pal, A., Roy, B., Pal, S., Saha, J. 1998c. Clinical manifestations of chronic Arsenic toxicity, its natural history and therapy: Experience of study in West Bengal, India. International Conference on Arsenic pollution of ground water in Bangladesh: Causes, effects and remedies, Dakha, Bangladesh 8–12 Feb. p. 91.

Guha Mazumder, D.N., Ghoshal, U.C., Saha, J., Santra, A., De, B.K., Chatterjee, Dutta, S., Angle, C.R., Centeno, J.A. 1998d. Randomized placebo-controlled trial of 2,3-dimercaptosuccinic acid in therapy of chronic arsenicosis due to drinking arsenic-contaminated subsoil water. *Clin. Toxicol.*, **36** (7), 683–690.

Huang, Y.Z., Qian, X.C., Wang, G.Q. et al. 1985. Endemic chronic arsenicism in Xinjiang. *Chin. Med. J. (Engl.)*, **98**, 219–222.

Kew, J., Morris, C., Aihie, A., Fysh, R., Jones, S., Brooks, D. 1993. Arsenic and mercury intoxication due to Indian ethnic remedies. *Br. Med. J.*, **306**, 506–507.

Kreppel, H., Paepacke, U., Thiermann, H., Szinicz, L., Reichl, F.X., Singh, P.K., Jones, M.M. 1993. Therapeutic efficacy of new dimercaptosuccinic acid (DMSA) analogues in acute arsenic trioxide poisoning in mice. *Arch. Toxicol.*, **67**, 580–585.

Leuz, K., Hruby, K., Druml, W., Eder, A., Gaszner, A., Kleinberger G., Pichler, M., Weiser, M. 1981. 2,3-Dimercapto succinic acid in human arsenic poisoning. *Arch. Toxicol.*, **47**, 241–243.

Mondal, B.K., Roychowdhury, J.R., Samanta, G. et al. 1996. Arsenic in ground water in seven districts of West Bengal, India—The biggest arsenic calamity in the world. *Curr. Sci.*, 70 (II), 976–986.

Shum, S., Whitehead, J., Vanghn, L.R.N., Surrey Shum, R.N., and Hale, T. 1995 Chelation of organo arsenate with Dimercaptosuccinic acid (DMSA). *Vet. Human Toxicol.*, **37**, 239–242.

Tseng, W.P., Chu, H.M., Hoiv, S.W. 1968. Prevalance of skin cancer in an endemic area of chronic arsenicism in Taiwan. *J. Natl. Cancer Inst.*, **40**, 453–463.

Tseng, W.P., Chu, H.M., How, S.W., Fong, J.M., Lin, C.S., Yeh, S. 1968. Prevalence of skin cancer in an endemic area of chronic arsenicism in Taiwan. *J. Natl Cancer Inst.*, **40**, 453–463.

Tseng, W.P. 1977. Effects and dose–response relationships of skin cancer and blackfoot disease with arsenic. *Environ. Health Perspect.*, **19**, 109–119.

UNICEF, 1998. Plan of action to combat the situation assessing arsenic contamination in drinking water: Plan to assist Government of West Bengal by UNICEF, UNICEF East India office, Calcutta, 1998, p. 6.

Arsenic Exposure and Health Effects
W.R. Chappell, C.O. Abernathy and R.L. Calderon (Editors)
© 1999 Elsevier Science B.V. All rights reserved.

Clinical Approaches to the Treatment of Chronic Arsenic Intoxication: From Chelation to Chemoprevention

Michael J. Kosnett

ABSTRACT

The dimercapto chelating agent, dimercaprol (British Anti-Lewisite, BAL) was developed in the 1940s as a treatment for acute poisoning by the vesicant organoarsenical chemical warfare agent, lewisite (dichloro [2-chlorovinyl] arsine). Water soluble analogs of dimercaprol, dimercaptosuccinic acid (DMSA, succimer) and dimercaptopropanesulfonic acid (DMPS, Unithiol, Dimaval) were introduced in the 1950s as chelators of arsenic and other heavy metals that offered the advantage of higher therapeutic index, and both oral and intravenous routes of administration. Animal models have demonstrated that BAL, DMSA, and DMPS are efficacious in averting morbidity and mortality if administered within minutes to hours of *acute* arsenic exposure. In patients with *chronic* arsenic exposure, administration of the chelating agents may result in a transient increase in urinary arsenic excretion. However, the ability of chelation to avert or reverse the clinical effects of chronic arsenic intoxication, such as anemia, neuropathy, portal hypertension or hyperkeratoses, or to decrease the future risk of arsenic-induced cancer, has not been established. Limited case series, supported by recent insights into the potential mechanisms of arsenic-induced carcinogenesis, suggest that oral treatment with retinoids, (Vitamin A analogs), may have promise in the treatment of chronic cutaneous arsenicism, and may also have impact on the development of neoplasia. Selenium, an antioxidant nutrient that antagonizes many of the effects of arsenic in biological systems, also merits attention as a potential therapeutic agent for patients with a history of chronic arsenic exposure. Nonspecific supportive care, including tricyclic antidepressants for the painful dysesthesias of peripheral neuropathy, and topical keratolytics for palmar-plantar hyperkeratoses, may offer short-term symptomatic relief.

Keywords: treatment, chelation, retinoids, selenium

INTRODUCTION

Optimal treatment of chronic arsenic intoxication should be directed at three key objectives: (1) cessation of ongoing arsenic exposure; (2) provision of specific drugs or nutrients that might hasten recovery and/or avert disease progression, and (3) administration of nonspecific supportive care that will minimize existing symptomatology. Limitations in public health resources and infrastructure have sometimes slowed implementation of the first goal, and success in the second and third goals has been constrained by a paucity of controlled clinical studies. An overview of clinical data pertaining to the treatment of chronic arsenic intoxication is instructive to highlight the promise and uncertainties of potential therapeutic modalities.

CHELATORS

Dithiol chelating agents were originally developed as specific pharmaceutical treatments of _acute_ arsenic poisoning. The first such chelator, dimercaprol ("British Anti-Lewisite" or "BAL"), was originally introduced in the 1940s as a topical and parenteral (intramuscular) antidote for acute poisoning by the vesicant organoarsenical chemical warfare agent known as "lewisite" (dichloro [2-chlorovinyl] arsine) (Peters et al., 1945; Stocken and Thompson, 1946). Water soluble analogs of dimercaprol, dimercaptopropanesulfonic acid (DMPS, "Unithiol", "Dimaval"), and dimercaptosuccinic acid (DMSA, succimer, "Chemet"), were developed as heavy metal chelators in the 1950s, (Petrunkin, 1956; Liang et al., 1957) and offered the advantage of higher therapeutic index, and both oral and intravenous routes of administration (Aposhian et al., 1984). Controlled animal experiments have demonstrated that dimercaprol, DMSA, and DMPS increase survival when administered within minutes to hours after _acute_ poisoning with lethal doses of organic or inorganic arsenicals (Stocken and Thompson, 1946; Tadlock and Aposhian, 1980) (Table 1). However, the efficacy of these agents declined in proportion to the length of time after acute arsenic exposure that

TABLE 1

Animal experiments indicate that dithiol chelating agents are protective against acute arsenic poisoning

Rat survival after topical lewisite (dichloro[2-chlorovinyl]arsine) ≈ 30 mg/kg [1]			
No chelator	0/6		
BAL (50–70 mg/kg topical inunction)	8/8		

Protection of mice against lethal effects of sodium arsenite (0.14 mmol/kg s.c.)	Thiol compound (mmol/kg) [2]	Cumulative 21 Day Survival (# surviving/start)	% Survival
	[saline]	0/24	0
	0.25 DMSA	24/24	100
	0.14 DMSA	20/24	83
	0.07 DMSA	16/24	67
	0.25 DMPS	24/24	100
	0.14 DMPS	21/24	88
	0.07 DMPS	19/24	79
	0.25 BAL	22/24	92
	0.14 BAL	2/24	8

(1) Data from Stocken and Thompson (1946).
(2) Injected i.p. immediately after NaAsO$_2$. Data from Tadlock and Aposhian (1980).

treatment was begun. In studies of the effect of dimercaprol on experimental organoarsenical poisoning in rabbits, Eagle et al. (1946) noted that all animals survived when a single injection of dimercaprol was administered 5 minutes after the arsenical, compared to no survival if treatment were delayed for 6 hours. Data obtained by Tadlock and Aposhian (1980) on the efficacy of single dose of DMSA (0.25 mmol/kg, i.p.) against a lethal dose of sodium arsenite (0.14 mg/kg, s.c.) in mice suggests that beneficial effects on survival may begin to diminish when treatment is delayed for 2 or more hours.

For the first five decades following their introduction into medical formularies, the capacity of chelating agents to improve clinical outcome in cases of *subacute or chronic* arsenic exposure had never been investigated in animal experiments or in carefully controlled human clinical trials. Case series of patients undergoing dimercaprol treatment for subacute arsenical dermatitis and other complications of syphilis treatment with organoarsenical medication suggested that chelation may accelerate clinical improvement, but these early studies were conducted without rigorous controls, and their clinical relevance to long-term ingestion of inorganic arsenic is questionable (Eagle and Magnuson, 1946; Carleton et al., 1948).

Two recent studies have utilized controlled animal experiments or human trials to examine the potential benefit of chelating agents in subacute or chronic arsenic poisoning. Flora et al. (1995) exposed rats to 1 mg/kg oral arsenate for three weeks, followed by 5 days of oral treatment with either DMSA (50 mg/kg po tid), DMPS (50 mg/kg po tid), or normal saline. The chelators partially reversed arsenic-associated inhibition of blood delta amino-levulinic dehydratase, but had no effect on an arsenic-associated increase in blood zinc protophorphyrin. An impact, if any, on anemia or leukopenia was not reported. The chelators increased hepatic glutathione; however the chelators, but not the arsenic alone, appeared to increase serum transaminases (ALT and AST). The authors stated that the chelators had a salutary effect on arsenic-associated histopathological changes in the kidney and liver, but data regarding the magnitude of arsenic effect and chelation reversal in all animals was not provided. No information on the effect of arsenic or the chelators on renal function was reported. Extrapolation of these limited findings to chelation treatment of chronic arsenic exposure in humans is further constrained by the poor suitability of the rat as an animal model for arsenic toxicity (Vahter, 1994), and the use of chelator doses 5 fold higher than customary human regimens.

Guha Mazumder et al. (1998) conducted a randomized, placebo controlled trial of DMSA in adult men with overt chronic arsenicism, including hyperkeratosis and hyperpigment-ation, attributed to long-term ingestion of arsenic in drinking water. The 11 subjects randomized to DMSA had consumed arsenic in drinking water ($660 \pm 390 \,\mu g/L$) for a mean of 15.25 ± 10.7 years, up until 0.54 ± 1.37 months before entry into the trial. The 10 subjects randomized to matched placebo had comparable arsenic exposure, i.e. consumption of arsenic in drinking water ($650 \pm 340 \,\mu g/L$) for a mean of 21.6 ± 11.95 years, up until 1.67 ± 1.22 months prior to study entry. All subjects were hospitalized during the study regimen, which in the active drug arm consisted of an initial course of DMSA, 350 mg po qid × 7 days, followed by 350 mg po tid × 14 days. This regimen was then repeated following a 3 week drug-free hiatus. Pre- and post intervention measurements included a single blinded (patient blinded) standardized, graded assessment of signs and symptoms, urinary arsenic excretion, liver function tests, and double blinded, graded histological assessment of skin biopsies. Results indicated that clinical assessment scores underwent a similar degree of improvement in both the DMSA and the placebo group. In like manner, there were no differences between the DMSA and placebo group in laboratory parameters, and dermal histopathology, at the beginning and at the end of treatment. The authors concluded that their study "has conclusively shown that DMSA is globally ineffective in the therapy of chronic arsenic toxicity in man".

In a recent study (Aposhian et al., 1997), the administration of DMPS to subjects with very recent, long-term ingestion of arsenic in drinking water was associated with a prompt increase in the excretion of arsenic in the urine that was several fold above pre-chelation levels. In 13 subjects consuming arsenic in drinking water (528 μg/L) up until one day prior to the administration of a single oral 300 mg dose of DMPS, total urine arsenic increased from a baseline of 605 $\pm$ 81 μg/g creatinine (Cr) to a peak of 2325 $\pm$ 258 μg/g Cr in the first two hours post chelator. In 11 control subjects chronically consuming water containing arsenic at a concentration of 21 μg/L, DMPS resulted in the baseline urine arsenic concentration of 91 $\pm$ 17 μg/g Cr transiently increasing to 305 $\pm$ 79 μg/g Cr. The data are consistent with chelation accelerating the decorporation of arsenic in chronically exposed humans. However, animal experiments suggest that compared to cessation of exposure alone, DMPS chelation may predominantly affect the rate of arsenic excretion, rather than long-term net excretion (Maiorino and Aposhian, 1985). At the present time, there are no follow-up data available to determine whether a short-term increase in urinary arsenic excretion associated with chelation will result in a lower risk of long-term adverse outcomes, such as cancer.

RETINOIDS

More than 50 years ago, Hall (1946) and colleagues described a beneficial effect of oral supplementation with Vitamin A (retinol) in the treatment of cutaneous arsenicism. In that report, oral Vitamin A, 150,000 USP units per day for 3 months resulted in a partial regression of palmar hyperpigmentation and hyperkeratoses in a 39-year-old male who had taken Fowler's solution (potassium arsenite) for treatment of childhood chorea. More recently, Thianprasit (1984) presented a case series of 9 patients with cutaneous arsenicism who were treated for 2 to 7 months with oral etretinate, a synthetic aromatic retinoid. Clinical and histopathological improvement was noted in arsenical hyperkeratoses, but not in hyper-pigmentation. Other case reports of regression of arsenical keratoses with etretinate treatment have been published (Biczo et al., 1986; Sass et al., 1993). It is noteworthy that etretinate and other retinoids have been reported to have antikeratinizing effects in other disorders of keratinization, such as hereditary palmoplantar keratoderma, pityriasis rubra pilaris, and certain ichthyoses (Fritsch, 1992).

In addition to causing regression in arsenical keratoses, retinoids may offer significant promise in the chemoprevention of arsenic-related cancers. The interaction of endogenous and exogenous retinoids with nuclear receptors influences the expression of genes that effect cell differentiation, proliferation, and induction of apoptosis (Miller, 1998). Some clinical trials, recently reviewed by Lotan (1996) and Hong and Sporn (1997) suggest a beneficial role for retinoids in chemoprevention of cancer in multiple organs. For example, Bouwes Bavinck et al. (1995) reported a prospective, double-blind, placebo controlled trial of acitretin in renal transplant patients that resulted in decreased occurrence of cutaneous squamous cell carcinoma and keratotic skin lesions. A prospective randomized, controlled trial of retinoids in patients with chronic cutaneous arsenicism is clearly indicated at this point in time. However, the therapeutic challenge will lie in selecting the right drug, at the proper dosage, at the correct stage of carcinogenesis. In addition, because retinoids and high dose retinol may have adverse effects, including teratogenesis, such trials will require careful attention to patient selection and surveillance.

SELENIUM

Considering the size and demographics of the large populations at risk of arsenic-induced cancer throughout the world, it would also be advantageous if readily available, non-proprietary nutrients were found to have a role in the chemoprevention of arsenic-induced cancers. Selenium, an antioxidant mineral that appears to antagonize several cellular effects of arsenic in biological systems, merits particular attention in this regard (Schrauzer, 1992). Several mechanisms may account for a modulating effect of selenium on chronic arsenic toxicity, including, but not limited to, the *in vivo* formation of arsenic–selenium compounds, an impact of selenium and selenoenzymes on arsenic-associated oxidative stress, effects of selenium on arsenic methylation, and the influence of selenium on cell growth. It is noteworthy that a recent multi-center double-blind randomized trial of selenium supplementation was associated with a marked reduction in cancers of the lung, prostate and colon (Clark et al., 1996). Interestingly, there was no reduction in the incidence of basal cell carcinoma and squamous cell carcinoma of the skin (the study's primary endpoints), however the trial was terminated early due to the major impact on other cancer incidence.

NONSPECIFIC SUPPORTIVE CARE

Nonspecific supportive care of the ambulatory patient with chronic arsenic intoxication has not been examined with controlled studies, but several approaches might be considered for certain adverse symptoms. Topical keratolytic agents, such as salicylic acid in petrolatum have been proposed as treatment for discomfort resulting from palmar and plantar hyperkeratoses (Saha, 1995; Guha Mazumder, 1996), which may persist long after arsenic exposure ends. Tricyclic antidepressants such as amitryptiline may have utility in relieving the painful dysesthesias of arsenic peripheral neuropathy (Wilner and Low, 1993). Fortunately, however, this symptom will often resolve spontaneously within several months following cessation of arsenic exposure.

REFERENCES

Aposhian H. V., Carter, D.E., Hoover, T.D., Hsu, C.A., Maiorino R.M., and Stine, E. 1984. DMSA, DMPS, DMPA—as arsenic antidotes. *Fund. Appl. Toxicol.*, **4**, S58–S70.

Aposhian, H.V., Arroyo A., Cebrian, M.E., Del Razo, L.Z., Hurlbut, K.M., Dart, R.C., Gonzalez-Ramirez, D., Kreppel, H., Speisky H., Smith, A., Gonsebatt, M.A., Ostrosky-Wegman, P., Aposhian, M. 1997. DMPS-Arsenic challenge test. I. Increased urinary excretion of monomethylarsonic acid in humans given dimercaptopropane sulfonate. *J. Pharm. Expt. Ther.* **282**, 192–200.

Biczo, Z., Berta, M., Szabo, M., Nagy, G.Y. 1986. Traitement de l'arsenicisme chronique par l'etretinate. *La Presse Medicale*, **15**, 2073.

Bouwes Bavinck, J.N., Tieben, L.M., Van Der Woude, F.J., Tegzess, A.M., Hermans, J., Ter Schegget J., Vermeer B.J. 1995. Prevention of skin cancer and reduction of keratotic skin lesions during acitretin therapy in renal transplant recipients: a double-blind, placebo-controlled study. *J. Clin. Oncol.*, **13**, 1933–1938.

Carleton A.B., Peters, R.A., Stocken, L.A., Thompson, R.H.S., and Williams, D.I. Clinical uses of 2,3-dimercaptopropanol (BAL). VI. The treatment of complications of arseno-therapy with BAL (British Anti-Lewisite). *J. Clin. Invest.*, **25**, 497–527.

Clark L.C., Combs, G.F., Turnbull, B.W., Slate, E.H., Chalker, D.K., Chow, J. Davis, L.S., Glover, R.A., Graham, G.F., Gross, E.G., Krongrad, A., Lesher, J.L., Park, K.H., Sanders, B.B., Smith, C.L., Taylor J.R. 1996. Effects of selenium supplementation for cancer prevention in patients with carcinoma of the skin. A randomized controlled trial. *JAMA*, **276**, 1957–1963.

Eagle, H., Magnuson, H.J. 1946. The systemic treatment of 227 cases of arsenic poisoning (encephalitis, dermatitis, blood dyscrasias, jaundice, fever) with 2,3-dimercaptopropanol (BAL). *Am. J. Syph. Gon. V.D.*, **30**, 420–441.

Eagle, H., Magnuson, H.J., Fleischman, R. 1946. Clinical uses of 2,3 dimercaptopropanol (BAL). I. The systemic treatment of experimental arsenic poisoning (marphasen, lewisite, phenyl arsenoxide) with BAL. *J. Clin. Invest.*, **25**, 451–66.

Flora, S.J.S., Dube, S.N., Arora U, Kannan, G.M., Shukla, M.K., Malhotra, P.R. 1995. Therapeutic potential of meso 2,3,-dimercaptosuccinic acid or 2,3,-dimercaptopropane 1-sulfonate in chronic arsenic intoxication in rats. *BioMetals*, **8**, 111–116.

Fritsch, P.O. 1992. Retinoids in psoriasis and disorders of keratinization. *J. Am. Acad. Derm.*, **27**, S8–S14.

Guha Mazumder D.N. 1996. Treatment of chronic arsenic toxicity as observed in West Bengal. *J. Ind. Med. Assoc.*, **94**, 41–42.

Guha Mazumder, D.N., Ghoshal, U.C., Saha, J., Santra, A., De, B.K., Chatterjee, A., Dutta, S., Angle, C.R., Centeno, J.A. 1998. Randomized, placebo-controlled trial of 2,3-dimercaptosuccinic acid in therapy of chronic arsenicosis due to drinking arsenic-contaminated subsoil water. *Clin. Toxicol.*, **36**, 683–690.

Hall, A.F. 1946. Arsenical keratoses disappearing with vitamin A therapy. *Arch. Derm. Syph.*, **53**, 154.

Hong W.K., and Sporn, M.B. Recent advances in chemoprevention of cancer. *Science*, **278**, 1073–1077.

Liang, Y., Chu, C., Tsen, Y., Ting, K. 1957. Studies on antibilharzial drugs. VI. The antidotal effects of sodium dimercaptosuccinate and BAL-glucoside against tartar emetic. *Acta Physiol. Sin.*, **21**, 24–32.

Lotan R. 1996. Retinoids in cancer chemoprevention. *FASEB J.*, **10**, 1031–1039.

Maiorino, R.M., Aposhian, H.V. 1985. Dimercaptan metal binding agents influence the biotransformation of arsenite in the rabbit. *Toxicol. Appl. Pharmacol.*, **77**, 240–250.

Miller W.H. 1998. The emerging role of retinoids and retinoic acid metabolism blocking agents in the treatment of cancer. *Cancer* **83**, 1471–82.

Peters, R.A., Stocken, L.A., Thompson, R.H.S. 1945. British Anti-Lewisite. *Nature*, **156**, 616–619.

Petrunkin, V.E. 1956. Synthesis and properties of dimercapto derivatives of alkylsulfonic acids. *Ukr. Khem. Zh.*, **22**, 603–607.

Saha K.C. 1995. Chronic arsenical dermatoses from tube-well water in West Bengal during 1983–87. *Ind. J. Dermatol.*, **40**, 1–12.

Sass, U., Grosshans, E., Simonart, J.M. 1993. Chronic arsenicism: criminal poisoning or drug-intoxication? Report of two cases. *Dermatology*, **186**, 303–305.

Schrauzer G.N. 1992. Selenium. Mechanistic aspects of anticarcinogenic action. *Biol. Trace Element Res.*, **33**, 51–62.

Stocken L.A., Thompson, R.H.S. 1946. British Anti-Lewisite. 2. Dithiol compounds as antidotes for arsenic. *Biochem. J.*, **40**, 535–548.

Tadlock, C.H., Aposhian, H.V. 1980. Protection of mice against the lethal effects of sodium arsenite by 2,3-dimercapto-1-propane-sulfonic acid and dimercaptosuccinic acid. *Biochem. Biophys. Res. Comm.*, **94**, 501–507.

Thiaprasit, M. 1984. Chronic cutaneous arsenism treated with aromatic retinoid. *J. Med. Assoc. Thailand*, **67**, 93–100.

Vahter M. 1994. Species differences in the metabolism of arsenic compounds. *Appl. Organometallic Chem.*, **8**, 175–182.

Wilner, C., Low, P.A., 1993. Pharmacological approaches to neuropathic pain. In: P.J. Dyck (ed.), *Peripheral Neuropathy*, pp. 1709–1720. W.B. Saunders, Philadelphia.

Arsenic Exposure and Health Effects
W.R. Chappell, C.O. Abernathy and R.L. Calderon (Editors)
© 1999 Elsevier Science B.V. All rights reserved.

Major Interventions on Chronic Arsenic Poisoning in Ronpibool District, Thailand—Review and Long-Term Follow Up

Chanpen Choprapawon, Sirinporn Ajjimangkul

ABSTRACT

This study aimed to assess the outcome, impact and problems of the major interventions during the past ten years to control chronic arsenic poisoning in Ronpibool District, Nakorn Sri Thammarat, the southern Province of Thailand, in order to find better strategies to combat the illnesses. In early 1998, the principal investigator conducted four focus group discussions in the district. The results from the focus group discussions showed that the toxicity from arsenic contamination in natural water sources in Ronpibool District had been recognized by the Thai government and the residents of this district for more than 10 years. Since 1987, the government sectors have launched several interventions. Major measures include: providing of alternative water sources (e.g. distribution of big water jars to collect rain water, construction of deep tube wells, and construction of village pipe water systems), periodic monitoring of drinking water and environmental samples, case detection and supportive treatment for patients with severe skin manifestations. Also included were health education for the villagers; and site remediation to clean up the 3,000 tons of high-grade mine waste piles. However, the major interventions during these ten years show less success than expected. All four group discussions agreed that they still lack a good understanding of the situation and desire continuous communication with the government sectors. Villagers are concerned about their economic status and some old life styles inhibit their participation in solving this problem. The investigators conducted a survey of all households (approximately total 5,000 households) in the district. Information was collected on demographics, sources of water used and suspected cases of skin lesions. After the survey was completed, household mapping was done in a manner that the villagers could easily utilize for long-term follow up and design the appropriate interventions for the specific problem of each village. It is proposed that a well-planned, long-term study should be established in order to control the problem and to provide a rich source of scientific knowledge to help us understand the specific disease occurrence and changing pattern of illnesses in this area.

Keywords: chronic arsenic poisoning, Thailand, intervention, long-term follow up

INTRODUCTION

The first skin cancer case in the area associated with chronic exposure to arsenic was detected in 1987. An epidemiological survey was done in 1994 and found that the prevalence of skin manifestation from chronic arsenic poisoning was 26.3 percent (Ekpalakorn and Rodcline, 1994). From this figure it can be estimated that there would be around 5,000 cases of skin lesions and more than 10,000 people in Ronpibool District had arsenic levels higher than 50 μg per 100 mg in hair and nails (Table 1). This study also found that more than two-thirds of the drinking water supplies were contaminated with arsenic and about 2% had an arsenic level higher than the safety limit (Table 2). Another study found high arsenic intake from cooked food with the level of 726.8 μg per person per day (Boriboon et al., 1996).

Potential sources of arsenic contamination were from high grade arsenopyrite waste piles in a bedrock mining area and wastes from an ore-dressing plant located in the mountainous area drained by surface water (Williams et al., 1996). The amount of mining waste was estimated at more than 3,000 tons of wastes in this area and the contamination extended for more than 20 km from the sources. Not only was high arsenic contamination found in the

TABLE 1

Prevalence (%) of chronic arsenic poisoning by village and stage, Ronpibool District, Nakorn Sri Thammarat Province, Thailand, 1994

Village no.	High As/No Skin Lesions		High As/Wiith Skin Lesions		All (%)
	%	95% CI	%	95% CI	
1	40.6	23.7–59.4	53.1	34.7–70.9	93.7
2	62.7	48.1–75.9	28.5	15.9–41.7	90.2
3	72.5	56.1–85.4	17.5	7.3–32.8	90.0
4	34.3	19.1–52.2	22.9	10.4–40.1	57.2
5	57.5	40.9–73.0	35.0	20.6–51.7	92.5
6	73.3	54.1–87.7	23.3	9.9–42.3	96.6
7	60.9	49.9–71.2	18.4	10.9–28.1	79.3
8	29.7	15.9–47.0	2.7	0.1–14.2	32.4
9	31.7	18.1–48.1	22.0	10.6–37.6	53.7
10	75.6	60.5–87.1	13.3	5.1–26.8	88.9
11	83.3	62.6–95.3	16.7	4.7–37.4	100.0
12	44.1	33.8–54.8	49.5	38.9–60.0	93.6
13	73.2	57.1–85.8	17.1	7.2–32.1	90.3
14	47.8	26.8–69.4	26.1	10.2–48.4	73.9
All	55.8	51.8–59.8	26.3	22.9–30.0	82.1

TABLE 2

Percentage of arsenic contamination in water samples, Ronpibool District, Nakorn Sri Thammarat Province, Thailand, 1994

Water samples	No.	Median As(mg/kg)	% of As (0.05 mg/kg)+
Rain	86	0.001	not detected
Shallow Ground	70	0.002	2.8
Pipe	20	0.003	5.0
Canal	7	not detected	not detected
Deep Ground	3	not detected	not detected
Bottle	2	not detected	not detected
Total	188	0.006	1.5

environment, but also the copper and cadmium contamination in surface waters was detected especially in the downstream area (Arrykul et al., 1987). During this decade, several interventions have been implemented. This study describes experiences from the past and present interventions and proposes long-term research and development as a tool to empower community involvement as well as to get more knowledge on the health effects of chronic arsenic poisoning in the Southern district of Thailand.

METHOD

This study consisted of two parts. The first part was to review past and present experiences, regarding the interventions to control chronic arsenic poisoning, from all working documents and minutes of the meetings during 1987–1995. The second part was the assessment of knowledge, perception and practice of the samples by conducting four focus group discussions among villagers in the district. Each focus group consisted of 7 to 16 members of the same ages or social status. A total household survey was conducted by the VHVs (Village Health Volunteer) in every village. The VHVs collected demographic data and interviewed a house member about the sources of water used in the household. The VHVs located and drew a map of the village. Water resource locations for each house were illustrated in the map using different legends to differentiate among the main sources. After finishing the survey, the VHVs presented the survey results and the map in the District Committee meeting for data verification and proposing major problems of the villages. This data will be the baseline for the long-term study which will be conducted by a multi-institutional research team.

RESULTS

The review of past and present interventions could be divided into three main periods.

First Phase

This was first few years after the detection of skin cancer and arsenic poisoning during 1987–1988. This period might be called "the emergency and panic phase". It started when the mass media disclosed to the public the occurrence of cancer and poisoning in this area. The Prime Minister, whose motherland is in the south near this province, visited the district and ordered every ministry to take action to stop poisoning in this district. Major objectives of all interventions (Silaparassami, 1994) during 1987–1989 were to:
1. conduct a quick study to identify potential sources of contamination,
2. test water sources to find the contaminated sites and to close the contaminated wells and label each with a warning sign,
3. provide several thousand big water jars and tanks to store rain water to substitute for the shallow ground water which has been normally used for several decades,
4. construct deep ground water wells as alternative safer sources in the district,
5. construct small irrigation ditches in other subdistricts to be used for community pipe systems,
6. provide treatment for severe cases of arsenicism and educate the villagers through a local network,
7. test and provide small water treatment systems.
However, because of public panic and the lack of good management, the villagers could not sell their agriculture products after the massive announcement in the national newspapers and television broadcasts. In addition, confusing and incorrect information was disseminated to the public even among the government officers. So the government tried to keep all information from the public in order to stop national panic and local conflict (Choprapawon, 1994).

Even though the government had launched many interventions during the first few years, the situation did not improve much. Most villagers continued to drink contaminated water. Of particular concern was the lack of community involvement which led to the second phase of the control.

Second Phase

After all the movement had slowed down during 1989–1993, in the early period there was no clear focal point for long-term interventions. There were only periodic meetings in the Ministry of Public Health to follow up some interventions. Off and on there were some debates in parliament. In this phase the major activity was among the technical and academic communities to review and discuss research questions which might be important for further actions. Several workshops and national conferences were held to identify and discuss priority research topics among the experts in various fields. The priority research (The National Epidemiology Board of Thailand and Prince Sonkhla University, 1989) included: (a) the characteristics and distribution of arsenic contamination in soil and water, (b) the epidemiological study of chronic arsenic poisoning and the risk factors, (c) the effect of arsenic on child development (including intelligence), and (d) the development of a social program to change the behavior of the villagers. The forums presented the recommendations directly to the Prime Minister but there was no prompt action from the government sectors.

The technical forum has continued to collaborate by establishing a network in order to follow the situation and identify opportunities to initiate effective interventions. With this strategy, some actions in the government sector were initiated. A major one was the remediation plan for 3,000 tons of mining waste which had been dumped in this area for more than 70 years. Another plan was a medical study of the skin lesions. Following the actions in this second phase, three important projects were initiated, the first was to study the effect of arsenic in child development and intelligence (Siripittayakunkit et al., 1997), the second study was the search for a geochemical solution (Williams et al., 1996) and the third one was the geo-hydrological study to assess the spreading of the heavy metal contamination along the natural water resources to the Pak-Panang Bay (Arrykul et al., 1996).

Third Phase

In the past, community leaders and the folk sectors had been the recipient of all government interventions. In late 1993, a community committee was established to raise community participation (Choprapawon, 1994). The District Committee for Environmental Management consists of community leaders, public health personnel, officers from the Department of Public Administration, Department of Mineral Resources, Department of Irrigation, Department of Health, Department of Medical Sciences, Department of Pollution Control, and invited members of the local political party. From the Committee meeting, several strategies were implemented as follows:

1. A first cartoon book for the lay person was written and illustrated by a primary school teacher living and working in this area. Thirty thousand copies were produced by the Department of Health, MoPH and distributed to every household and primary school pupils, in order to educate the whole district and all stakeholders.

2. The first epidemiological survey was conducted by the public health ministry which showed that the prevalence of skin diseases was 26.3% (Ekpalakorn and Rodcline, 1994). Another study, supported by the WHO, assessed arsenic intake from cooked food in a sample of households (Boriboon et al., 1996).

3. Department of Mineral Resources conducted the survey of contaminated areas and started the process of remediating 3,000 tons of mining wastes (Paijitprapapon and Pluemarrom, 1997).

4. The Department of Health budgeted the construction of 3 community pipe systems using the new irrigation unit constructed by the Department of Irrigation, Ministry of Agriculture (Department of Health, 1996).
5. The Department of Environmental Quality Promotion, Ministry of Science, Technology and Environment, proposed JICA support a feasibility study on environmental management (Environmental Research and Training Center, Department of Environmental Quality Promotion, 1998).

Beginning in 1993, the District Committee held regular meetings every two months (Kongsawas, 1995–1997). Through this forum the committees slowly learned about the problem as it was quite difficult for them to understand the process of arsenic poisoning. The symptoms of chronic arsenic poisoning are often difficult to perceive, especially early on in the exposure process. In addition, some intervention, especially the finding of alternative water sources, has to deal with the allocation of resources among the households. This sometimes led to arguments and fighting among the villagers. In the process of five years of these meetings, the community began to take an active role.

The investigators completed four focus group discussions among a group of 12 secondary school children aged 15–17 years, a group of 7 village leaders, a group of 8 low-risk village residences , and a group of 7 arsenicism cases and their family members. When comparing the results of the focus group discussion with previous in-depth interviews on the perception and knowledge of the villagers (Ajjimangkul, 1992), we found that the villagers and the students have better understanding and knowledge about causes, clinical manifestation of arsenic poisoning, prevention strategies and previous intervention efforts.

From May to June 1998, the VHVs completed the total household survey which showed that 48.1 percent of the residents were male. Only village 16 has a higher number of males than female (Table 3). The pattern of water use has changed from previously as rain water is the major source of drinking water in every village and piped water is used for bathing and other purposes. Bottled water was frequently used in villages 7 and 13 (Table 4). When

TABLE 3

Number and percent of population by sex from district survey in Ronpibool District, Nakorn Sri Thammarat Province, Thailand, 1998

Village no.	Male		Female		Total	
	No.	%	No.	%	No.	%
1	274	47.6	296	52.4	576	3.4
2	830	47.7	902	52.3	1,738	10.4
3	667	49.4	679	50.6	1,351	8.1
4	577	48.1	621	51.9	1,200	7.2
5	491	46.4	563	53.6	1,057	6.3
6	489	48.6	501	51.4	1,007	6.0
7	857	49.2	884	50.8	1,744	10.4
8	680	48.9	705	51.1	1,389	8.3
9	448	47.1	501	52.9	951	5.7
10	571	46.3	659	53.7	1,233	7.4
11	361	51.6	336	48.4	700	4.2
12*	414	47.0	464	53.0	881	5.3
13	453	44.7	558	55.3	1,013	6.1
14	317	49.2	324	50.8	644	3.9
15	158	47.0	174	53.0	336	2.0
16	445	50.3	436	49.7	884	5.3
Total	8,032	48.1	8,672	51.9	16,704	100.0

*Not completed. Changing number of villages from the previous year due to the resettlement of the district.

TABLE 4

Pattern of water use by village in Ronpibool District, Nakorn Sri Thammarat Province, Thailand, 1998 (in %)

Village no.	Drinking water			Other uses		
	Rain	Well	Other	Rain	Well	Other
1	95.2	18.6	6.6	78.8	61.7	8.0
2	78.3	18.8	38.2[1]	66.9	22.4	35.0[1]
3	75.0	19.0	55.1[1]	30.4	30.7	63.5[1]
4	79.3	30.1	13.1	37.2	62.4	16.0
5	46.0	51.5	8.5	20.1	77.6	6.7
6	78.7	40.2	7.6	30.0	68.5	23.5
7	82.8	20.3	42.3[2]	38.2	41.8	37.7[3]
8	97.4	27.3	23.7[4]	33.0	32.4	68.2[4]
9	88.9	27.6	27.0[1]	55.2	37.9	39.8[1]
10	79.3	37.5	51.2[1]	36.0	44.7	55.4[1]
11	82.8	54.0	51.6[4]	38.9	52.1	66.3[1]
12	74.0	13.1	41.5[1]	60.9	22.1	44.4[1]
13	71.8	13.1	46.7[2]	56.6	15.3	51.7[2]
14	75.5	32.3	4.2	25.6	72.8	2.1
15	93.0	26.9	28.1[1]	70.0	26.9	44.9[1]
16	86.3	40.3	10.9[1]	71.3	49.7	19.7[1]
Total	78.7	28.2	30.9	44.6	43.2	38.4

(1) Mainly pipe; (2) mainly bottle; (3) mainly ground; (4) mainly canal.

TABLE 5

Change in pattern of water use (%), 1986 and 1997, Ronpibool District, Nakorn Sri Thammarat Province, Thailand

Type of water	Drinking		Cooking		Bathing	
	1986	1997	1986	1997	1986	1997
Shallow ground	80.0	10.8	82.2	24.2	80.0	76.6
Rain	17.7	81.7	10.0	62.3	0.2	1.2
Pipe	4.5	8.7	6.7	21.3	16.8	49.1
Deep Ground	0	0	0.6	0	2.8	3.6
Bottled	0	10.5.	0	4.2	0	0

Source: Oshikawa (1998).

compared with another study (Oshikawa, 1998), the pattern of water use among known arsenic poisoning cases changed from shallow ground water to rain water (Table 5) which is similar to the results from the recent household survey. The study of Oshikawa also found changes in the stage of skin manifestations. There were some improvements in the stage of skin lesions and about one third remained unchanged (Table 6).

DISCUSSION

During the past 10 years, the Thai government has expended many resources to improve the condition of chronic arsenic poisoning in Ronpibool district, Nakorn Sri Thammarat Province. Major interventions include the big water jars for storing rain water. From focus group discussions and observation, most households increase consumption of drinking rain water when it is available. However, during dry seasons, they may return to using shallow ground water. Drinking bottled water is found more in the town but still reduces exposure to arsenic. Even though some patients showed improvement of skin lesions, the villagers have

TABLE 6

Changes in stages of skin lesions and other outcomes over 10 years, Ronpibool District, Nakorn Sri Thammarat Province, Thailand, 1998

Stage 1987	Stage in 1997				Dead	Missing	Total
	0	1	2	4			
1	62	173	12	0	18	359	624
	(9.9)	(27.7)	(1.9)	(0.0)	(2.9)	(57.5)	(100.0)
2	4	30	45	4	15	90	188
	(2.1)	(15.9)	(23.9)	(2.1)	(7.9)	(47.9)	(100.0)
4	0	1	1	2	2	0	6
	(0.0)	(16.7)	(16.7)	(33.3)	(33.3)	(0.0)	(100.0)
Total	66	204	58	6	35	449	818
	(8.1)	(24.9)	(7.1)	(0.7)	(4.3)	(54.9)	(100.0)

Source: Oshikawa (1998).

reverted to using contaminated water so the situation might get worse unless alternative sources of drinking water can be found.

From past experiences, the villagers were treated as beneficiaries in every intervention without active involvement. The lack of responsibility for the solutions and continuity of the effort were the main problems. In addition, there was a lack of strong and continuous political commitment of the local and central governments, a lack of movement by local authorities, misunderstanding by the villagers and a lack of good public information.

In order to increase awareness and understanding about the situation, more involvement of the folk sector is needed both in the research and interventions. The household survey data on the pattern of water use and the prevalence of skin diseases should consolidate their efforts and raise concerns about the problem. A well-planned, long-term study would be one important strategy to combat the problem on one hand, and on the other hand it should be the source of scientific knowledge to help us understand the occurrence of specific diseases and the changing pattern of illnesses in this area.

REFERENCES

Ajjimangkul, S. 1992. The study on pattern of water used among mothers with children aged under 5 in Ronpibool District, Nakorn Sri Thammarat Province: their perception, believe, and behavior. Ms.C. thesis. Mahidol University.

Arrykul, S. et al. 1987. Arsenic in water and soil in Ronpibool District. Unpublished report, Faculty of Engineering, Prince of Songkhla University.

Arrykul, S. et al. 1996. Contamination of arsenic, cadmium and lead in Pakpanang River Basin, Nakorn Sri Thammarat Province, Thailand. International Symposium on Geology and Environment, 31 January–2 February 1996.

Boriboon, P. et al. 1996. The study on arsenic intake from daily cooked food in Ronpibool District, Nakorn Sri Thammarat Province. Choprapawon, C. 1994. The review of chronic arsenic poisoning in Ronpibool District, Nakorn Sri Thammarat Province. Report submitted to the National Committee on Environmental Health Problems.

Department of Health, Ministry of Public Health of Thailand. 1996. Summary of the construction of community pipe system to control arsenic poisoning in Ronpibool District, Nakorn Sri Thammarat Province.

Ekpalakorn, V., Rodcline A. 1995. The prevalence survey of chronic arsenic poisoning in Ronpibool District, Nakorn Sri Thammarat Province. Unpublished report, Division of Epidemiology, Office of the Permanent Secretary, Ministry of Public Health.

Environmental Research and Training Center, Department of Environmental Quality Promotion, Ministry of Sciences, Technology and Environment. 1998. Project formation study on environmental management planning survey for arsenic contaminated area of Nakorn Sri Thammarat Province.

Kongsawas, S. 1995–1997. Report of the committee meetings, Ronpibool District Committee on Environmental Control.

Oshikawa, S. 1998. Re-examination of a cohort of subjects with arsenical skin lesions ten years after inception. Ms.C. thesis, Prince of Songkhla University.

Paijitprapapon, A., Pluemarrom N. 1997. Arsenic waste remediation in Ronpibool District: Problems and Limitation. Presented in the National Symposium on the Control of Arsenic Contamination, Ronpibool District, Nakorn Sri Thammarat Province. 2 July 1997. Nikko Hotel, Bangkok.

Silaparassami, Y. 1994. Summary of major interventions to control chronic arsenic poisoning in Ronpibool District, Nakorn Sri Thammarat Province. Report of the Provincial Health Office, Nakorn Sri Thammarat Province.

Siripittayakunkit, A. et al. 1997. Association between chronic arsenic exposure and schoolchildren's growth and intellectual ability at Ronpibool District, Nakorn Sri Thammarat Province. Report submitted to the Thailand Health Research Institute, National Health Foundation.

The National Epidemiology Board of Thailand and Prince of Songkhla University. 1989. Arsenic: Silent harm threat to Pakpanang Bay.

Williams, M., Fordyce, F., Paijitprapapon, A., Charoenchaisri, P. 1996. Arsenic contamination in surface drainage and groundwater in part of the southeast Asian tin belt, Nakorn Sri Thammarat Province, southern Thailand. *Environ. Geol.*, **27**, 16–33.

Arsenic Exposure and Health Effects
W.R. Chappell, C.O. Abernathy and R.L. Calderon (Editors)

Rapid Action Programme: Emergency Arsenic Mitigation Programme in Two Hundred Villages in Bangladesh

Quazi Quamruzzaman, Shibtosh Roy, Mahmuder Rahman, Selim Mia,
Ashraf Islam Arif

ABSTRACT

The presence of arsenic in tubewells in Bangladesh has been known since 1984. The impact on public health from this exposure to arsenic is only just being understood. Between November 1997 and April 1998, Dhaka Community Hospital was authorised by the Bangladesh Government (with UNDP support) to conduct a Rapid Action Programme (RAP) to establish a baseline on the extent of the problem in a sample population. Data are collected from rural locations thought to be situated in vulnerable areas. Presence/absence of arsenic in tubewells, according to field kit testing and number and gender of clinically symptomatic adults and children are recorded. It is impossible for the survey to be conducted in isolation of the affected communities' needs, so follow-up clinics are arranged for patients via the Hospital's own programmes. Social awareness campaigns are mounted because of the social stigma attached to those affected. Contaminated wells are marked and villagers advised about alternative safe water sources. The RAP has highlighted various practical problems: reliability of field testing kits, mapping of villages, unknown number of privately sunk tubewells. The extent of the contamination and number of arsenicosis victims may be far greater than many organisations currently acknowledge.

Keywords: arsenicosis, tubewells, water field test kits

INTRODUCTION

It was known as early as 1984 that arsenic was present in the tubewells of one area of Bangladesh. This information was withheld from public knowledge and the policy of tubewell sinking continued and continues even today. In 1993, more contaminated tubewells were identified by the Government but again there was no publicity about the findings; no remedial action was taken.

As part of its integrated health care programme, dermatologists from Dhaka Community Hospital (DCH) who were visiting rural sites to see referred patients were reporting unusual numbers of cases of hyperpigmentation, melanosis and keratosis. They were suspicious that the cause may be arsenic poisoning and looked for further cases in the areas. Patients were produced from neighbouring villages who had been to Calcutta, India for treatment of their dermatological conditions, which local health workers had been unable to diagnose or treat. They confirmed that they had been diagnosed as suffering from the effects of chronic arsenic toxicity. Alarmed by this, DCH contacted The School of Environmental Sciences, Jadevpur University, Calcutta (SOES) who proposed that from their studies in West Bengal, it was highly probable that significant areas of Bangladesh aquifers would also be contaminated with arsenic.

DCH lobbying campaigns led to Government action. With funding from the United Nations Development Programme (UNDP) and under the co-ordination of the World Bank, DCH because of the experience and knowledge it had already gained about arsenic and because of its known capacity to work at the field level throughout Bangladesh, was appointed by the Government to conduct a Rapid Action Programme (RAP). The aim was to gain an understanding of the extent and magnitude of the problem. The programme was carried out between November 1997 and April 1998.

The primary objective of this emergency programme was to provide immediate relief to the areas known to be the most affected in Bangladesh. Patients were to be identified and all tubewells located and marked as contaminated or uncontaminated. Based upon the recommendations of the local residents, the best options for arsenic-free water were to be identified and provided. It was thus hoped to prevent these people from further exposure to arsenic. The results of the RAP were to be used to provide a basis for future actions.

METHODOLOGY

Two hundred villages were selected based on the West Bengal hydro-geological studies of ground water contamination in 34 districts of the south western part of Bangladesh. These consisted of villages that had already been identified by various organisations as having contaminated tubewells. The adjacent villages were also included in the sample. They were all situated in twenty southern districts of Bangladesh and thought to be highly affected.

The programme was implemented by groups of community health workers, trained to conduct face-to-face interviews and to identify the various skin lesions associated with arsenicosis. They also received training in mapping techniques. One member in each team was trained to use field test kits for the detection of arsenic in water. A physician supported each group.

Two structured, partially open-ended interview schedules by face-to-face interview were used: one to determine the population and health of the household and arsenicosis status of the members; the other to determine water sources used by the household for different purposes, water uses and sanitation arrangements. Household members who were present when the community worker visited were screened for arsenicosis. A return visit was made to screen those absent at work during the first visit and information on those not present but members of the household, was obtained from the respondent. The doctor in each group confirmed the identification of patients by physical examination.

The following steps were taken:
- Mapping of the village.
- Registration of households.
- Screening of household members for arsenicosis.
- Testing of the tubewell by field kit method: 10% of samples were re-tested in the Government Department of Public Health Engineering laboratory to validate the findings.
- Distribution of health cards to households and water testing cards to tubewell owners.
- Colouring tubewells: contaminated tubewells were painted red.
- Management of arsenicosis patients: the physician provided symptomatic treatment and this was followed by a DCH health visit when, as well as attending patients, the local health workers were given information about arsenicosis symptoms and care.
- Spotting tubewells on the village map.
- Spotting other water sources on the village map: including deep tubewells, rivers, dug wells, ponds, canals, 'beals' etc.
- Community awareness programmes.
- National database: all data collected was submitted to the National Arsenic Mitigation and Information Centre to be used to form a national database.

During the three days or so that a group was in a village a series of educational programmes were undertaken to promote knowledge e.g. that skin lesions were the result of arsenic poisoning and not contagious; that drinking or using tubewell water for cooking could cause arsenicosis but using it for washing or cleaning purposes was not injurious; contaminated water needs to be managed as a community issue with responses worked out locally etc.

RESULTS

1. The survey covered a population of 469,424 in 91,787 households.
2. A total of 32,651 tubewells were tested, of which approximately 11% were Government owned; the remainder were sunk either privately or by non-government organisations.
3. The field kit test detected the presence of arsenic only above the limit of 0.1 mg/l, while the Bangladesh Government limit is currently 0.05 mg/l.
4. Approximately 62% of the tubewells that were tested (damaged tubewells (<2%) were not tested) were arsenic positive (i.e. ≥0.1 mg/l).
5. A total of 1,802 patients were detected with skin lesions: 54.99% were male, 35.96% were female and 9.05% were children below the age of 15 years. In similar studies in West Bengal only 1.7% of patients were found below the age of 15 years.
6. There is still a large number of the population using surface water for their cooking. 47% of people are using pond water for washing and 45% for household work, and 31% are using pond water for cooking.
7. No village was found without contamination. The contamination range was from 5% to 100% of the tubewells.

DISCUSSION

In a pilot phase, maps were obtained from Government sources but when these were used at field level, it was found that some villages had expanded so much that they had designated themselves as two villages e.g. Old X and New X, the latter not mapped. Some villages had been reconstructed and bore no resemblance to the original maps. Similarly, Government sunk tubewell locations were recorded and the existing maps were used to locate them but non-government tubewells had to be found by using local knowledge.

Previous estimates of the total number of tubewells in Bangladesh have been based on Government figures. This survey found, on average, 160 tubewells per village, which if applied to the whole of Bangladesh, would mean approximately 11 million tubewells nation-wide, almost doubling the previously estimated number of tubewells in the country. Since the RAP was conducted, more districts have been identified with arsenic-contaminated water, making the task of testing each well formidable, yet all negative wells will need periodic testing to ensure the well-being of the users.

The water field test kits had limitations. The one used in the RAP only tested positive above arsenic levels of 0.1 mg/l but it was reliable. The original programme design was to test for levels above 0.05 mg/l in accordance with the Government's recommendation. Some kits do have lower detection levels and are reliable but they have the disadvantage of not being user-friendly. Bromide burns were reported and arsenic gas emission proved hazardous for field testers. The Bangladesh climate is hot and humid and not all kits perform reliably under these conditions.

Mass training in patient identification for all health workers is required. More awareness exists now than before the RAP but, equally, dermatological patients are panicking that any lesion is from arsenic poisoning. Clear diagnostic guidelines are needed especially as biological sample testing is only available in Dhaka and too expensive for most patients.

From the RAP, the prevalence of arsenicosis appears to be 3 in 1000 population. The total population is approximately 120 million and while at the time of the RAP, it was thought that only a limited number of geographical areas were affected, it is now reported that the major part of the country is contaminated. The number of pediatric patients is comparatively high. Many of those seen had passed the 'reversible' symptom stage and their prognosis is poor.

Certain unexpected results were found. In some villages, one tubewell would have a positive result while the neighbouring tubewell, only a few metres distant, would register as negative, even when the water was laboratory tested by atomic absorption spectro-photometer (<0.05 mg/l). In some households, all members were using the same water sources but only one member showed symptoms. These anomalies could not be investigated further as they were outside the scope of the RAP.

CONCLUSIONS

Further investigation is obviously required to understand the effects of arsenic poisoning. Many hypotheses are being proposed about why the population of Bangladesh is apparently so susceptible to arsenic; why it is suffering such advanced symptoms of arsenic toxicity. The effects of a poor nutritional status or possible genetic factors are frequently mentioned. Whatever the reasons are, the consequences of prolonged ingestion of even low doses of arsenic have profound implications for the country.

Research takes time and the concern now has to be to provide safe drinking and cooking water to the population. A reliable, safe, easy-to-use field test kit is needed that may be distributed to households so that they are able to monitor their own tubewells and water sources. It was previously thought that 97% of the population relied solely on tubewell water for drinking and cooking. The RAP revealed that in reality other traditional water sources are still being used even if in small numbers. For the immediate future a return to traditional water sources with care being taken about water quality, along with the use of arsenic-free tested tubewells may be the only practical solution.

ACKNOWLEDGEMENTS

The authors wish to acknowledge the following for their help in this study: the 200 communities, for their co-operation; the field workers and physicians and Abdur Rashid Khan.

Arsenic Exposure and Health Effects
W.R. Chappell, C.O. Abernathy and R.L. Calderon (Editors)

Public Health Hazard Surveillance and Response to Arsenic Contamination

Henry Anderson, Lynda Knobeloch, Charles Warzecha

ABSTRACT

In 1987, it was discovered that a large number of private drinking water wells in two northeastern Wisconsin counties contained arsenic levels that exceeded the federal drinking water standard of 50 micrograms per liter. The source of this contamination is a natural vein of arsenic in the bedrock underlying this region. A year earlier, arsenic-contaminated surface soils were identified in apple and cherry orchards located in a neighboring county. This contamination was attributed to the historical use of lead arsenate as an agricultural insecticide. As a result of these findings, which demonstrated the potential for significant human exposure to toxic forms of arsenic, the Wisconsin Bureau of Environmental Health issued soil contact and drinking water advisories, developed public information materials, organized community meetings, and conducted a health survey.

Keywords: public health, arsenic, drinking water, soil, orchards

BACKGROUND

Surveillance of environmental quality and related human health effects is an important public health function. In the past, public health surveillance consisted primarily of tracking the incidence of communicable diseases, especially those that can be prevented through immunization, improved sanitation, or behavior modification. During the past decade, public health agencies have expanded their activities to include monitoring the prevalence of chronic diseases such as heart disease and diabetes, as well as conditions like asthma, birth defects, and cancer that may be caused or exacerbated by environmental factors.

In Wisconsin, the identification and surveillance of environmental hazards is a collaborative, multi-agency effort. The Department of Natural Resources is responsible for overseeing the quality of our water, soil, and air. In addition, the Department of Agriculture, Trade and Consumer Protection and researchers at the University of Wisconsin conduct a variety of soil and water quality studies. Department of Health and Family Services' Bureau of Environmental Health uses environmental quality data generated by these agencies to evaluate potential health risks. When a significant health risk is identified, it is the Bureau's responsibility to ensure protection of public health. If an imminent hazard exists, the Bureau has authority to order emergency intervention strategies, such as water supply replacement or environmental remediation. In some cases, the Bureau may also conduct exposure and health outcome studies.

In the late 1980s the Wisconsin Department of Natural Resources discovered that surface soils in apple and cherry orchards located in Door County contained high levels of lead and arsenic. This contamination was traced to the historical use of lead arsenate as an insecticide. In an effort to reduce the potential for human exposure to these toxic metals, the Bureau of Environmental Health issued soil contact advisories for lead and arsenic. Bureau staff also prepared educational materials that were distributed to landowners and concerned citizens and presented information on health risks and exposure reduction strategies at community meetings.

In 1987, Department of Natural Resources' water quality experts found that an ancient bedrock formation that extends beneath more than 20,000 private water supplies in northeastern Wisconsin was slowly releasing arsenic into the groundwater (see Figure 1). Widespread water quality testing in the area confirmed that arsenic concentrations in many of these wells exceeded the federal drinking water standard. In response to this discovery, the Wisconsin Bureau of Environmental Health took action to reduce human ingestion of arsenic-contaminated drinking water and conducted a study to assess the health effects of past exposure. This article summarizes these cases and discusses effective strategies for dealing with environmental health emergencies.

METHODS

Surface Soil Study

Soil samples were collected at 34 sites in Door County by Department of Natural Resources staff and consultants. Sampling sites were selected primarily to include pesticide mixing areas, but also included orchards and background locations. Soil borings extended to a depth of five feet or to bedrock, and were collected continuously in approximately 0.5-ft increments. Samples were submitted to a state-certified laboratory for lead and arsenic analysis.

Private Well Study

Press releases that encouraged private well owners to submit a water sample for arsenic analysis were published in local newspapers. Families that requested sample kits were asked to complete a health survey and sign a release that authorized investigators to access their

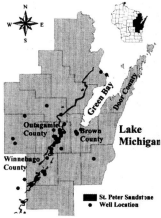

Fig. 1. Illustration of Northeastern Wisconsin showing St. Peter Sandstone formation and the location of private drinking water wells that contained >50 μg/L arsenic.

arsenic well test result. Most of the surveys were returned before the families received their water test results minimizing potential recall bias. Arsenic analyses were performed by state-certified laboratories. Survey and water test data were analyzed using EpiInfo 6.01 software developed by the U.S. Centers for Disease Control and Prevention.

RESULTS

Soil Contamination

Surface soil samples collected from existing and abandoned orchards in Door County had arsenic concentrations ranging from <20 mg/kg (limit of detection) to 140 mg/kg (RMT, 1987). Arsenic levels were significantly higher in apple orchards than in cherry orchards, and were highest in pesticide mixing areas (see Table 1). Most lead levels ranged from 25 mg/kg to 520 mg/kg, however a concentration of 48,000 mg/kg was detected in one isolated hot spot. Inorganic forms of arsenic and lead are relatively immobile in soil. To evaluate the potential for these metals to leach to groundwater, lead and arsenic were measured at varying depths. It was found that lead levels were highest in the top six inches of surface soil, while arsenic levels were highest at a depth of 6 to 12 inches below the surface (see Table 1).

TABLE 1

Summary of lead and arsenic levels in Door.County orchards (RMT, 1987)

	Depth (ft)	No. samples	Lead	Arsenic
Mixing areas	0–0.5	136	388	84
	0.5–1.0	7	103	103
	1.0–1.5	9	85	21
	1.5–2.0	17	21	<20
Apple orchards	0–0.05	16	360	62
	0.5–1.0	5	81	35
	1.0–1.5	4	<20	<20
	1.5–2.0	6	<20	<20
Cherry orchards	0–0.5	8	51	<20
	0.5–1.0	0	—	—
	1.0–1.5	0	—	—
	1.5–2.0	1	<20	<20

All values are geometric means in mg/kg.

In response to concerns about exposure to lead and arsenic through intermittent contact with surface soil in and around orchards, the Bureau of Environmental Health issued soil contact advisories of 400 mg/kg for lead and 100 mg/kg for arsenic (Bureau of Environmental Health, 1997). The advisory for lead was based on the U.S. EPA's determination that chronic exposure to soil that contains less than 400 mg lead per kg does not significantly alter blood lead levels in preschool-aged children. The arsenic advisory assumes a maximum soil intake rate of 100 mg/day and allows daily ingestion of 10 micrograms of arsenic. This intake level is lower than the average daily intake arsenic estimate from fruits and vegetables and is assumed to be safe for people of all ages.

After these advisories were issued, concerns were raised about conversion of orchards to other land uses, such as residential, commercial and recreational property. In response to this concern, the Bureau recommended that arsenic levels not exceed site-specific background levels (usually <10 mg/kg) on land that was being converted to residential use. Although a lower level could be justified based on the toxicity of inorganic arsenic, it was judged impractical and unnecessary to require cleanup to levels that were significantly below those that exist naturally in the area.

Groundwater Contamination

In 1987, a groundwater study conducted by the Wisconsin Department of Natural Resources, Green Bay Headquarters revealed a large vein of arsenic in a bedrock layer found at the interface of the St. Peter Sandstone and Sinnipee Dolomite (Stoll et al., 1995). This geologic formation stretches from southern Brown County into Outagamie and Winnebago Counties and lies beneath more than 20,000 private water supply wells. Water samples from 1943 private wells in the Fox River Valley, one of the fastest growing regions of the state, contained arsenic concentrations that ranged from 1 to 1,200 μg/L. Levels exceeded 5 μg/L in 622 (32%) of these wells, and 68 (3.5%) of the wells had arsenic concentrations that exceeded the federal standard of 50 μg/L.

In response to this finding, the Bureau of Environmental Health developed a public education campaign and conducted a family health survey. One component of the education campaign was the development of a fact sheet on the health effects of arsenic-contaminated drinking water. Water testing laboratories mailed these fact sheets to families along with their arsenic test result and county health officials distributed them to residents in the affected communities. In addition to this effort, Bureau staff participated in a series of public meetings and worked with the Department of Natural Resources and county health officials to coordinate water testing and the distribution of bottled water.

To assess the health impact of past exposure to arsenic-contaminated drinking water, the Bureau conducted an area-wide health and drinking water study. Self-administered surveys were used to collect information about individual water use habits and health status. These surveys were mailed to each family that requested a water sample kit. To minimize recall bias, surveys were sent along with water collection kits and were, in most cases, returned before families received their water test result. Surveys were returned by 637 families (64% response rate). These surveys provided water intake and health status information for 1,623 individuals who had lived in their homes for more than one year. Daily arsenic intakes were calculated by multiplying the self-reported water intake rates by the arsenic concentration in the well water. Using this method, it was estimated that 1,233 respondents had ingestion rates of less than 5 μg of arsenic per day from water, 390 had intakes of 5–49 μg/day, and 45 had intakes that exceeded 49 μg/day (Haupert, 1994; Haupert et al., 1996).

Comparison of daily arsenic intake levels and illness rates found that people who ingested more than 49 micrograms of arsenic per day were significantly more likely to report skin cancer, kidney problems, tremors, and unexplained hair loss. Non-cancerous skin changes and gastrointestinal illness rates were not correlated with arsenic exposure (see Table 2).

TABLE 2

Relative risk of adverse health effects related to arsenic intake

Health effect	Arsenic Intake	
	5–49 μg/day No. cases (RR, 95% CI)	≥50 μg/day No. cases (RR, 95% CI)
Skin cancer	5 (0.77, 0.13–1.42)	5 (3.28, 2.17–4.40)
Kidney problems	6 (0.64, 0.19–1.09)	2 (1.98, 1.23–2.73)
Unexplained tremors	1 (0.30, 0.00–1.44)	1 (2.66, 1.50–3.72)
Unexplained hair loss	10 (0.31, 0.00–0.72)	3 (4.68, 1.47–14.83)
Stomach upset	21 (0.57, 0.32–0.82)	1 (0.22, 0.00–1.23)
Diarrhea	14 (0.65, 0.35–0.75)	1 (0.36, 0.00–1.38)
Skin discoloration	7 (0.73, 0.31–1.15)	1 (1.16, 0.13–2.19)

These findings must be interpreted cautiously, however, since the number of exposed individuals and illnesses is quite small.

DISCUSSION

Over the past twenty years, environmental health has become an increasingly important part of the public health agenda. It has expanded beyond the traditional food sanitation and radiation environmental health programs to include natural and man-made chemical contamination. Private water supplies and indoor environments have been of increasing importance.

Apple and cherry production have been important to the Door County economy since the late 1800s. At its peak, Door County had 10,500 acres of cherry orchards, and 2,200 acres of apple orchards. This industry was heavily dependent on pesticides to control insects that threaten these crops. Lead arsenate was virtually the sole insecticide used until World War II when DDT and other organic insecticides became available. The recent popularity of Door County as a vacation destination and summer retreat for the metropolitan areas of Milwaukee and Chicago has increased pressure to convert agricultural land to other uses. Many abandoned orchards are now the sites of single-family homes, condominiums, restaurants and parks. This evolving land use scenario created the potential for public exposure to arsenic-contaminated soils.

Door County's case is not unique. Across the United States, population has grown disproportionately in rural and suburban areas, many of which are not served by municipal water systems. When that occurs, and the area has naturally-contaminated water (arsenic, radon, fluoride), or contamination due to human activity (nitrate, pesticides, VOCs, gasoline) it is imperative that the contamination be identified and measures taken to prevent exposures. Public health hazard surveillance using computerized environmental databases is important in identifying potential exposures and targeting a public health response.

Unfortunately, public health has had to focus upon reacting to contamination rather than being actively involved in zoning and land use planning to prevent the exposures from occurring. Once homes are built, the remedy options are limited. Costs of whole-home water treatment or the drilling of a new well are often prohibitive and long-term maintenance of a bottled water program difficult. While public health advisories are an effective way to inform the public, compliance is often less than desired. Compliance is especially difficult when the contaminant does not result in a taste or odor problem or acute health effects. Public health programs need to become more proactive in primary environmental exposure prevention planning.

REFERENCES

Bureau of Environmental Health. 1997. Memo from Charles Warzecha to Kathy Erdman/Rick Stoll, Door County Lead Arsenate Consultation, September 25, 1997.

Haupert, T.A. 1994. Arsenic in Drinking Water Supply Wells in Outagamie and Winnebago Counties, WI: Study of Possible Health Effects. Masters Thesis, UW-Green Bay, May 1994.

Haupert, T., Wiersma, J.H. and Goldring, J. 1996. Health effects of ingesting arsenic-contaminated groundwater. *Wisc. Med. J.*, Feb. 1996.

RMT, Inc. 1987. Door County Lead Arsenate Investigation.

Stoll, R., Burkel, R., and LaPlant, N. 1995. Naturally occurring arsenic in sandstone aquifer water supply wells of Northeastern Wisconsin. Wisconsin Groundwater Research and Monitoring Project Summaries. WI Dept Natural Resources PUBL-WR-423-95.

Arsenic Exposure and Health Effects
W.R. Chappell, C.O. Abernathy and R.L. Calderon (Editors)
© 1999 Elsevier Science B.V. All rights reserved.

Full-scale Application of Coagulation Processes for Arsenic Removal in Chile: A Successful Case Study

Ana María Sancha

ABSTRACT

The presence of arsenic in some drinking water supplies in the North of Chile has a natural origin due to the hydrogeologic characteristics of the area, which has a predominance of quaternary volcanism. The impacts on the health of the population supplied with this water during the sixties pushed the Chilean authorities to study the problem and develop a solution. In 1970 the first of the four water treatment plants for arsenic removal presently in operation was built. The arsenic is removed with coagulation processes. Coagulation converts soluble As into insoluble reaction products facilitating their subsequent removal from water by sedimentation and filtration. The general treatment process involves oxidation (pre and post treatment), pH adjustment and ferric or alum coagulant. Operational considerations that are important for removing arsenic include effluent turbidity and filter run length. Turbidity removal is a prerequisite for efficient arsenic removal. The arsenic level in finished water is 0.04 mg/L. The surface raw water is characterized by high hardness, salinity, and alkalinity, and a low turbidity. Orthophosphate and natural organic matter (NOM) have not been currently detected. Arsenic occurs mainly in the As(V) and As (III) oxidation states with As(V) dominant. Organic species (methylated arsenic) are rarely present and are considered of little significance compared with inorganic species. The As concentration in the raw water is in the range 0.40–0.60 mg/L. The new WHO guidelines for As, 0.01 mg/L, presents several challenges to Chile. Our work with coagulation treatment processes for their potential to satisfy a more stringent standard concludes that it is not possible to lower arsenic concentrations to below 0.03 mg/L using this conventional technology. We face the challenge of developing a technology that clearly protects the public's health but, at the same time, is not prohibitively costly. The paper will highlight and analyze the background and key issues in the Chilean arsenic removal technology including some technical details of Chilean water utilities.

Keywords: arsenic removal, water treatment, Antofagasta-Chile, hydroarsenicism

TABLE 1

Salar del Carmen treatment utility, Chile. Representative raw water quality

Parameter	Value (mg/L)
pH	8.0–8.4
Alkalinity ($CaCO_3$)	100.0–120.0
Hardness ($CaCO_3$)	130.0–150.0
Chloride (Cl)	120.0–140.0
Sulfate (SO_4)	80.0–100.0
Total Dissolved Solids	700.0–800.0
Silica (SiO_2)	40.0–50.0
Boron (B)	3.0–4.0
Fluoride (F)	0.30–0.50
Arsenic (As)	0.40–0.60

INTRODUCTION

Drinking water supplies for the population of the north region of Chile contain arsenic as a result of the geochemical characteristics of this zone, since rivers which originate in the Andes springs have developed their hydrographic basins in volcanic materials. (Henriquez, 1968, Henriquez, 1978). Arsenic occurs, principally, as an inorganic form with predominance of the pentavalent specie (Sancha et al., 1992). The arsenic-contaminated water presents, in general, a low level of turbidity and natural organic materials and high levels of alkalinity, chloride, hardness, sulfate, boron fluoride and silica (Table 1).

Before water treatment utilities for arsenic removal were set up in Chile, skin lesions were detected especially in children, also some serious cardiovascular cases were reported (Merino, 1965; Borgoño et al., 1971; Puga et al., 1973; Rosenberg, 1974; Zaldivar 1974; Moran et al., 1977; Borgoño et al., 1977).

The health effects reported during the sixties forced the Chilean authorities to worry about the problem and search for a solution with the participation, in the first stage, of German researchers. Thus in the seventies the first two utilities for arsenic removal were built in Chile. Later on, in the eighties, based on studies carried out by Chilean researchers, two new utilities were built and the two existing ones were improved to produce water with a maximum arsenic level in the range of 0.040–0.050 mg As/L. (Sancha, Ruiz et al., 1984).

TECHNOLOGIES FOR ARSENIC REMOVAL

The technologies available for arsenic removal from water supplies include chemical precipitation through coagulation–filtration and softening, activated alumina or carbon adsorption, ion exchange and processes with membranes such as reverse osmosis, electrodialysis and nanofiltration. Most of these technologies have not been tested at full-scale, with present experience only at the laboratory and pilot-plant scale. When these advanced technologies are tested in a more realistic scenario with complex matrix water such as hardness, sulfate, total dissolved solids, selenium, nitrate, fluoride, chloride, orthophosphate natural organic materials and suspended materials, some of these technologies might present important limitations and the removal efficiency may drop significantly (Hathaway and Rubel, 1987; Joshi and Chandhuri, 1996; Waypa et al., 1997; Clifford et al., 1986; Shen, 1993; Gupta and Chen, 1978; Gulledge et al., 1973; McNeill et al., 1995; Bellack, 1971).

The only technology for which there is experience at full scale is coagulation–filtration; these processes have been used in Chilean water treatment utilities for arsenic removal since the seventies. Coagulation–filtration processes remove arsenic through chemical sorption and particle removal. During coagulation the addition of Al or Fe coagulants, facilitates the conversion of soluble inorganic As species into insoluble reaction products that are formed by mechanisms of precipitation, coprecipitation or adsorption. The formation of these insoluble products facilitates subsequent removal by means of sedimentation and filtration processes. In this way, the efficiency of arsenic removal by coagulation will depend on the formation of insoluble products, arsenic adsorption on them and the removal of the resulting material. Any problem that may arise in these processes will limit the efficiency of the arsenic removal process.

FULL-SCALE APPLICATION OF COAGULATION PROCESSES FOR ARSENIC REMOVAL IN CHILE

Surface water resources for main cities in the North of Chile like Antofagasta and Calama with an approximate population of 400,000 inhabitants are naturally contaminated with arsenic. The raw water—a mixture of the waters of the Toconce, Lequena, Quinchamale and Siloli-Polapi rivers—has an As concentration in the range of 0.40–0.45 mg/L. At the present time, the four water treatment utilities for arsenic removal existing in Chile treat 1730 L/s (Table 2). In these utilities arsenic is removed from water using a technology based on the coagulation– filtration processes with preoxidation, double filtration and post-oxidation (Figure 1).

The purpose of preoxidation or prechlorination is to ensure that all the arsenic to be removed is in the pentavalent form because this specie is more efficiently removed (Sorg and Logsdon, 1978; Sancha et al., 1992). The post oxidation or post chlorination ensures that residual arsenic is As(V) a less toxic specie (WHO, 1981). The purpose of double filtration, where the latter is like a post-treatment, is to improve the removal of particulates formed during the coagulation processes because residual arsenic is directly proportional to residual turbidity which is formed by the particulate matter and metallic hydroxides on which arsenic has been sorbed.

Experience has demonstrated that the most important parameters in the arsenic removal processes are: arsenic speciation, raw water pH, oxidant and coagulant doses, adsorption time, gradient and flocculation time, floc sedimentation and filtration rate, filtering media and filter run length (Sancha, Ruiz et al., 1984). Aluminum sulfate was initially used as the coagulant and now ferric chloride (30 mg/L) is used. The coagulant based on Fe salts is much more effective in removing arsenic than aluminum-based coagulants. With aluminum salts, the dose required to remove As from water is higher than with iron salts, the resulting flocs with Al are smaller, slighter and they take longer to settle (70–75 m^3/m^2/day) in a large pH range (6.5–8.0). No lime addition is necessary since the water is already hard. During some

TABLE 2

Water treatment utilities for arsenic removal, Chile

Utility	Capacity (L/s)	Water sources	As (mg/L)
Complejo Salar del Carmen		R. Toconce	0.600–0.900
Planta Antigua (1970)	500	R. Lequena	0.150–0.350
Planta Nueva (1988)	520	Quinchamale	0.100–0.250
		Siloli Polapi	< 0.050
Cerro Topater (1978)	500	R. Toconce	0.600–0.900
		R. Lequena	0.150–0.350
		Quinchamale	0.100-0.250
Chuquicamata (1989)	210	R. Colana	0.070–0.090
		R. Inacaliri	0.080–0.090

Fig. 1. Arsenic removal from surface water, Salar del Carmen Water Treatment Plant, Antofagasta, Chile.

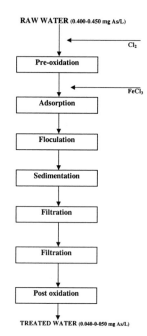

RAW WATER (0.400-0.450 mg As/L)

Cl₂

Pre-oxidation

FeCl₃

Adsorption

Floculation

Sedimentation

Filtration

Filtration

Post oxidation

TREATED WATER (0.040-0-050 mg As/L)

periods of the year it is necessary to adjust the pH because raw water presents daily and hourly pH variations in the range 8–9.

Chilean utilities for As removal operate at a filtration rate of 140–180 m^3/m^2/day. This rate is lower than the values used in utilities that remove turbidity, 3% of the finished water is used in washing coagulants and filter units. The treatment of raw water with As content within a range of 0.400–0.450 mg/L produces drinking water with residual As in the range 0.040–0.050 mg/L.

A recent study carried out in Chile (Sancha et al., 1997) showed that a careful operation of these water utilities, consisting of an appropriate adjustment of raw pH and coagulant dosage would allow As residual levels to reach 0.03–0.02 mg/L. With coagulation technology it is impossible to reduce this residual level any further.

COST OF ARSENIC REMOVAL IN CHILE

The last arsenic removal water utility of Complejo Salar del Carmen (Q = 520 L/s) built in 1988 in Chile had a cost of approximately 20 million US dollars. In this utility the current cost of producing drinking water with 0.040 mg/l residual As is 0.04 US $/m^3$. This cost does not take into consideration the cost of handling and final disposal of the arsenic residual produced in the treatment.

Recent studies made in Chile (González, 1997) concluded that if this country adopts the new WHO recommendation for Maximum Contaminant Level for Arsenic in drinking water (0.010 mg/L), it will be necessary to use reverse osmosis as post-treatment to the system that is currently being used in Chile, which could result in a higher cost for the population of the North of Chile.

DISCUSSION AND CONCLUSIONS

In Chile arsenic is removed from water supplies with coagulation/filtration processes using $Al_2(SO_4)_3$ or $FeCl_3$. Arsenic removal has a better performance using $FeCl_3$ as coagulant. The dependence of pH in the efficiency of As removal is greater when Al salts are used as coagulants than when $FeCl_3$ is used. The As (V) is removed from water more effectively than As (III). The oxidation of arsenite to arsenate is a critical process to obtain effective arsenic removal.

The key factors in arsenic removal at Chilean utilities are the pH of the water to be treated, the coagulant dose used and the processes of separation of the flocs that are formed. There exists a good correlation between particulate (turbidity) removal and arsenic removal. The efficiency of the arsenic removal process can be improved by pre-conditioning the raw water pH and optimizing the addition of chemical agents (Cl_2 and Fe Cl_3) and the separation of the flocs.

Arsenic removal from drinking water sources is not the end of the contamination problem. It is necessary to have a system to adequately dispose of the sludge generated in the treatment without producing a new contamination problem. In the Chilean case, the sludge and washing water are disposed of in the desert in ponds that are especially conditioned for this purpose.

For the maximum level of As in drinking water recommended by organizations such as WHO and USEPA the feasibility of reaching the proposed value at a reasonable cost for the countries affected by this problem should be considered. Countries require solutions that, together with protecting the people's health are at the same time technically effective and economically and socially adequate. In the case of Chile, recent studies show that to fulfill the WHO quality goals would mean to double the current water price for the consumers in the North of the country.

ACKNOWLEDGMENTS

This work was supported by the Universidad de Chile and the Comisión Nacional de Investigación Científica y Tecnológica (CONICYT/FONDEF). The author thanks her students for all their efforts in the development of the arsenic removal tests.

REFERENCES

Bellack, E. 1971. Arsenic removal from potable water. *JAWWA*, **62** (7), 454.

Borgoño, J.M. and Greiber, R. 1971. Estudio Epidemiológico del Arsenicismo en Antofagasta. *Rev. Med. Chile*, **9**, 702–709.

Borgoño, J.M., Vincent, P., Venturino, H., and Infante, A. 1977. Arsenic in the drinking water of the city of Antofagasta: epidemiological and chemical study before and after the installation of the treatment plant. *Environ. Health Perspect.*, **19**, 10–105.

Clifford, D., Subramanian, S., and Sorg, T.D. 1986. Removing dissolved inorganic contaminants from water. *Environ. Sci. Technol.*, **20** (11), 1072.

González, N. Karla. 1997. Alternativas de Remoción de Arsénico desde Fuentes de Agua Potable en la II Región Antofagasta y Costos Asociados. Memoria para optar al título de Ingeniero Civil, Universidad de Chile.

Gulledge, J.H. and O'Connor, J.T. 1973. Removal of arsenic (V) from water by adsorption on aluminum and ferric hydroxides. *JAWWA*, **65** (8), 547.

Gupta, S.K. and Chen, K.Y. 1978. Arsenic removal by adsorption. *JWPCF*, 493.

Hathaway, S.W. and Rubel, F. Jr. 1987. Removing arsenic from drinking water. *JAWWA*, **79** (8), 161.

Henríquez A.H. 1968. Causas del Alto Contenido de Arsénico en los Ríos Toconce y Hojalar. Instituto de Investigaciones Geológicas. Santiago, Chile.

Henriquez A.H. 1978. Misión relativa al Programa de Cooperación Subregional sobre el Arsénico y otros Contaminantes en el Agua en relación con el Volcanismo Cuaternario (Argentina, Bolivia, Chile y Peru). UNESCO. Montevideo. Documentos Técnicos en Hidrología.

Joshi, A. and Chandhuri, M. 1996. Removal of arsenic from ground water by iron oxide-coated sand. *J. Environ. Eng.*, **122** (8).

McNeill, L.S. and Edwards, M. 1995. Soluble arsenic removal at water treatment plants. *JAWWA*, **87** (4) 105.

Merino, R. 1965. Consecuencias Epidemiológicas de Aguas Contaminadas con Arsénico en México referidas al Agua Potable de Antofagasta. *Cátedra de Ingeniería Sanitaria*. Escuela de Salubridad. Universidad de Chile.

Moran, S., Maturana, G., Rosenberg, H., Casanegra, P., and Dubernet, J. 1977. Occlusions coronariennes lives a une intoxication arsenicale chronique. *Arch. Mal. Coeur*, **70** (10), 1115–1120.

Puga, F., Olivos, P., Greinber, R. et al. 1973. Hidroarsenicismo crónico. Intoxicación Arsenical Crónica en Antofagasta. Estudio Epidemiológico y Clínico. *Rev. Chilena Pediatría*, Vol. 44 N. 3.

Rosenberg, H.G. 1974 Systemic arterial disease and chronic arsenicism in infants. *Arch. Pathol.*, **97**, 360–365.

Sancha A.M., Ruiz G. et al. 1984. Estudio del Proceso de Remoción de Arsénico de Fuentes de Agua Potable empleando Sales de Aluminio. *XIX Congreso Interamericano de Ingeniería Sanitaria y Ambiental. Chile, Tema I Agua Potable, Vol II.*

Sancha, A.M., Vega, F., and Fuentes, S. 1992. Speciation of Arsenic present in Water Inflowing to the Salar del Carmen Treatment Plant in Antofagasta, Chile and its Incidence on the Removal Process. *International Seminar Proceedings (183-186). Arsenic in the Environment and its Incidence on Health. Universidad de Chile. Santiago, Chile.*

Sancha, A.M., Vega, F., Fuentes, S. 1992. Efficiency in Removing Arsenic from Water Supplies for Large Towns. Salar del Carment Plant. Antofagasta, Chile. *International Seminar Proceedings (159-163). Arsenic in the Environment and its Incidence on Health. Universidad de Chile. Santiago, Chile.*

Sancha, A.M., González, K. and Pérez, O. 1997. Reduciendo la Exposición de la Población Chilena a Contaminantes. Opciones para Remover Arsénico de Fuentes de Agua Potable de la II Región. *XII Congreso Chileno de Ingeniería Sanitaria y Ambiental. Copiapó-Chile.*

Shen, Y.S. 1993. Study of arsenic removal from drinking water. *JAWWA*, **65** (8), 543.

Sorg, T.J. and Logsdon, G.S. 1978. Treatment technology to meet the interim primary drinking water regulations for inorganics: Part 2. *JAWWA*, **70** (7), 379.

Waypa, J., Elimelech, M., and Hering, J. 1997. Arsenic removal by RO and NF membranes. *JAWWA*, October.

World Health Organization, 1981. *Arsenic Environmental Health Criteria 18.*

World Health Organization. 1993. *Guidelines for Drinking Water Quality.* Second Edition Vol 1. Recommendations.

Zaldivar, R. 1974. Arsenic contamination of drinking water and foodstuffs causing endemic chronic poisoning. *Beitr. Pathol.*, **151**, 384–400.

Arsenic Exposure and Health Effects
W.R. Chappell, C.O. Abernathy and R.L. Calderon (Editors)
© 1999 Elsevier Science B.V. All rights reserved.

Development of an Anion Exchange Process for Arsenic Removal from Water

Dennis A. Clifford, Ganesh Ghurye, Anthony R. Tripp

ABSTRACT

Arsenate (As(V)) anions are readily removed from drinking water by anion exchange with chloride-form strong-base resins. However, the process is not simple ion exchange, because anion resins can convert monovalent arsenate ($H_2AsO_4^-$) to divalent arsenate ($HAsO_4^{2-}$) with the expulsion of a proton and a lowering of pH. A similar but more significant pH change results from the conversion of bicarbonate (HCO_3^-) to carbonate (CO_3^{2-}) within the resin, and these pH changes can affect process design. Another surprising aspect of the chloride-for-arsenate ion-exchange process is that spent NaCl brine, heavily contaminated with arsenic, may be reused directly to regenerate the exhausted ion-exchange bed. Finally, in spite of the low level ($\leq 2\,\mu g/L$) to which arsenic must be treated, empty bed contact times (EBCTs) as short as 1.5 minutes may be employed in the process. In lab studies at the University of Houston and in field studies carried out in McFarland, California, Hanford, California, and Albuquerque, New Mexico, several versions of the anion-exchange process for arsenic removal have been tested at bench and field scale. Currently, the optimal process comprises the use of a type 2 polystyrene divinylbenzene, strong-base anion resin, 1.5 minutes EBCT, 1 M NaCl regenerant, and regenerant reuse up to twenty five times with the chloride concentration maintained at approximately 1.0 M. Typically, 400 to 800 bed volumes of water can be treated before arsenic breakthrough, and the arsenic concentration in the process effluent is consistently below $2\,\mu g/L$. The process will be tested at full scale in Albuquerque in the near future.

Keywords: arsenic, ion exchange, anion exchange, water treatment

INTRODUCTION

MCL for Arsenic in Water

Based on recent studies that suggest that arsenic is a more serious carcinogen than previously thought, USEPA is in the process of revising the current arsenic MCL from 0.05 mg/L to a more conservative MCL in the range of 0.002 to 0.020 mg/L. Because USEPA has categorized arsenic as a Class A Carcinogen, its MCL goal (MCLG) will be set at zero.

Water Quality and Arsenic Speciation in Relation to Removal

Although arsenic can exist in both organic and inorganic forms, only inorganic arsenic in the +III or +V valence state has been found to be significant where potable water supplies are concerned (Andreae, 1977; Irgolic, 1981).

Depending on the redox condition of the groundwater, either arsenite (As(III)) or arsenate (As(V)) forms will be predominant. The pH of the water is also very important in determining the arsenic speciation. The primary arsenate (As(V)) species found in groundwater in the 6 to 9 pH range are monovalent $H_2AsO_4^-$ and divalent $HAsO_4^{2-}$. These anions result from the dissociation of arsenic acid, H_3AsO_4, which exhibits pK_a values of 2.2, 7.0 and 11.5. Uncharged arsenious acid, H_3AsO_3 is the predominant species of trivalent arsenic found in natural waters. Only at pH values above its pK_a of 9.2 does the monovalent arsenite anion, $H_2AsO_3^-$, predominate.

The Chloride Ion-Exchange Process

In the mid 1980s, chloride-form ion exchange with strong-base anion (SBA) resin was shown to be an effective treatment technology for removing 80–100 $\mu g/L$ arsenic from well water in Hanford, California (Clifford, 1990). Because only As(V) is removed by anion exchange in the pH range of natural waters, preoxidation to convert As(III) to As(V) is necessary but pH adjustment is not. The chlorinated and filtered source water is passed downflow through a 1–2 m deep bed of chloride-form, strong-base anion exchange resin, and the chloride–arsenate ion-exchange reaction takes place in the pH 8 to 9 range. (See Eq. (1) in which R represents a positively charged resin exchange site). Regeneration with excess NaCl according to Eq. (2) is readily accomplished, and returns the resin to the chloride form, ready for another exhaustion cycle.

Exhaustion $2\,RCl + HAsO_4^{2-} = R_2HAsO_4 + 2\,Cl^-$ \hfill (1)

Regeneration $R_2HAsO_4 + 2NaCl = 2\,RCl + Na_2HAsO_4$ \hfill (2)

Research by Horng and Clifford (1997) into the behavior of polyprotic anions in strong-base anion (SBA) resins has shown that SBA resins tend to convert monovalent ions into divalent ions, as shown in Figure 1. The example conversion of monovalent $H_2AsO_4^-$ into $HAsO_4^{2-}$ produces one micromole of HCl for each micromole of monovalent arsenic converted to

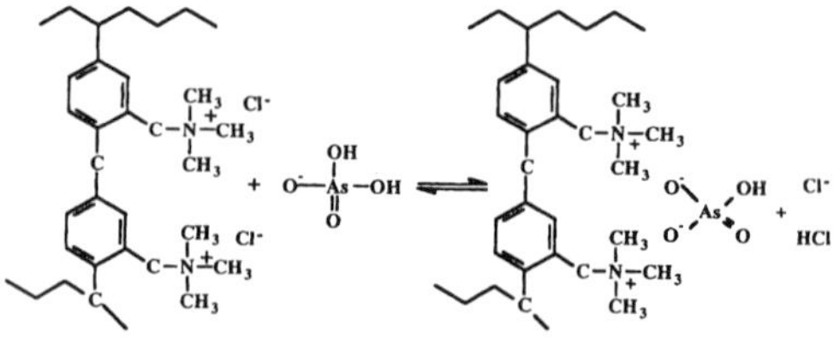

Fig. 1. Arsenic (V) ion exchange with production of acid due to the conversion of monovalent $H_2AsO_4^-$ to divalent $HAsO_4^{2-}$ within the resin.

$$2\,RCl + H_2AsO_4^- = R_2HAsO_4 + HCl + Cl^-$$

divalent arsenic within the resin. The HCl produced diffused out of the resin into the water. Because the concentration of arsenic is so small, the resulting pH reduction is not significant in a typical bicarbonate-buffered ground water. However, when a similar resin-phase reaction occurs during bicarbonate (HCO_3^-) conversion to carbonate (CO_3^{2-}), the pH reduction can be quite significant during the early portions of an ion-exchange run when pHs below 5.0 may be observed (Horng and Clifford, 1997).

Figure 2 is a sketch of a simple ion exchange process for removing arsenic (V) anions from water. Because sulfate is present in all ground waters and is highly preferred by anion resins, the sulfate concentration of a water is a critical water quality consideration when applying an ion-exchange process. For the Hanford, California ground water, which contained only 5 mg/L sulfate, the run length to arsenic breakthrough was very long, about 4,000 BV, when using polystyrene SBA resin and an MCL of 50 μg/L (Clifford, 1990). It was also shown that (a) exhaustion EBCTs as short as 1.4 min could be used, and (b) the spent resin could easily be regenerated with 2–3 bed volumes of 1 M NaCl. The run lengths were progressively (3–5%) shorter, however, with succeeding regenerations, and this was attributed to resin fouling by fine mica particles in the feed water, which was not filtered. A 10-year-later retrospective look at the Hanford results posed several questions: (a) Could ion exchange achieve effluent concentrations consistently below 2 μg/L, (b) could the new nitrate-selective resins be used for combined arsenic and nitrate removal, (c) could waters containing up to 150 mg/L sulfate be treated to reasonable (>350 bed volumes) run lengths, and (e) could the spent regenerant be reused to conserve salt and minimize waste?

To answer these questions, further ion exchange research for arsenic removal was carried out in McFarland, California in 1995 on a 180 mg/L TDS water contaminated with 50 mg/L sulfate and 16 μg/L arsenic. During the same study, water from Hanford, California containing 50 μg/L arsenic was trucked to the Mobile Lab in McFarland for some experiments. In the McFarland–Hanford studies, a conservative MCL of 2 μg/L arsenic was used to determine run length. It was found that effluent arsenic concentration was always below 2 μg/L for typical run lengths, which were in the range of 250–900 BV depending on sulfate concentration. (With sulfate-spiked McFarland water, the arsenic run lengths were 900 and 250 BV for sulfate concentrations of 50 and 220 mg/L, respectively.) The sulfate-selective resins, especially ASB-2, were far superior to the nitrate-selective variety, which is reasonable in light of the fact that arsenic was apparently removed as a divalent ion similar to sulfate. The McFarland arsenic-removal research (Ghurye et al., 1998) led to the confirmation of the selectivity sequence shown in Figure 3, which had been suggested by Horng (1983) based on his lab studies of arsenic (V) removal by ion exchange. The significant influence of sulfate concentration in the ground water on arsenic ion exchange is demonstrated by Figure 4, which is based on equilibrium multicomponent chromatography theory (Helfferich and Klein, 1970) and computer predictions (Horng, 1983; Tirupanangadu, 1996) of arsenic run length (BV) as a function of sulfate concentration in world average ground water. The results

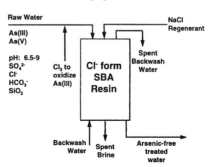

Fig. 2. Schematic diagram of a simple chloride-ion-exchange process for the removal of arsenic (V) from water using strong-base anion resin.

Fig. 3. Typical strong-base-anion resin selectivity sequence for common anionic and neutral constituents of natural ground water.

SO_4^{2-} > $HAsO_4^{2-}$ > CO_3^{2-}, NO_3^- >

Cl^- > $H_2AsO_4^-$, HCO_3^- >>

$Si(OH)_4$, H_3AsO_3

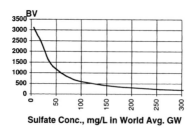

Sulfate Conc., mg/L in World Avg. GW

Fig. 4. The effect of sulfate concentration in ground water on the run length (bed volumes throughput) of a typical ion exchange process for arsenic (V) removal from world average ground water.

are conservative and are based on achieving only 60% of the theoretical run length after allowing for arsenic leakage caused by mass transfer limitations. For this water, which is relatively low in TDS, acceptable run lengths approaching 400 BV may be attained with up to about 150 ppm sulfate in the feed water. As anticipated, as sulfate concentration increases, the run length decreases dramatically.

After the McFarland arsenic studies, the main unanswered questions dealt with ways to improve the efficiency of the NaCl regeneration step, which is the focus of this paper. All subsequent experiments described here were carried out using Ionac ASB-2, type 2, SBA resin with an advertised exchange capacity of 1.4 meq/mL. (No endorsement is intended; similar resins are available from other manufacturers.)

METHODOLOGY

Sampling and Analysis

All experiments were carried out in the University of Houston/U.S. Environmental Protection Agency (UH/EPA) Mobile Lab (Clifford and Bilimoria, 1984) located at the West Mesa Reservoir in Albuquerque, New Mexico. Arsenic samples were preserved with nitric acid and shipped to the University of Houston for analysis by hydride generation atomic absorption spectroscopy using a Perkin Elmer Flow Interruption Analysis (FIAS) System with a detection limit of 0.1 μg/L.

During the course of the research, the Albuquerque West Mesa ground water exhibited the following average concentrations: As (21 μg/L), NO_3–N (1.2 mg/L), Cl^-(11 mg/L), sulfate (70 mg/L), total alkalinity (137 mg/L as $CaCO_3$), total hardness (53 mg/L as $CaCO_3$), pH 8.2, and total dissolved solids (328 mg/L).

Small Column Tests and Buret Regenerations

Ion exchange column exhaustion tests were performed in 2.5-cm (1-in) i.d. glass columns with a resin bed volume of 375 mL and a resin bed depth of 74 cm (29 in). Exhaustion flow rate was set at 250 mL/min, corresponding to an empty bed contact time (EBCT) of 1.5 minutes. To find the "optimum" range of regeneration conditions, the bed of ASB-2 resin was exhausted three times and regenerated twice. The exhausted (to 600 BV, i.e., 2 μg/L arsenic breakthrough) resin was then mixed and divided into 20-mL aliquots and regenerated under various conditions in a 25-mL burette. These regenerations, referred to as "buret-regeneration" tests, were analyzed to determine the "optimum" range of regenerant concentration and amount for further 2.5-cm column exhaustion-regeneration tests.

Following the regeneration optimization testing, the 2.5-cm i.d. columns were exhausted and regenerated at levels of 0.5, 1 or 2 eq chloride/eq resin with NaCl concentrations of 0.5

and 1.0 M at a superficial linear velocity of at least 2 cm/min. Upon completion of a series of exhaustion cycles, the resins were exhaustively regenerated at a regeneration level of 4 eq chloride/eq resin (20.4 lb NaCl/ft³) with 1.0 M NaCl. This was done to ensure that each new set of experiments would be performed with an essentially arsenic- and sulfate-free resin.

Brine reuse experiments were performed with 1 M Cl⁻ at a regeneration level of 2 eq chloride/eq resin. During regeneration, the first 0.5 BV was wasted as displacement rinse. Regeneration was followed by 5 BV of slow rinse using product water at the same flow rate used for regeneration. The first 0.5 BV of slow rinse was collected and added to the spent brine. Make-up salt was added to the spent brine to maintain the chloride concentration at 1.0 M.

Large Column Tests

After the arsenic ion-exchange process with brine reuse had been developed using the small 2.5-cm (1-in) i.d. columns and buret regenerations, the optimum process operating conditions were verified in a series of exhaustions and regenerations carried out in a much larger column. The 15.2-cm (6-in) column tests were performed in a clear acrylic-plastic column with a resin volume of 13.9 L (3.67 gal) and a resin bed depth of 76 cm (30 in). The flow rate through the column was maintained at 9.27 L/min (2.45 gal/min) corresponding to an EBCT of 1.5 min. Regenerant flow rate was 400 mL/min corresponding to a SLV of 2.2 cm/min. During regeneration, the first 0.5 BV was wasted as displacement rinse. Make-up salt was added to the spent brine to maintain the chloride concentration at 1.0 M.

RESULTS AND DISCUSSION

Buret Regeneration Test Results

Effect of Regenerant Concentration

Dilute regenerants (0.5–1.0 M) were more effective than concentrated regenerants (2.0–4.0 M). The amount of salt required to remove 95% of the total arsenic eluted from the resin increased from an average of 1.99 eq chloride/eq resin for 0.25 M NaCl to 5.45 eq chloride/eq resin for 4 M NaCl. The total arsenic eluted from the resin also decreased with an increase in the concentration of regenerant brine. From an average of 110% arsenic removal for 0.5 M NaCl, the arsenic removal dropped to 80% when 4.0 M NaCl was used.

It is known that divalent ion elution from a conventional resin is enhanced by an increase in the concentration of regenerant. This is referred to as electroselectivity reversal (Helfferich, 1962).

The opposite appeared to be true in the buret regeneration tests. Hence, besides regenerant concentration, there seemed to be another factor, regenerant flow rate, that influenced arsenic elution. A more dilute regenerant, while delivering the same amount of salt, also delivered more water through the resin at a higher flow rate. In other words, a more dilute regenerant solution produced greater rinsing effect than a stronger regenerant solution.

To determine the effect of rinsing, three more regenerations were performed with 0.5, 1.0, and 2.0 M NaCl regenerant using a total of only 2 eq chloride/eq resin. In these tests arsenic removal decreased from 102% for 0.5 M NaCl to 41% for 2.0 M NaCl. Once again, dilute regenerants yielded higher arsenic removal.

Effect of Regenerant Flow Rate

Regarding the effect of flow rate on arsenic elution from the exhausted resin, 80% arsenic removal was attained at a regenerant loading of 2 eq chloride/eq resin and a superficial linear velocity (SLV) of 2.0 cm/min while only 36% recovery was attained under the same conditions at a SLV of 0.5 cm/min. The lower arsenic recovery at the slower flow rate suggested that serious channeling was occurring during the slow-flow regenerations, which were undertaken to keep the total regenerant contact time at approximately one hour while reducing the regenerant loading from 8 to 2 eq chloride/eq resin.

Effect of Regenerant Amount

Regeneration studies showed that 0.5 and 1.0 M regenerant solutions at 2 eq chloride/eq resin gave essentially complete arsenic elution. Hence, in subsequent 1-inch column exhaustion tests, both 0.5 and 1.0 M regenerant solutions at 0.5, 1.0, and 2.0 eq chloride/eq resin were chosen for further investigation to actually quantify the effect of regeneration variables on arsenic elution and leakage in the product water and run length to arsenic breakthrough during cyclic column operation.

Small Column Exhaustion Test Results

Once a range of optimum regeneration conditions was developed using minicolumn studies, the range of regeneration conditions was verified using cyclic exhaustion-regenerations in 2.5-cm columns.

Arsenic Leakage

For the same regeneration level, comparable arsenic leakages were obtained with 0.5 and 1.0 M NaCl. At a regeneration level of 0.5 eq chloride/eq resin, excessively high arsenic leakage was produced for both 0.5 and 1.0 M NaCl. At regeneration levels of 1 and 2 eq chloride/eq resin, the arsenic leakage remained below the target MCL of 2 μg/L at all times for both 0.5 and 1.0 M regenerants.

Arsenic Run Length

Once again, for the same regeneration level, there was no significant difference in arsenic run length between 0.5 and 1.0 M NaCl regenerants. However, run length to arsenic breakthrough increased significantly with increasing regeneration level for both 0.5 and 1.0 M regenerants.

Using 1.0 M NaCl, run length to arsenic breakthrough increased from 340 to 640 BV when the regeneration level was increased eight-fold from 0.5 to 4.0 eq chloride/eq resin. It is important to note that increasing the regeneration level caused a less-than-proportionate increase in the run length. That is, the number of bed volumes to arsenic breakthrough per eq chloride used decreased as the regeneration level increased, leading to increasing in-efficiency in salt usage.

Spent Brine Reuse

In an effort to improve the economics of the ion-exchange process by conserving salt and reducing brine discharge, spent brine was compensated to 1.0 M chloride and reused. Brine reuse was surprisingly successful; it produced low arsenic leakage in the product water and only slightly decreased the run length to arsenic breakthrough.

Although a regeneration level of 1 eq chloride/eq resin produced low arsenic leakage and 460 BV of run length during 1-inch column exhaustion tests for both 0.5 and 1.0 M regenerants, brine reuse experiments were performed at a higher regeneration level of 2 eq chloride/eq resin to ensure a minimum run length of 400 BV in the event that higher sulfate concentrations were encountered in the feed water. In order to minimize regenerant volume and regeneration time, 1.0 M NaCl was chosen as the regenerant. A total of twenty-six exhaustion/regeneration cycles were performed with and without make-up chloride addition.

Fig. 5. Typical performance of the ion-exchange process with direct reuse of spent brine without prior treatment to remove arsenic. Brine fortified with NaCl each cycle to maintain chloride concentration at 1.0 M.

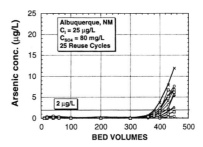

Brine Reuse With Chloride Make-up

The results of the brine reuse experiments are shown in Figure 5. Eighteen exhaustion/ regeneration cycles were performed (17 reuse cycles) with make-up chloride addition. Arsenic leakage in the product water was unexpectedly low; it never exceeded 0.6 μg/L which was equal to or better than the performance of Ionac ASB-2 using fresh brine as regenerant. The run length to arsenic breakthrough decreased slightly from 450 BV for the first exhaustion cycle to 400 BV for the 18th exhaustion, i.e., a 11% reduction in run length to arsenic breakthrough after 17 reuse cycles. For some exhaustion cycles, the run length to arsenic breakthrough was shortened due to unusually high sulfate concentration in the feed water (up to 103 mg/L). After 17 recycles, significant concentrations of arsenic, sulfate and bicarbonate had accumulated in the spent brine to levels of 17,040 μg/L, 151,200 mg/L (1.58 M, 3.15 N), and 24,400 mg/L (0.39 M), respectively. The pH of the spent brine varied in the range of 8.9–9.3.

Thus, spent brine reuse proved extremely effective in substantially reducing salt consumption with no adverse effect on arsenic leakage in the product water. Arsenic leakage in the product water did not exceed 0.6 μg/L for any sample during the 18 exhaustion cycles. However, substantial arsenic concentrations were present in the first 0.5 BV of slow rinse, which will have to be treated along with the spent brine before eventual disposal.

Brine Reuse Without Chloride Make-up

In order to determine if the accumulated bicarbonate (0.39 M) in the spent regenerant could effectively regenerate the resin, four exhaustion cycles were performed without make-up salt addition. Due to excess chloride remaining in the reuse brine, there was no adverse effect on arsenic leakage or run length to arsenic breakthrough for the first exhaustion cycle following regeneration without make-up chloride addition. However, the run length to arsenic breakthrough decreased rapidly thereafter to 200 BV accompanied by an increase in arsenic leakage which reached 11.3 μg/L for the fourth exhaustion without chloride make up. The chloride concentration, as expected, kept decreasing after make-up chloride addition was stopped, and the bicarbonate in the spend brine was not a good regenerant although its concentration varied between 0.39 and 0.25 M.

When make-up chloride addition was resumed, the run length to arsenic breakthrough was immediately restored to 400 BV, and the arsenic leakage decreased to an average of 0.1 μg/L with an arsenic leakage range of 0.0–0.2 μg/L

Large Column Test Results

The successful 2.5 cm (1-inch) column exhaustion tests were scaled up in a 15.2-cm (6-inch) column with spent brine reuse and make-up chloride addition to 1.0 M. Effluent arsenic concentrations remained below the target MCL of 2 μg/L at all times during the fifteen exhaustion cycles performed. The average arsenic leakage during the first 100 BV of these cycles was 0.1 $\pm$ 0.1 μg/L (range of 0–0.9 μg/L). Although arsenic and sulfate concentrations

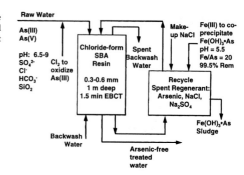

Fig. 6. Schematic of the arsenic (V) ion-exchange process with direct brine reuse and eventual removal of arsenic from spent brine (after about 25 or more cycles) by coprecipitation with Fe(III).

reached 26,600 μg/L and 120,500 mg/L (1.26 M), respectively, arsenic concentration in the ion-exchange product water remained low, and the run length did not decrease below 400 BV.

Process Schematic for Arsenic Ion Exchange with Brine Reuse

Figure 6 is a schematic of a suggested full-scale arsenic ion-exchange process with brine reuse. Shown also in the figure is the addition of ferric iron (Fe(III)) for the eventual coprecipitation and removal of arsenic from the spent recycle brine before it is disposed of. The Fe(OH)$_3$ sludge contaminated with arsenic would be dried and placed in a hazardous waste land fill for final disposal. The arsenic-free brine would be metered into a sanitary sewer or sent to an evaporation pond for final disposal.

CONCLUSIONS

An arsenic ion-exchange process was developed during research in Hanford and McFarland, California and Albuquerque, New Mexico. The research led to the following conclusions:

1. Arsenic removal to below 2 μg/L was achieved on a consistent basis for over 400 bed volumes using (a) chloride-form ion exchange with commercially available sulfate-selective strong-base resins (especially the type 2 resin ASB-2), (b) EBCT of 1.5 min, and (c) a typical bed depth of 76 cm (30 in)
2. The special nitrate-selective resins did not perform as well as the conventional sulfate-selective resins for arsenic removal based on (a) shorter runs to arsenic breakthrough, and (b) greater arsenic peaking after its breakthrough.
3. Spent brine, fortified after each regeneration to maintain the chloride concentration at 1.0 M, was reused without arsenic removal, without any adverse effect on arsenic leakage or run length.
4. Sulfate levels in ground water as high as 114 mg/L gave run lengths beyond 400 BV even when using recycled brine.

ACKNOWLEDGMENTS

This research was funded by the US Environmental Protection Agency, the City of Albuquerque, and the University of Houston. The authors are grateful for the technical and administrative support of Tom Sorg, and Darren Lytle, USEPA research engineers; Kelly Uhlrich, Paul Henderson, Ray deLeon, and Philip Holderness of the McFarland Mutual Water Company; John Stomp, Larry Blair, and Barbara Gastian, Public Works Department, City of Albuquerque; and Norman Gaume, Engineer Advisor, NM Interstate Stream Commission.

REFERENCES

Andreae, M.O. 1978. Distribution and speciation of arsenic in natural waters and some marine algae. *Deep Sea Res.*, **25** (4), 391–402.

Clifford, D.A. 1990. Ion exchange and inorganic adsorption. In: F. Pontius (ed.), *Water Quality and Treatment*, 4th Edn., Chap. 9, pp. 561–639, McGraw Hill, New York.

Clifford, D.A. and Bilimoria, M.R. 1984. A Mobile Drinking-Water Treatment Research Facility for Inorganic Contaminants Removal, PB 84-145 507, NTIS, Springfield, VA 22161, 75 pp., Summary report EPA-600/2-84-018, 6 pp., U.S. EPA, Cincinnati, OH.

Clifford, D.A. and Lin, C.C. 1986. Arsenic Removal From Drinking Water in Hanford, California, Summary Report, Univ. of Houston Dept. of Civil and Environmental Engineering, Houston, TX.

Ghurye, G.L., Clifford, D.A., and Tripp, A.R. 1999. Combined arsenic and nitrate removal by ion exchange and KDF media. *J. Am. Water Works Assoc.*, **9.**

Helfferich, F.G. and Klein, G. 1970. *Multicomponent Chromatography: Theory of Interference*, Marcel Dekker, New York.

Helfferich, F.G. 1962. *Ion Exchange*, McGraw-Hill, New York.

Horng, L.L. and Clifford, D.A. 1997. The behavior of polyprotic anions in ion-exchange resins. *J. React. Funct. Polym.*, **35** (1/2), 41–54.

Horng, L.L. 1983. Reaction Mechanisms and Chromatographic Behavior of Polyprotic Acid Anions in Multicomponent Ion Exchange, PhD. Dissertation, University of Houston, University Park, Houston, TX.

Irgolic, K.J. 1982. Speciation of Arsenic Compounds in Water Supplies, U.S. Environmental Protection Agency. Project Summary. EPA-600/S1-82-010.

Tirupanangadu, M.S. 1996. Development of a Multicomponent Chromatography Program for Predicting Effluent Concentration Histories. M.S. Thesis, University of Houston, Houston, TX.

Arsenic Exposure and Health Effects
W.R. Chappell, C.O. Abernathy and R.L. Calderon (Editors)
© 1999 Elsevier Science B.V. All rights reserved.

Subterranean Removal of Arsenic from Groundwater

U. Rott, M. Friedle

ABSTRACT

In some regions of the world arsenic, as also iron and manganese, is a natural component of the aquifer. In contrast to iron and manganese, which are not very toxic for human beings, arsenic has a high toxicity so that the arsenic contamination of pumped groundwater is affecting the health of millions of people. Much of the total arsenic consists of As(III) which is more toxic than As(V). The maximum admissible concentration for arsenic according to the German guideline for drinking water had been 0.04 mg/L As until 1990. Since then the limit has been 0.01 mg/L As (Rott and Meyerhoff, 1996). In Pabna e.g., a northern district of Bangladesh, a very high arsenic contamination of 14 mg/L was found in the pumped groundwater. In this paper the results of three field studies of large scale plants for *in situ* treatment of groundwater with elevated concentrations of iron, manganese and arsenic are presented. The parameters arsenic and iron, measured in the pumped groundwater, fell below the guideline limits of 0.01 mg/L As and 0.2 mg/L Fe respectively, within the first few treatment cycles. On the other hand, the period of ripening of the manganese removal normally lasts several weeks or months. The reason for the delayed start of the demanganization is the dependence on bacteria which must first adapt to the changed environment. As the duration of treatment continues, the concentration of Mn can fall below the guideline value of 0.05 mg/L. *In situ* treatment of groundwater can be a cost-efficient and reliable alternative for conventional above-ground water treatment. Because of the use of the aquifer as a natural reactor no filter sludge is produced and no above-ground buildings are necessary. In the case of new building or extension of an existing treatment plant, *in situ* processing should always be taken into account.

Keywords: arsenic, iron, manganese, ammonia, water treatment, ground-water supply, *in situ* treatment, groundwater, mobilisation

INTRODUCTION

In connection with the planning of the water supply for three communities in Germany, different variants for the treatment of groundwater with elevated contents of iron, manganese and arsenic are discussed. Field experiments have been conducted from 1994 to the present. The main aim of the experiments that have been carried out by the Institute of Sanitary Engineering, Water Quality and Solid Waste Management of the University of Stuttgart were to prove the transferability of the practical experience with subterranean removal of iron and manganese from groundwater to a similar elimination of arsenic (Rott and Meyerhoff, 1996).

After a short description of the application of *in situ* treatment and the general structure of a treatment plant, the results of field experiments for removal of arsenic under the specific conditions, concerning the raw water quality and the character of the wells and the aquifer, are presented.

METHODOLOGY

By the subterranean removal of iron, manganese and arsenic, the oxidation and filtration processes of conventional above-ground water treatment plants are transferred into the aquifer. Therefore the underground is used as a natural bio-chemical reactor. In this technology, a part of the pumped groundwater is recirculated back into the aquifer carrying an oxidising agent, generally atmospheric oxygen. A simple approach to introduce oxygen into the water is the application of a water jet air pump (Rott and Friedle, 1998).

After the pump, a degasification container is used to purge out the excessive gas. As an alternative, technical oxygen can be used as oxidising agent. Because of the high concentration of technical oxygen, chemical reactions are accelerated compared to air-oxygen.

The oxygen-enriched water is reinfiltrated into the aquifer, using the filter pipes of the production wells. The ratio of the delivered volume and the recharged water volume is called the "efficiency coefficient". This coefficient usually reaches values between 2 and 12, depending on the aquifer- and raw water conditions. The basic configuration of an *in situ* treatment plant is shown in Figure 1.

Figure 2 shows the structure of an *in situ* treatment plant near Paderborn, in the north of Germany. The drinking water supply has a total capacity of about 3.75 Mio. m^3/a. The treatment plant consists of four horizontal filter wells with different aquifer characters. The structure of a horizontal filter well is given in Figure 3. In this case horizontal filter wells are used because of the greater yielding capacity of this kind of well. Because of the great distance to the other wells and the enrichment station of more than 4 km, well IV is equipped with its own oxygenation station.

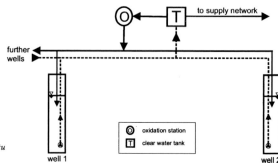

Fig. 1. Basic configuration of an *in situ* treatment plant.

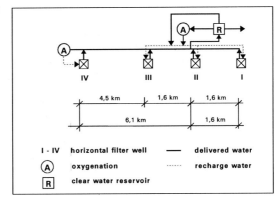

Fig. 2. Scheme of *in situ* plant near Paderborn.

This plant shows that applying *in situ* treatment of groundwater can lower the concentrations of iron, manganese, ammonia, nitrite, nitrate, sulphurhydrogen and organic substances far beyond the drinking water standards. A drinking water supply is thus possible directly from the aquifer without any further above-ground treatment.

Because of the input of oxygen, the redox potential of the water is increased. A number of different physical, chemical and biological processes in the surrounding area of the well screen section, the so-called oxidation-zone, start or are intensified. The alternate operation of the wells for delivering groundwater and infiltration of oxygen-rich water induces alternating oxidation- and adsorption-periods on the surface of the solid material in the aquifer.

During the groundwater delivering period (discharge) Fe(II), Mn(II) and As(III) are adsorbed to the surface of soil grains which are partially coated by previously deposited oxidation products and bacteria. In the following recharge period the bivalent ions are oxidised to relatively insoluble ferric hydroxides and manganese oxides by the oxygen transported with the infiltration water into the pores of the aquifer.

The oxidation processes are accelerated by autocatalytic effects of the oxidation products and by autotrophic micro-organisms utilising energy from the oxidation process. Additionally, dissolved iron and manganese are adsorbed on the bacteria sheaths by the bio-film.

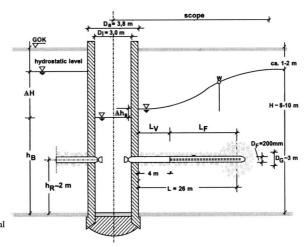

Fig. 3. Structure of a horizontal filter well.

Fig. 4. Scheme of the oxidation zone.

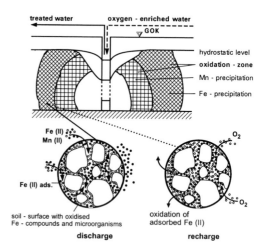

discharge   recharge

Arsenic(III) requires first an oxidation to Arsenic(V) before it can be precipitated by iron or adsorbed to iron-hydroxide and manganese-oxide, which are known as remarkable adsorbers for arsenic ions. Furthermore, other ions such as cadmium, copper, zinc and other micro-pollutants can be removed (Gulledge and O'Conner, 1973; Pierce and Moore, 1982).

Some specific bacteria are also able to oxidise ammonia in a two step process, the so-called nitrification. This process is very important for the *in situ* treatment, because of the high oxygen-consumption of more than 3.55 mg O_2/mg NH_4^+. Figure 4 shows the oxidation zone and the preparation process of the groundwater.

Because of the different oxidation-reduction potential, the removal of iron and manganese can only take place in spatial isolated regions of the aquifer. An increase of the redox potential from 40 to 160 mV reduces the solubility of iron in water from 10 mg/L to 0.1 mg/L. On the other hand, the stability-range of manganese-oxides starts with a higher pH-value compared to a lower pH-value for iron(III)-hydroxides. This means that the oxidation of manganese requires a higher redox-potential and a higher pH-value, respectively, than the iron oxidation.

The Eh-pH-Diagram for selected chemical combinations which are often part of the aquifer is shown in Figure 5 (Rott and Friedle, 1998).

A great advantage of *in situ* treatment is the retention of the oxidation products of iron, manganese and arsenic. While in above-ground treatment plants voluminous, arsenic-

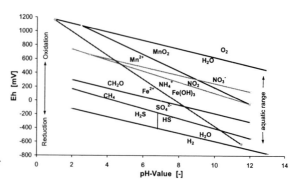

Fig. 5. Eh-pH-Diagram for aquatic systems.

TABLE 1

Raw water quality

Parameter	Raw water (A)	Raw water (B)	Raw water (C)	Guideline limit
Iron: Fe (mg/L)	0.94	1.97	0.94	0.20
Manganese: Mn (mg/L)	0.20	0.35	0.15	0.05
Arsenic (III): As(III) (mg/L)	–	0.024	–	–
Arsenic, total: As (mg/L)	0.015	0.038	0.015	0.010

containing sludge is produced, which requires an ecologically compatible and cost-efficient disposal, the oxidation products remain in the aquifer because of subterranean groundwater treatment.

Although the oxidation products are deposited in the aquifer, a blockage of the underground system does not occur. The first reason is the proportionality of the volume of the oxidation-zone to the volume of the infiltrated water. This means, that the oxidation zone increases with a decreasing pore volume. The second reason is the deposit of the oxidation products in so-called "dead-end-pores" and the aging of the voluminous hydrous hydroxides to less voluminous oxides and oxide hydrates. Last but not least, the reduction of chemical iron combinations decreases because of the application of *in situ* treatment.

RESULTS

The initial parameters of the raw water quality found in the three case studies are shown in Table 1. In all three cases the raw water was almost free of oxygen.

Treatment Results of Plant A

The processing aim of the field experiment was a permanent falling short of the parameters iron, manganese and arsenic below the valid guideline limits. The limiting values for drinking water are 0.2 mg/L Fe, 0.05 mg/L Mn and 0.01 mg/L As.

The scheme of Plant A is shown in Figure 6. In this case of application, groundwater from well 1 is enriched with air oxygen and infiltrated into well 2.

Well 2 has a depth of ca. 115 m and consists of four filter pipes with a common length of about 50 m. The aquifer at the well location is fissured. In spite of these bad ancillary conditions, very good treatment results could be achieved. The treatment results of the experiment are given in Figure 7. As the illustration shows, the removal of iron, manganese and arsenic began within the first few treatment cycles. One cycle consists of a recharge and

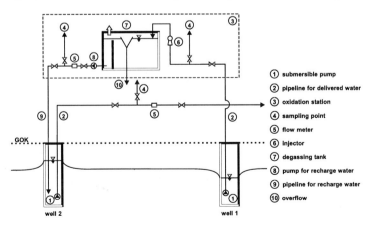

① submersible pump
② pipeline for delivered water
③ oxidation station
④ sampling point
⑤ flow meter
⑥ injector
⑦ degassing tank
⑧ pump for recharge water
⑨ pipeline for recharge water
⑩ overflow

Fig. 6. Scheme of Plant A.

well 2   well 1

delivery period and two short pauses of ca. 30 minutes between the two phases. The iron concentration fell below the limit value of 0.2 mg/L after three days of treatment, while the arsenic concentration of the delivered groundwater had already decreased below the guideline limit of 0.01 mg/L As after the first infiltration of oxygen-enriched water.

The results of the arsenic elimination are in a close relation to the removal of iron, because iron(II) and iron(III) is known as an excellent floccing agent for arsenic. The increased iron concentrations between a total delivery volume of ca. 22,000 and 24,000 m³ are explicable with an entry of ferrous particles which are detached from the filter pipe of the well when the submersible pump was activated.

Contrary to expectations, the removal of manganese also started within the first treatment cycle and reached ca. 50% of the raw water concentration at the end of the field experiment. It can be assumed that the treatment results improve with a further application of *in situ* treatment.

To investigate a potential remobilization of the deposited arsenic oxidation products precipitated in the aquifer as ferric arsenate ($FeAsO_4$) or As(V) adsorbed to ferric and manganese-hydroxides, the experiment at plant A was terminated with a kind of "crash test", which means a continuous delivery was realised for four weeks without any infiltration of oxygen-rich water. The discharge time of about one month can also be expressed in the form of the efficiency coefficient of 23 as in Figure 8 (Rott and Friedle, 1998).

Whereas the manganese concentration increased immediately because of remobilization of manganese hydroxides or -oxides, the concentrations of iron and arsenic were nearly constant over the total delivery time. This result verifies the assumption of a high adsorption capacity in the oxidation zone for Fe(II) and As(V) with the consequence of a stable operation of *in situ* treatment for the removal of iron and arsenic (Rott and Meyerhoff, 1996).

Treatment Results of Plant B

As generally shown in Figure 1, Plant B consists of two wells of 5" diameter which are operated alternately for production and recharge with a flow of 3 L/s. A complete oxidation from As(III) to As(V) was obtained within the first few days of treatment. Figure 9 shows the total arsenic and iron concentrations always at the end of the delivery periods.

Corresponding to the removal of iron, the arsenic concentration decreased after several cycles of the *in situ* treatment. After approximately 20 treatment cycles the arsenic concentrations were continuously lower than the guideline limit of 0.01 mg/L (Rott and Meyerhoff, 1996).

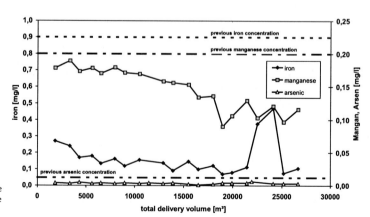

Fig. 7. Iron, manganese and arsenic in dependence of the delivery volume.

Fig. 8. As, Fe and Mn in delivered groundwater of plant A (long-term experiment).

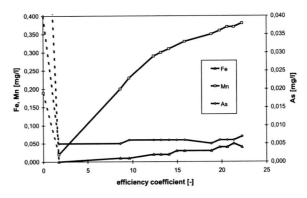

Fig. 9. Iron and arsenic concentrations at the end of the delivery period.

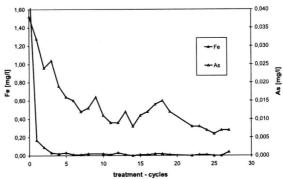

Treatment Results of Plant C

In the third case (Plant C) the application of *in situ* treatment is different to the generally normal technique represented in Fig. 10 because the experiment is practised with only one well. The infiltration water is taken from a clear water reservoir by gravity and is recharged by technical oxygen with a concentration of 12 mg/L. The water flows into the well by the natural hydrostatic pressure. Because of that, there are no additional energy costs for pumping. This technique variant requires a sufficiently sized drinking water tank for the provision of the enrichment water. Furthermore the delivery of drinking water is only possible from the reservoir during the infiltration phases.

As described in the second example (Plant B) the oxidation of arsenic(III) also took place in the first days of treatment, comparable with the removal of iron. Simultaneously to the removal of iron, the arsenic concentrations decrease from the beginning of *in situ* treatment. After 16 treatment cycles the arsenic values reached the guideline limit of 0.01 mg/L with deviation of ± 0.005 mg/L. After the starting period of some weeks, this plant was operated with recharge water from the reservoir and oxygen from the air only.

DISCUSSION

The three examples demonstrate the capability of *in situ* treatment. All field experiments which have been carried out by the Institute of Sanitary Engineering, Water Quality and Solid Waste Management achieved very good treatment results. While the removal of iron and arsenic normally starts after a few treatment cycles, the removal of manganese requires several weeks or months. The delayed beginning of the manganese removal is due to the

Fig. 10. Scheme of Plant C.

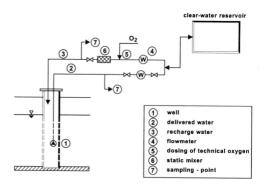

adjustment time of the micro-organisms which have to adapt to the varied surrounding conditions. However, it is confirmed that the removal of manganese and other oxidable substances improves with an increasing duration of *in situ* treatment and that by applying this method, drinking water can be supplied directly from the aquifer without any further above-ground treatment.

In order to obtain further findings of the chemical and biological mechanisms of *in situ* treatment a current research project titled "Analysis of physical, chemical and micro-biological processes in order to optimise *in situ* treatment of reduced groundwater" is being carried out at the Institute of Sanitary Engineering, Water Quality and Solid Waste Management of the University of Stuttgart.

CONCLUSIONS

Assuming appropriate hydrogeological and geochemical conditions, *in situ* treatment using oxygen as the only reagent can be an alternative low-cost technique for drinking water treatment. The technique of subterranean treatment can be used for the removal of iron, manganese, arsenic, ammonia and organic substances. *In situ* treatment makes use of the aquifer as a natural reactor for physical, chemical and microbiological processes. In compari-son to conventional treatment processes such as filtration and flocculation, *in situ* processes are often less expensive, both in investment and operating costs. In addition, wells and submersible pumps are protected against encrustations of ferric and manganese hydroxides and oxides (Rott and Meyerhoff, 1996). A great advantage of *in situ* treatment is the avoidance of any waste products and the resulting disposal problems as well as the use of only natural reactions without any treatment chemicals.

How far the described positive findings can be transferred to other habitats with differing aquifer- and raw water characteristics, particularly with different pH-values or iron-, manganese- and arsenic-concentrations, has to be checked with comparative field tests before a full-scale plant is built.

REFERENCES

Gulledge, J.H. and O'Connor, J.T. 1973. Removal of arsenic (V) from water by adsorption on aluminium and ferric hydroxides. *JAWWA*, **65**, S. 548–552.

Pierce, M. and Moore, C.B. 1982. Adsorption of arsenite and arsenate on amorphous iron hydroxide. *Water Res.*, **16** (1982), S. 1247–1253.

Rott, U. and Meyerhoff, R. 1996. *In situ* treatment of arsenic in groundwater. Workshop on Natural Origin In-organic Micropollutants. IWSA, Wien, 06–07 May 1996.

Rott, U. and Friedle, M. 1998. Drinking Water Supply based on Groundwater Protection and Treatment in the Aquifer, 3. Int. Water Technology Conference, Alexandria, Egypt, 20–23 May 1998.

Rott, U. and Meyerhoff, R. 1996. *In situ* treatment of groundwater. International Conference on Urban Engi-neering in Asian Cities in the 21st Century. Bangkok, 20–23 November 1996.

Arsenic Exposure and Health Effects
W.R. Chappell, C.O. Abernathy and R.L. Calderon (Editors)
© 1999 Elsevier Science B.V. All rights reserved.

Mode of Action Studies for Assessing Carcinogenic Risks Posed by Inorganic Arsenic

Melvin E. Andersen, Harvey J. Clewell, III, Elizabeth T. Snow,
Janice W. Yager

ABSTRACT

Mode of action (MOA) is emphasized as a unifying concept in new U.S. EPA carcinogen risk assessment guidelines. Optimally, MOA hypotheses relate carcinogenicity to obligatory precursor effects, link cancer and non-cancer responses through common pathways, and predict dose–response relationships via biologically-based dose–response (BBDR) models. Inorganic arsenic (As_i) increases skin lesions, cardiovascular disease, and several types of cancers in humans. The MOA or MOAs for As_i toxicity/carcinogenicity is poorly understood. Multiple effects may be idiosyncratic, each with a distinct MOA. Alternatively, only a limited number of precursor steps may be involved in all tissues. This paper outlines proposed MOAs of As_i carcinogenesis—impaired DNA repair, altered DNA methylation, increased growth factor synthesis, and increased oxidative stress. Increasingly, MOA hypotheses are suggesting that concentrations of critical gene products, including growth factors, redox-sensitive proteins, and DNA repair/DNA methylating enzymes, may be altered by As_i. These alterations would enhance tumor promotion or progression. A potential MOA for As_i acting as a late-stage tumor progressor is evaluated in relation to specific data needs for an As_i risk assessment and to the development of a BBDR model for As_i-induced internal tumors in humans. MOA studies of transcriptional processes, measurements of As_i dosimetry in humans, and dose–response evaluations for precursor endpoints appear important for supporting public health decisions about the risks posed by human As_i exposures. Studies of the transcriptional/ post-translational activities of arsenite and metabolites are likely to prove especially valuable for both cancer and non-cancer risk assessments.

Keywords: mode of action, arsenic carcinogenesis, BBDR modeling, tumor progression, cancer risk assessment

INTRODUCTION

Ingestion of water containing inorganic arsenic (As_i) has been associated with increases in tumors of the skin and several internal organs. Tumor incidence increases when water concentrations exceed 600 $\mu g/L$ (Tseng, 1977; Tseng, et al., 1968). The current drinking U.S. EPA drinking water standard is 50 $\mu g/L$. Thus, there is a relatively small margin of exposure (MOE) between carcinogenic concentrations of As_i (600 $\mu g/L$) and the drinking water standard (50 $\mu g/L$). To assess the true human risks from arsenic at ambient drinking water concentrations requires clarification of the manner in which arsenic compounds interact with cells to cause cancer. The shape of the tumor dose response should be a reflection of the dose response curve for these interactions and should influence the calculation of cancer risks from the compound.

In 1996, U.S. EPA developed and circulated revisions to their guidelines for carcinogen risk assessment (EPA, 1996). The newly proposed guidelines focus on 'mode of action' as an integrating concept to guide both evaluation of the tumor dose response curve in the observable range and the extrapolation of the expected dose-response to low doses. Mode of action (MOA) is a description of key events and processes starting with the interaction of an agent with a cell, through operational and anatomical changes, resulting in cancer formation. In this definition, a key event is an empirically observable precursor step that is a necessary element of the carcinogenic process. Mode of action may include direct DNA-reactivity or other predisposing conditions, such as enhancements of cell replication in the absence of DNA-reactivity. The latter cell proliferation responses may arise from mitogenic stimuli or from recurrent cytotoxicity followed by reparative hyperplasia.

In the new guidelines the dose–response analysis has two parts. A curve fitting routine or biological model is used in the region of observation to determine a point of departure, usually an effective dose causing a 10% increase in incidence (i.e., an ED_{10}) or a lower bound on the ED_{10} (i.e., an LED_{10}) for tumors. The second part is analysis in the region of extrapolation. This extrapolation may follow linear, threshold, or non-linear procedures depending on mode of action and on the availability of biologically based dose response (BBDR) models. When mode of action data support a non-linear extrapolation, a margin of exposure (MOE) is calculated. The MOE is the ratio of the point of departure (i.e., the ED_{10} or LED_{10}) divided by the human exposure. The decision regarding the appropriate value of the MOE is left with the risk manager and could vary for different populations of exposed individuals or for different use scenarios for a particular carcinogen.

The mode of action by which a chemical causes tumors may also be related to the mode of action by which the chemical causes non-cancer effects as well. Both nasal toxicity and carcinogenicity of vinyl acetate are related to its metabolism to acetaldehyde and acetic acid within the sustentacular cells of the nasal olfactory epithelium. These irritant compounds lead to cell toxicity, recurrent regeneration, and neoplastic transformations. Risk assessment for both the cancer and non-cancer endpoints should utilize the dose response of the olfactory degeneration lesions as a precursor step (Bogdanffy, et al., 1999). Another example is the receptor-mediated interactions of compounds such as the dioxins. The binding of dioxins to the Ah receptor and transcriptional activation by the Ah receptor–ligand complex appears to be an obligatory step for toxic and carcinogenic effects of these compounds. Receptor activation process could serve as precursor step for aiding in assessing risks of neoplastic and other toxic endpoints. Thus, mode of action may serve as an integrating concept to organize risk assessments for effects on diverse organ systems based on some common obligatory step. In the absence of evidence for common steps, each effect caused by a compound has to be fully evaluated to independently determine the risks of each of the toxic effects for humans (Barton et al., 1998).

Mode of action information serves several purposes in risk assessments. This knowledge can organize a risk assessment based on compelling evidence of a specific mode of action.

Secondly, a group of mode of action hypotheses can be considered in order to determine the expected low dose behavior of each of the modes of action. Lastly, considerations of modes of action can serve to structure ongoing research studies to enhance the eventual application of these study results in risk assessments. It is this last application that appears to have potential for guiding on-going studies with As_i. This paper evaluates the state of the science for As_i in relation to some proposed modes of action for its carcinogenic and non-cancer effects. After outlining the toxic effects and various proposed modes of action, we examine a particular mode of action that links internal tumors and specific molecular interactions of arsenite with tumor progression. In addition, we discuss data needs required for validating this mode of action and assessing its potential impact on the dose–response curve for As_i-induced carcinogenesis in humans.

BACKGROUND

Arsenic Carcinogenesis

As_i exposures have been associated with increased incidence of cancers in multiple tissues, including skin, lung, kidney, liver and bladder. In populations in Taiwan exposed via drinking water, the increased incidence of bladder cancer appears to be highly non-linear with sharp increases occurring at water concentrations above 500 $\mu g/L$ (Guo et al., 1994). Mathematical analysis of the increased lung cancer incidence in arsenic-exposed workers (Mazumdar et al., 1989) indicated that As_i acted at a late stage in the carcinogenic process. Despite these observations in human populations, arsenic has not been shown to be an animal carcinogen in conventional bioassays. The discrepancy between the lack of carcinogenicity in test animals and As_i's action as a risk factor for cancer at multiple sites in humans is a distinctive aspect of the carcinogenic potential of this compound.

Non-Cancer Endpoints

There are associations between increased As_i and peripheral vascular disease in Taiwanese populations (Tseng, 1977; Wu et al., 1989). Blackfoot disease, a syndrome in which there is thickening of arterial walls, leading to hardening of the arteries and loss of elasticity, causes gangrene in the digits and limbs. The incidence of Blackfoot in the Taiwanese populations was 9/1000 and increased with age and dose of As_i. Increases in hyperpigmentation and keratosis were also noted in this population. Chen et al. (1996) found an association between As_i and hypertension. Other studies have demonstrated a dose-dependent relationship between diabetes mellitus (Lai et al., 1992), ischemic heart disease (Chen et al., 1996), and peripheral vascular disease (Tseng et al., 1996). Several epidemiological studies attempted to determine the relationship between cancer and other endpoints to see if there are conditions that might serve as useful biomarkers of cancer risk. Cancer incidence was higher among those with Blackfoot disease after adjusting for dose (Chiou et al., 1995). As_i-associated skin disease also appears to be an indicator of higher cancer risk (Cuzick et al., 1992; Tsuda et al., 1995). However, there are inadequate data to determine whether any of these noncancer effects are direct precursors to tumors.

Metabolism

The most common form of arsenic in the environment is arsenate. This pentavalent form is reduced to trivalent arsenite and methylated to methyl and dimethylarsonic acids (*i.e.*, MMAA and DMAA) in multiple animal species (Thompson, 1993). Arsenite is the most reactive of these compounds. The pathways of arsenate metabolism are surprisingly complex (Thompson, 1993). Other metabolic intermediates, including methyl arsonous and dimethylarsinous acids, are also present in the body after arsenate or arsenite exposures. Trivalent, methylated compounds are also present. In addition, glutathione conjugates of As_i

are believed to be important intermediates in DMAA production. Any of these various metabolites individually or in combination may be involved in the molecular interactions, toxicity and carcinogenicity of As_i. Biological interactions of these various compounds with the carcinogenic process could occur at any of several points in the conversion of normal cells to a malignant phenotype, including the initiation, promotion, or progression phases of carcinogenesis.

Some Proposed Modes of Action at the Cellular/Molecular Level

Oxidative Stress and Tumor Promotion by DMAA

Dimethylarsonic acid (DMAA) is a major metabolite of arsenate. Methylation enhances urinary excretion and has been regarded as a detoxification pathway for inorganic arsenic in the body. When DMAA was administered to mice at high doses, single strand DNA breaks were noted in lung tissue (Yamanaka and Okada, 1994). The nature of these DNA alterations were examined *in vitro* and assumed to be due to generation of a DMAA peroxy radical produced from DMAA. DMAA promotes tumors in rats initiated with a treatment regimen that included diethylnitrosamine, N-methyl-N-nitrosourea, N-butyl-N-(4-hydroxybutyl)-nitrosamine, 1,2-dimethylhydrazine and N-*bis*-(2-hydroxyproplyl) nitrosamine. Increased tumor incidences in bladder, kidney, liver and thyroid (Yamamoto et al., 1995) were observed in rats treated with these initiators and with high doses of DMAA for 24 weeks (100 or 400 μg/ml). In a second study, tumor promotion in the bladder by DMAA was examined following initiation with 0.05% N-butyl-N-(4-hydroxybutyl)-nitrosamine (Wanibuchi et al., 1996). In these studies, tumor multiplicity was increased at 10 μg/ml DMAA. In the DMAA control rats without treatment with the initiator, there was toxicity in bladder cells, measured by increased cell proliferation, but no evidence for any initiating activity of DMAA.

Inhibition of DNA Repair

Inorganic arsenic compounds are not mutagenic at single gene loci (Rossman et al., 1980). However, arsenite is co-mutagenic in several assay systems. Arsenite enhanced the mutagenic effects of UV-irradiation in *E. coli* (Rossman, 1981) and in mammalian cells (Lee et al., 1985). Arsenite and UV-irradiation caused a greater than additive increase in mutation frequency in a pZ189 shuttle vector system in DNA repair proficient GM 637 human fibroblasts (Wiencke et al., 1997). The clastogenic interactions between UV irradiation and arsenite were greatest during the G1 to S phase of the cell cycle. The co-mutagenesis of sodium arsenite with N-methyl-N-nitrosourea in intact cell systems was observed with treatments of 10 μM arsenite for 3 hours or 5 μM for 24 hours (Li and Rossman, 1989a). While the target for arsenite's effects on co-mutagenicity are unclear, its ability to inhibit completion of DNA excision repair has been associated with effects on DNA ligase II (Li and Rossman, 1989b). The sensitivity of DNA ligase to direct inhibition by arsenite was also tested in nuclear extracts obtained from V79 cells 3 hours after induction with 4 μM N-methyl-N-nitrosourea. Nuclear extracts prepared in this fashion primarily contain DNA ligase II. Arsenite inhibition of the DNA ligase in this nuclear extract occurred only at concentrations in the millimolar range, much higher than the concentration active in the intact cells. Work presented at this conference (Snow and Hu, 1998) indicated that the effects of arsenite on DNA repair capability and on oxidation-reduction status in human keratinocytes are probably associated with post-translational or transcriptional mechanisms rather than with direct inhibition of the ligase enzyme.

Altered DNA Methylation

Exposure of human lung adenocarcinoma A549 cells to arsenite (0.08–2 μM) or arsenate (30–300 μM) caused hypermethylation of the promoter region of the p53 tumor suppressor gene

(Mass and Wang, 1997). Limited evidence was also provided for hypermethylation throughout the genome. Hypermethylation of a promoter region could silence the tumor suppressor genes, effectively serving to 'initiate' cells to an altered phenotype by biochemical effects on the genome. In contrast, other investigators have proposed a role of pan-genomic hypomethylation in As_i induced cell transformation (Zhao et al., 1997). Arsenite was used to transform a rat epithelial cell line (TRL1215) by growing the cells in the presence of arsenite for 18 weeks at concentrations of 0.125 to 0.5 μM. These transformed cells gave rise to aggressive, malignant tumors when inoculated into nude mice. DNA hypomethylation occurred concurrently with malignant transformation. These transformed cells had reduced levels of S-adenosyl-methionine (19% reduction at 0.5 μM). Arsenite-induced DNA hypomethylation was related to dose and duration and persisted after cessation of arsenite treatment. Hyperexpressibility of the metallothionine gene, which is increased by DNA hypomethylation, was also detected in the transformed cells. Although the mRNA for DNA methyltransferase (MeTase) was increased two-fold, the enzymatic activity was reduced to half in the arsenite-treated cells.

The interactions of arsenite that are responsible for these effects on epithelial cell DNA methylation are not presently understood. The authors proposed that methylation is reduced due to persistent decreases in the S-adenosyl methionine (SAM) pool related to the metabolism of arsenite to MMA and DMAA (Zhao et al., 1997). Calculations of methyl pool size and SAM consumption by arsenite metabolism could be used to bolster the case for cofactor depletion as the basis for these effects.

Human Bladder Cancer

The association between drinking water exposures to inorganic arsenic compounds and internal tumors is strongest for bladder cancer. Epithelial cells in this tissue are potentially exposed to exogenous As_i and its metabolites via systemic exposures from the blood and via arsenate and its metabolites that are excreted into the urine. As with several other tissues, including colon and brain, there are ongoing efforts to determine the most prevalent pathways leading to neoplastic changes in the bladder (Spruck et al., 1994). Bladder tumors tend to follow two divergent pathways, one in which p53 gene mutations occur early and a second where loss of heterozygosity (LOH) of chromosome 9 is an early event (Spruck et al., 1994). This latter pathway produces papillary transitional cell carcinomas (TCCs) that are often multifocal with little tendency to progress (Figure 1). These tumors had a high frequency of LOH on chromosome 9. In contrast, carcinomas *in situ* (CIS) have a high frequency of p53 mutations and frequently progress (Figure 1). Chromosome 9 alterations in CIS lesions were associated with more invasive tumors. This model for bladder cancer from Spruck and colleagues (Spruck et al., 1994) can be converted into a quantitative, multi-pathway BBDR model of bladder carcinogenesis. The quantitative model then serves to organize data on background incidence of bladder cancer and to show possible mechanisms by which As_i may influence human bladder tumor progression.

Multipathway Tumor Progression Model for As_i Exposures

The multipathway BBDR cancer schematic (Figure 2) proposes a late stage effect of As_i, acting either directly or via a biotransformation product, on completion of the carcinogenic process. Each transition is represented by a mutation rate (μ) and each cell type in the circles has particular birth (α) and death (β) rates. Here As_i increases mutation rates (μ) after loss of function of both alleles of the p53 gene. When tumors from the endemic Black Foot area in Taiwan were analyzed for p53 mutations, two CGA $\rightarrow$ CAC changes at location 175 and three tumors with double mutations were noted (Shibata, *et al.*, 1994). These observations were interpreted as indicating a mechanism where there was an increase in DNA mutation per damage-inducing event. Stimulated cell division with inefficient repair was proposed as a

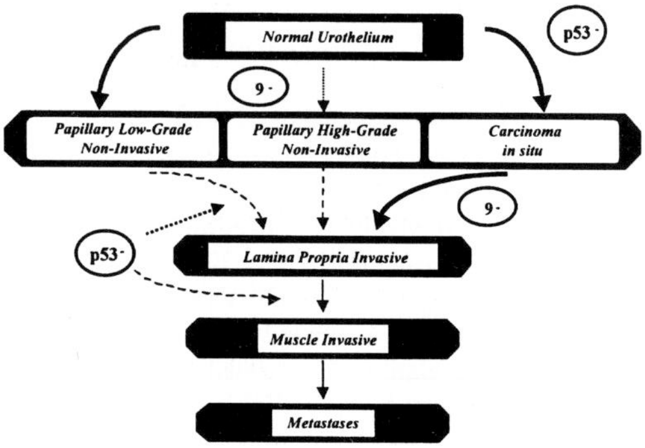

Fig. 1. The proposed association of specific mutations, loss of p53 and loss of heterozygosity on chromosome 9, with progression of bladder carcinogenesis in humans. The pathway has been described by Spruck et al. (1994). The papillary lesions do not tend to progress while carcinoma *in situ* have a more dire prognosis.

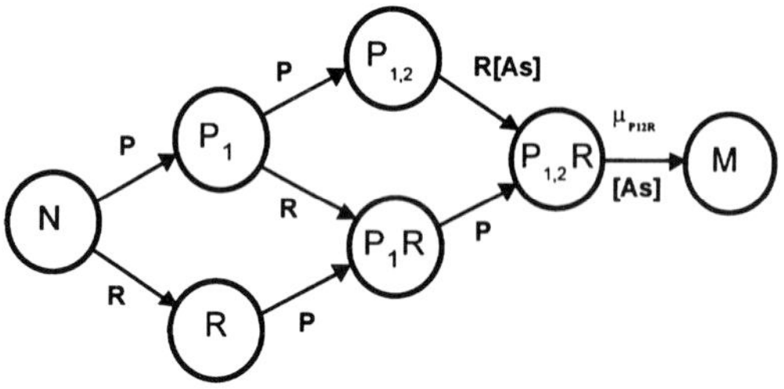

Fig. 2. Schematic of a multipathway model for bladder cancer. The cell type $P_{1,2}R$ has loss of function of both p53 alleles and LOH of chromosome 9. M is the invasive, malignant tumor cell. The transitions at which arsenic compounds are proposed to alter mutation rates are shown with [As] on the connecting line. P represents p53 mutational events: R represents growth stimulatory mutational events, N is the normal cell type, μ_{P12R} is the mutation frequency for the $P_{1,2}R$ cell.

Fig. 3. Hypothesis for some of the toxic and carcinogenic effects of arsenic. Linkage between cellular effects and dosimetry of inorganic arsenic with a biologically based dose response model for As_i-induced carcinogenesis model. Arsenic/arsenic metabolites may primarily affect cancer induction at a late stage after several mutational events create a "mutation-prone" phenotype.

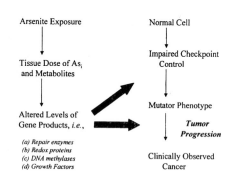

possible mechanism for these observations in As_i-exposed individuals. This observation is consistent with our multipathway cancer model in which As_i leads to higher mutation frequencies by altering levels of critical DNA-repair enzymes either before or after loss of checkpoint control (Figure 3). This BBDR model cannot be completed with all appropriate parameters at the present time. The process of model building, however, needs to begin early and help shape the collection and interpretation of experimental results.

The relative importance of the different modes of action might vary from tissue to tissue. Altered fidelity of DNA replication related to impaired DNA repair competence or altered methylation may be more related to progression with tumors of the internal organs. Growth factor control may be more important in determining the responses of the skin (both carcinogenic and non-cancer responses) and vascular tissues. The skin responses do seem to be qualitatively different from those in internal organs. The action of arsenic on skin may be more related to promotion than late-stage effects on progression.

Molecular Mechanisms of Regulation

Potential Transcriptional Control Molecules

Consistent with the U.S. EPA's focus on common precursor events, our working MOA for As_i emphasizes common molecular events in different tissues that then lead to diverse responses based on different tissue sensitivities for downstream events in the toxicity cascade. This molecular event is the control of a range of gene products by As_i. This concept ties together disparate effects into a common picture where initial molecular steps give rise to a complex set of final responses. Increasingly, study results indicate that As_i exposures of cells in culture lead to alterations in concentrations of protein products. These proteins include growth factors in keratinocytes, methyltransferases in rat liver epithelial cells, and DNA-repair enzymes and glutathione-dependent enzymes in human keratinocytes. These observations raise the question of the manner in which As_i can regulate these proteins. Would the action be direct or indirect occurring secondary to toxicity? One possibility is that the active signaling form of As_i may be associated with a biotransformation product. Among the metabolites that are implicated in arsenite biotransformation are several glutathione (GSH) derivatives. Glutathione conjugates of arsenite itself or of mono- and dimethylated forms of trivalent arsenic have been discussed as intermediates in the biotransformation pathways for As_i (Thompson, 1993).

Arsenite enhances glutathione levels in human fibroblasts (Lee and Ho, 1995), Chinese hamster V79 cells (Ochi, 1997), and human keratinocytes (Snow and Hu, 1998). Most compounds that react with glutathione cause GSH-depletion. A transient rebound increase in GSH occurs only after exposure ceases. Among organic compounds, 2-nitropropane actually enhances GSH levels rapidly and persistently after treatment (Zitting et al., 1981). 1-Nitropropane, another substrate for glutathione conjugation, does not cause increases in glutathione (Haas-Jobelius et al., 1992). The initial conjugate formed with 2-nitropropane, isopropyl-S-glutathione, is broadly similar in structure to the dimethylated trivalent arsenic conjugate with GSH. The space-filling structures of cysteine derivatives of these conjugates were optimized with a chemical drawing program (Figure 4). Perhaps, the GSH conjugates of the various forms of trivalent arsenic serve as inappropriate signals for the regulation of critical proteins whose concentrations are altered by As_i treatment. This hypothesis will have to be examined from a mode of action perspective and from the point of view of tissue dosimetry in exposed humans. Studies would evaluate the mechanisms of the control of gene product concentrations by As_i and its metabolites. Other studies would assess the tissue concentrations of these compounds in humans exposed under conditions that increase the incidence of specific arsenic-related health conditions.

 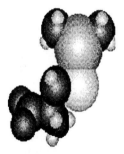

Fig. 4. Cysteine derivative structures derived from glutathione conjugates of 2-nitropropane or dimethylarsinous acid.

2-Nitropropane Displacement Product

Dimethyl Arsinous Acid Addition Product

Assessing the Hypothesis that a Mutator Phenotype and Enhanced Progression Underlie As$_i$ Carcinogenicity in Humans

Research Needs

The proposed MOA for As$_i$ as a tumor progressor acting through transcriptional processes is plausible, although still speculative. In addition, there is no direct evidence for the form of As$_i$ that might be involved with gene regulation. GSH conjugates of either arsenite or of methylated arsenic derivatives are attractive candidate compounds since GSH is an important natural cofactor. However, no data are available in humans or experimental animals assessing tissue concentrations of these proposed metabolites. Despite these difficulties, this hypothesis is consistent with a series of observations and focuses on the possibility that a common precursor action of As$_i$ is associated with most or all of its toxic properties. The next step in testing this hypothesis is development of the database on tissue exposure, human tumor development, and gene product regulation. Continuation of dosimetry and epidemiology studies in human populations and molecular and transcriptional studies with specific cell types *in vitro* will be especially informative. The following types of studies should be considered.

Human Studies

- Analyze existing dosimetry studies in human volunteers and in populations in areas with high water As$_i$ concentrations in order to estimate urinary tract and target tissue exposures (for arsenite, DMAA, and glutathione conjugates) associated with increased incidences of adverse effects.
- Extend mutational analysis of background and As$_i$-associated tumors to assess the range of mutations in tumors of the bladder, skin, and other sites and to assess the progression portion of the current hypothesis.

In vitro DNA Repair/Co-mutagenesis Studies

- Studies are still needed to identify the manner in which arsenite interferes with DNA repair to act as a co-mutagen. What is the basis of the higher sensitivity of the whole cells versus isolated ligase proteins? Are the ligase genes regulated by As$_i$ and is the regulation at the level of transcription or at the level of post-translational processes? Do GSH-conjugates act as signaling molecules? These same questions should be posed for gene products found altered in the cell transformation studies (Zhao, *et al.*, 1997) and in the growth factor studies (Germolec, *et al.*, 1996).
- Extend studies of DNA-repair competence to evaluate dose–response relationships in normal cells and in cells from p53 deficient transgenic mice or in human cells transformed with SV40 large T-antigen to block p53 action. Do these cells show enhanced responsiveness to As$_i$?
- Consider toxicity/tumor promotion studies in transgenic mice lacking one or both p53 alleles; these transgenics may be a better model for As$_i$ carcinogenesis.

BBDR Model Building:

- Continue efforts to refine progression/promotion models and efforts to work with investigators to insure that laboratory studies on mode of action contain quantitative dose–response components that aid in risk assessments and hypothesis testing.

The importance of the data derived from these mode of action studies is in confidently predicting the shape of the dose–response curve at low levels of As$_i$, i.e., is the dose response curve highly non-linear and, if so, what MOE should be proposed for regulation? The shape of the curve could be probed by studies on the dose–response relationships of DNA-repair,

DNA-methylation, or enhancement of growth factors. Conversely, studies of gene product regulation by As_i and its biotransformation products might eventually be used as a precursor effect for these extrapolations.

SUMMARY

Epidemiological results in human populations show unequivocal association between As_i exposures and cancer. The increased incidences occur at multiple tumor sites when arsenic concentrations are within an order of magnitude of the ambient groundwater concentrations found in some regions in the U.S. Responsible public health positions must rely on specific expectations related to the shape of the dose–response curve in the dose region immediately adjacent to the region of observation. The shape of the curve in this region depends on the mode of action of As_i as a human carcinogen. Enhanced progression/tumor promotion due to alterations in concentrations of gene products related to DNA repair, DNA methylation, and cell proliferation appears to be a promising mode of action hypothesis to explain the cancer and non-cancer effects. An outline and rationale for applying a mode-of-action-based risk assessment approach to human bladder cancer is provided here. This outline demonstrates the elements in this mode-of-action-based approach and is intended to generate discussion regarding optimal strategies for combining mechanistic studies of the DNA effects of As_i with the more quantitative approaches required for conducting risk assessments with this important element.

ACKNOWLEDGMENTS

We gratefully acknowledge support from the Electric Power Research Institute (EPRI).

REFERENCES

Barton, H.A., Andersen, M.E., Clewell, H.J., III. 1998. Harmonization: Developing consistent guidelines for applying mode of action and dosimetry information to cancer and noncancer risk assessment. *Hum. Ecol. Risk Assess.*, **4**, 75–115.

Bogdanffy, M.S., Sarangapani, R., Plowchalk, D.R., Jarabek, A., Andersen, M.E. 1999. A biologically-based risk assessment for vinyl acetate-induced cancer and non-cancer inhalation toxicity. *Toxicol. Sci.*, in press.

Chen, C., Chiou, H., Chiang, M., Lin, L., Tai, T. 1996. Dose–response relationship between ischemic heart disease mortality and long-term arsenic exposure. *Arterioscler. Thromb. Vasc. Biol.*, **16**, 504–510.

Chiou, H., Hsueh, Y., Liaw, K., Horng, S., Chiang, M., Pu, Y., Lin, J., Huang, C., Chen, C. 1995. Incidence of internal cancers and ingested inorganic arsenic: a seven-year follow-up study in Taiwan. *Cancer Res.*, **55**, 1296–1300.

Cuzick, J., Sasieni, P., Evans, S. 1992. Ingested arsenic, keratoses, and bladder cancer. *Am. J. Epidemiol.*, **136**, 417–421.

EPA. 1996. *Proposed Guidelines for Carcinogen Risk Assessment*. EPA 600-P-92-003C, Office of Research and Development, Washington, DC.

Germolec, D.R., Yoshida, T., Gaido, K., Wilmer, J.L., Simeonova, P.P., Kayama, F., Burleson, F., Dong, W., Lange, R.W., Luster, M.I. 1996. Arsenic induces overexpression of growth factors in human keratinocytes. *Toxicol. Appl. Pharmacol.*, **141**, 308–318.

Guo, H.-R., Chiang, H.-S., Hu, H., Lipsitz, S.R., Monson, R.R. 1994. Arsenic in drinking water and urinary cancers: a preliminary report. In: W.R. Chappell, C.O. Abernathy and C.R. Cothern (eds.), *Arsenic Exposure and Health*, pp. 119–128. Science and Technology Press, Northwood, England.

Haas-Jobelius, M., Coulston, F., Korte, F. 1992. Effects of short-term inhalation exposure to 1-nitropropane and 2-nitropropane on rat liver enzymes. *Ecotoxicol. Environ. Saf.*, **23**, 253–259.

Lai, M.S., Hsueh, Y.M., Chen, C.J., Shyu, M.P., Chen, S.Y., Kuo, T.L., Wu, M.M., Tai, T.Y. 1992. Ingested inorganic arsenic and prevalence of diabetes mellitus. *Am. J. Epidemiol.*, **139**, 484–492.

Lee, T.C., Ho, I.C. 1995. Modulation of cellular antioxidant defense activities by sodium arsenite in human fibroblasts. *Arch. Toxicol.*, **69**, 498–504.

Lee, T.-C., Oshimura, M., Barrett, J.C. 1985. Comparison of arsenic-induced cell transformation cytotoxicity, mutation and cytogenetic effects in Syrian hamster embryo cells in culture. *Carcinogenesis*, **6**, 1421–1426.

Li, J., Rossman, T. 1989a. Mechanism of comutagenesis of sodium arsenite with n-methyl-n-nitrosourea. *Biol. Trace Elem. Res.*, **21**, 373–381.

Li, J.-H., Rossman, T.G. 1989b. Inhibition of DNA ligase activity by arsenite: a possible mechanism of its comutagenesis. *Mol. Toxicol.*, **2**, 1–9.

Mass, M.J., Wang, L. 1997. Arsenic alters cytosine methylation patterns of the promoter of the tumor suppressor gene P53 in Human lung cells: a model for a mechanism of carcinogenesis. *Mutat. Res.*, **386**, 263–277.

Mazumdar, S., Redmond, C.K., Enterline, P.E., Marsh, G.M., Costantino, J.P., Zhou, S.Y.J., Patwardhan, R.N. 1989. Multistage modeling of lung cancer mortality among arsenic-exposed copper-smelter workers. *Risk Anal.*, **9**, 551–563.

Ochi, T. 1997. Arsenic compound-induced increases in glutathione levels in cultured Chinese hamster V79 cells and mechanisms associated with changes in gamma-glutamylcysteine synthetase activity, cystine uptake and utilization of cysteine. *Arch. Toxicol.*, **71**, 730–740.

Rossman, T.G. 1981. Enhancement of UV-mutagenesis by the low concentrations of arsenite in *E. coli*. *Mutat. Res.*, **91**, 207–211.

Rossman, T.G., Stone, D., Molina, M., Troll, W. 1980. Absence of arsenite mutagenicity in *E. coli* and Chinese hamster cells. *Environ. Mutagen.*, **2**, 317–379.

Shibata, A., Ohneseit, P.F., Tsai, Y.C., Spruck, C.H., III, Nichols, P.W., Chiang, H.-S., Lai, M.-K., Jones, P.A. 1994. Mutational spectrum in the P53 gene in bladder tumors from the endemic area of black foot disease in Taiwan. *Carcinogenesis*, **15**, 1085–1087.

Snow, E.T., Hu, Y. 1998. Modulation of DNA repair and glutathione levels in human keratinocytes by micromolar arsenite. *Third International Conference on Arsenic Exposure and Health Effects*, pp. 33. San Diego, CA.

Spruck, C.H., III, Ohneseit, P.F., Gonzalez-Zulueta, M., Esrig, D., Mijao, N., Tsai, Y.C., Lerner, S.P., Schmutte, C., Yang, A.S., Cote, R., Dubeau, L., Nichols, P.W., Hermann, G.G., Steven, K., Horn, T., Skinner, D.G., Jones, P.A. 1994. Two molecular pathways to transitional cell carcinoma of the bladder. *Cancer Res.*, **54**, 784–788.

Thompson, D.J. 1993. A chemical hypothesis for arsenic methylation in mammals. *Chem.-Biol. Interact.*, **88**, 89–114.

Tseng, C.H., Chong, C.K., Chen, J.I., Tai, T.Y. 1996. Dose–response relationship between peripheral vascular disease and ingested inorganic arsenic among residents in blackfoot disease endemic villages in Taiwan. *Atherosclerosis*, **120**, 125–133.

Tseng, W.P. 1977. Effects and dose–response relationships of skin cancer and blackfoot disease with arsenic. *Environ. Health Perspect.*, **19**, 109–119.

Tseng, W.P., Chu, H.M., How, S.W., Fong, J.M., Lin, C.S., Yeh, S. 1968. Prevalence of skin cancer in an endemic area of chronic arsenicism in Taiwan. *J. Natl. Cancer Inst.*, **40**, 453–462.

Tsuda, T., Babazono, A., Yamamoto, E., Kurumatani, N., Mino, Y., Ogawa, T., Kishi, Y., Aoyama, H. 1995. Ingested arsenic and internal cancer: a historical cohort study followed for 33 years. *Am. J. Epidemiol.*, **141**, 198–209.

Wanibuchi, H., Yamamoto, S., Chen, H., Yoshida, K., Endo, G., Hori, T., Fukushima, S. 1996. Promoting effects of dimethylarsinic acid on N-butyl-N-(4-Hydroxybuty) nitrosamine-induced urinary bladder carcinogenesis in rats. *Carcinogenesis*, **17**, 2435–2437.

Wiencke, J., Yager, J., Varkonyi, A., Hultner, M., Lutze, L. 1997. Study of arsenic mutagenesis using the plasmid shuttle vector pZ189 propagated in DNA repair proficient human cells. *Mutat. Res.*, **386**, 335–344.

Wu, M.-M., Kuo, T.-L., Hwang, Y.-H., Chen, C.-J. 1989. Dose–response relationship between arsenic concentration in well water and mortality from cancers and vascular diseases. *Am. J. Epidemiol.*, **130**, 1123–1132.

Yamamoto, S., Konishi, Y., Matsuda, T., Murai, T., Shibata, M.-A., Yuasa, I.M., Otani, S., Kuroda, K., Endo, G., Fukushima, S. 1995. Cancer induction by an organic arsenic compound, dimethylarsinic acid (cacodylic acid), in F344/DuCrj rats after pretreatment with five carcinogens. *Cancer Res.*, **55**, 1271–1276.

Yamanaka, K., Okada, S. 1994. Induction of lung-specific DNA damaged by metabolically methylated arsenics via the production of free radicals. *Environ. Health Perspect.*, **102**, 37–40.

Zhao, C.Q., Young, M.R., Diwan, B.A., Coogan, T.P., Waalkes, M.P. 1997. Association of arsenic-induced malignant transformation with DNA hypomethylation and aberrant gene expression. *Proc. Natl. Acad. Sci. USA*, **94**, 10907–10912.

Zitting, A., Savolainen, H., Nickels, J. 1981. Acute effects of 2-nitropropane on rat liver and brain. *Toxicol. Lett.*, **9**, 237–246.

Arsenic Exposure and Health Effects
W.R. Chappell, C.O. Abernathy and R.L. Calderon (Editors)
© 1999 Elsevier Science B.V. All rights reserved.

Observations on Arsenic Exposure and Health Effects

Kenneth G. Brown

ABSTRACT

The current EPA risk assessment for ingestion of inorganic arsenic was published in 1988, based on analysis of epidemiologic data from the Blackfoot disease region of Taiwan. The components of the exposure–response assessment basically consist of a health effect endpoint (skin cancer), an exposure metric (arsenic concentration in well water used for drinking), a model (the Multistage-Weibull), and parameters for scaling risk to the U.S. population (body weight, water consumption rate). These form what might be called the *current paradigm* for arsenic risk assessment. The EPA risk assessment identifies several information gaps, some of which have been narrowed by research in the last ten years. While improved information will enhance risk estimation under the current paradigm, the emphasis in this article is on observations that suggest broadening the paradigm itself. In particular, it is proposed that the regions of the U.S. with the highest arsenic concentrations in drinking water supplies be surveyed for skin signs of arsenicism.

Keywords: arsenic toxicity, drinking water, skin cancer, risk assessment

INTRODUCTION

As part of its mission to protect the public from unsafe levels of harmful environmental substances, the U.S. Environmental Protection Agency is reviewing its current standard of 50 μg/L of inorganic arsenic in drinking water (called the MCL: maximum contaminant level). The current EPA risk assessment of inorganic arsenic (U.S. EPA, 1988) is based on what will be called the *current paradigm* which consists of statistically fitting a mathematical model (exposure–response model) to data showing the rate of occurrence of a health effect at different exposure levels (exposure–response data). The model-fitting is a way of summarizing the relationship between exposure level and response rate, thus providing a means of predicting the response rate at an arbitrary exposure level including values below the observed range as typically required for setting an MCL. From the regulatory standpoint, it may be acceptable to set the MCL at an arsenic concentration where the increased risk of the health effect is a prescribed value, such as one per million lifetimes. The exposure–response curve serves as a way of determining what exposure value corresponds to that level of added risk. An added step is extrapolating risk estimates from the observed population to the target population, taking into account any differences that may modify risk (such as adjusting for weight, age, lifestyle, or other discernible factors).

The resultant risk estimates for the target population depend on how reliably and accurately each of the above steps can be accomplished, specifically on: quality of data for health effects and arsenic exposure, the accuracy of the mathematical model chosen for fitting the data (accuracy as a correct theoretical description of the relationship between exposure and response), and adequate adjustment for differences in risk modifiers between the study population and the target population.

To relate the components listed above specifically to the U.S. EPA risk assessment of arsenic, the following may be observed. The health effect endpoint is skin cancer in the observed population of Taiwan. Numerous other health effects endpoints, some cancer and some non-cancer, have been attributed to arsenic with varying degrees of evidential support (NRC, 1999, Chap. 4). Exposure–response data for these endpoints is virtually non-existent, however, except for some internal cancer sites (Chen et al., 1992; Wu et al., 1989; NRC, 1999, Chap. 10). The exposure data are arsenic concentrations in well water used for drinking water in the Blackfoot disease region of Taiwan. In general, epidemiologic data are highly preferable to animal data for evaluating health effects in humans, but a disadvantage is that intake can be controlled in animal experiments but not in humans, so human exposure is not known as accurately. The correct exposure–response model is unknown. Different models typically give different predictions of risk at low arsenic concentrations considered for an MCL. In extrapolating predicted risk from the Taiwan study population to the U.S. population, differences in average weight and intake of drinking water can be taken into account, but the potential effects of differences between the populations in intake of inorganic arsenic from dietary sources and in general nutrition and other socio-economic factors are difficult to take into account because of limited data and inadequate scientific understanding of their potential relevance.

By broadening the current paradigm we have in mind surveying portions of the U.S. population most highly exposed to arsenic in drinking water and food for early signs of arsenicism (signs of arsenic toxicity from chronic exposure). This approach is suggested as an alternative, not a substitute, for the current one, as the term "broaden" would indicate. It also has limitations, but they are *different limitations*, related to our premise. The premise of this article, stated more precisely below, is that it is unlikely that severe arsenic-induced health effects (e.g., cancer of the skin or elsewhere) will occur in a chronically exposed population in which there are no cases of cutaneous manifestations of arsenic toxicity (skin signs). One or more cases of arsenic-induced skin signs has typically identified populations at risk in parts of the world where arsenic in drinking water is several hundred micrograms/liter, but evidence is lacking at much lower arsenic concentrations. Thus, the most tenuous aspect of the premise is that the risk of skin signs exceeds the risk of severe health effects, and by a nontrivial amount, even at low arsenic concentrations (such as <50 μg/L).

The technical description of the premise is as follows:

1. Early skin signs, diagnosable as specific to arsenic toxicity, will tend to occur in higher numbers than severe health effects in an arsenic-exposed population (i.e., the probability of skin signs in an exposed person is higher than the probability of a severe health effect).
2. Persons with skin signs are at greater risk of developing a severe health effect than those without skin signs.
3. Some exposed persons may develop severe health effects without experiencing skin signs.
4. In an arsenic-exposed population of size n, the probability that k (≥ 1) persons develop arsenic-induced severe health effects and 0 persons develop skin signs, rapidly decreases toward zero as k or n increases.

The next two sections describe what we consider to be the two most important areas of uncertainty in extrapolation of risk from Taiwan to the U.S., included here to motivate the need to broaden the current risk assessment paradigm for arsenic. The section thereafter describes the somewhat limited data on the relationship between occurrence of skin signs and more severe health effects.

ARSENIC IN FOOD

Data on intake of inorganic arsenic from food in Taiwan are limited and the data are highly variable (Schoof at al., 1998). It appears, however, that average intake of inorganic arsenic from food sources may be higher in Taiwan than in the U.S., perhaps on the order of 50 μg/day and 14 μg/day, respectively. When food sources are taken into account, a differential rate of dietary intake between Taiwan and the U.S. causes a shift in the exposure–response curve for arsenic in water in the U.S., as shown in Figure 1. The long-dash line in Fig. 1 is the exposure–response curve from the EPA risk analysis, which implicitly assumes that the dietary intake of arsenic is the same for Taiwan and the U.S. The intakes are shown as zero in the figure legend, but the same curve would apply at any nonzero value so long as it is the same for both Taiwan and the U.S. The short-dash curve is the same as the long-dash curve except that it is shifted to the right. The shift results from dietary arsenic intake being higher in Taiwan than in the U.S., the magnitude of the shift being proportional to the difference in dietary intakes (the mathematics is explained in Brown and Abernathy, 1997). The figure shows a threshold for the short-dash curve, but there could also be a gradual but

EFFECT OF ARSENIC IN FOOD (microgram/day)

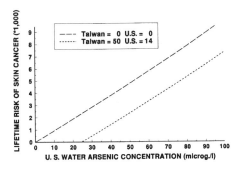

Fig. 1. The excess lifetime risk from arsenic in U.S. drinking water for two different assumptions of dietary arsenic intake in Taiwan and the U.S.

undetectable increase in risk at the low-exposure region shown as a threshold. The point to be made is that extrapolation of risk estimates from Taiwan to the U.S., for arsenic in drinking water, appear quite sensitive to whatever difference may exist in intake from arsenic in food, which is not very well known.

RISK MODIFIERS

Another difficulty in extrapolating risk estimates from one population to another is the potential difference in what may well be collectively called risk modifiers. For example, on average, a Taiwanese in the study population weighs less than the average American in the target population, and the mortality rates differ, affecting lifetime risk estimates. Both of these factors can be addressed quantitatively and were taken into account in the EPA risk assessment for skin cancer. The extent to which other risk modifiers may exist is partly speculative. For example, it is difficult to know if nutrition, and more generally, socio-economic status (SES), is a factor. This issue is raised in the EPA risk assessment, but scientific knowledge is still insufficient to address it definitively.

The people of the Blackfoot disease region of Taiwan tend to be very poor, subsisting on a diet high in rice and yams. Hsueh et al. (1995) reports an exposure–response relationship between duration of consumption of dried sweet potatoes and prevalence of skin cancer, suggesting that poor nutrition, or a correlate, may be related to susceptibility. There is other observational evidence to support the hypothesis of a nutrition-susceptibility link. Mandal et al. (1996) observed that poor people of West Bengal are more affected by arsenical melanosis and that nutritional status appears to play an important role in arsenic toxicity. Mazumder et al. (1998) found increased prevalence of keratoses (1.6 fold) in persons below 80% of standard body weight, further suggesting that malnutrition may play a role. Similarly, field examinations in Bangladesh led to the observation that poverty stricken rural populations were most affected and that those with adequate nourishment could withstand even a moderate amount of arsenic without skin manifestations. It was also observed that at the preliminary stage of skin lesions, like diffuse melanosis, victims recovered if they discontinued use of contaminated water and got nourishing food (SOES, 1996).

Genetics may play a role in susceptibility to arsenic toxicity, and this may be a factor within and/or between populations. Consider the observations of A. P. Arroyo, Secretary of Health, Second Region (Chile), posed in the form of the following questions (paraphrased): (1) Why do the people of Atacameño, who drank water with high levels of total arsenic (600 μg/L) for many decades, not have arsenic-associated diseases? (2) Why are there only *some* families in Antafagosta with arsenic-related diseases? (Why do some people develop arsenic-related disease and others do not, when both are exposed to the same levels of arsenic?) (Arroyo, 1998).

The point of this section is that factors such as nutrition/SES, genetics, and perhaps others, may act to modify risk to different degrees in different populations, including the study population and the target population. For example, extrapolation of risk estimates from the Blackfoot disease region to the Atacameño population of Chile, or vice versa, would probably be quite erroneous, but for reasons not understood. It is difficult to know what risk modifiers may be affecting extrapolation of risk estimates from Taiwan to the U.S., whether they may be significant, and how they would alter the exposure–response curve for the U.S.

EARLY STAGES

We now turn to the evidence that skin signs may be correlated with the development of more severe health effects and that screening the U.S. population in regions where arsenic levels are highest may be a useful empirical supplement to the current risk assessment paradigm. It seems to be a common finding among populations where arsenic-related health effects have been observed, at much higher concentrations than found in the U.S., that within the population as a whole the earliest manifestations of arsenicism are distinctive changes in skin pigmentation. The prevalence study by Tseng et al. (1968, 1977) in the Blackfoot disease region of Taiwan found hyperpigmentation, keratosis, and skin cancer in subjects of age 5, 15, and 25 years, respectively, at the youngest. Further, there were 18 times as many cases of skin signs (hyperpigmentation or keratosis) as skin cancer, skin signs were present in 94% of the skin cancer cases, and skin cancer was 70 times more prevalent among those with skin signs than those without skin signs (see also Yeh, 1973). Figure 2 shows a diagram of the relative proportions.

If the same relative proportions in Figure 2 hold in the U.S., at *the relatively low exposure concentrations in the U.S.*, then the probability that an arsenic-induced skin cancer would be *without* concomitant skin signs is about 0.06. This value quickly diminishes as the number of cases increases: 2 cases ($0.06^2 = 0.004$), 3 cases (0.0002), etc. The chance of seeing these numbers of skin cancer cases without concomitant skin signs *and* without skin signs in any other people with similar exposure (assuming there are some) would be still lower. This suggests that screening populations in the U.S. for skin signs (that could also include questions about arsenic exposure, urinalysis for arsenic content, etc.), starting with regions where the arsenic concentration in water is highest, may provide a better idea of the extent to which skin cancer from arsenic may be a risk in the U.S. Obviously the whole U.S. population could not be surveyed, but it is reasonable to consider the small percentage with exposures above some limit (say 10 or 25 μg/L). It should be noted at this point that previous epidemiologic studies in the U.S. have not found an excess risk of skin cancer in the regions studied, but that may be due to small sample sizes. Also, an ongoing study in Utah by the U.S. EPA is reporting evidence of health effects related to arsenic (Lewis et al., 1999).

It is not so clear that early signs typically occur concomitantly with arsenic-induced cancer at internal sites, although the data are limited compared to the large study of Tseng and colleagues. Cuzick et al. (1992) followed a cohort of 478 patients treated with Fowler's solution (potassium arsenite) in England during the period 1945–1969. In a subcohort of 142 patients examined for signs of arsenicism in 1970, all eleven subsequent cancer deaths occurred in those with signs of arsenicism. The experience of Tsuda et al. (1995), however, seems to be a little different.

Tsuda et al. (1995) followed an historical cohort of 454 residents in Niigata Prefecture, Japan, from 1959 until 1992, who were exposed to drinking water polluted by a small factory nearby where arsenic trisulfide had been produced for more than 40 years. Exposure was estimated to be only about five years, however, from 1954, when the concentration in wells used for drinking was inferred to have increased substantially, to 1959, when the wells were tested and the plant was closed. The problem was discovered when an 11-year old boy, with hyperpigmentation, white spots, and hyperkeratosis of the skin, was diagnosed with

TSENG STUDY - OUTCOMES PROPORTIONAL TO AREA

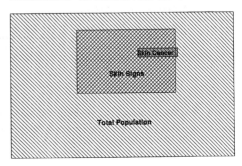

Fig. 2. Venn diagram with areas proportional to prevalence of skin signs and skin cancer in the study population of Tseng et al. (1968).

arsenicism. The government undertook measurements of wells in the region in 1959, with values ranging from undetectable to over 3,000 $\mu g/L$. Medical examinations of the town residents were also conducted in 1959, with 383 participants. Of those, 97 had at least one of the dermal signs of arsenicism: hyperkeratosis, hyperpigmentation, and white spots. The criteria for diagnosis of arsenicism, however, was based on hyperpigmentation, hepato-megaly, and pancytopenia. Thus, 88 persons were included in the "chronic arsenicism" group, 13 of which had no skin signs of arsenicism, and 97 were in the "skin signs" group, with 75 persons in both.

The objective of the study was to elucidate the effect of ingested arsenic on the development of internal malignancies, and to identify the relation between skin signs of arsenicism and successive cancer deaths to see if skin signs are a useful predictor of future cancer development. They concluded that "mortality from cancer was correlated with the severity of chronic arsenicism...However, among the group exposed to the highest con-centrations of arsenic ($\geq 1,000 \mu g/L$), excesses of cancer mortality were observed among those with both positive and negative skin signs. The results demonstrate that negative skin signs are no assurance of low risk for cancer development."

As the authors note, this study is limited by its small size and the potential for nondifferential misclassification of exposure because well-arsenic concentrations were measured only once. There were a total of 32 cancers deaths (out of a total of 91 deaths among the 383 persons examined), but they included only 9 lung cancers and 3 urinary cancers, sites at which cancer has been linked to arsenic in other studies. The lung cancer mortalities appear to be related to positive skins signs and to severity of arsenicism while the urinary cancer mortalities do not (all 3 occurred in persons without skin signs), but the numbers are small and some caution is warranted. This study would lead us to question whether the chance that an arsenic-induced malignancy would be accompanied by skin signs in a given individual is as high as indicated for skin cancer in the study of Tseng et al., but the high prevalence of skin signs (25%) reinforces the notion that skin signs may be a reliable indicator for excessive exposure to arsenic in even small communities.

DISCUSSION

The association of arsenic with severe health conditions, e.g., cancer, peripheral cardio-vascular disease, ischemic heart disease, to name a few, are difficult to study in the U.S. because they are not specific to arsenic exposure. Other countries, with very high arsenic concentrations in drinking water compared to the U.S., have typically observed skin signs and other symptoms as an indicator of arsenic toxicity within a population. Screening for early signs and symptoms of arsenic toxicity within water supply systems with the highest arsenic levels in the U.S. could provide empirical evidence to consider in conjunction with the current risk assessment paradigm that projects risk of cancer in the U.S. If arsenic toxicity is found, then it indicates that arsenic intake is too high (due to the local water supply or otherwise). People should be protected against the risk of any health effect from arsenic, including skin signs, and not just severe health effects.

A limitation is that we do not know if a safe level of arsenic exposure for one health effect (e.g., skin signs) is also a safe level for another effect (e.g., cancer or heart disease), but that is an argument for expanding the risk assessment paradigm to include all avenues of investigation. Surveying exposed subpopulations of the U.S. for skin signs, combined with personal exposure histories to arsenic (that may include urinalysis), is one such avenue. Skin signs can occur relatively early and have been reported at relatively low concentrations (<50 μ/L, Tsuda (1995), Mazumder et al. (1998)). There is also some evidence that people with skin signs may be at higher risk for development of at least some of the more severe effects, although further study is needed.

REFERENCES

Arroyo, A.P. 1997. The arsenic problem in North of Chile: past, present, and future. Program abstract. Arsenic: Health Effects, Mechanisms of Action and Research Issues. An NCI/EPA symposium. September 22–24, 1997. Hunt Valley, MD.

Brown, K.G. and Abernathy, C.O. 1997. The Taiwan skin cancer risk analysis of inorganic arsenic ingestion: effects of water consumption rates and food arsenic levels. In: Abernathy C.O., Calderon R.L., Chappell W.R. (eds.), *Arsenic Exposure and Health Effects*, Vol/ 2, pp. 260–271. Chapman and Hall. New York.

Chen C.-J., Chen C.-W., Wu M.-M., and Kuo T.-L. 1992. Cancer potential in liver, lung, bladder and kidney due to ingested inorganic arsenic in drinking water. *Br. J. Cancer*, **66**, 888–892.

Cuzick, J., Sasieni, P., and Evans, S. 1992. Ingested arsenic, keratoses, and bladder cancer. *Am. J. Epidemiol.*, **136**, 4, 417–421.

Hsueh, Y.-M., Cheng, G.-S., Wu, M.-M., Yu, H.-S., Kuo, T.-L., and Chen, C.-J. 1995. Multiple risk factors associated with arsenic-induced skin cancer: effects of chronic liver disease and malnutritional status. *Br. J. Cancer*, **71**, 109–114.

Lewis, D.R., Southwick, J.W., Ouellet-Hellstrom, R., Rench, J., Calderson, R.L. 1999. Drinking water arsenic in Utah: a cohort mortality study. *Environ. Health Perspect.*, **107**, 5, (in press).

Mandal, B.K., Chowdhury, T.R., Samanta, G., Basu, G.K., Chowdhury, P.P., Chanda, C.R., Lodh, D., Karan, N.K., Dhar, R.K., Tamili, D.K., Das, D., Saha, K.C., and Chakraborti, D. 1996. Arsenic in groundwater in seven districts of West Bengal, India—The biggest arsenic calamity in the world. *Current Sci.*, **70**, 976–986.

Mazumder, D.N.G., Haque, R., Gosh, N., De, B.K., Santra, A., Chakraborty, D., and Smith, A.H. 1998. Arsenic levels in drinking water and the prevalence of skin lesions in West Bengal, India. *Inter. J. Epidemiol.*, **27**, 871–877.

NRC (National Research Council). 1999. *Arsenic in Drinking Water*. National Academy Press. Washington, D.C.

Schoof, R.A., Yost, L.J., Crecelius, E., Irgolic, K., Goessler, W., Guo H.-R., Greene, H. 1998. Dietary arsenic intake in Taiwanese districts with elevated arsenic in drinking water. *Hum. Ecolog. Risk Assess.*, **4**, 117–135.

SOES (School of Environmental Studies). 1996. Bangladesh's arsenic calamity may be more serious than West Bengal. A report from the School of Environmental Studies, Jadavpur University, Calcutta, India.

Tseng, W.-P., Chu, H.-M., How, S.-W., Fong, J.-M., Lin, C.-S., and Yeh, S. 1968. Prevalence of skin cancer in an endemic area of chronic arsenicism in Taiwan. *J. Nat. Cancer. Inst.*, **40** (3), 453–463.

Tseng, W.-P. 1977. Effects and dose–response relationships of skin cancer and blackfoot disease with arsenic. *Environ. Health Perspect.*, **19**, 109–119.

Tsuda, T., Babazono, A., Yamamoto, E., Kurumatani, N., Mino, Y., Ogawa, T., Kishi, Y., Aoyama, H. 1995. Ingested arsenic and internal cancer: a historical cohort study followed for 33 years. *Am. J. Epidemiol.*, **141**, 198–209.

U.S. EPA (U.S. Environmental Protection Agency). 1988. *Special Report on Ingested Inorganic Arsenic. Skin Cancer; Nutritional Essentiality.* EPA/625/3-87/013.

Wu M.-M., Kuo T.-L., Hwang Y.-H., and Chen C.-J. 1989. Dose–response relation between arsenic concentration in well water and mortality from cancers and vascular diseases. *Am. J. Epidemiol.*, **130**, 1123–1232.

Yeh, S. 1973. Skin cancer in chronic arsenicism. *Human Pathol.*, **4** (4), 469–485.

Keyword Index

SOCIETY FOR ENVIRONMENTAL GEOCHEMISTRY AND HEALTH

ANNOUNCING

Fourth International Conference on Arsenic Exposure and Health Effects
June 18–22, 2000
San Diego, California

FOR FURTHER INFORMATION:
Please e-mail, fax, or mail your name, affiliation, address, phone, fax, and e-mail address.

International Conference on Arsenic Exposure and Health Effects
Dr. Willard R. Chappell
Campus Box 136
University of Colorado at Denver
P.O. Box 173364
Denver, CO 80217-3364
USA

E-mail: as98@carbon.cudenver.edu
Fax: (303) 556-4292
Phone: (303) 556-4520

Visit our website: www.cudenver.edu/as98

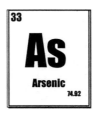